# RNA Interference and Viruses

## Current Innovations and Future Trends

Edited by

Miguel Angel Martínez

Fundacio irsiCaixa
Hospital Universitari Germans Trias i Pujol
Badalona
Spain

Caister Academic Press

Caister Academic Press
Norfolk, UK

www.caister.com

British Library Cataloguing-in-Publication Data
A catalogue record for this book is available from the British Library

ISBN: 978-1-904455-56-1



Cover image: Adapted from Figure 8.1.

Printed and bound in Great Britain

# Contents

# Contributors

**Xabier Abad**
Department of Hepatology and Gene Therapy
Center for Applied Medical Research (CIMA)
University of Navarra
Pamplona
Spain

xabad@progenika.com

**Patrick Arbuthnot**
Antiviral Gene Therapy Research Unit
Department of Molecular Medicine and Haematology
University of the Witwatersrand Medical School
South Africa

patrick.arbuthnot@wits.ac.za

**Sailen Barik**
Department of Biochemistry and Molecular Biology
University of South Alabama College of Medicine
Mobile, AL
USA

sbarik@jaguar1.usouthal.edu

**Ben Berkhout**
Laboratory of Experimental Virology
Department of Medical Microbiology
Center for Infection and Immunity Amsterdam (CINIMA)
Academic Medical Center
University of Amsterdam
Amsterdam
The Netherlands

b.berkhout@amc.uva.nl

**Ben Berkhout**
Laboratory of Experimental Virology
Department of Medical Microbiology
Center for Infection and Immunity Amsterdam (CINIMA)
Academic Medical Center
University of Amsterdam
Amsterdam
The Netherlands

b.berkhout@amc.uva.nl

**Olivier ter Brake**
Laboratory of Experimental Virology
Department of Medical Microbiology
Center for Infection and Immunity Amsterdam (CINIMA)
Academic Medical Center
University of Amsterdam
Amsterdam
The Netherlands

olivierterbrake@gmail.com

**Abdullah Ely**
Antiviral Gene Therapy Research Unit
Department of Molecular Medicine and Haematology
University of the Witwatersrand Medical School
South Africa

abdulla.ely@wits.ac.za

**Puri Fortes**
Department of Hepatology and Gene Therapy
Center for Applied Medical Research (CIMA)
University of Navarra
Pamplona
Spain

pfortes@unav.es

**Gabriele Fuchs**
Stanford University
Department of Microbiology and Immunology
Stanford, CA
USA

gfuchs@stanford.edu

**Jovanna González-Rojas**
Department of Hepatology and Gene Therapy
Center for Applied Medical Research (CIMA)
University of Navarra
Pamplona
Spain

sjgonzalez@unav.es

**Ana Giner**
Centre for Research in Agricultural Genomics (CRAG)
CSIC-IRTA-UAB
Barcelona
Spain

agrgmy@cid.csic.es

**Bertsy Goic**
Institut Pasteur
Viruses and RNA interference
Paris
France

bertsy.goic-figueroa@pasteur.fr

**Joost Haasnoot**
Laboratory of Experimental Virology
Department of Medical Microbiology
Center for Infection and Immunity Amsterdam (CINIMA)
Academic Medical Center
University of Amsterdam
Amsterdam
The Netherlands

p.c.haasnoot@amc.uva.nl

**Luc J.W. van der Laan**
Erasmus MC-University Medical Centre
Department of Surgery
Rotterdam
The Netherlands

l.vanderlaan@erasmusmc.nl

**Lóránt Lakatos**
Agricultural Biotechnology Center ABC
Plant Biology Institute
Gödöllõ
Hungary

lakatos@abc.hu

**Juan José López-Moya**
Centre for Research in Agricultural Genomics (CRAG)
CSIC-IRTA-UAB
Barcelona
Spain

jlmgmy@cid.csic.es

**Maria Nevot**
Fundacio irsiCaixa
Hospital Universitari Germans Trias i Pujol
Badalona
Spain

mnevot@irsicaixa.es

**Miguel Angel Martínez**
Fundacio irsiCaixa
Hospital Universitari Germans Trias i Pujol
Badalona
Spain

mmartinez@irsicaixa.es

**Cara T. Pager**
Stanford University
Department of Microbiology and Immunology
Stanford, CA
USA

ctpager@stanford.edu

**Qiuwei Pan**
Department of Gastroenterology & Hepatology
Erasmus MC-University Medical Centre
Rotterdam
The Netherlands

q.pan@erasmusmc.nl

**Sébastien Pfeffer**
Architecture et Réactivité de l'ARN
Université de Strasbourg
Institut de Biologie Moléculaire et Cellulaire du CNRS
Strasbourg
France

s.pfeffer@ibmc.u-strasbg.fr

**Maria-Carla Saleh**
Institut Pasteur
Viruses and RNA interference
Paris
France

carla.saleh@pasteur.fr

**Mélanie Tanguy**
Architecture et Réactivité de l'ARN
Université de Strasbourg
Institut de Biologie Moléculaire et Cellulaire du CNRS
Strasbourg
France

m.tanguy@ibmc.u-strasbg.fr

# Other Books of Interest

# Preface

Fire and Mello discovered that introducing long double-stranded RNA into the nematode *Caenorhabditis elegans* led to the targeted degradation of homologous mRNA. That revealed the existence of a fundamental mechanism for gene expression regulation, now known as RNA interference (RNAi) (Fire *et al., Nature* 1998). This discovery has given rise to the advent of novel tools for biological research and drug discovery. RNA interference is emerging as a powerful modality for battling some of the most notoriously challenging viral clinical targets (hepatitis C virus, human immunodeficiency virus, and other human viruses). However, several critical issues associated with this novel technology must be resolved before it can progress to testing in human clinical trials. The two central topics of this book are the latest findings on RNAi–virus interactions and progress in the development of RNA interference-based antiviral therapeutics. Some chapters emphasize general concepts concerned with the role of RNA interference in natural antiviral defence mechanisms; others suggest how to improve the efficacy and safety of RNA interference-based antiviral drugs; still others describe how this technology has been used as a new therapeutic tool for fighting specific viruses, including human immunodeficiency virus, hepatic viruses, or respiratory viruses. The book provides an in-depth review of RNA interference and viruses and discusses potential new avenues for research. I thank all the authors, who have contributed their time and expertise to produce a book that should provide stimulus for further research. I would like also to acknowledge Horizon Scientific Press for inviting me to serve as guest editor for this book.

Miguel Angel Martinez

# RNAi: an Antiviral Defence System in Insects

1

Bertsy Goic and Maria-Carla Saleh

## Abstract

Multicellular organisms evolved sophisticated defence systems to confer protection against pathogens. An important characteristic of these immune systems is their ability to act both locally at the site of infection and at distal uninfected locations (Baulcombe, 2004; Dorner and Radbruch, 2007; Roitt *et al.*, 2001; Voinnet, 2005). Insects rely on multiple immune responses to combat infection; one of them is RNA interference (RNAi). Here we review the current knowledge on the general mechanism of RNAi and summarize what is known about the antiviral role of RNAi in the insect model *Drosophila melanogaster*. We also discuss the strategies evolved by viruses to suppress the RNAi response. Finally, we briefly describe the RNAi mechanism in other insects of economic and/or health relevance.

## Introduction

### *Drosophila melanogaster* as an animal model

*Drosophila melanogaster* was introduced as an experimental animal model at the beginning of the twentieth century in the context of evolutionary biology (Kohler, 1994). Around 1908, Thomas H. Morgan, an embryologist of some renown at the time, began growing *Drosophila melanogaster* in large numbers with an interest in exploring the existence of 'macromutations' (Bellen, 1998). Some time during the course of 1910, Morgan came across a fly with white eyes. This was the first allele of the *white* gene; its linkage to the sex chromosome triggered a revolution in our understanding of heredity and led to the establishment of genetics as a subject. Through the activities of Morgan and his students, *Drosophila melanogaster* did more than any other model to lay down the foundations of Genetics as a discipline and a tool for Biology. Throughout the twentieth century, *Drosophila melanogaster* has been broadly used as a model in developmental biology, benefiting from rapid generation time, ease and robustness of culturing and low maintenance costs compared with other animal models. One hundred years onward, the basic knowledge about this insect, its genetics and the tools to manipulate it have considerably grown. Moreover, the similarities in genetic make up and the developmental programming logic between flies and humans reveal *Drosophila melanogaster* itself as a consummated model organism (Bier, 2005; Reiter and Bier, 2002). Of particular significance to human biology, over 70% of the proteins involved in human disease exist in flies, suggesting that *Drosophila melanogaster* may act as a model system for human diseases (Bier, 2005; Reiter and Bier, 2002), learning, behaviour (Margulies *et al.*, 2005) and even alcoholism and other drug addictions (Bellen, 1998).

> *When with foresight and luck Morgan selected this species for studies in heredity and together with Sturtevant, Bridges and Muller derived from it the evidence for the existence, arrangement, and complex transmission of genes in the chromosomes, the significance of the results was not owing to* Drosophila melanogaster *as*

*a unique organism, but as representative of all organisms.*

Sturtevant (1932)

Nowadays, *Drosophila melanogaster* is also an important experimental model in the field of Immunology. In addition to the aforementioned characteristics, it offers the advantage that immunity can be studied at the level of the whole organism by monitoring the resistance of multiple mutant flies to infection. The success of this model largely relies on the possibility of applying genetic and genomic approaches to this organism (Adams and Sekelsky, 2002) (Box 1.1). Furthermore, another reason for studying immunology in insects in general, and *Drosophila melanogaster* as a safe representative model, is that many arthropods are vectors for human pathogens and a mastery of their immune system could lead to better vector and/or disease control.

In this chapter, we will focus on the immune response that insects mount against viruses, the underlying mechanisms and the strategies by which viruses suppress this response. We will mainly describe what is known in *Drosophila melanogaster*, but will also discuss relevant information from other arthropod models.

## General aspects of the immune system in insects

A competition exists between a host's defence responses and a pathogen's need to control and use the host cell, or organism, for its replication cycle. Having coexisted for thousands of years, both insects and microorganisms have influenced each other's evolution. Thus, insects have developed sensitive mechanisms for recognition of pathogens and an array of strategies to defend themselves against attacks by bacteria, fungi, parasites and viruses.

To combat infection, the fruit fly, *Drosophila melanogaster* relies on multiple defence responses, many of which are shared with higher organisms (Hultmark, 2003; Tzou *et al.*, 2002). These responses include the use of physical barriers together with local and systemic innate immune responses. First, epithelia, such as that beneath the cuticle, in the alimentary tract and in the tracheae, act both as a physical barrier and local defence against pathogens by producing antimicrobial peptides and reactive oxygen species. Second, specialized cells, haemocytes,[1] participate in phagocytosis and encapsulation of foreign intruders in the haemolymph.[2] Finally, the fat body, a functional equivalent of the mammalian liver, produces humoral response molecules including antimicrobial peptides (Lemaitre and Hoffmann, 2007). In the sophisticated relationship between host and pathogen, the adult fruit fly has developed different ways to defend itself against a broad spectrum of aetiological agents (Fig. 1.1). These are briefly summarized below.

### Bacteria and fungi

A hallmark of *Drosophila melanogaster* defence system is the synthesis and secretion of potent antimicrobial peptides (AMP) that accumulate in the haemolymph where they oppose invading microorganisms, generating a systemic immune response (Lemaitre and Hoffmann, 2007). This response persists for several days, even weeks, and can confer protection against a second challenge by the same pathogen (Boman *et al.*, 1972). AMP production is mainly produced by the fat body. Due to its large size and its location inside the open circulatory system, the fat body represents a powerful organ for the synthesis and secretion of immune effectors.

Injection of Gram-positive or Gram-negative bacteria and fungi induces different patterns of AMP expression, indicating that microbial recognition mechanisms can distinguish between various classes of invaders (Lemaitre *et al.*, 1997) (Fig. 1.1). Whereas the recognition of bacteria is achieved through sensing of species specific forms of peptidoglycan by the peptidoglycan recognition proteins, the fungal detection seems to rely also on a member of the Gram-positive bacteria binding proteins (Bangham *et al.*, 2006).

1 Cells found in the haemolymph of insects that are equivalent to the blood cells in vertebrates.

2 Insects have an open circulatory system. The haemolymph is the blood-like fluid that bathes tissues and is circulated throughout the body by the dorsal vessel, a functional equivalent of the heart.

**Box 1.1 Mutagenesis in fruit flies**

In *Drosophila*, there are two main ways to induce mutagenesis: ethyl methane sulphonate (EMS) and the P-transposable elements. The first one is the most common because is easy to administer and causes the highest frequency of mutations, mainly point mutations. The disadvantages of EMS are the difficulty in mapping point mutations and that often the progeny of mutagenized males are mosaic and the mutation is not transmitted to the next generation, unless the germ cells are also mutated (St Johnston, 2002). The second one is to use P-element induced mutations, because the mutated gene can be rapidly and easily identified, however is very inefficient. The P-element is a class II transposon, which means that its movement within the genome is made possible by a transposase. The wild type P-element is a sequence of 2,907 bp that includes 31 bp inverted repeat at each end and a gene for P-transposase, which consists of four exons. To be competent to response to transposase, a P-elements construct requires only a few hundred base pairs of P-elements sequence on each end, including inverted repeats. Almost any sequence can be placed between these ends and usually a marker gene is included for following the insertion phenotypically. The mutagenesis system typically uses an autonomous (with a functional transposase) but immobile element, and a mobile non-autonomous (lack a functional transposase gene due to mutation) element. Flies from subsequent generations can then be screened by phenotype or PCR. Other examples for common transposable elements are *mariner*, *Hermes* and *piggyBac* (Adams and Sekelsky, 2002).

After pathogen recognition, fungi, yeast and Gram-positive bacteria activate the Toll pathway. Gram-negative bacteria activate the Imd (immune deficiency) pathway, producing differential gene expression in response to the infection. For instance, Toll pathway activation produces drosomycin, whereas diptericyn is produced by the Imd pathway.

The barrier epithelia that are in constant contact with large numbers of microorganisms, is lined with a chitinous matrix, producing a hostile environment to microbial colonization with the help of lysozyme secretions. Additionally, in the gut lumen there is a local production of AMP and reactive oxygen species (ROS) that generate complementary inducible defences (Lemaitre and Hoffmann, 2007).

## Viruses and parasitic nucleic acids

Viruses are important natural pathogens in *Drosophila melanogaster*, more than 40% of all flies in natural environments are infected. More than 25 *Drosophila melanogaster* viruses have been identified, all RNA viruses (Ashburner, 2005; Habayeb *et al.*, 2006). The immune response against viruses is distinct from the response involved in the microbial defence. Recent studies suggest the existence of three types of antiviral response: activation of the JAK/STAT pathway to induce putative antiviral genes by undefined cytokine(s) (Dostert *et al.*, 2005), degradation of viral RNAs by the RNAi machinery (Galiana-Arnoux *et al.*, 2006; van Rij *et al.*, 2006; Wang *et al.*, 2006), and, very recently, a role for the DExD/H-box helicase of Dicer-2 regulating the induction of genes like *Vago* that can control viral load in some *Drosophila melanogaster* tissues (Deddouche *et al.*, 2008) (see below for details).

The JAK/STAT pathway, a cascade that plays an important role in the antiviral response in mammals, has also been identified in the fruit fly. It has four main cellular components, the ligand Unpaired (upd), the receptor Domeless, the Janus Kinase (JAK) Hopscotch and the STAT transcription factor (Agaisse and Perrimon, 2004). JAK/STAT deficient flies are as resistant to bacterial and fungal infections as wild type flies and express normal AMP profiles; however, they are more sensitive to infection with *Drosophila* C virus (DCV), a virus infection model in fruit flies (Dostert *et al.*, 2005). The hypothesis was proposed that during virus infection, the JAK/STAT pathway responds to tissue injury through upd, a haemocyte-released cytokine. Upd activates the pathway by binding to the Domeless receptor in the fat body (Agaisse and Perrimon, 2004), but the precise role of this pathway in the insect immune response remains unsolved.

During DCV infection, some specific responsive genes are regulated by the JAK/STAT pathway. One of them, *vir-1* (virus-induced RNA 1), is induced 24 h after viral infection in

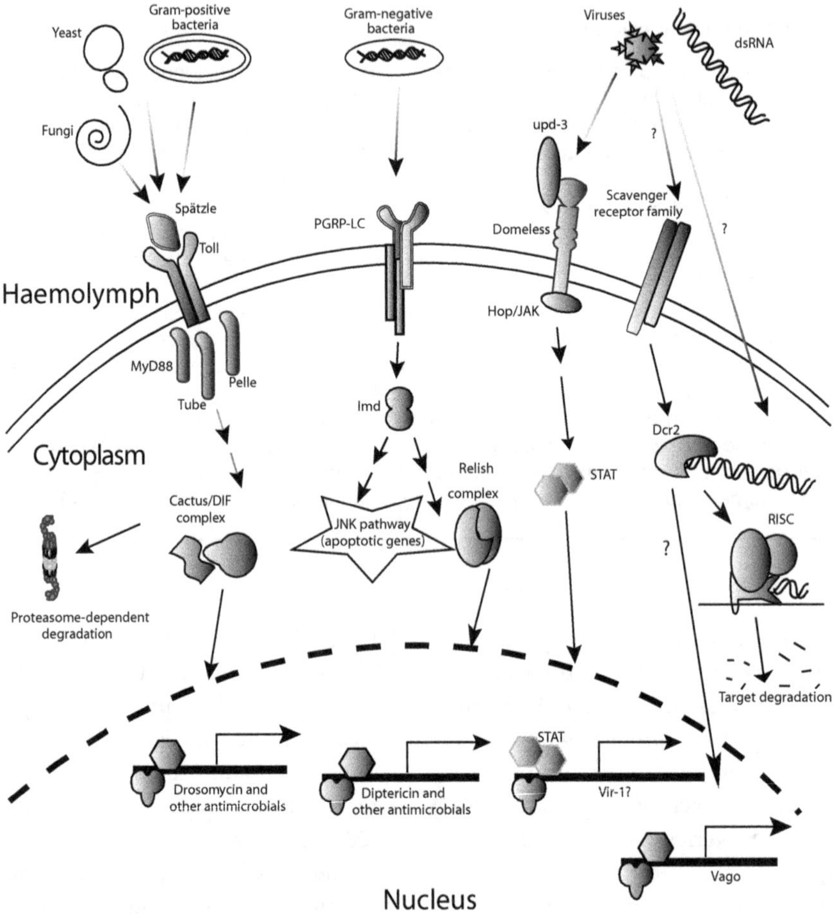


**Figure 1.1** Immune response of *Drosophila melanogaster.* Fruit flies have developed multiple immune responses and these mechanisms depend on the nature of the pathogens. The recognition and the defence strategies vary depending if the invader is a Gram-positive or Gram-negative bacteria, fungi, yeast or virus. Described here are the main and simplified models for each pathway. The antimicrobial peptide genes are regulated by a balance between two signalling pathways: the Toll pathway, which is largely activated by fungi, yeast and Gram-positive bacteria; and the Imd pathway, which is mainly activated by Gram-negative bacteria. **Toll pathway** is activated by the recognition of cleaved form of Spätzle by Toll receptor. This causes the recruitment of three intracellular proteins, MyD88, Tube and Pelle. By a mechanism not yet characterized, Cactus is phosphorylated and then degraded by the proteasome. As a consequence, transcription factors are released and move to the nucleus for transcription of antimicrobial genes. The **Imd pathway** is activated during Gram-negative infection (predominantly, but not solely). The microbial sensing occurs by the transmembrane receptor PGRP-LC, which recruits the adaptor Imd. Downstream of Imd activation, Relish is phosphorylated and cleaved, that allows the Rel domain to translocate to the nucleus for transcription, whereas the inhibitory domain remains stable in the cytoplasm. On the other hand, to sense viral and parasite nucleic acids, there are two pathways described: JAK/STAT and RNAi pathways. The *Drosophila melanogaster* **JAK/STAT pathway** has four main components: JAK, encoded by hopscotch (hop); STAT; a ligand, encoded by unpaired (upd); a receptor, encoded by Domeless. The current model of the JAK/STAT signal cascade is the following: the extracellular upd protein binds and activates the receptor Domeless, which in turn activates the intracellular Hop/JAK; Hop/JAK then phosphorylates and stabilizes STAT. The phosphorylated STAT dimerizes and translocates to the nucleus to transcribe antiviral factors. For the systemic **RNAi pathway**, the recognition of viral dsRNA beyond the infected site is important. There is evidence that a Scavenger receptor is involved in the dsRNA recognition. This dsRNA triggers the RNAi silencing pathway. Dcr-2, a component of the RNAi machinery, might also have a role in transcription of antiviral factors, such as Vago protein. It is also possible that dsRNA reaches the uninfected cells by other undescribed models. Adapted from Lemaitre and Hoffmann (2007).

the ventral epidermis, but not in the fat body, in a JAK/STAT-dependent manner. The data suggest that *vir-1* is not induced in the main tissue in which DCV replicates; instead *vir*-1 would be induced elsewhere after a signal generated by DCV-infected cells. This role as antiviral defence is supported by the fact that flies deficient in the JAK Hopscotch are more susceptible to DCV infection. Even though *Drosophila melanogaster* produces molecules in a JAK/STAT-dependent manner, none of the identified and tested genes regulated by DCV infection has shown a direct antiviral effect (Dostert *et al.*, 2005).

The best-characterized antiviral response in insects is the small interfering RNA (siRNA) pathway or RNA interference (RNAi) pathway. This silencing RNA-dependent mechanism belongs to the small RNA pathways and makes reference to a type of gene silencing mechanism conserved among various species from different kingdoms (fungi, animals and plants). Indeed, RNAi-related pathways (siRNA, endo-siRNAS, miRNA and piRNA) play roles in several aspects of cell life, which range from gene expression, to epigenetic modification and regulation of heterochromatin, to resistance to pathogens. Whether these two responses to viral infection, the RNAi and JAK/STAT pathways, can act in a synergistic way is an open question.

For the host, RNAi is a defence mechanism against both viruses and genomic parasites (such as transposable elements) and also can regulate other components of innate immunity by modulating gene expression (Ding and Voinnet, 2007). Immune system genes and the parasite molecules interact closely and often evolve rapidly. It is thought that this results from a co-evolutionary arms race, in which there is a reciprocal process of adaptation and counter adaptation between parasites and host. So RNAi is a primary defence of many organisms against pathogens.

Almost all RNA silencing pathways are triggered by double-stranded RNA (dsRNA) precursors that vary in length and origin. This dsRNA is rapidly converted into small RNAs ranging from 21 to 32 nucleotides, depending on the silencing pathway involved. Subsequently, small RNAs are incorporated into the multiprotein complex RISC (RNA induced silencing complex), where they guide the recognition of RNAs through an Argonaute/Piwi family member. Argonaute/Piwi proteins lie at the heart of the small RNA pathways as they bind the small RNAs and mediate transcriptional or post-transcriptional gene silencing.

Arthropods in general, encode a functional RNAi pathway induced by dsRNA. The first report was made in *Drosophila melanogaster* (Kennerdell and Carthew, 1998; Pal-Bhadra *et al.*, 1997), followed by many other insects, such as the red flour beetle (Bucher *et al.*, 2002), mosquito (Billecocq *et al.*, 2000; Gaines *et al.*, 1996), tick (Aljamali *et al.*, 2002, 2003) and honeybee (Amdam *et al.*, 2003; Beye *et al.*, 2002). See the section 'RNAi in other insect models' in this chapter for more details.

In a general way, RNAi provides an antiviral defence in plants and animals that is able to seek out and destroy viral RNAs in a sequence specific manner. As a counter-defence, many of the insect RNA viruses, like DCV, Flock House virus (FHV) or Cricket Paralysis virus (CrPV), encode RNAi suppressors. The real significance of RNAi as antiviral response in *Drosophila melanogaster* is supported by two main observations. First, mutant flies in Dicer-2 (*Dcr-2*) or Argonaute-2 (*Ago-2*), which affect the RNAi pathway, have increased susceptibility to viral infections (Galiana-Arnoux *et al.*, 2006; van Rij *et al.*, 2006; Wang *et al.*, 2006) (Box 1.2). Second, some genes involved in RNAi response against viruses, but not their housekeeping miRNA paralogues, are the fastest evolving genes in the *Drosophila melanogaster* genome, which is a signature of a host pathogen arms race (Obbard *et al.*, 2006).

It was generally believed that RNAi evolved specifically as a means of protection against viruses and transposable elements in the genome. However, recent discoveries show that RNAi is involved in many other cellular processes, suggesting that the defence mechanism, while obviously important, might actually be a derivative process rather than the evolutionary origin of the mechanism.

Here we present an overview of our current knowledge on the RNAi antiviral response in the context of two fundamental questions: (a) How is the antiviral RNAi response in insects? and (b) What strategies have viruses developed to evade it?

## RNA interference in insects

Sequence-specific gene silencing was discovered in transgenic plants under the form of co-suppression between introduced transgenes or between a transgene and its homologous endogenous genes (Matzke *et al.*, 1989; Napoli *et al.*, 1990; van der Krol *et al.*, 1990). The first indication of RNA silencing as an antiviral response came from observations in transgenic plants, where virus infection triggered RNA silencing of homologous virus-derived transgenes (Lindbo *et al.*, 1993). Subsequent studies confirmed that RNA silencing is a natural antiviral response in plants that specifically targets the viral genome and homologous RNAs for degradation (chapter 2 of this book).

The early evidence implicating viruses as targets of RNA silencing in insects came from the use of alphavirus-based expression vectors. Mosquitoes infected with a recombinant alphavirus carrying a heterologous viral sequence also developed RNA-mediated, homology-dependent virus resistance (Adelman *et al.*, 2001; Billecocq *et al.*, 2000; Gaines *et al.*, 1996; Olson *et al.*, 1996). Specific targeting of insect viruses by RNAi was confirmed in *Drosophila melanogaster* cells transfected with dsRNA corresponding to part of the FHV genome, which become resistant to FHV (Li *et al.*, 2002). The first discovered viral suppressor of RNAi system in invertebrates was the FHV B2 protein, essential to suppression but not involved in replication (Li *et al.*, 2002).

With the exception of retroviruses, all other RNA viruses produce dsRNAs as replication intermediates or genome products. Additionally, many DNA viruses generate dsRNA as a result of convergent transcription of their compact genomes. Thus, dsRNA is often a molecular signature of virus infection and as such, is a trigger for a range of host responses, including RNA interference (Marques and Carthew, 2007).

Several studies established that the core components of the RNAi machinery and the siRNA pathway, e.g. Ago-2, Dcr-2, R2D2, are essential for antiviral defence (Galiana-Arnoux *et al.*, 2006; van Rij *et al.*, 2006; Wang *et al.*, 2006; Zambon *et al.*, 2006). Mutant flies deficient in these components are hypersensitive to infection by insect pathogens –DCV, CrPV, FHV and Sindbis virus. The lack of RNAi components in these flies is associated with higher mortality and with a significant increase of viral titres. Thus, the enhanced viral susceptibility of RNAi defective mutant flies is due to their inability to control virus replication, underscoring the necessity of a functional RNAi pathway for efficient antiviral defence (Ding and Voinnet, 2007).

A fundamental feature of RNA silencing is the production of siRNA from dsRNA by the endoribonuclease Dcr-2. These siRNAs loaded onto the RISC complex guide the Ago-2 component protein to cut target RNAs in a homology-dependent manner (Fig. 1.2). RNA silencing in fungi, worms and plants involves a cellular RNA-dependent RNA polymerase (RdRP), which amplifies the template RNA used to cut the target RNA. In *Drosophila melanogaster* and mammals however, RNA silencing remains functional, although no cellular RdRP have been described.

## The main components of the siRNA pathway

### Dicer

In the fruit fly genome there are two paralogues of Dicer, *Dcr-1* and *Dcr-2*. Dcr-1 is primarily involved in the biogenesis of micro RNAs (miRNAs) (Box 1.3), while Dcr-2 is involved in the generation of

**Box 1.2 Viral infection in *Drosophila melanogaster***

To analyze the antiviral response or host-virus interaction in *Drosophila*, adult flies (3-5 days) are inoculated by microinjection with the viral stock intra-thoraxically or in the abdominal cavity. The virus, via haemolymph, reaches distinct host tissues for its replication. Flies isolated in different parts of the world display varying susceptibility to virus infection. Some of those viruses are horizontally transmitted through contact or ingestion. Even when injections are not the natural way for a viral spread, this procedure has been used to ensure reproducible infection and lethality under laboratory conditions (Lemaitre and Hoffmann, 2007).

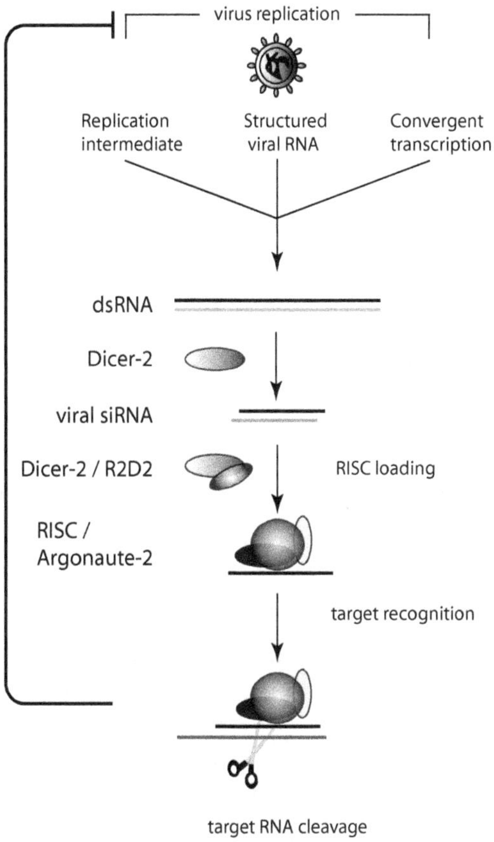


**Figure 1.2** RNA silencing in virus infection. Double-stranded RNA (dsRNA) is commonly generated during viral replication and is considered a molecular pattern associated with viral infection. In the *Drosophila melanogaster* RNAi pathway, dsRNA molecules are processed by the RNase III enzyme Dicer 2 (Dcr-2) to form 21–23 nucleotide siRNA (small interfering RNAs) duplexes. Dicer-2-associated protein R2D2 binds to siRNAs and facilitates its loading onto the effector complex RISC (RNA-induced silencing complex). The siRNA duplex associates with the Argonaute 2 (Ago-2) protein and is unwound with one strand retained in RISC. The single-stranded siRNA in RISC guides sequence-specific degradation of complementary or near-complementary target mRNAs. After cleavage, the messages are degraded. Adapted from van Rij (2008).

siRNAs. The structural characteristics of Dicer proteins include a C-terminal dsRNA binding domain, two RNase III domains, a PAZ domain (Dcr-2 lacks it), a domain of unknown function (DUF-283) and an N-terminal DExH-box helicase domain (Kavi *et al.*, 2008).

Based on studies in X-ray structure, it is proposed that Dicer forms an intramolecular dimer such that two RNase domains come together to form a single catalytically active dsRNA processing centre (Zhang *et al.*, 2004). The RNase III domains also participate in other protein–protein interactions that are important for the overall efficiency of the RNAi process. The *Drosophila melanogaster* Dcr-2 cleaves the dsRNA substrate from the termini in an ATP-dependent manner (Jiang *et al.*, 2005; Lee *et al.*, 2004). These siRNAs are 21 nucleotides long. They originate from exogenous dsRNA arising from either viral genomes or their replicative intermediates, fold back RNAs expressed by the host or virus, or overlapping pairs of sense and antisense transcripts.

It was recently shown that Dcr-2 plays an important role in RISC assembly and the selection of bona fide siRNA. In *Drosophila melanogaster*, there is a second type of ribonuclease III, Drosha, but this enzyme is involved in the generation of precursor miRNA molecules, short RNAs molecules naturally expressed by the cell that regulate a wide variety of genes.

A binding partner for Dcr-2 is R2D2. R2D2 belongs to a growing family of dsRNA-binding proteins that function in tandem with specific RNase III enzymes. R2D2 contains two tandem dsRNA-binding domains at the N-terminus and forms a heterodimeric complex with Dcr-2. Experimental approaches with *R2D2* mutant flies show that this protein is dispensable for siRNA production *in vivo* (Liu *et al.*, 2006). Some studies suggest that the Dcr-2/R2D2 complex not only binds long dsRNA, but also binds duplex siRNA, forms the RISC loading complex and facilitates the transfer of siRNA onto Ago-2, the catalytic component of RISC. Hence, even if R2D2 is not necessary for siRNA production, it is required for loading siRNA onto RISC (Liu *et al.*, 2006; Pham *et al.*, 2004; Tomari *et al.*, 2004). Photo-cross-linking experiments suggest that R2D2 preferentially binds the more stable end of the siRNA, whereas Dcr-2 is often found at the less stable end (Tomari *et al.*, 2004). Thus, R2D2 could be a sensor for the thermodynamic asymmetry of siRNA, thereby orienting Dcr-2 to the end that is easier to unwind.

The role of Dcr-2/R2D2 complex in loading siRNA into RISC complex depends on its ability to bind siRNA. Therefore, Dcr-2 and R2D2 coordinately bind siRNA and constitute the

**Box 1.3 Small RNA related pathways**

The small RNAs that provide target specificity to the silencing machinery can be distinguished by their origin, function and their association to the Argonaute/Piwi proteins family. Below we summarize the major pathways present within the animal kingdom.

**siRNAs** (small interfering RNAs) are 21 nucleotides long. They originate from exogenous perfectly based-paired dsRNA arising from either viral genomes or replicative intermediates; fold back RNAs expressed by the host or virus; or overlapping pairs of sense and antisense transcripts. Their main function is in defense against parasitic nucleic acids, such as transposons and viruses (Figure 2).

**endo-siRNAs** (endogenous small interfering RNAs) are 21 nucleotides in length. They are generated by genomic sources of dsRNAs that included inverted repeat structures, sites of bidirectional transcription and antisense transcripts from various loci, including retrotransposons and pseudogenes. *Drosophila* endo-siRNA production requires not only canonical RNAi factors (Dcr-2 and Ago-2), but some endo-siRNAs also depend on the canonical microRNA biogenesis factor Loquacious. Endo-siRNAs are produced in both gonadal and somatic tissues.

**miRNAs** (micro RNAs) are ~22 nucleotides long. They originate from a single, long transcript of endogenous origin that forms imperfectly base-paired hairpin structures. miRNAs represent a large class of evolutionarily conserved RNAs that were discovered because of their critical roles in development and cellular regulation. They guide translational inhibition or cleavage of cellular RNAs.

**piRNAs** (Piwi-interacting RNAs) are slighty longer RNAs from ~24 to 32 nucleotides in length and they form RNA-protein complexes with Piwi proteins. They are believed to be processed from single-stranded primary transcripts that are transcribed from defined genomic regions. Their biogenesis is not completely understood, but it is Dicer independent. They contribute to the silencing of transposable elements in the animal germ line. rasiRNAs (repeat-associated small interfering RNAs) belong to the piRNA class because of their association with members of the Piwi family of proteins. They originate from retrotransposon sequences and from certain repeat-rich genome regions. They play a crucial role in controlling the expression of homologous sequences dispersed throughout the genome.

gateway for siRNA loading and RISC activation (Liu *et al.*, 2006). But how do Dcr-2 and R2D2 facilitate siRNA loading onto RISC? The newly synthesized siRNA is dsRNA, but the siRNA exists as a ssRNA in an active RISC complex. Thus, the nascent siRNA duplex must be unwound during (or after) RISC assembly. Recent studies strongly suggest a model for separation of siRNA strands and activation of RISC complex. After the Dcr-2/R2D2 complex recruits duplex siRNA to Ago-2, the Piwi domain of Ago-2 cleaves the 'passenger strand' and facilitates the formation of active RISC complex containing only the 'guide strand' (Matranga *et al.*, 2005; Rand *et al.*, 2005). However, this model cannot account for the elimination of the passenger strand of miRNAs, since their precursors are not fully complementary and would not be cleaved by Ago-2. A presumptive explanation is that Dcr-2/R2D2 complex facilitates unwinding of the siRNA duplex, since Dcr-2 carries a DExH helicase domain and is in physical contact with the siRNA end. However, this hypothesized mechanism does not seem to be correct, since mutations in the Dcr-2 helicase motif do not affect RISC assembly. Alternatively, the Dcr-2/R2D2 complex may recruit an unknown helicase, as is the case in *Caenorhabditis elegans*, DRH-1 and DRH-2, but there are not yet evidence for this (Liu *et al.*, 2006).

One of the best-understood functions of Dcr-2 is endoribonucleic cleavage of dsRNA, such as described above. However, it seems to have other functions in the antiviral response in *Drosophila melanogaster*, to either directly or indirectly control the expression of some genes. During DCV infection the expression of more than 150 genes is triggered (Dostert *et al.*, 2005) together with the induction of one or more as yet unknown cytokines that activate the Domeless receptor and the JAK Hopscotch, which leads to activation of the *Drosophila melanogaster* STAT factor and gene induction. Hopscotch mutant flies have higher viral loads than wild type flies, which suggest that some of the induced genes encode antiviral molecules. One such gene induced by viral infection, Vago, participates in the control of viral load in the fat body. The function of Vago is not yet clear, but is postulated to act either as an antiviral molecule targeting virions or as a cytokine triggering an antiviral state in neighbouring cells. The induction of this gene is dependent on Dcr-2, and more specifically, on the DExD/H-box helicase domain (Deddouche *et al.*, 2008). The DExD/H-box helicase domain present in Dcr-2 is also part of other closely related proteins involved in the sensing of viral RNA by the RIG-like receptors (RLR) helicases, RIG-1, Mda5 and LGP2 in mammals. The sequence data suggest that the helicase domains of Dicer proteins and RLRs belong to the same family of helicases that are essential in sensing viral infection in multicellular organisms, so it is not unexpected that Dcr-2 is acting as a sensor that triggers the antiviral inducible response in flies (Deddouche *et al.*, 2008). Thus, an evolutionary conserved set of DExD/H-box helicases directs antiviral response in *Drosophila melanogaster* and mammals.

Dcr-2, which is known to interact with dsRNA, could be a sensor that activates Vago expression. The finding that neither Ago-2 nor R2D2 was required for Vago induction suggests that the sensing function of Dcr-2 and its ribonuclease function in RNA silencing are uncoupled. The signalling downstream of Dcr-2 probably differs from that of RLR pathway, as Dcr-2 does not contain caspase-recruitment domains and the *Drosophila melanogaster* genome does not contain orthologues of either the signal transducer IPS-1 (interferon-β promoter stimulator 1) or the IRF (Interferon regulatory factor) transcription factors (Deddouche *et al.*, 2008).

Then, Dcr-2 seems to have a dual function in fruit fly, as a component of the RNAi antiviral immunity and as an RLR-like helicase regulating the induction of molecules that can control viral load in some tissues.

## Small interfering RNAs (siRNAs)

The siRNAs generated in *Drosophila melanogaster* are 21 nt in length with a characteristic 5′-$PO_4$ group and a 3′-OH group with 2 nt overhangs at the 3′ ends (Elbashir *et al.*, 2001). These functional groups are indicative of ribonuclease III activity (Dcr-2). The 5′-$PO_4$ group stabilizes the RISC complex and is important for RISC fidelity in determining the correct cleavage site on the target RNA. In absence of this group, the RNAi response cannot be initiated. It is believed that 5′-$PO_4$ groups are recognized at multiple steps in RISC assembly. The presence of the 3′-OH group is important, but not essential as mutations in this end are relatively well tolerated (Kavi *et al.*, 2008). For a correct gene silencing, the siRNA pathway requires an exact complementarity between the guide strand and the target RNA.

Dcr-2 acts on a long dsRNA molecule to produce siRNA, which is the hallmark of RNA silencing mechanisms. The two strands of the siRNA duplex have different thermodynamic stabilities at their ends, an important property to identify the guide strand of the siRNA. The guide strand is incorporated into the RISC complex and mediates sequence-specific degradation of complementary or near-complementary target mRNAs.

## RNA-induced silencing complex (RISC)

RISC is a multi-subunit ribonucleoprotein complex that plays a key role in RNAi with a reported size in the range of 200 to 500 kDa. RISC shows RNase activity when exposed to RNAs homologous to input dsRNAs, but lacks activity in the presence of non-targeted RNAs.

Being a multiprotein complex, RISC assembly is critical for generating an effective RNAi response. Unlike Ago-2 loaded via the RISC assembly pathway, affinity-purified (Liu *et al.*, 2004) or recombinant Ago-2 (Rivas *et al.*, 2005) can only be programmed with single, not double-stranded siRNA, suggesting that the essential function of the RISC loading complex is to facilitate the directional loading of a double-stranded siRNA into Ago-2. The idea that Ago-2 mediates passenger strand cleavage triggering siRNA strand dissociation also accounts for previous observation where one passenger strand appears to be destroyed for every cycle of assembly of a guide strand into target-cleaving RISC (Schwarz *et al.*, 2003). Passenger-strand cleavage would strongly reinforce siRNA functional asymmetry by coupling passenger-strand destruction to RISC assembly.

As the guide strand is produced, additional proteins are loading in the holo-RISC (active RISC), which presumably help in the formation of RISC. It is postulated that various protein components found *in vivo* in RISC might play an important role in assembly, target cleavage, formation of a distinct effector complex, or all of them. Biochemical purification of the *Drosophila melanogaster* embryonic lysate has led to the identification of the following components: (i) Ago-2, (ii) dFXR (*Drosophila melanogaster* orthologue of fragile X mental retardation protein), (iii) VIG (vasa intronic gene), (iv) Tudor-SN (a nuclease with a tudor domain and bearing 5 nuclease domains homologous to the Staphylococcus nuclease domains), (v) R2D2, (vi) Aubergine, (vii) Armitage-RNA helicase, (viii) siRNAs and potentially other unidentified factors (Kavi *et al.*, 2008). Although these proteins are present when purifying RISC, it is not known whether all of them are essential partners for the Dicer family (Dcr-1 and Dcr-2 in fruit flies) and/or for the holo-siRISC.

The active RISC is directed to its target RNA via the guide strand, which then results in a selective degradation mediated by Ago-2. The specificity of this reaction is conferred by the complementarity between the guide strand and the target RNA. The accessibility of the target is important for the efficacy of the RNAi response, because target RNAs adopting a complex secondary structure hinder cleavage by RISC (Kavi *et al.*, 2008).

## Systemic RNA silencing

### RNAi amplification

*Drosophila melanogaster* is a complex multicellular organism that must generate an efficient antiviral state in the whole body. This antiviral state requires more than a cellular and local antiviral response, which although important, may not be sufficient. In this sense, the spread of an RNAi antiviral signal is important for the establishment of systemic immunity. The spreading of RNAi signal in the fruit fly has been controversial, since RNA silencing, persistence and spread in fungi, plants and worms involves a cellular RNA-dependent RNA polymerase (RdRP). However, the multiple turn over of RISC may mediate RNA silencing in the absence of a cellular RdRP in *Drosophila melanogaster* and mammalian cells. In addition to their work as guides for RNA target degradation in the absence of RdRP (Schwarz *et al.*, 2002), other studies suggest a 'primer' function for siRNAs amplifying the signal (Lipardi *et al.*, 2001). The fact that the *Drosophila melanogaster* genome does not contain orthologues of RdRP, which may amplify silencing, argues against the existence of systemic silencing. Other results that point in this direction are that the 3′-OH group of the siRNA guide is not essential for RNAi, an observation that argues against the 'primer' model, which would be required for systemic silencing. On the other hand, Paterson and co-workers proposed that siRNAs function as primers and not guides, thus arguing in favour of the presence of systemic silencing in *Drosophila melanogaster.* They showed that the 3′-OH is essential for RNAi and to trigger the amplification (Lipardi *et al.*, 2001). It may also be possible that molecules other than RdRP foster systemic silencing (Kavi *et al.*, 2008) since we recently showed a systemic and active RNAi spread in virus-infected flies (Saleh *et al.*, 2009) (discussed in more detail below in spreading section). In accordance with this set of data, a study from Chang and Taylor showed that in absence of an identifiable RdRP in mammals, DNA-directed RNA polymerases or other enzymes may perform a similar role to RdRPs (Chang and Taylor, 2002) and therefore an RNAi signal amplification supporting post-transcriptional gene silencing could happen.

### Spreading

Once a local virus infection is initiated, the distant cells of the organism remain susceptible unless a global antiviral state is generated. The RNAi machinery is one of the best understood candidates capable of producing this state in the fruit fly. Because the trigger of antiviral RNAi is an exogenous dsRNA molecule of viral origin and because Dcr-2 has the ability to sense dsRNAs, it seems reasonable to think that dsRNA is actually the molecule responsible for the systemic antiviral state. In this sense an interesting question arises: how does the dsRNA enter uninfected cells to generate an effective antiviral RNAi state? We showed in *Drosophila melanogaster* S2 cells that dsRNA reaches the RNAi machinery via a specific pathway that involves clathrin-mediated endocytosis and scavenger-like pattern-recognition receptors (Saleh *et al.*, 2006). Furthermore, the down-regulation of several components of vesicular intracellular transport and lipid metabolism blocked the RNAi response not only in *Drosophila melanogaster* but also in *Caenorhabditis elegans*. Thus, the uptake and transport of dsRNA to initiate an RNAi response is an active process that involves vesicle-mediated intracellular trafficking and depends on lipid modifications and cytoskeleton guidance (Saleh *et al.*, 2006). Interestingly, mutant flies defective in the exogenous dsRNA uptake machinery are hypersensitive to infection with DCV and Sindbis virus. Mortality in dsRNA-uptake defective flies is accompanied by higher levels of viral RNA and $10^2$–$10^5$-fold increases in viral titres. Therefore, the dsRNA uptake mechanism based on receptor-mediated endocytosis contributes to arthropod antiviral defence through the control of the dsRNA dissemination (Saleh *et al.*, 2009). It is appealing to invoke the endocytic pathway in the uptake of dsRNA, as endocytosis is the major mechanism by which cells absorb material from their 'external' environment. The implications of endocytic pathway genes in *Drosophila melanogaster* and *Caenorhabditis elegans* might also imply a more general and important role for endocytosis in RNAi in other organisms (Whangbo and Hunter, 2008). Several questions remain to be answered; one of the most intriguing is how the systemic immunity is produced/induced in insects?

In all multicellular organisms, the communication between cells is crucial to a proper coordination of the system. Cellular communication is important for development, function (nutrition, homeostasis, etc) and defence. Such communication often involves small molecules or proteins being transported or translocated from the signalling cells to the receptor cells. The recognition of the signals normally activates specific pathways to generate a correspondent response. The RNAi pathway does not seem to be an exception. Silencing in plants and *Caenorhabditis elegans* is non-cell autonomous and can be transmitted between cells; however, the mechanism to transport silencing information seems to be different in these organisms. For instance, plants use distinct mechanisms for cell-to-cell and long distance transport of silencing information. The long distance transport of silencing signals is through the phloem, comprised of companion cells that support sieve elements. Cell to cell transport in leaves occurs though plasmodesmal connection, which connect the cytoplasm of almost all cells [with the exception of guard cells that surround sites of gaseous exchange (stomata)] (Jose and Hunter, 2007).

Like in plants, in *Caenorhabditis elegans* the antiviral spread and the systemic RNAi nature is well-studied. The silencing signals can cross cellular boundaries and spread between cells and tissues. Furthermore, *Caenorhabditis elegans* and some other organisms can perform environmental RNAi, that is, sequence-specific gene silencing in response to environmentally encountered dsRNA (Whangbo and Hunter, 2008). dsRNA ingestion produces a specific and durable silencing effect, although generally not as potent as the effect of dsRNA injection. Importantly, these silencing signals can be generated in the absence of target gene expression in the intestine (Winston *et al.*, 2007). A critical component to perform systemic RNAi response in *Caenorhabditis elegans* is SID-1 (systemic RNAi defective 1) (Winston *et al.*, 2002). SID-1 mutant worms remain competent in cell autonomous RNAi, but cannot induce a systemic RNAi in response to dsRNA. A heterologous SID-1 expression system in *Drosophila melanogaster* S2 cells that lack a SID-1 homologue confers rapid dsRNA import from the growth media by an apparently passive mechanism. Together, these observations suggest that SID-1 might function as a dsRNA channel or as part of a multiprotein channel (Feinberg and Hunter, 2003).

Environmental RNAi in multicellular organisms involves the uptake of dsRNA by a primary group of cells, followed by the systemic spread of gene silencing into a secondary group of cells or tissues. This environmental RNAi has also been documented in other invertebrates including planaria, parasitic nematodes and the honeybee (Geldhof *et al.*, 2007; Newmark *et al.*, 2003; Orii *et al.*, 2003; Patel *et al.*, 2007). In the fruit fly, environmental RNAi seems to function under certain conditions, such as soaking embryos in dsRNA solutions. However, RNAi in response to dsRNA feeding has not been reported in larvae or adults flies.

In *Drosophila melanogaster,* the spreading model suggests that infected cells release viral dsRNA that is subsequently taken up by uninfected cells through the dsRNA uptake pathway, thereby eliciting an antiviral RNAi response. It was previously thought that *Drosophila melanogaster* is unable to systemically spread an RNAi response, based on observations that endogenously expressed RNA hairpins did not spread from cell to cell (Roignant *et al.*, 2003). However, our work demonstrated that upon virus infection, infected cells spread a silencing signal systemically that elicits protective RNAi-dependent immunity throughout the organism. While uninfected *Drosophila melanogaster* cells appear to lack a constitutive mechanism for RNAi systemic spread, as seen in plant and worm cells, they do have an active and highly efficient mechanism for dsRNA uptake that we showed to be essential for antiviral immunity. Accordingly, dsRNA is normally not released from uninfected cells, but virus infection may induce dsRNA release either through lysis of infected cells or through a virally induced shedding mechanism. We propose that these virally derived dsRNAs are taken up into uninfected cells to generate virus-specific intracellular immunity that prevents virus spread (Fig. 1.3). A direct prediction of this model is that during viral infection, specific pathogen dsRNAs are spread to induce a systemic silencing. The spread of virus-derived specific RNAi signals from thorax to head in infected flies has been recently demonstrated (Saleh *et al.*, 2009). This specific antiviral response is dependent on the RNAi core machinery and the dsRNA uptake pathway, as flies defective in those components were hypersensitive to infection with DCV and Sindbis virus. It is remarkable that blocking the spread of the RNAi signal has such a profound effect on antiviral immunity. This suggests that the cell-autonomous RNAi response is insufficient in controlling a viral infection.

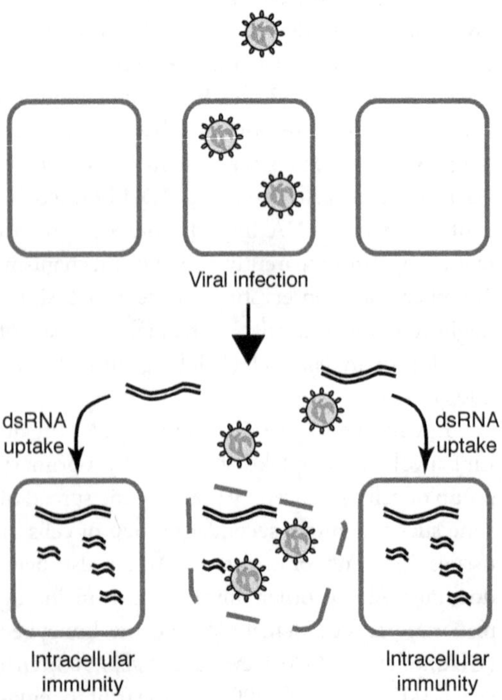


**Figure 1.3** Model for systemic RNAi viral immunity in *Drosophila melanogaster*. Upon viral infection, virus-specific dsRNAs (e.g. replication intermediates) are generated during the initial rounds of virus replication. Following cell death or lysis, dsRNAs are taken up and processed by uninfected cells to protect them from subsequent infection, thereby preventing virus spread. Reprinted from Saleh *et al.* (2009) with permission.

## Viral escape from RNAi in insects

The finding in 1998 that plant viruses encode proteins that are suppressors of RNA silencing (Anandalakshmi *et al.*, 1998; Brigneti *et al.*, 1998) was a milestone in the RNAi research, providing one of the strongest experimental support for a role of RNAi in antiviral defence. Today, more than 50 viral suppressor proteins from over 30 virus genera have been described and many of them are dsRNA binding proteins (Li and Ding, 2006). Viral suppressors of RNA silencing (VSRs) are encoded by taxonomically diverse viruses with (+)RNA, (−)RNA, dsRNA, ssDNA or dsDNA genomes infecting plants, invertebrates and vertebrates. While sequestering virus-specific siRNA or dsRNA molecules away from the RNAi pathway is apparently the primary mechanism for some suppressors, it is now clear that there is a diverse range of suppression mechanisms (Li and Ding, 2006).

VSRs encoded by viruses of different families often share no homology at the primary amino acid sequence level. However, homologues or counterparts (by location in the viral genome) of a known VSR from the same virus family always encode silencing suppression activity, even if they may share minimal sequence homology (Li and Ding, 2006). Most of the known viral suppressor genes overlap either the N-terminal or C-terminal region of viral RdRPs (Li and Ding, 2006; Pfeffer *et al.*, 2002). The first animal VSRs, B2 of FHV, was identified in an RNA silencing assay established in plants. These broad-spectrum activities of VSRs have been demonstrated for NS1 of influenza A virus and p19 of tombusviruses. That these VSRs are mostly dsRNA binding proteins may explain why they are active in both the animal and plant kingdom (Li and Ding, 2006). The varying mechanisms by which different suppressors inhibit RNAi can affect different steps in the pathway, such as (i) suppression of siRNA production (nodavirus B2 suppression protein) (Lu *et al.*, 2005; Sullivan and Ganem, 2005), (ii) sequestration of siRNA (tombusvirus P19 suppression protein) (Vargason *et al.*, 2003; Ye *et al.*, 2003), (iii) targeting Argonaute protein by protein–protein interaction (polerovirus P0 suppression protein) (Bortolamiol *et al.*, 2007) or (cucumovirus 2b suppression protein) (Zhang *et al.*, 2006), or (iv) inhibition of systemic silencing by sequestration of the 'spreading' signal (potexvirus P25) (Bayne *et al.*, 2005; Voinnet *et al.*, 2000). These different levels of mechanistic or functional inhibition of the RNAi pathway have been mainly founded in plants and up to now only a few of them in animals.

Nowadays the importance of the RNAi pathway in the antiviral response is clear; however, recent reports suggest that viruses also exploit RNA silencing mechanisms for their own benefit. For example, cleavage of the early SV40 mRNAs by its own miRNAs led to reduced expression of viral T antigens and reduced sensitivity to lysis by cytotoxic T cells without reducing the yield of infectious virus (Sullivan *et al.*, 2005). Whether insect viruses could manipulate the siRNA or miRNA pathways to improve their interaction with the host and what the physiological implications would be are open questions.

Some of the best characterized insect RNAi viral suppressors are summarized below.

## Flock house virus (FHV)

FHV is a member of the family Nodaviridae, which contains a positive-strand RNA bipartite genome and includes pathogens for insects and fishes. The genome of FHV consists of RNA1 that encodes an RdRP catalytic subunit while RNA2 encodes capsid-protein precursor. The RdRP of RNA1 templates the synthesis of a subgenomic RNA (RNA3) that encodes two small non-structural proteins, protein B1 and B2 (Fig. 1.4). B2 is synthesized at high levels early in the replication cycle (Li *et al.*, 2002).

FHV infection in *Drosophila melanogaster* cells induces the formation of membranous vesicles called 'spherules' on the outer mitochondrial membrane in which viral replication occurs (Kopek *et al.*, 2007; Miller *et al.*, 2001). Infection of *Drosophila melanogaster* induces the RNAi response in the Dcr-2/Ago-2 canonical pathway and B2 expression is essential to establish the infection. B2 acts as structural component of the viral RNA replication complex to inhibit production of siRNAs, including 5′-terminal siRNA, even though B2 is dispensable for RNA replication (Aliyari *et al.*, 2008).

B2 is a potent RNAi suppressor of 106 amino acids. It contains a dsRNA binding domain located at the N-terminal region and this domain binds dsRNA as an homodimer (Chao *et al.*, 2005; Lingel *et al.*, 2005). B2 binds tightly ($K_d$ ~1 nM) to dsRNA in a sequence-independent manner, but does not bind ssRNA, DNA or DNA–RNA hybrids. RNA binding is not selective for siRNAs, as altering the length and termini (5′-$PO_4$ and 3′ 2-nt overhangs) did not disrupt binding. B2 also inhibits the cleavage of a 35-bp RNA substrate by Dicer *in vitro*. An *in silico* assay shows that B2 can be modelled onto longer dsRNAs, suggesting that B2 may prevent Dicer processing by protectively coating the FHV genome (Fig. 1.4).

When *Drosophila melanogaster* S2 cells are infected with a B2-deficient mutant of FHV, (+) and (−) siRNA are produced in approximately the same ratio, indicating that the viral replicative intermediates, rather than structured regions of (+) or (−) RNAs, serves as the substrate of Dcr-2. Moreover, these siRNA are targeting predominantly the 5′-terminal region of the genomic RNA (5′-terminal 400-nt region of RNA1). Thus, the initiation of the progeny (+) RNA synthesis triggers processing of the 5′-terminal nascent dsRNA intermediates into siRNA. However, when S2 cells were infected with wild type FHV, the 5′-terminal siRNAs were not more abundant in comparison with other regions (van Rij and Berezikov, 2009). This indicates that the massive production of viral siRNAs targeting the nascent

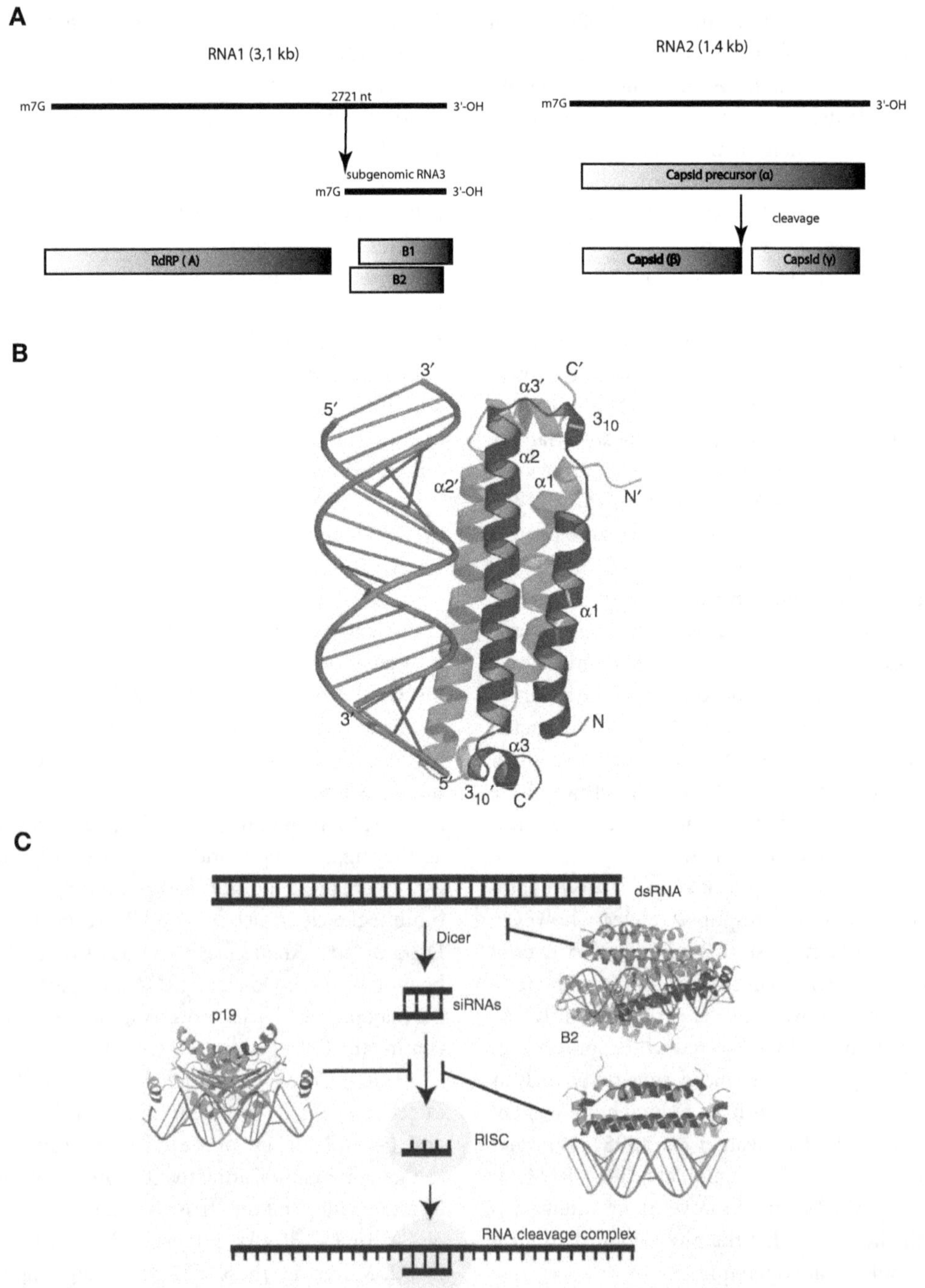


**Figure 1.4** Flock house B2 protein, an example of RNAi suppressor. (A) Flock house virus (FHV) is a segmented, bipartite and positive ssRNA virus. The genome is composed of RNA1 = 3.1 kb and RNA2 = 1.4 kb. The 5′ end of each segment is capped. The 3′ end has no poly-A tract. RNA1 encodes directly the RNA-dependent RNA polymerase (RdRP), and RNA2, the capsid protein. A subgenomic RNA (RNA3) is also produced and encode proteins B1 and B2, the RNAi suppressor. (B) Scheme of dsRNA-B2 interaction. B2 binds dsRNA as a dimmer and recognizes two successive minor grooves and the major groove that separates them. (C) Model of suppression of RNAi pathway by B2 or P19. B2 can inhibit both Dicer cleavage and siRNA incorporation into RISC. P19 acts by preventing siRNA incorporation into RISC. Adapted from Chao *et al.* (2005).

viral replicative intermediates was inhibited by B2 in FHV infected cells. Albeit the production of siRNA was markedly reduced by B2 expression, it was not completely eliminated. These 'remaining' siRNAs, however, became inactive in antiviral silencing and unable to inhibit viral replication. This result indicates that B2 suppresses the antiviral activity of siRNA in addition to inhibiting siRNA production (Aliyari *et al.*, 2008).

Even when B2 does not selectively bind siRNAs, it can bind them with high affinity. Thus, it is also possible that B2 inhibits RNA silencing by preventing siRNA loading into RISC. Indeed, detection of FHV deficient B2 mutants in *Drosophila melanogaster* cells is only possible after depletion of Ago-2, indicating that FHV infection triggers antiviral silencing in an Ago-2-dependent manner (Chao *et al.*, 2005). The ability to bind dsRNA as well as siRNA allows B2 to suppress RNA silencing in at least two steps in the RNAi pathway (Fig. 1.4).

### Drosophila C virus (DCV)

DCV infection is one of the best established models of viral infection in *Drosophila melanogaster.* DCV is a non-enveloped, (+) ssRNA virus belonging to the Dicistroviridae family. The 9264-nt genome contains two large open reading frames that encode the non-structural and structural polyproteins. Non-structural proteins encoded by ORF1 include helicase, protease and RdRP. The structural proteins encoded by ORF2 are VP0, VP1, VP2, VP3 and VP4, with VP0 a precursor of VP3 and VP4, which are assembled in the capsid.

DCV can infect *Drosophila melanogaster* cell lines and cause cell death. Injection of the virus into adult flies also causes death within 4–10 days, depending on the dose. The route of natural infection for DCV is not clear, but ingestion of the virus neither causes illness nor induces a substantial immune response (Roxstrom-Lindquist *et al.*, 2004). It was demonstrated recently that several genes controlled by the JAK-STAT pathway are specifically up-regulated upon DCV infection. Injection of DCV into adult flies induces a substantial change in gene expression profile. Approximately 140 genes are up-regulated by more than two-fold, but only one-third overlap with the 230 genes up-regulated during bacterial and fungal infection. Notably, the well characterized antimicrobial peptide genes are not induced by DCV infection (Dostert *et al.*, 2005). Thus, there is a specific antiviral pathway. It is known that DCV induces the RNAi pathway, since flies that lack a functional RISC are rapidly killed by DCV (Galiana-Arnoux *et al.*, 2006; van Rij *et al.*, 2006).

DCV encodes a suppressor which inhibits RNAi silencing initiated by long dsRNA, but does not affect the miRNA pathway. This suppressor, named DCV-1A, acts upstream of RISC specifically inhibiting Dcr-2 function. DCV-1A is translated as part of a long polyprotein that includes the viral RdRP and is encoded at the N-terminus of ORF1. DCV-1A has a canonical active dsRNA-binding domain that inhibits Dicer processing. The binding to dsRNA is sequence independent; however, length dependence was observed. Thus, the affinity of DCV-1A for the different lengths of dsRNA closely correlates with efficiency of DCV to mediate suppression of RNAi (van Rij *et al.*, 2006).

It was proposed that because DCV-1A is translated as part of the ORF that includes the RdRP, RNA replication would be coupled to the silencing suppression function. In this way, a viral genome block is created that is perhaps most effective since productive viral RNA replication requires suppression of RNA silencing (Li and Ding, 2006).

### Cricket paralysis virus (CrPV)

CrPV is a member of a group of insect picorna-like viruses. CrPV, which is closely related to DCV, encodes a RNA suppressor that maps to the same genomic region as DCV (Wang *et al.*, 2006). However, a canonical double-stranded RNA binding domain for RNAi suppression such as the one in DCV could not be predicted in CrPV (and other members of the Dicistroviridae family). In agreement with this, CrPV did not inhibit dsRNA processing in S2 cells, suggesting a different mechanism of action for the CrPV RNAi suppressor (van Rij *et al.*, 2006).

### Other viruses

The main models of virus–host interactions in the *Drosophila melanogaster* RNAi field are (+) ssRNA viruses that replicate in the cytoplasm. To carry out studies with (−) ssRNA, dsRNA or

DNA viruses (such as Sigma virus, DXV, etc.), will give us important and valuable data to understand the antiviral mechanism(s), in view that these viruses present different dsRNA intermediates and might produce alternative mechanisms to induce the RNA silencing response.

## RNAi in other insect models

Although *Drosophila melanogaster* is the best characterized insect model to study RNAi, other invertebrate models (the majority of them in economically or medically relevant arthropods, like honeybees, mosquitoes, etc.) have been studied for the presence of an active RNA silencing mechanism.

RNAi-mediated gene knockdown has been reported in several insect orders, including Diptera, Coleoptera, Hymenoptera (Araujo *et al.*, 2006; Cruz *et al.*, 2006; Dong and Friedrich, 2005; Turner *et al.*, 2006). Although most of these studies have used injected dsRNA, these results show a functional and active mechanism of RNA silencing in the insect class, suggesting a role in the defence against exogenous RNAs (see Table 1.1 for a list of insects shown to generate an RNAi response).

### Mosquitoes

Mosquitoes are insects in the family Culicidae. Both male and female mosquitoes are nectar feeders, but the females of many species are also capable of haematophagy (blood meals).

Because of this, mosquitoes can be vectors for viruses and parasites that can cause disease in humans. Infection of humans occurs when an infected mosquito feeds on a person. This cycle could be interrupted by killing the mosquitoes, isolating infected people from all mosquitoes while they are infectious or vaccinating the exposed population. All three techniques have been used, often in combination but without great success, to control mosquito-transmitted diseases (Ribeiro and Francischetti, 2003). Viruses such as Western, Eastern and Venezuelan equine encephalitis viruses, Chikungunya virus (alphaviruses);, Japanese, St. Louis and tick-borne encephalitis viruses, yellow fever virus, Dengue virus, West Nile virus (flaviviruses), Lacrosse virus and Rift Valley virus (bunyaviruses) (Weaver, 2006) are all transmitted by mosquitoes bites.

In recent years it has been demonstrated that mosquito cells support RNAi and homologues of the *Drosophila melanogaster* RNAi components are identified in the sequenced genome of *Anopheles gambiae*. Also, Ago-2, Dcr-2 and TSN (Tudor Staphylococcal nuclease) have been identified in *Aedes aegypti* by homology to *Drosophila melanogaster* and the *Anopheles Gambiae* orthologues (Campbell *et al.*, 2008b). As in *Drosophila melanogaster*, it has been demonstrated that viral replication triggers the RNAi silencing response in mosquitoes. This response is Ago-2 dependent and sensitive to the B2 suppressor. Variations in the activity of this antiviral pathway in mosquitoes might explain in part why some mosquitoes are competent human vectors and others are not. It is also possible that the methods to evade, escape or suppress the RNAi response will be effective only in the natural vectors (Li *et al.*, 2004).

When *Aedes aegypti* was artificially infected with two different Sindbis virus genotypes and the small viral RNAs produced were analysed, interesting differences were found in replication kinetics of these two virus. They could indicate a correlation between replication efficiency in the mosquito and susceptibility to the antiviral RNAi defence (Campbell *et al.*, 2008b). In the same work Campbell and co-workers showed that Sindbis infected mosquitoes did not enhance significantly Dcr-2 or Ago-2 at transcriptional level, but there was an enrichment for TSN transcripts (Campbell *et al.*, 2008b). The TSN transcript enrichment in *Aedes Aegypti* during Sindbis infection contrasts with reports of TSN depletion during DCV infection in *Drosophila melanogaster* (Dostert *et al.*, 2005). However, DCV is an insect-specific pathogen whereas Sindbis infects insects and vertebrates, establishing a persistent, non-lethal infection in the insect vector.

By comparative genomics of small RNA regulatory pathways in mosquito vectors it was found that Ago-2 protein is significantly more diverse among mosquito species than the miRNA pathway nuclease, Ago-1. This finding suggests that antiviral defence effectors are evolving at a faster rate than those involved in housekeeping functions. Simultaneously, some entomopathogenic viruses may evolve into arboviruses (Campbell *et al.*, 2008a). When considering these pathways, it is important to keep in mind that evolutionary

**Table 1.1** Insects shown to generate an RNAi response

| Insect | dsRNA delivered by | Silencing occurs in |
|---|---|---|
| Beetle (*Tribolium* sp.) | Intra-abdominal injection of pupae | Next generation (Bucher *et al.*, 2002) |
| | Intra-abdominal injection of larvae | Adult bristles (Tomoyasu *et al.*, 2008) |
| Cockroach (*Blatella periplaneta*) | Intra-abdominal injection | Ovariole (Ciudad *et al.*, 2006) |
| | Injection into cercus of larvae | Neurons and skin (Marie *et al.*, 2000) |
| Cricket (*Gryllus* sp.) | Intra-abdominal injection | Next generation (Mito *et al.*, 2005) |
| Fleshfly (*Sarcophaga* sp.) | Injection into larval haemolymph | Pupal haemocytes (Nishikawa and Natori, 2001) |
| Fruitfly (*Drosophila* sp.) | Intra-abdominal injection | Neurons in the head (Dzitoyeva *et al.*, 2003) |
| | Intra-thoracic injection | Systemic antiviral state (Saleh *et al.*, 2009) |
| Giant silkmoth (*Hyalophora cecropia*) | Injection into pupae | Next generation (Bettencourt *et al.*, 2002) |
| Grassopper (*Schistocerca* sp.) | Injection into heart vessel of larvae | Developing eye (Dong and Friedrich, 2005) |
| Honeybee (*Apis* sp.) | Injection into haemolymph | Fat body (Amdam *et al.*, 2003) |
| | Feeding | Antiviral response against Israeli acute paralysis virus (Maori *et al.*, 2009) |
| | Embryo injection | E30 homeobox motif (Beye *et al.*, 2002) |
| Kissing bug (*Rhodnius sp.*) | Thoracic injection of larvae | Salivary gland (Araujo *et al.*, 2006) |
| Locust (*Locusta* sp.) | Intra-abdominal injection | Next generation (He *et al.*, 2006) |
| Milkweed bug (*Oncopeltus*) | Intra-abdominal injection | Next generation (Liu and Kaufman, 2004) |
| Mosquito (*Anopheles gambiae*) | Thoracic injection of larvae | Midgut, fat body (Blandin *et al.*, 2002) |
| Mosquito (*Aedes egyptae*) | Injection into haemocele | Gut (Bartholomay *et al.*, 2004) |
| | Injection dsRNA/viral infection | Antiviral response against Sindbis virus (Campbell *et al.*, 2008b) |
| Moth (*Spodoptera* sp.; *Manuka* sp.; *Epiphyas* sp.) | Injection into haemocoele | Midgut (Rajagopal *et al.*, 2002) |
| | Intra-abdominal injection | Fat body (Eleftherianos *et al.*, 2006) |
| | Fed droplets to larvae | Antennae of adults (Turner *et al.*, 2006) |
| Silkworm (*Bombyx sp.*) | Injection into haemocoele | Midgut, silk glands (Tabunoki *et al.*, 2004) |
| Termite (*Reticulitremes* sp.) | Injection into side of the thorax | Fat body (Zhou *et al.*, 2006) |
| Tick (*Amblyomma* sp.; *Ixodes* sp.) | Injection into haemolymph | Salivary gland (Aljamali *et al.*, 2003; Sukumaran *et al.*, 2006) |
| | Fed droplets to larvae | Salivary gland (Soares *et al.*, 2005) |
| Wasp (*Nasonia* sp.) | Intra-abdominal injection | Next generation (Lynch *et al.*, 2006) |

Adapted from Jose and Hunter (2007).

pressure exists on both the arbovirus and the mosquito vector to modulate the immune response.

There are multiple potential alternatives to develop new tools in the control of viral diseases using the RNAi pathway. For instance, if genes involved in the antiviral response are identified, one could look for variations in these genes in the vector population and associate them with vector competence and transmission. In this way, it would be possible to have a rapid screening to identify vector populations. These vector populations can be targeted for control or elimination (Sanchez-Vargas *et al.*, 2004). It is also possible to think of other control strategies that induce a more robust antiviral RNAi pathway in mosquitoes. Mosquitoes could be modified to be poorly competent vectors or as vectors that completely eliminates the virus by producing specific fragments of viral dsRNA that quickly trigger an RNAi response (Sanchez-Vargas *et al.*, 2004). A rapid RNAi response could be a key in viral clearance, since virus may replicate in and escape the midgut before the RNAi response is under way. For example, there are some efforts to develop transgenic mosquitoes, in order to introduce anti-dengue genes into dengue competent mosquito population. The idea is to block the replication and transmission of the viruses by producing anti-dengue effector molecules (dsRNA) throughout the whole mosquito or preferentially in the salivary glands or midgut, the site of the primary encounter with the pathogen or site of transmission respectively. This approach would prevent the insect viral infection by destruction of the viral genome before the virus could evade the RNAi response (Travanty *et al.*, 2004).

### *Tribolium castaneum* (red flour beetle)

Red flour beetles attack stored grain products and seeds (flour, cereals, pasta, biscuits, beans, nuts, etc.) causing damage and economic loss. This species is important for RNAi pathway research since *Tribolium castaneum* is becoming an accepted model for the study of systemic RNAi in insects. It was shown that injection of dsRNA into the mother's haemocoele results in knock-down of zygotic genes in offspring embryos ('parental RNAi') (Bucher *et al.*, 2002). Even though the systemic RNAi has been demonstrated in this beetle, a recent genomic comparison of *Caenorhabditis elegans* and *Tribolium castaneum* reveals a lack of conservation of a systemic RNAi mechanism (Tomoyasu *et al.*, 2008). For example, *Tribolium castaneum* lacks the *Caenorhabditis elegans*-like RdRP, then the signal amplification observed in this insect must be based on a different gene with a similar activity or possibly by a different mechanism. RdRP-like activity has been demonstrated in cell-free extracts from *Drosophila melanogaster* embryos, even though the RdRP gene is not present in these insects (Araujo *et al.*, 2006).

### *Epiphyas postvittana* (light brown apple moth)

The light brown apple moth (*Epiphyas postvittana*) is a leaf-roller moth belonging to the Lepidoptera. The insect is regarded as an herbivorous generalist, and the larvae feed on numerous horticultural crops with limited natural predators. There is interesting work carried out in the light brown apple moth that shows RNAi effects after dsRNA feeding in larval stage. In this work, dsRNA against carboxyestereses was incorporated into an artificial diet. Gene repression was observed after 2 days of feeding while maximal effect occurred after 7 days. These genes are thought to be gut-expressed and thus, only a local RNAi effect was required for repression. However, the most interesting observation is the knock-down of a gene expressed in the adult antenna through feeding dsRNA to larvae, demonstrating the persistence of the RNAi signal throughout the larval and adult stages with a systemic spread of RNAi signalling from the gut to the antennae (Price and Gatehouse, 2008; Turner *et al.*, 2006). Although not all insect systems have been successfully tested in gene down-regulation by dsRNA feeding, this result is very encouraging since it opens the possibility of pest control in agriculture and human health.

### *Apis mellifera* (honeybee)

The honeybee, *Apis mellifera*, is a key model for social behaviour and ecologically essential as a pollinator. Compared with other sequenced insect genomes, the *Apis mellifera* genome evolves more slowly than that of the fruit fly and malaria mosquito, and is more similar to vertebrates for

RNA interference and DNA methylation genes, among others (Honeybee Genome Sequencing Consortium, 2006). Honeybees live in highly crowded hives, providing favourable demographic conditions for infectious diseases. Curiously, given the predicted disease pressures in honeybee colonies, the honeybee genome encodes fewer proteins implicated in insect immune pathways when compared to other insect genomes. Although the Toll, Imd and JAK/STAT pathways seem to be intact, paralogue counts for gene families implicated in these pathways are reduced by two-thirds (Evans *et al.*, 2006). This reduction spans every step in the immune response from pathogen recognition to immune effectors, and implies a reduced flexibility in the abilities of honeybees to recognize and resist pathogens. The results suggest that honeybees use novel immune pathways, are poorly defended against pathogens at the individual level, and/or have immune systems that are narrowly focused on a relatively small group of coevolved pathogens.

The honeybee genome harbours only a single homologue of SID-1, a gene essential for systemic RNAi (Winston *et al.*, 2002), whereas multiple copies of this gene occur in the moth and draft *Tribolium castaneum* genomes (Honeybee Genome Sequencing Consortium, 2006). The honeybee SID-1 homologue clusters with one of three beetle and two of three moth SID homologues, as well as two vertebrate paralogues, one of which was lost in fish but preserved in birds and mammals. This disparity in SID-1-like gene numbers may relate to honeybee RNAi efficacy. RNAi in honeybees was first demonstrated by embryo injection of dsRNAs (Beye *et al.*, 2002), and later experiments showed that, as in other invertebrates, injection of dsRNA can reduce target gene expression away from the site of injection (Amdam *et al.*, 2003). However, the molecular details of honeybee gene silencing have not been clearly elucidated (Honeybee Genome Sequencing Consortium, 2006) and little is known about the role of SID-1 proteins in insect systemic RNAi.

The honeybee genome also encodes many other proteins similar to the core RNAi machinery of other species, including two Dicer (Dcr) enzymes, a Drosha homologue and other RISC components, like Ago-1 and Ago-2, as well as a full suite of dsRNA binding proteins – R2D2, Pasha and Loquacious – all of which have yet to be confirmed as components of the RNAi pathway in honeybees.

Israeli acute paralysis virus (IAPV) has been characterized as a bee-affecting dicistrovirus (Maori *et al.*, 2007a) and is one of the most interesting examples of dsRNA-related antiviral immunity. It has been reported that by feeding bees with dsRNA of IAPV sequences, IAPV infection is silenced and bee mortality is prevented. Furthermore, the presence of IAPV-specific siRNA (21, 22 and 25 bp in length) was reported in bees fed IAPV-dsRNA, indicating that the antiviral response had been initiated by the ingested dsRNA (Maori *et al.*, 2009). These results are promising for antiviral therapies in honeybees used for honey production and pollination of many major fruit crops.

Recently, DNA versions of genomic segments of non-retro RNA viruses have been found in their respective host genomes, and the reciprocal exchange of genome sequence between host and viruses has been demonstrated (Maori *et al.*, 2007b; Tanne and Sela, 2005). This is the case of IAPV and bees, with beneficial results for bees, since insects carrying segments of the viral genome are resistant to IAPV (Maori *et al.*, 2007a).

## Concluding remarks

The recent discovery of several families of small, regulatory, non-coding RNAs and their implication in fundamental biological processes, from microbial immunity to development, has opened a fast-paced field of research that continues to provide unexpected insights into biology. The study of RNA silencing in insects has unravelled a new, sophisticated and coordinated immune system that relies on recognition of specific nucleic acids and is selectively activated by exogenous dsRNA. In face of this discovery, important questions arise about the origin and the evolutionary destiny of the RNAi immune system. One of the most relevant, being if RNAi is a natural antiviral mechanism in mammals. Since information is contradictory in this field, one would like to ask why evolution would select against an antiviral mechanism that is very efficient in invertebrates. Two outstanding characteristics of the mammalian adaptive immune system are immunological memory, i.e. the pathogen-specific response maintained in the absence of the original stimulant; and recognition, i.e. target-specific destruction of the pathogen. When comparing RNAi as an immune system with the adaptive immune system, can memory be the central point to favour the adaptive immune system against the nucleic-acid-based immune system? Is there memory in the RNAi immune system of insects and is this memory based on the half-life of RISC preloaded with virus specific siRNAs? Another possibility is that the existence of memory could be linked to the existence of an RdRP. In the presence of amplification of the silencing signal, a 'memory-like state' could be generated and spread throughout the organism. It is also possible that long-term 'immunological' memory would not confer an advantage to short-lived species such as insects, as it would to higher animals. If antiviral RNAi is vestigial in mammals, it should also be possible to find animals or possibly, specific developmental stages in vertebrates, that exhibit a double response: cellular immunity-protein-based and RNAi immunity nucleic-acid based.

Besides these mechanistic considerations, it is also important to see if RNAi can live up to the promise as a great therapeutic tool. When considering the impact of RNAi as an antiviral in insects, we think immediately of arboviruses and mosquitoes. Future directions of research should address if RNAi modulates the outcome of viral infection in mosquitoes, how virus–vector interactions are modulated, whether arboviruses have suppressors of RNAi, whether RNAi acts as a factor to restrict the host range of certain arboviruses (i.e. does arbovirus X infect *Anopheles* and not *Culex* due to an RNAi-mediated response?), and finally, if the vector-RNAi response can be used as a strategy to control arboviruses.

Finally, it is tempting to consider the role of RNAi as an immune system as much more involved than to only control viral infections. In the work from Navarro *et al* (Navarro *et al.*, 2008), infection of *Arabidopsis* with Turnip mosaic virus, which produces a suppressor of both siRNA and microRNA pathways, promotes infection by non-pathogenic bacteria, which may explain the observed synergy between viral and bacterial pathogens seen in the field. This observation point to the fact that RNAi may be a central controller with a wide-spectrum of effect on the immune response, not only involved in antiviral responses, and that bacteria have evolved to suppress RNA silencing in order to escape from this response. Once again, plants lead the way on the discovery of RNAi and show us where to direct our steps. Let's follow them.

## Acknowledgements

We are grateful to members of the Saleh lab for their help in preparing this book chapter. Christophe Antoniewski, Marco Vignuzzi and Ronald van Rij for fruitful discussions and critical reading of the manuscript. We apologize to those colleagues whose work has not been discussed. Bertsy Goic is a Manlio Cantarini fellow. Maria Carla Saleh is funded by the Institut Pasteur.

## References

Adams, M.D., and Sekelsky, J.J. (2002). From sequence to phenotype: reverse genetics in *Drosophila melanogaster*. Nat. Rev. Genet. *3*, 189–198.

Adelman, Z.N., Blair, C.D., Carlson, J.O., Beaty, B.J., and Olson, K.E. (2001). Sindbis virus-induced silencing of dengue viruses in mosquitoes. Insect Mol. Biol. *10*, 265–273.

Agaisse, H., and Perrimon, N. (2004). The roles of JAK/STAT signaling in *Drosophila* immune responses. Immunol. Rev. *198*, 72–82.

Aliyari, R., Wu, Q., Li, H.W., Wang, X.H., Li, F., Green, L.D., Han, C.S., Li, W.X., and Ding, S.W. (2008). Mechanism of induction and suppression of antiviral immunity directed by virus-derived small RNAs in *Drosophila*. Cell Host Microbe *4*, 387–397.

Aljamali, M.N., Bior, A.D., Sauer, J.R., and Essenberg, R.C. (2003). RNA interference in ticks: a study using histamine binding protein dsRNA in the female tick *Amblyomma americanum*. Insect Mol. Biol. *12*, 299–305.

Aljamali, M.N., Sauer, J.R., and Essenberg, R.C. (2002). RNA interference: applicability in tick research. Exp. Appl. Acarol. *28*, 89–96.

Amdam, G.V., Simoes, Z.L., Guidugli, K.R., Norberg, K., and Omholt, S.W. (2003). Disruption of vitellogenin gene function in adult honeybees by intra-abdominal injection of double-stranded RNA. BMC Biotechnol. *3*, 1.

Anandalakshmi, R., Pruss, G.J., Ge, X., Marathe, R., Mallory, A.C., Smith, T.H., and Vance, V.B. (1998). A viral suppressor of gene silencing in plants. Proc. Natl. Acad. Sci. U.S.A. *95*, 13079–13084.

Araujo, R.N., Santos, A., Pinto, F.S., Gontijo, N.F., Lehane, M.J., and Pereira, M.H. (2006). RNA interference of the salivary gland nitrophorin 2 in the triatomine bug *Rhodnius prolixus* (Hemiptera: Reduviidae) by dsRNA ingestion or injection. Insect Biochem. Mol. Biol. *36*, 683–693.

Ashburner, M., Golic, K.G., and Hawley, R.S. (2005). Drosophila: a Laboratory Handbook, 2nd edn (Cold Spring Harbor, NY, Cold Spring Harbor Laboratory Press.).

Bangham, J., Jiggins, F., and Lemaitre, B. (2006). Insect immunity: the post-genomic era. Immunity 25, 1–5.

Bartholomay, L.C., Fuchs, J.F., Cheng, L.L., Beck, E.T., Vizioli, J., Lowenberger, C., and Christensen, B.M. (2004). Reassessing the role of defensin in the innate immune response of the mosquito, *Aedes aegypti*. Insect Mol. Biol. *13*, 125–132.

Baulcombe, D. (2004). RNA silencing in plants. Nature *431*, 356–363.

Bayne, E.H., Rakitina, D.V., Morozov, S.Y., and Baulcombe, D.C. (2005). Cell-to-cell movement of potato potexvirus X is dependent on suppression of RNA silencing. Plant J. *44*, 471–482.

Bellen, H.J. (1998). The fruit fly: a model organism to study the genetics of alcohol abuse and addiction? Cell 93, 909–912.

Bettencourt, R., Terenius, O., and Faye, I. (2002). Hemolin gene silencing by ds-RNA injected into *Cecropia pupae* is lethal to next generation embryos. Insect Mol. Biol. *11*, 267–271.

Beye, M., Hartel, S., Hagen, A., Hasselmann, M., and Omholt, S.W. (2002). Specific developmental gene silencing in the honey bee using a homeobox motif. Insect Mol. Biol. *11*, 527–532.

Bier, E. (2005). *Drosophila*, the golden bug, emerges as a tool for human genetics. Nat. Rev. Genet. *6*, 9–23.

Billecocq, A., Vazeille-Falcoz, M., Rodhain, F., and Bouloy, M. (2000). Pathogen-specific resistance to Rift Valley fever virus infection is induced in mosquito cells by expression of the recombinant nucleoprotein but not NSs non-structural protein sequences. J. Gen. Virol. *81*, 2161–2166.

Blandin, S., Moita, L.F., Kocher, T., Wilm, M., Kafatos, F.C., and Levashina, E.A. (2002). Reverse genetics in the mosquito *Anopheles gambiae*: targeted disruption of the Defensin gene. EMBO Rep. *3*, 852–856.

Boman, H.G., Nilsson, I., and Rasmuson, B. (1972). Inducible antibacterial defence system in *Drosophila*. Nature *237*, 232–235.

Bortolamiol, D., Pazhouhandeh, M., Marrocco, K., Genschik, P., and Ziegler-Graff, V. (2007). The Polerovirus F box protein P0 targets ARGONAUTE1 to suppress RNA silencing. Curr. Biol. *17*, 1615–1621.

Brigneti, G., Voinnet, O., Li, W.X., Ji, L.H., Ding, S.W., and Baulcombe, D.C. (1998). Viral pathogenicity determinants are suppressors of transgene silencing in *Nicotiana benthamiana*. EMBO J. *17*, 6739–6746.

Bucher, G., Scholten, J., and Klingler, M. (2002). Parental RNAi in *Tribolium* (Coleoptera). Curr. Biol. *12*, R85–86.

Campbell, C.L., Black, W.C.t., Hess, A.M., and Foy, B.D. (2008a). Comparative genomics of small RNA regulatory pathway components in vector mosquitoes. BMC Genomics *9*, 425.

Campbell, C.L., Keene, K.M., Brackney, D.E., Olson, K.E., Blair, C.D., Wilusz, J., and Foy, B.D. (2008b). *Aedes aegypti* uses RNA interference in defense against Sindbis virus infection. BMC Microbiol. *8*, 47.

Chang, J., and Taylor, J. (2002). *In vivo* RNA-directed transcription, with template switching, by a mammalian RNA polymerase. EMBO J. *21*, 157–164.

Chao, J.A., Lee, J.H., Chapados, B.R., Debler, E.W., Schneemann, A., and Williamson, J.R. (2005). Dual modes of RNA-silencing suppression by Flock House virus protein B2. Nat. Struct. Mol. Biol. *12*, 952–957.

Ciudad, L., Piulachs, M.D., and Belles, X. (2006). Systemic RNAi of the cockroach vitellogenin receptor results in a phenotype similar to that of the *Drosophila* yolkless mutant. FEBS J *273*, 325–335.

Cruz, J., Mane-Padros, D., Belles, X., and Martin, D. (2006). Functions of the ecdysone receptor isoform-A in the hemimetabolous insect *Blattella germanica* revealed by systemic RNAi *in vivo*. Dev. Biol. *297*, 158–171.

Deddouche, S., Matt, N., Budd, A., Mueller, S., Kemp, C., Galiana-Arnoux, D., Dostert, C., Antoniewski, C., Hoffmann, J.A., and Imler, J.L. (2008). The DExD/H-box helicase Dicer-2 mediates the induction of antiviral activity in drosophila. Nat. Immunol. *9*, 1425–1432.

Ding, S.W., and Voinnet, O. (2007). Antiviral immunity directed by small RNAs. Cell *130*, 413–426.

Dong, Y., and Friedrich, M. (2005). Nymphal RNAi: systemic RNAi mediated gene knockdown in juvenile grasshopper. BMC Biotechnol. *5*, 25.

Dorner, T., and Radbruch, A. (2007). Antibodies and B cell memory in viral immunity. Immunity *27*, 384–392.

Dostert, C., Jouanguy, E., Irving, P., Troxler, L., Galiana-Arnoux, D., Hetru, C., Hoffmann, J.A., and Imler, J.L. (2005). The Jak-STAT signaling pathway is required but not sufficient for the antiviral response of drosophila. Nat. Immunol. *6*, 946–953.

Dzitoyeva, S., Dimitrijevic, N., and Manev, H. (2003). Gamma-aminobutyric acid B receptor 1 mediates behavior-impairing actions of alcohol in *Drosophila*: adult RNA interference and pharmacological evidence. Proc. Natl. Acad. Sci. U.S.A. *100*, 5485–5490.

Elbashir, S.M., Martinez, J., Patkaniowska, A., Lendeckel, W., and Tuschl, T. (2001). Functional anatomy of siRNAs for mediating efficient RNAi in *Drosophila melanogaster* embryo lysate. EMBO J. *20*, 6877–6888.

Eleftherianos, I., Millichap, P.J., ffrench-Constant, R.H., and Reynolds, S.E. (2006). RNAi suppression of recognition protein mediated immune responses in the tobacco hornworm *Manduca sexta* causes increased susceptibility to the insect pathogen *Photorhabdus*. Dev. Comp. Immunol. *30*, 1099–1107.

Evans, J.D., Aronstein, K., Chen, Y.P., Hetru, C., Imler, J.L., Jiang, H., Kanost, M., Thompson, G.J., Zou, Z., and Hultmark, D. (2006). Immune pathways and defence mechanisms in honey bees *Apis mellifera*. Insect Mol. Biol. *15*, 645–656.

Feinberg, E.H., and Hunter, C.P. (2003). Transport of dsRNA into cells by the transmembrane protein SID-1. Science *301*, 1545–1547.

Gaines, P.J., Olson, K.E., Higgs, S., Powers, A.M., Beaty, B.J., and Blair, C.D. (1996). Pathogen-derived resistance to dengue type 2 virus in mosquito cells by expression of the premembrane coding region of the viral genome. J. Virol. *70*, 2132–2137.

Galiana-Arnoux, D., Dostert, C., Schneemann, A., Hoffmann, J.A., and Imler, J.L. (2006). Essential function *in vivo* for Dicer-2 in host defense against RNA viruses in drosophila. Nat. Immunol. *7*, 590–597.

Geldhof, P., Visser, A., Clark, D., Saunders, G., Britton, C., Gilleard, J., Berriman, M., and Knox, D. (2007). RNA interference in parasitic helminths: current situation, potential pitfalls and future prospects. Parasitology *134*, 609–619.

Habayeb, M.S., Ekengren, S.K., and Hultmark, D. (2006). Nora virus, a persistent virus in *Drosophila*, defines a new picorna-like virus family. J. Gen. Virol. *87*, 3045–3051.

He, Z.B., Cao, Y.Q., Yin, Y.P., Wang, Z.K., Chen, B., Peng, G.X., and Xia, Y.X. (2006). Role of hunchback in segment patterning of *Locusta migratoria manilensis* revealed by parental RNAi. Dev. Growth Differ. *48*, 439–445.

Honeybee Genome Sequencing Consortium. (2006). Insights into social insects from the genome of the honeybee *Apis mellifera*. Nature *443*, 931–949.

Hultmark, D. (2003). *Drosophila* immunity: paths and patterns. Curr. Opin. Immunol. *15*, 12–19.

Jiang, F., Ye, X., Liu, X., Fincher, L., McKearin, D., and Liu, Q. (2005). Dicer-1 and R3D1-L catalyze microRNA maturation in *Drosophila*. Genes Dev. *19*, 1674–1679.

Jose, A.M., and Hunter, C.P. (2007). Transport of sequence-specific RNA interference information between cells. Annu. Rev. Genet. *41*, 305–330.

Kavi, H.H., Fernandez, H., Xie, W., and Birchler, J.A. (2008). Genetics and biochemistry of RNAi in *Drosophila*. Curr. Top. Microbiol. Immunol. *320*, 37–75.

Kennerdell, J.R., and Carthew, R.W. (1998). Use of dsRNA-mediated genetic interference to demonstrate that frizzled and frizzled 2 act in the wingless pathway. Cell 95, 1017–1026.

Kohler, R.E. (1994). Lords of the Fly (Chicago: The University of Chicago Press).

Kopek, B.G., Perkins, G., Miller, D.J., Ellisman, M.H., and Ahlquist, P. (2007). Three-dimensional analysis of a viral RNA replication complex reveals a virus-induced mini-organelle. PLoS Biol. 5, e220.

Lee, Y.S., Nakahara, K., Pham, J.W., Kim, K., He, Z., Sontheimer, E.J., and Carthew, R.W. (2004). Distinct roles for *Drosophila* Dicer-1 and Dicer-2 in the siRNA/miRNA silencing pathways. Cell *117*, 69–81.

Lemaitre, B., and Hoffmann, J. (2007). The host defense of *Drosophila melanogaster*. Annu. Rev. Immunol. *25*, 697–743.

Lemaitre, B., Reichhart, J.M., and Hoffmann, J.A. (1997). *Drosophila* host defense: differential induction of antimicrobial peptide genes after infection by various classes of microorganisms. Proc. Natl. Acad. Sci. U.S.A. *94*, 14614–14619.

Li, F., and Ding, S.W. (2006). Virus counterdefense: diverse strategies for evading the RNA-silencing immunity. Annu. Rev. Microbiol. *60*, 503–531.

Li, H., Li, W.X., and Ding, S.W. (2002). Induction and suppression of RNA silencing by an animal virus. Science *296*, 1319–1321.

Li, W.X., Li, H., Lu, R., Li, F., Dus, M., Atkinson, P., Brydon, E.W., Johnson, K.L., Garcia-Sastre, A., Ball, L.A., *et al.* (2004). Interferon antagonist proteins of influenza and vaccinia viruses are suppressors of RNA silencing. Proc. Natl. Acad. Sci. U.S.A. *101*, 1350–1355.

Lindbo, J.A., Silva-Rosales, L., Proebsting, W.M., and Dougherty, W.G. (1993). Induction of a highly specific antiviral state in transgenic plants – implications for regulation of gene expression and virus resistance. Plant Cell 5, 1749–1759.

Lingel, A., Simon, B., Izaurralde, E., and Sattler, M. (2005). The structure of the flock house virus B2 protein, a viral suppressor of RNA interference, shows a novel mode of double-stranded RNA recognition. EMBO Rep. *6*, 1149–1155.

Lipardi, C., Wei, Q., and Paterson, B.M. (2001). RNAi as random degradative PCR: siRNA primers convert mRNA into dsRNAs that are degraded to generate new siRNAs. Cell *107*, 297–307.

Liu, J., Carmell, M.A., Rivas, F.V., Marsden, C.G., Thomson, J.M., Song, J.J., Hammond, S.M., Joshua-Tor, L., and Hannon, G.J. (2004). Argonaute2 is the catalytic engine of mammalian RNAi. Science *305*, 1437–1441.

Liu, P.Z., and Kaufman, T.C. (2004). Hunchback is required for suppression of abdominal identity, and for proper germband growth and segmentation in the intermediate germband insect *Oncopeltus fasciatus*. Development *131*, 1515–1527.

Liu, X., Jiang, F., Kalidas, S., Smith, D., and Liu, Q. (2006). Dicer-2 and R2D2 coordinately bind siRNA to promote assembly of the siRISC complexes. RNA *12*, 1514–1520.

Lu, R., Maduro, M., Li, F., Li, H.W., Broitman-Maduro, G., Li, W.X., and Ding, S.W. (2005). Animal virus replication and RNAi-mediated antiviral silencing in *Caenorhabditis elegans*. Nature *436*, 1040–1043.

Lynch, J.A., Brent, A.E., Leaf, D.S., Pultz, M.A., and Desplan, C. (2006). Localized maternal orthodenticle patterns anterior and posterior in the long germ wasp *Nasonia*. Nature *439*, 728–732.

Maori, E., Lavi, S., Mozes-Koch, R., Gantman, Y., Peretz, Y., Edelbaum, O., Tanne, E., and Sela, I. (2007a). Isolation and characterization of Israeli acute paralysis virus, a dicistrovirus affecting honeybees in Israel: evidence for diversity due to intra- and inter-species recombination. J. Gen. Virol. *88*, 3428–3438.

Maori, E., Paldi, N., Shafir, S., Kalev, H., Tsur, E., Glick, E., and Sela, I. (2009). IAPV, a bee-affecting virus associated with colony collapse disorder can be silenced by dsRNA ingestion. Insect Mol. Biol. *18*, 55–60.

Maori, E., Tanne, E., and Sela, I. (2007b). Reciprocal sequence exchange between non-retro viruses and hosts leading to the appearance of new host phenotypes. Virology *362*, 342–349.

Margulies, C., Tully, T., and Dubnau, J. (2005). Deconstructing memory in *Drosophila*. Curr. Biol. *15*, R700–713.

Marie, B., Bacon, J.P., and Blagburn, J.M. (2000). Double-stranded RNA interference shows that Engrailed controls the synaptic specificity of identified sensory neurons. Curr. Biol. *10*, 289–292.

Marques, J.T., and Carthew, R.W. (2007). A call to arms: coevolution of animal viruses and host innate immune responses. Trends Genet. *23*, 359–364.

Matranga, C., Tomari, Y., Shin, C., Bartel, D.P., and Zamore, P.D. (2005). Passenger-strand cleavage facilitates assembly of siRNA into Ago2-containing RNAi enzyme complexes. Cell *123*, 607–620.

Matzke, M.A., Primig, M., Trnovsky, J., and Matzke, A.J. (1989). Reversible methylation and inactivation of marker genes in sequentially transformed tobacco plants. EMBO J. *8*, 643–649.

Miller, D.J., Schwartz, M.D., and Ahlquist, P. (2001). Flock house virus RNA replicates on outer mitochondrial membranes in *Drosophila* cells. J. Virol. *75*, 11664–11676.

Mito, T., Sarashina, I., Zhang, H., Iwahashi, A., Okamoto, H., Miyawaki, K., Shinmyo, Y., Ohuchi, H., and Noji, S. (2005). Non-canonical functions of hunchback in segment patterning of the intermediate germ cricket *Gryllus bimaculatus*. Development *132*, 2069–2079.

Napoli, C., Lemieux, C., and Jorgensen, R. (1990). Introduction of a chimeric chalcone synthase gene into petunia results in reversible co-suppression of homologous genes in trans. Plant Cell *2*, 279–289.

Navarro, L., Jay, F., Nomura, K., He, S.Y., and Voinnet, O. (2008). Suppression of the microRNA pathway by bacterial effector proteins. Science *321*, 964–967.

Newmark, P.A., Reddien, P.W., Cebria, F., and Sanchez Alvarado, A. (2003). Ingestion of bacterially expressed double-stranded RNA inhibits gene expression in planarians. Proc. Natl. Acad. Sci. U.S.A. *100* Suppl. *1*, 11861–11865.

Nishikawa, T., and Natori, S. (2001). Targeted disruption of a pupal hemocyte protein of *Sarcophaga* by RNA interference. Eur. J. Biochem. *268*, 5295–5299.

Obbard, D.J., Jiggins, F.M., Halligan, D.L., and Little, T.J. (2006). Natural selection drives extremely rapid evolution in antiviral RNAi genes. Curr. Biol. *16*, 580–585.

Olson, K.E., Higgs, S., Gaines, P.J., Powers, A.M., Davis, B.S., Kamrud, K.I., Carlson, J.O., Blair, C.D., and Beaty, B.J. (1996). Genetically engineered resistance to dengue-2 virus transmission in mosquitoes. Science *272*, 884–886.

Orii, H., Mochii, M., and Watanabe, K. (2003). A simple 'soaking method' for RNA interference in the planarian *Dugesia japonica*. Dev. Genes Evol. *213*, 138–141.

Pal-Bhadra, M., Bhadra, U., and Birchler, J.A. (1997). Cosuppression in *Drosophila*: gene silencing of Alcohol dehydrogenase by white-Adh transgenes is Polycomb dependent. Cell *90*, 479–490.

Patel, A., Fondrk, M.K., Kaftanoglu, O., Emore, C., Hunt, G., Frederick, K., and Amdam, G.V. (2007). The making of a queen: TOR pathway is a key player in diphenic caste development. PLoS ONE *2*, e509.

Pfeffer, S., Dunoyer, P., Heim, F., Richards, K.E., Jonard, G., and Ziegler-Graff, V. (2002). P0 of beet Western yellows virus is a suppressor of posttranscriptional gene silencing. J. Virol. *76*, 6815–6824.

Pham, J.W., Pellino, J.L., Lee, Y.S., Carthew, R.W., and Sontheimer, E.J. (2004). A Dicer-2-dependent 80s complex cleaves targeted mRNAs during RNAi in *Drosophila*. Cell *117*, 83–94.

Price, D.R., and Gatehouse, J.A. (2008). RNAi-mediated crop protection against insects. Trends Biotechnol. *26*, 393–400.

Rajagopal, R., Sivakumar, S., Agrawal, N., Malhotra, P., and Bhatnagar, R.K. (2002). Silencing of midgut aminopeptidase N of *Spodoptera litura* by double-stranded RNA establishes its role as *Bacillus thuringiensis* toxin receptor. J. Biol. Chem. *277*, 46849–46851.

Rand, T.A., Petersen, S., Du, F., and Wang, X. (2005). Argonaute2 cleaves the anti-guide strand of siRNA during RISC activation. Cell *123*, 621–629.

Reiter, L.T., and Bier, E. (2002). Using *Drosophila melanogaster* to uncover human disease gene function and potential drug target proteins. Expert Opin. Ther. Targets *6*, 387–399.

Ribeiro, J.M., and Francischetti, I.M. (2003). Role of arthropod saliva in blood feeding: sialome and post-sialome perspectives. Annu. Rev. Entomol. *48*, 73–88.

Rivas, F.V., Tolia, N.H., Song, J.J., Aragon, J.P., Liu, J., Hannon, G.J., and Joshua-Tor, L. (2005). Purified Argonaute2 and an siRNA form recombinant human RISC. Nat. Struct. Mol. Biol. *12*, 340–349.

Roignant, J.Y., Carre, C., Mugat, B., Szymczak, D., Lepesant, J.A., and Antoniewski, C. (2003). Absence of transitive and systemic pathways allows cell-specific and isoform-specific RNAi in *Drosophila*. RNA *9*, 299–308.

Roitt, I., Brostoff, J., and Male, D. (2001). Immunology, 6th edn (London: Mosby).

Roxstrom-Lindquist, K., Terenius, O., and Faye, I. (2004). Parasite-specific immune response in adult *Drosophila melanogaster*: a genomic study. EMBO Rep. *5*, 207–212.

Saleh, M.C., Tassetto, M., van Rij, R.P., Goic, B., Gausson, V., Berry, B., Jacquier, C., Antoniewski, C., and Andino, R. (2009). Antiviral immunity in *Drosophila* requires systemic RNA interference spread. Nature *458*, 346–350.

Saleh, M.C., van Rij, R.P., Hekele, A., Gillis, A., Foley, E., O'Farrell, P.H., and Andino, R. (2006). The endocytic pathway mediates cell entry of dsRNA to induce RNAi silencing. Nat. Cell Biol. *8*, 793–802.

Sanchez-Vargas, I., Travanty, E.A., Keene, K.M., Franz, A.W., Beaty, B.J., Blair, C.D., and Olson, K.E. (2004). RNA interference, arthropod-borne viruses, and mosquitoes. Virus Res. *102*, 65–74.

Schwarz, D.S., Hutvagner, G., Du, T., Xu, Z., Aronin, N., and Zamore, P.D. (2003). Asymmetry in the assembly of the RNAi enzyme complex. Cell *115*, 199–208.

Schwarz, D.S., Hutvagner, G., Haley, B., and Zamore, P.D. (2002). Evidence that siRNAs function as guides, not primers, in the *Drosophila* and human RNAi pathways. Mol. Cell *10*, 537–548.

Soares, C.A., Lima, C.M., Dolan, M.C., Piesman, J., Beard, C.B., and Zeidner, N.S. (2005). Capillary feeding of specific dsRNA induces silencing of the isac gene in nymphal *Ixodes scapularis* ticks. Insect Mol. Biol. *14*, 443–452.

St Johnston, D. (2002). The art and design of genetic screens: *Drosophila melanogaster*. Nat. Rev. Genet. *3*, 176–188.

Sturtevant, A.H. (1932). The use of mosaics in the study of the developmental effects of genes. Proc. 6th Int. Congr. Genet. *1*, 304–307.

Sukumaran, B., Narasimhan, S., Anderson, J.F., DePonte, K., Marcantonio, N., Krishnan, M.N., Fish, D., Telford, S.R., Kantor, F.S., and Fikrig, E. (2006). An *Ixodes scapularis* protein required for survival of *Anaplasma phagocytophilum* in tick salivary glands. J. Exp. Med. *203*, 1507–1517.

Sullivan, C.S., and Ganem, D. (2005). A virus-encoded inhibitor that blocks RNA interference in mammalian cells. J. Virol. *79*, 7371–7379.

Sullivan, C.S., Grundhoff, A.T., Tevethia, S., Pipas, J.M., and Ganem, D. (2005). SV40-encoded microRNAs regulate viral gene expression and reduce susceptibility to cytotoxic T cells. Nature *435*, 682–686.

Tabunoki, H., Higurashi, S., Ninagi, O., Fujii, H., Banno, Y., Nozaki, M., Kitajima, M., Miura, N., Atsumi, S., Tsuchida, K., *et al.* (2004). A carotenoid-binding protein (CBP) plays a crucial role in cocoon pigmentation of silkworm (Bombyx mori) larvae. FEBS Lett. *567*, 175–178.

Tanne, E., and Sela, I. (2005). Occurrence of a DNA sequence of a non-retro RNA virus in a host plant genome and its expression: evidence for recombination between viral and host RNAs. Virology *332*, 614–622.

Tomari, Y., Du, T., Haley, B., Schwarz, D.S., Bennett, R., Cook, H.A., Koppetsch, B.S., Theurkauf, W.E., and Zamore, P.D. (2004). RISC assembly defects in the *Drosophila* RNAi mutant armitage. Cell *116*, 831–841.

Tomoyasu, Y., Miller, S.C., Tomita, S., Schoppmeier, M., Grossmann, D., and Bucher, G. (2008). Exploring systemic RNA interference in insects: a genome-wide survey for RNAi genes in Tribolium. Genome Biol. *9*, R10.

Travanty, E.A., Adelman, Z.N., Franz, A.W., Keene, K.M., Beaty, B.J., Blair, C.D., James, A.A., and Olson, K.E. (2004). Using RNA interference to develop dengue virus resistance in genetically modified Aedes aegypti. Insect Biochem. Mol. Biol. *34*, 607–613.

Turner, C.T., Davy, M.W., MacDiarmid, R.M., Plummer, K.M., Birch, N.P., and Newcomb, R.D. (2006). RNA interference in the light brown apple moth, *Epiphyas postvittana* (Walker) induced by double-stranded RNA feeding. Insect Mol. Biol. *15*, 383–391.

Tzou, P., De Gregorio, E., and Lemaitre, B. (2002). How *Drosophila* combats microbial infection: a model to study innate immunity and host–pathogen interactions. Curr. Opin. Microbiol. *5*, 102–110.

van der Krol, A.R., Mur, L.A., Beld, M., Mol, J.N., and Stuitje, A.R. (1990). Flavonoid genes in petunia: addition of a limited number of gene copies may lead to a suppression of gene expression. Plant Cell *2*, 291–299.

van Rij, R.P. (2008). Virus meets RNAi. Symposium on antiviral applications of RNA interference. EMBO Rep. *9*, 725–729.

van Rij, R.P., and Berezikov, E. (2009). Small RNAs and the control of transposons and viruses in *Drosophila*. Trends Microbiol. *17*, 163–171.

van Rij, R.P., Saleh, M.C., Berry, B., Foo, C., Houk, A., Antoniewski, C., and Andino, R. (2006). The RNA silencing endonuclease Argonaute 2 mediates specific antiviral immunity in *Drosophila melanogaster*. Genes Dev. *20*, 2985–2995.

Vargason, J.M., Szittya, G., Burgyan, J., and Hall, T.M. (2003). Size selective recognition of siRNA by an RNA silencing suppressor. Cell *115*, 799–811.

Voinnet, O. (2005). Non-cell autonomous RNA silencing. FEBS Lett. *579*, 5858–5871.

Voinnet, O., Lederer, C., and Baulcombe, D.C. (2000). A viral movement protein prevents spread of the gene silencing signal in *Nicotiana benthamiana*. Cell *103*, 157–167.

Wang, X.H., Aliyari, R., Li, W.X., Li, H.W., Kim, K., Carthew, R., Atkinson, P., and Ding, S.W. (2006). RNA interference directs innate immunity against viruses in adult *Drosophila*. Science *312*, 452–454.

Weaver, S.C. (2006). Evolutionary influences in arboviral disease. Curr. Top. Microbiol. Immunol. *299*, 285–314.

Whangbo, J.S., and Hunter, C.P. (2008). Environmental RNA interference. Trends Genet. *24*, 297–305.

Winston, W.M., Molodowitch, C., and Hunter, C.P. (2002). Systemic RNAi in *C. elegans* requires the putative transmembrane protein SID-1. Science *295*, 2456–2459.

Winston, W.M., Sutherlin, M., Wright, A.J., Feinberg, E.H., and Hunter, C.P. (2007). *Caenorhabditis elegans* SID-2 is required for environmental RNA interference. Proc. Natl. Acad. Sci. U.S.A. *104*, 10565–10570.

Ye, K., Malinina, L., and Patel, D.J. (2003). Recognition of small interfering RNA by a viral suppressor of RNA silencing. Nature *426*, 874–878.

Zambon, R.A., Vakharia, V.N., and Wu, L.P. (2006). RNAi is an antiviral immune response against a dsRNA virus in *Drosophila melanogaster*. Cell. Microbiol. *8*, 880–889.

Zhang, H., Kolb, F.A., Jaskiewicz, L., Westhof, E., and Filipowicz, W. (2004). Single processing center models for human Dicer and bacterial RNase III. Cell *118*, 57–68.

Zhang, X., Yuan, Y.R., Pei, Y., Lin, S.S., Tuschl, T., Patel, D.J., and Chua, N.H. (2006). Cucumber mosaic virus-encoded 2b suppressor inhibits *Arabidopsis* Argonaute1 cleavage activity to counter plant defense. Genes Dev. *20*, 3255–3268.

Zhou, X., Oi, F.M., and Scharf, M.E. (2006). Social exploitation of hexamerin: RNAi reveals a major caste-regulatory factor in termites. Proc. Natl. Acad. Sci. U.S.A. *103*, 4499–4504.

# RNA Silencing in Plants and the Role of Viral Suppressors

2

Ana Giner, Juan José López-Moya and Lóránt Lakatos

Abstract

The term RNA silencing refers to several pathways present in eukaryotic organisms that lead to the sequence-specific elimination or functional blocking of RNAs with homology to double-stranded RNAs (dsRNAs) that have previously triggered the mechanism. Besides playing important roles in developmental control, RNA silencing forms part of the defence against viruses in plants, acting as a potent antiviral mechanism. To escape from the RNA silencing-based defence, most plant viruses make use of different strategies, the most common relying in the action of viral proteins with the capacity to suppress RNA silencing. The characterization of these viral suppressors is providing useful insights to understand how RNA silencing works, revealing components and steps in the silencing pathways.

## Introduction

In plants, the phenomenon of RNA silencing, also known as post-transcriptional gene silencing (PTGS), or RNA interference (RNAi), is a conserved process that plays an important role in the regulation of gene expression, and also is an essential component of the defence responses (Tomari and Zamore, 2005; Voinnet, 2005). In particular, silencing mechanisms are rather important for the control of transposons and molecular pathogens such as viruses (Wang and Metzlaff, 2005; Girard and Hannon, 2008). Indeed, viruses are parasites that make extensive use of the molecular machineries of their host for many processes, including for translation of their own genetic information, what makes them susceptible to be targeted by this posttranscriptional defence pathway. Furthermore, most plant viruses have RNA genomes, which are easily recognized by RNA silencing, as it also occurs with the essential RNA transcripts derived from viruses having DNA genomes. Consequently, plant viruses have been forced to develop responses against RNA silencing in order to survive and replicate in their hosts. This is true independently of the variability in terms of genome composition [single-stranded (ss) or double-stranded (ds) RNA or DNA] and genome structure that plant viruses exhibit (Hull, 2002).

The presence of multifunctional proteins with different roles during their cycles is common for viruses, as a direct consequence of having all the information required for their replication, maintenance and spreading in relatively small genomes. A clear example of this can be found in the many functions played by viral capsid proteins (Callaway *et al.*, 2001). Among the most representative elements contained in plant virus genomes, and essential for viral functions, the following deserve to be mentioned: proteins for encapsidation into virions, proteins acting as determinants of pathogenicity, proteins involved in transmission, proteins assisting the movement within the different parts of the plant, and proteins for replication (Nagy and Pogany, 2008; Taliansky *et al.*, 2008; Ziegler-Graff and Brault, 2008). At the same time, it has been recognized recently that this list needs to be completed with proteins required to interfere with host defences at different steps of the PTGS pathways, in order to avoid silencing (Voinnet, 2005; Burgyan,

2008). Collectively, these viral proteins are known as suppressors of silencing because they can interfere effectively with RNA silencing, and they are the focus of the present review. As mentioned above, it is common that viral gene products with the silencing suppression function are also involved in other essential processes. For instance, the first viral suppressor of silencing characterized was the HCPro protein of potyviruses (Anandalakshmi *et al.*, 1998; Brigneti *et al.*, 1998; Kasschau and Carrington, 1998), also involved in vector transmission and proteolytic processing of viral precursors (Maia *et al.*, 1996). Since then, a large number of phylogenetically unrelated suppressors have been identified, exhibiting an extraordinary variability, and being present in practically all plant virus species studied (Voinnet *et al.*, 1999; Carrington *et al.*, 2001; Baulcombe, 2002; Voinnet, 2005).

As a consequence of the action of viral suppressors in the silencing pathways, viruses might cause alterations in development and other essential processes that are endogenously controlled, at least partially, by RNA silencing mechanism (Chapman *et al.*, 2004; Palukaitis *et al.*, 2008). It is therefore possible that these viral suppressors of silencing are responsible for the appearance of certain phenotypes and symptoms in virus infected plants, as an important component of the pathogenic action. This explanation is particularly relevant for the pathogenicity of viroids, molecular pathogens devoid of protein encoding capacity (Tabler and Tsagris, 2004; Wang *et al.*, 2004; Flores *et al.*, 2005). However, these phenotypic alterations are often limited, due to evolutionary adaptation, because well adapted parasites should cause only damage compatible with survival of their hosts, and therefore the viral silencing suppressors should not provoke excessive effects. In other words, the action of viral silencing suppressors must have been evolutionarily shaped just to maintain an adequate balance between the replication of the virus surpassing the defence responses, and the survival of the host (Torres-Barcelo *et al.*, 2008).

In this chapter, the recent findings about how the different types of viral suppressors found in plant viruses exert their action will be presented and discussed.

## Small RNAs involved in RNA silencing pathways in plants

It is well established that the initial reaction in any RNA silencing response in eukaryotes is triggered by the presence of double-stranded RNAs (dsRNAs) in the cell. Taking into account the different possible origins of dsRNAs (Tomari and Zamore, 2005), a first division of RNA silencing pathways can be established between the endogenous pathways based on microRNAs (miRNAs) deriving from specific nuclear genes, and the pathways depending on cis-acting short interfering RNAs (siRNAs) from other origins (Bartel, 2004; Kidner and Martienssen, 2005a; Jones-Rhoades *et al.*, 2006; Xie and Qi, 2008). In both cases, and also in plants and animals systems, the activation of RNA silencing results in the appearance of characteristic small RNAs (Ramachandran and Chen, 2008). These small RNAs ranged from 21 to 30 nucleotides (nt) in length, and in fact their presence constitutes a diagnostic distinctive feature of RNA silencing.

There are several types of small RNAs that can be found in plants. First, as indicated above, miRNAs are endogenous small regulatory RNAs, originated from nuclear genes leading to hairpin precursors after transcription (Mallory and Vaucheret, 2006; Meyers *et al.*, 2008). In their mature form miRNAs have near perfect complementarity, only four or fewer mismatches, with the sequence of the target genes that they control (Mallory *et al.*, 2004). They are considered essential to regulate the expression of a significant number of genes, playing key roles both in development and in responses to the environment (Jones-Rhoades *et al.*, 2006); to better illustrate their importance as regulatory elements, it is estimated that about two-thirds of the known plant miRNAs can modulate the expression of transcription factors, placing them upstream of many key-regulatory pathways (Mallory and Vaucheret, 2006). Although the miRNA main function of gene regulation is common to animal and plants, their biogenesis has some differences in both types of organisms, and they even differ in the way they are commonly arranged in their genomes, with the long RNA precursors known as primary miRNAs (pri-miRNAs) often containing a cluster of distinct miRNAs in animals or a single miRNA in plants (Millar and Waterhouse, 2005). Also the localization in the cell during maturation differed among animals and plants: whereas in animals the imperfect stem–loop sequences of pri-miRNAs are first recognized and cleaved in the nucleus, generating intermediate 60- to 70-nt-long precursors (pre-miRNAs) that are specifically transported to the cytoplasm to continue their maturation, in plants the whole process is considered to take place in the nucleus (Bartel, 2004; Vaucheret, 2006). Pre-miRNAs are recognized by a type of RNase III-like ribonuclease known as Dicer (Park *et al.*, 2002; Carmell and Hannon, 2004), to give rise to miRNA/miRNA* duplexes. In the model plant *Arabidopsis thaliana*, the protein DCL1 (Dicer-like 1) is considered responsible of this step, although other elements specifically implicated in miRNA biogenesis and maturation have been identified, such as HYL1 (Han *et al.*, 2004; Vazquez *et al.*, 2004a; Hiraguri *et al.*, 2005) and HEN1 (Yu *et al.*, 2005), the later acting in the protection of the mature miRNA throughout methylation. Interestingly, DCL1 is itself regulated by a miRNA (Xie *et al.*, 2003). As we will see later, the miRNA pathway coincide with other RNA silencing routes in the use of effectors complexes denominated RISC (RNA induced silencing complex), where at least a member of the Argonaute family of proteins is implicated (Joshua-Tor, 2006; Tolia and Joshua-Tor, 2007; Hutvagner and Simard, 2008). In the case of the miRNA pathway in Arabidopsis, the AGO1 protein is the key component of RISC (Baumberger and Baulcombe, 2005). As in the case of DCL1, expression of AGO1 is also regulated by miRNA (Vaucheret *et al.*, 2004).

Instead of being generated from endogenous long hairpin precursors derived from specific genes as the miRNAs, other siRNAs can be originated from different sources of long dsRNAs cleaved by Dicer enzymes (Blevins *et al.*, 2006). A first category is constituted by the siRNAs, mostly 21–24 nt in length and depending on DCL2, DCL3 and DCL4 for their biogenesis. SiRNAs were first identified in plants undergoing posttranscriptional gene silencing as a result of a viral infection (Hamilton and Baulcombe, 1999). However, other small RNAs included in different categories of siRNAs might have endogenous origins not related with defence against viruses, such as the heterochromatin-associated siRNAs, the trans-acting siRNAs (ta-siRNAs), and the natural interfering cis-acting siRNAs (nat-siRNAs). Heterochromatin-associated siRNAs, predominantly 24 nt long, commonly derive from transposable elements and direct methylation either of DNA or of histones, leading to transcriptional gene silencing, an action depending on a different complex denominated RITS (RNA induced transcriptional silencing complex) (Ekwall, 2004; Verdel *et al.*, 2004). The 21 nt long ta-siRNAs are generated from endogenous RNAs by DCL4 and can regulate the activity of genes through the cleavage of non perfectly matching sequences (Peragine *et al.*, 2004; Vazquez *et al.*, 2004b; Allen *et al.*, 2005), but differed from miRNAs because they require components from both the siRNA suppression pathway (such as RDR6 and SGS3) as well as from the miRNA pathway (such as AGO1, HEN1, and HYL 1). Finally, the natural cis-acting siRNAs are derived from natural antisense transcripts, and have been described as a new system of cross-regulation between two *cis*-antisense gene pairs, requiring DCL1 and 2, RDR6, SGS3, and NRPD1A for their biogenesis (Borsani *et al.*, 2005).

## Core elements of the RNA silencing pathways in plants

There are several recent works that reviewed our current knowledge about the essential components of the RNA silencing pathways in plants and their specific activities (Brodersen and Voinnet, 2006; Chapman and Carrington, 2007). Although an extensive revision exceeds the focus of the present chapter, we can outline a brief summary of those elements, indicating in which steps they are more important. At least three core components Dicer-like (DCL) enzymes, Argonaute (AGO) proteins and RNA-dependent RNA polymerases (RDRP) deserve to be briefly mentioned.

DCL enzymes are multicomponent complexes with distinctive dsRNA binding, RNA helicase, RNase III and PAZ (denominated for being present in Piwi/Argonaute and/Zwille proteins) domains (Bernstein *et al.*, 2001; Hammond *et al.*, 2000). Four DCL homologues have been identified in *Arabidopsis,* each of them involved in the generation of different size classes of small RNAs from specific targets. As indicated, DCL1 processes 21 nt miRNAs from their imperfect stem–loop precursors and it is localized mainly in the nucleus, while DCL2, DCL3 and DCL4 process both replicating viral RNAs as well as other dsRNAs into 22-, 24- and 21-nt siRNAs, respectively (Borsani *et al.,* 2005; Bouche *et al.,* 2006; Fusaro *et al.,* 2006). Even though the roles of these DCLs have been determined in generation of discrete small RNAs, it has been also demonstrated, by using *Arabidopsis* mutants, a redundancy in their functions (Deleris *et al.,* 2006). For instance, in a dysfunctional DCL4 plant, the activity will be partially rescued by DCL2 (Blevins *et al.,* 2006; Bouche *et al.,* 2006; Deleris *et al.,* 2006; Fusaro *et al.,* 2006; Du *et al.,* 2007a; Qu *et al.,* 2008).

Argonaute proteins constitute the catalytic core components of active RISC, together with a single-stranded guide small RNA. The Argonaute proteins contain four different domains: the N-terminal region, a PAZ domain, a Middle domain and a PIWI domain. Crystallography techniques have allowed determining the structural basis for their RNA binding ability (Ma *et al.,* 2004; Song *et al.,* 2004; Yuan *et al.,* 2005). The number of argonaute genes varies in different species. In the *Arabidopsis* genome, ten Argonaute proteins have been identified so far (Hutvagner and Simard, 2008). The PAZ domain, also found in Dicer-like proteins, recognizes the characteristic protruding 3′ ends of miRNA and siRNAs, and therefore it is implicated on the process of selection of these small RNAs from other RNAs. This domain can bind to single-stranded RNAs and siRNA duplexes of random sequence, acting as siRNA binder and being theoretically implicated in the siRNA transfer from Dicer to Argonaute (Lingel *et al.,* 2004; Yuan *et al.,* 2005). The C-terminal 300-amino acid PIWI domain is highly conserved. The structure analysis revealed that PIWI domain has an RNase-H-like fold, carrying the endonuclease cleavage activity of the protein, and contains a 'binding pocket' (conserved Asp-Asp-Asp/Glu/His/Lys domain) that guides the strand 5′ end recognition of the guide RNA (Hutvagner and Simard, 2008).

The RDRPs constitute another important group of factors in both post-transcriptional and transcriptional silencing, contributing to the siRNAs biogenesis (Wassenegger and Krczal, 2006). One of the most important elements in this category is the RDRP6 (also called SDE1 and SGS2), known to be related to viral silencing and implicated in the formation of secondary 21nt siRNAs (Dalmay *et al.,* 2000; Mourrain *et al.,* 2000), that mediates RNA silencing signal spread in cell to cell and long distance movement in the plant (Diaz-Pendon *et al.,* 2007; Donaire *et al.,* 2008; Qu *et al.,* 2008).

## RNA silencing and virus infections in plants

As mentioned above, the dsRNAs triggering silencing responses can be derived from viruses, directly from the genome in the case of dsRNA viruses, or from replicative intermediates or secondarily structured regions in ssRNA viruses (Llave *et al.,* 2000; Voinnet *et al.,* 2000; Qu *et al.,* 2003; Thomas *et al.,* 2003; Voinnet, 2005). Viruses with DNA genomes are also targeted by RNA silencing mechanisms triggered by their transcripts (Bisaro, 2006). Apart from these virus-derived elements, dsRNA can appear after transcription of plant genes or transgenes, such as inverted repeat transgenes or aberrant genes that will be recognized by plant RDRPs.

Independently of the variable and often divergent origins of these dsRNAs that elicit the RNA silencing response, the molecular components that are involved in the pathway are basically common. Briefly, the dsRNAs are initially processed by Dicer enzymes, into short dsRNA fragments of about 20–25 nt long known as ds siRNAs. These elements have as hallmarks two typical structural features in the form of a 5′ phosphate and 2 nt overhanging 3′ ends (Elbashir *et al.,* 2001; Tomari and Zamore, 2005). After strand separation, one of the strands of the ds siRNA incorporates into a multi-subunit RISC (Matranga *et al.,* 2005). These complexes once assembled are the effectors that guide, in a sequence specific manner, either the degradation by slicing activity, or the translational arrest of messenger RNAs. The presence of a protein of the Argonaute family is a requisite of the effectors complexes, and they also participate in altering the chromatin to provoke transcriptional silencing (Hock and Meister, 2008). Although it was initially considered that plants and animals diverged in the mode of action of the RISC effectors complexes, with slicing being the most common mechanism in plants, recent results point towards the importance of translational arrest in plants (Brodersen *et al.*, 2008).

Together with these essential steps driven by Dicer and RISC, different additional elements participate in the RNA silencing pathways. For instance, upstream elements are required to produce and translocate RNA precursors leading to miRNAs with important roles in development and environmental responses (Mallory and Vaucheret, 2006). Other additional steps in the pathway include elements responsible of the movement of the RNA silencing signal from cell to cell and over long distances within the plant, through an amplification process dependent on the activity of endogenous RDRPs, allowing the systemic dispersion of the silencing responses all over the plant by *de novo* formation of siRNAs, in a process known as transitivity (Voinnet and Baulcombe, 1997; Mlotshwa *et al.*, 2002).

It is interesting to highlight that among the first studies leading to the discovery of the RNA silencing mechanisms in eukaryotes there are important contributions from plant virology, including works performed with pathogen derived resistant plants showing a negative correlation between levels of expression and resistance that have been recently summarized by their pioneering authors (Lindbo and Dougherty, 2005). This phenomenon reminded the co-suppression previously observed in transgenic petunias, and also was associated with the studies on cross-protection, where the level of sequence identity between the virus strains involved was determinant for the success of the protection. Nowadays, many of these intriguing phenomena have been explained by the RNA silencing responses triggered in plants. Furthermore, the phenomenon of synergism where the presence of a given virus serves to enhance the accumulation of other viruses was decisive to identify the first suppressor of gene silencing in the HCPro protein of potyviruses (Pruss *et al.*, 1997; Pruss *et al.*, 2004).

Even though several viral silencing suppressors have been since then identified among different virus families and reviewed quite extensively (Voinnet *et al.*, 1999; Li and Ding, 2001; Baulcombe, 2002; Roth *et al.*, 2004; Qu and Morris, 2005; Burgyan, 2008), the molecular

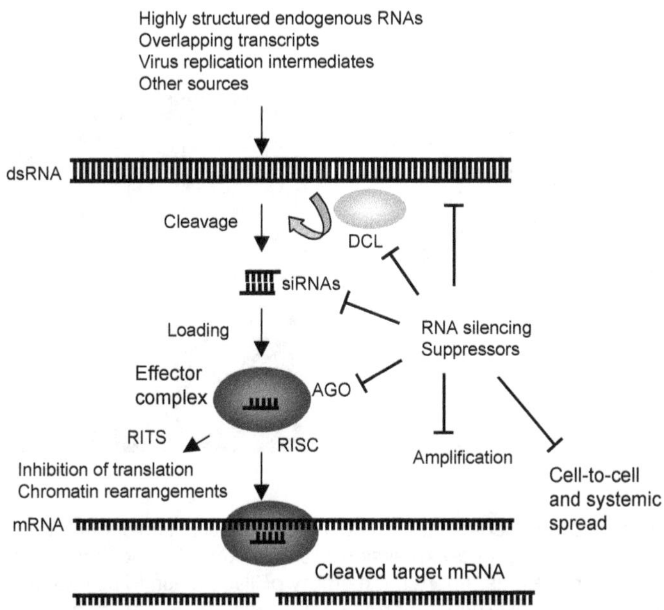


**Figure 2.1** Schematic representation of the common steps in the different RNA-silencing pathways in plants, with indication of the modes of interference by RNA silencing suppressors.

mechanisms underlying their specific activities are still in most cases under study, and only a subset of them are sufficiently known to allow their placement along discrete steps in the silencing pathways (Fig. 2.1). The variety of mechanism of action described is quite wide, and it is clear that there is not a single and unique way to interfere with RNA silencing. For instance, whereas some of these proteins act sequestering siRNAs, others seem to interfere with the RISC complexes. It is likely that a deeper characterization of the already known suppressors, and the finding of new ones, will lead to the identification of other modes of action. A first conclusion that can be derived from the observed variability in the sequence and mode of action of viral silencing suppressors is their likely recent origin in evolution. As mentioned, in many cases there is a functional overlap with other viral functions, which again argues in favour of a recent adoption of the suppression function.

## Experimental systems for identification of virus-derived RNA silencing suppressors

The identification of putative RNA silencing suppressor activities in proteins derived from plant viruses has been possible thanks to the design and use of different assays (Karjee *et al.*, 2008). Although their detailed description exceeds the scope of this article, a few indications about them might help to understand the next sections.

Among the most widely used systems available to identify silencing suppressors we can cite first the transient co-expression of the candidate element and a reporter element in plant leaves (Llave *et al.*, 2000; Voinnet *et al.*, 2000; Qu *et al.*, 2003; Thomas *et al.*, 2003). This well established assay is based on the use of the bacteria *Agrobacterium tumefaciens* carrying plasmids that allows transient expression of the proteins of interest. The system is based on the co-expression in the same plant leaf of the candidate viral gene and a reporter gene, usually green fluorescent protein (GFP) to allow visualization of the transient gene expression in transformed cells. Under standard conditions, protein expression will reach a peak around 2–3 days after the agroinfiltration, and in the case of GFP it can be easily visualized under UV light as a bright patch of green fluorescence, in contrast with the non-agroinfiltrated areas of the plant, that remains red under UV by the presence of chlorophyll. However, the overexpression of the foreign genes during these initial days generally results in the triggering of local RNA silencing, thus blocking further expression of the reporter gene and a gradual dimming of the green fluorescence (Johansen and Carrington, 2001; Voinnet *et al.*, 2003). In contrast to this, the

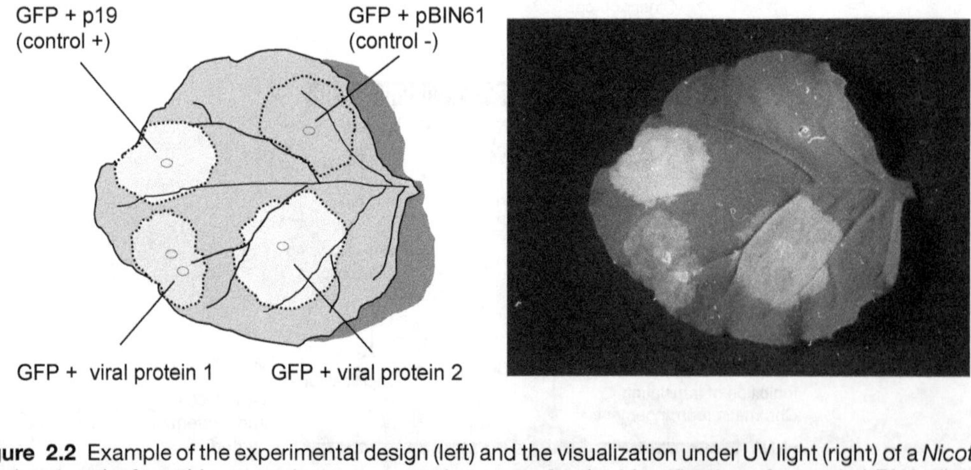


**Figure 2.2** Example of the experimental design (left) and the visualization under UV light (right) of a *Nicotiana benthamiana* leaf used in a transient co-expression assay for the identification of new viral RNA silencing suppressors, with the GFP protein as reporter, the empty pBIN61 plasmid as negative control, and the p19 of tombusvirus as positive control. The viral proteins assayed in this particular case were the P1b of the ipomovirus CVYV (Valli *et al.*, 2006) and a 1:10 dilution of the positive control for comparison purposes, and the picture was taken at 7 days post-agroinfiltration. Picture courtesy of A.M. Martín-Hernández (CRAG, Cabrils, Barcelona, Spain).

maintenance over time of the expression of the reporter gene indicates that the co-expressed viral protein has indeed silencing suppressor activity (Fig. 2.2). The most important advantages of the system as compared to the stable gene expression are in its simplicity and easiness to be performed, allowing the rapid assay of many candidate proteins without the need to generate transgenic plants.

Although the technique can be theoretically performed in different plant species, *Nicotiana benthamiana* represents a model of choice, because it was experimentally found that it allows high level expression of heterologous proteins in leaves (Goodin *et al.*, 2008). The reason for this peculiarity is still unclear, although it has been found that *N. benthamiana* lacks a functional salicylic acid inducible RDRP, which exists in other species such as tobacco and *Medicago truncatula*. Complementation studies showed that this RDRP is involved in antiviral defence. Since antiviral and transgene induced RNA silencing pathways are overlap in plants, this particular RDRP might account for this phenomenon (Yang *et al.*, 2004).

Different modification of this assay has been exploited to further dissect the activity of suppressors, as in the hairpin transcript-induced silencing assays, where the candidate and the reporter genes are co-agroinfiltrated with an inverted repeat GFP construct (GFP IR) capable to express a hairpin RNAs, which is a potent inducer of silencing rapidly processed into siRNAs by the silencing machinery. This system allows determining whether the suppressor is targeted against the hairpin, thus preventing siRNAs formation.

A further refinement of the system is constituted by the reversal of silencing assay, in which a transient expression assay is performed on transgenic plants expressing the reporter gene, as the constitutive GFP-expressing transgenic *N. benthamiana* line 16c initially produced in the laboratory of David Baulcombe (Ruiz *et al.*, 1998). The assay serves to determine whether the silenced phenotype can be reversed by the activity of a given viral protein. For instance, the 16c plants, once infiltrated with an *Agrobacterium tumefaciens* culture carrying a GFP construct, will trigger first local and later systemic silencing until the GFP will be completely silenced all over the plant. When candidate suppressors are co-infiltrated on the system, their effects can be comparatively evaluated. Furthermore, the possibility to investigate the effect of silencing suppressors on the mobile silencing signal can form part of the assay by periodic observation of the agroinfiltrated GFP-expressing plants under UV-light to find out differences in the spread of the signal to new tissues. Also virus vectors carrying the candidate suppressors from heterologous viruses can be tested in the reversal of silencing assay.

Another experimental possibility is represented by the use of transgenic plant lines where candidate silencing suppressors are expressed, and crossed to a series of well-characterized transgenic lines silenced for a reporter gene. This is more commonly used to better characterize the molecular activity of the candidate protein, allowing experimental designs including for instance grafting of different plant lines (Mallory *et al.*, 2001; Guo and Ding, 2002; Mallory *et al.*, 2003).

## Differences in viral RNA silencing suppression deriving from the variability of plant viruses

As mentioned before, the presence of suppressor of RNA silencing in viral genomes is nowadays considered a general feature in almost all plant viruses. However, the particularities of the different virus taxa, specifically those derived form their genome composition (such as RNA or DNA) could undoubtedly have a strong influence on the type of suppressor that each virus requires to deal with the RNA silencing mechanisms in their plant hosts. Here we will briefly summarize the major constraints that might have determined the adoption of the different strategies of RNA silencing suppression in plant viruses.

The wide majority of plant viruses present single-stranded positive sense RNA genomes. Once the virus enters the cell, the genomic RNA can be directly translated to produce viral proteins, and sometimes immediately after, or even simultaneously, the genome serves as template for RDRPs to synthesize the corresponding negative strand genomic RNA. These initial replicative processes, either performed by enzymes of viral, host or even both origins, will generate a replicative form, which might provide the first source of dsRNA able to induce a silencing response. Generally this process is considered to take place in the cytoplasm, and therefore this location can be determinant for the involvement of particular enzymes in the dsRNA processing (e.g., determine which would be the mainly involved DCL), and creates a first major difference between miRNAs pathways (occurring in the nucleus) and the siRNAs pathways that could operate against RNA viruses of cytoplasmic localization.

In the case of plant viruses with negative ssRNA genomes, generation of the positive strand will be necessary before the virus is able to produce any new viral protein, and thus these viruses might be even more susceptible to be early targeted by RNA silencing mechanisms. Nevertheless, one relevant feature that could pose differences between positive and negative ssRNA viruses is more related to their replicative cycles rather to their genomic source: unlike most positive ssRNA viruses, negative ssRNA plant viruses go through a life cycle alternating in both plants and their arthropod vectors (Wijkamp *et al.*, 1993; Falk and Tsai, 1998), which can force their evolution to adaptation to both systems. Silencing mechanisms have been also described in insects, but whether the same viral suppressors are able to act in both systems (plant and animals) has not been yet established as a requirement for the life cycle of a given virus. Interestingly, cross-kingdom activity of several viral suppressors of RNA silencing has been experimentally observed (Bucher *et al.*, 2004; Delgadillo *et al.*, 2004; Soldan *et al.*, 2005; Haasnoot *et al.*, 2007; Hemmes *et al.*, 2007; Schnettler *et al.*, 2008; Qian *et al.*, 2009).

The interest for studying plant viruses with ssDNA genomes has been boosted for the extraordinary economical importance of viruses belonging to the family Geminiviridae. In these viruses the presence of specific suppressors of silencing has been described (Vanitharani *et al.*, 2005; Bisaro, 2006). The replicative cycle of geminiviruses takes place in the nucleus of infected cells, through a rolling-circle mechanism (Gutierrez, 2000). As transcription can be a bidirectional process, intermediates of dsRNA may be generated, being the precursors for silencing.

Plant caulimoviruses and badnaviruses constitute examples of viruses with dsDNA genomes. As in other pararetroviruses, a RNA intermediate derived from transcription of the genome can be reverse transcribed also to produce the progeny dsDNA genomes. Although dsRNAs are not obligate replication intermediates for these viruses, they may nevertheless accumulate during infection (Blevins *et al.*, 2006) by forming secondary dsRNA structures. For instance, the large number of siRNAs deriving from the CaMV 35S leader region suggests that the secondary structure adopted by this element is a feature that triggers RNA silencing responses. A possible functional role for these leader-derived siRNAs has been postulated because they can regulate several endogenous genes in *Arabidopsis* (Moissiard and Voinnet, 2006).

## Representative examples of the different modes of action of viral RNA silencing suppressors

Searches for RNA silencing suppressors have become an essential part of the functional characterization of viral genomes. Although only a few viral suppressors have been completely characterized in their function, at least the molecular mechanism used in their suppression of RNA silencing activities are known to a level that allows elucidating whether they are acting at a particular step on the RNA silencing pathway (Fig. 2.3). In the present section, some of the different strategies that viruses can use to counteract the plant defence by interfering at different steps of the RNA silencing pathways will be analysed, using examples to illustrate how particular viral suppressor perform their actions.

### Interference by siRNA sequestering

Many unrelated plant viral proteins within the positive-strand RNA viruses have been shown to interfere with RNA silencing by sequestering siRNAs. Suppressors with that capacity have been described for instance in viruses belonging to the families Potyviridae, Tombusviridae, Bromoviridae and Closteroviridae. Consequently, this has been postulated as one of the most common strategies used by viruses to evade RNA silencing (Lakatos *et al.*, 2006).

**Main steps of the RNA silencing pathway in plants**

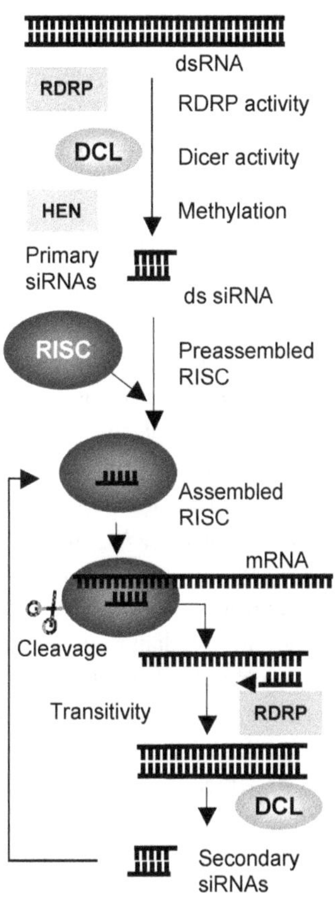


| Mechanism of suppression | Suppressor | Virus | Genus |
|---|---|---|---|
| Interference trhough dsRNA binding | p14 | PoLV | Aureusvirus |
| | p21 | PVX | Potexvirus |
| Interference with Dicer activity | p38 | TuCV | Carmovirus |
| siRNA sequestering | HCPro | TEV, PVY, TuMV & others | Potyvirus |
| | P1b | CVYV | Ipomovirus |
| | p19 | CRSV, TBSV | Tombusvirus |
| | p21 | BYV | Closterovirus |
| Interference with slicer activity | 2b | CMV, TAV | Cucumovirus |
| | P0 | BWYV, CABYV | Polerovirus |
| Interference with signal amplification | p69 | TYMV | Tymovirus |
| Other: inhibition of host factors | AL2 | TGMV | Begomovirus |

**Figure 2.3** Tabulated diagram of the different mechanisms of action of plant virus suppressors of RNA silencing, with the main steps of the silencing pathways in plants represented on the left, and some examples of viral proteins characterized for exerting their suppression function by interference at the indicated step (see text for details).

As mentioned above, after the dsRNAs are processed by Dicer enzymes into siRNAs, they need to be loaded into the RISC complex to guide the effect on homologous sequences. A wide variety of suppressors from divergent origins are able to interfere at this step, preventing siRNA incorporation into the RISC effector complex.

The protein HCPro of potyviruses such as Tobacco etch virus (TEV) has been characterized as a multifunctional product involved in aphid transmission, genome amplification, polyprotein processing, and virus movement within the plant (Maia *et al.*, 1996; Kasschau *et al.*, 1997). Evidence from studies involving the synergism phenomenon observed in co-infections between a potyvirus and unrelated viruses pointed to the suppression of a host defence mechanism, and the activity was mapped to the P1-HCPro region in the potyvirus genome (Pruss *et al.*, 1997). Soon, different groups at the same time concluded that the expression of the HCPro coding sequence alone was sufficient to suppress virus-induced RNA silencing (Anandalakshmi *et al.*, 1998; Brigneti *et al.*, 1998; Kasschau and Carrington, 1998), and diverse modes of action were proposed (Anandalakshmi *et al.*, 2000; Mallory *et al.*, 2001; Chapman *et al.*, 2004; Dunoyer *et al.*, 2004). As mentioned, the presence of the HCPro protein in plants is also related to the capability of the virus for cell to cell and long distance movements, a fact that might explain in part the conflicting results about its mode of action. Nevertheless, HCPro is not able to prevent completely the spread of the RNA silencing signal in the plant. Finally, *in vitro* and *in vivo* approaches to elucidate the molecular mechanism of suppression guiding the activity of HCPro proved that the strategy was the binding to small RNA duplexes of certain size (21 nt siRNAs), thus blocking the pathway (Lakatos *et al.*, 2006). Recently, an ability to interfere with the methylation of siRNAs and miRNAs was found for HCPro (Ebhardt *et al.*, 2005; Lozsa *et al.*, 2008), suggesting that methylation of small RNAs might occurs prior to RISC assembly, and in a cellular compartment compatible with the activity of HCPro as suppressor of silencing: the fact that small RNA methylation can be inhibited by viral RNA silencing suppressors localized in the cytoplasm indicated that methylation could take place in this compartment.

One interesting aspect of the analysis of the mechanisms of action of HCPro is related to the evolution of the silencing suppression functions in members of the family Potyviridae (Valli *et al.*, 2007). The fact that mutants of the closely related tritimoviruses remained fully infectious and capable to produce synergism after complete deletion of HCPro (Stenger *et al.*, 2005; Stenger *et al.*, 2007) opened the question if the RNA silencing suppression in members of other genera within the family might be shifted to a different gene product. Furthermore, the finding of related ipomoviruses naturally lacking the HCPro region, such as cucumber vein yellowing virus (CVYV) and squash vein yellowing virus (SqVYV) (Janssen *et al.*, 2005; Li *et al.*, 2008), prompted to reanalyse their genome structure. The existence of a duplicated P1 region was determined, and studies performed with CVYV revealed that in this virus the RNA silencing suppression activity was present in the P1b gene product, a protein completely unrelated to HCPro, although the mode of action was similar (Valli *et al.*, 2006; Valli *et al.*, 2008). Another interesting case of study will be a third ipomovirus, Sweet potato mild mottle virus (SPMMV), that retains the HCPro region and at the same time presents a long P1 region with a clear homology in its C-terminal region with the P1b of CVYV.

The p19 suppressor of tombusviruses also provides a well-characterized example of the mechanism of action based on siRNA sequestering. The ORF5-encoded 19 kDa protein (p19) of tombusviruses was first characterized as a pathogenic determinant, already implicated in plant defence (Dalmay *et al.*, 1993; Scholthof *et al.*, 1995). Later, it was identified as an RNA silencing suppressor (Voinnet *et al.*, 1999), and shortly after the first working model for RNA silencing suppression mechanism was proposed by showing that the p19 of carnation Italian ringspot virus (CIRV) can bind to siRNAs (Silhavy *et al.*, 2002). The siRNA sequestration theory was proven by *in vitro* and *in vivo* studies with cymbidium ringspot virus (CymRSV), which clearly showed that sequestering siRNAs prevents RISC assembly (Lakatos *et al.*, 2004).

The crystal structure of the p19 suppressor–siRNA complex was resolved, showing that p19 binds siRNA as a homodimer (Vargason *et al.*, 2003; Ye *et al.*, 2003). The beta sheets of the dimer contribute to the interaction with the helix of the siRNA and two tryptophan (Trp) amino acids situated on either side of the p19 homodimers bind the 5′ phosphate of the siRNA, acting as molecular callipers able to discriminate the size of the RNA molecule. The significance of the two Trps in siRNA binding was further proven by *in vitro* binding studies and by the inability of engineered viruses bearing mutations at these residues to suppress RNA silencing (Vargason *et al.*, 2003; Ye *et al.*, 2003). The ability of p19 to bind miRNAs has also been proved, a fact that implicates the protein in modulating of symptoms by interfering with endogenous miRNA pathways (Chapman *et al.*, 2004; Dunoyer *et al.*, 2004).

The third example of suppressor using a similar strategy is provided by the protein p21 encoded by the closterovirus Beet Yellows Virus (BYV) (Reed *et al.*, 2003), a viral product that was also defined as a small RNA binding protein inhibiting RISC assembly (Chapman *et al.*, 2004; Lakatos *et al.*, 2006). Although a crystal structure of an octameric form of p21 was resolved and found to be able to bind RNA (Ye and Patel, 2005), the details on how the suppression activity is exerted by this protein remains to be determined.

Apart from the referenced studies with HCPro, p19 and p21, several other viral proteins have been postulated to have adopted the 'siRNA sequestering' mechanism of action, such as the p14 of Pothos latent virus, the p38 of turnip crinkle virus, the p15 peanut clump virus, the B3 of barley stripe mosaic virus, the 2b protein of cucumber mosaic virus, and the P1b protein of cucumber vein yellowing virus, among others (Dunoyer *et al.*, 2002; Yelina *et al.*, 2002; Qu *et al.*, 2003; Merai *et al.*, 2005; Valli *et al.*, 2006; Goto *et al.*, 2007). Furthermore, a similar mechanism was also found when studying the B2 protein of the insect virus flock house virus (FHV) (Lingel *et al.*, 2005). The molecular basis underlying the siRNAs binding properties are likely to differ greatly from one viral protein to another, a fact that together with the consideration of the other roles that these viral suppressors may play during infection, reinforces the suggestion of their recent evolutionary acquisition.

## Interference through dsRNA binding

Besides the siRNA sequestering strategy, some viral proteins have been shown to interfere in the RNA silencing pathway upstream of the Dicer-mediated cleavage of dsRNA into siRNAs, therefore preventing the accumulation of siRNAs and the rest of subsequent steps. This is the presumed way of action of p14 from Pothos latent virus (PoLV) and the potato virus X (PVX) p25 protein.

The PoLV p14 presents an equivalent position in the genome to the tombusviral p19, both viruses belonging to Tombusviridae family, but despite this, sequence similarity between the two viruses is rather limited (Rubino and Russo, 1997). It has been shown that p14, as other silencing suppressors from the same families, are able to bind siRNAs, and also to longer (>100 nt) dsRNA. While p19 acts as strictly size-specific dsRNA binding protein, p14 seems to bind dsRNAs without any size preference. The presence of the protein in plant cells interferes with the accumulation of GFP mRNA and GFP derived siRNAs. Neither 21 nt nor 24 nt long GFP-specific siRNAs can be detected, pointing towards a mechanism of action based on preventing the formation of any siRNAs. When co-infiltrated with IR-GFP, neither hairpin transcripts nor the expected siRNAs deriving from those can be found. Nevertheless, p14 fails to prevent accumulation of viral siRNAs (Merai *et al.*, 2005).

The p25 from potato virus X (PVX) is responsible for movement of the virus into the plant, and also has been shown to be a suppressor of the systemic silencing response which interferes with an initiation step, possibly preventing dsRNA from being converted to siRNA (Voinnet *et al.*, 2000). Indeed, a link between movement and suppression of silencing was established throughout mutagenesis studies (Bayne *et al.*, 2005).

## Interference with Dicer activity

Other possibility of proteins acting as suppressors upstream of siRNA formation is the interference with the activity of Dicer enzymes. The carmoviral Turnip crinkle virus (TCV) encodes only five open reading frames (ORFs). The second one, from the 3 half of the genome, codes for the viral coat protein (CP), and represents the first characterized structural protein carrying silencing suppressor activity. Besides the important role in virion packaging and in virus–host interaction of the protein, studies with TCV CP reveal a strong suppressor function when overexpressed in the cytoplasm of the cell. Silencing suppressor assays showed that the protein is able to prevent siRNAs formation in co-agroinfiltration assays, what points that suppression occurs in a very early initiation step of the silencing pathway. Further characterization determined that even if the suppressor is not able to reverse established silencing, is able to block systemic spread of the signal in plants, and recently it has been determined that Dicer-like nuclease DCL4 is its primary target. The suppressor interferes with this DCL compromising antiviral activity in production of

21 nt. siRNA (Deleris *et al.*, 2006). Furthermore, the protein is also responsible of modulation of symptom development together with an RNA satellite from the virus.

As mentioned above for members of the Potyviridae family, an example of different mechanisms of action in suppressors deriving from closely related viruses has been recently reported between carmoviruses: in the case of pelargonium flower break virus (PFBV) the CP serves as RNA silencing suppressor through a presumably siRNA binding mechanism (Martinez-Turino and Hernandez, 2009).

Another example of inhibition of Dicer activity by an RNA silencing suppressor is the P6 protein of the cauliflower mosaic virus. Although P6 is known as a viral transactivator and forms cytoplasmic bodies, a fraction of P6 is exported into the nucleus, where P6 genetically and physically interacts with DRB4 protein. P6 sequesters DRB4 from the essential antiviral DCL4 protein complex thus inhibiting the siRNA processing step of RNA silencing (Haas *et al.*, 2008).

### Interference with slicer (RISC)

Some viruses have evolved in the development of strategies to interfere with the maintenance of the silencing signal rather than with the initiation steps of the pathway. For instance they have developed strategies to interfere downstream of siRNA formation with plant cell components of RISC complex.

One of the better-known examples is the 2b protein from cucumber mosaic virus (CMV). Shortly after the identification of 2b as an additional gene in the CMV genome (Ding *et al.*, 1994), works performed with CMV already postulated that it encoded a putative suppressor of host defences (Ding *et al.*, 1995). The 2b protein was characterized as silencing suppressor together with the PVY HCPro, and it was thought to act preventing initiation steps of silencing (Brigneti *et al.*, 1998), with little or no effect on miRNA functions when a mild strain (CMV Q) of the virus was used (Chapman *et al.*, 2004). It was also shown that the 2b protein of CMV suppresses systemic RNA silencing by interfering with the systemic delivery of the silencing signal (Guo and Ding, 2002). However, posterior studies with the most severe strain FNY-CMV revealed that *ago1* mutants exhibited hypersusceptibility to this virus (Morel *et al.*, 2002).

Plants infected with the FNY strain of CMV contain abundant 2b protein and elevate levels of miRNAs, which probably implies that the viral protein is also interfering with the miRNA pathways, eliciting developmental anomalies similar to *ago1* mutant alleles. Some studies recently propose that CMV 2b interacts directly with AGO1, inhibiting its cleavage activity (shown by RISC reconstitution assays) and that AGO1 recruits virus-derived siRNAs (*in vivo*) to attenuate RNA silencing and counteract host defence (Zhang *et al.*, 2006). CMV 2b interferes with miRNA pathways by a mechanism different from that used by P19. One of the reasons seems to be the nuclear localization of 2b, fact that encourage its interaction with AGO1, in contrast with the cytoplasmic localization of p19. This interaction has been studied in detail by *in vitro* reconstitution assays and mutant studies, resolving that CMV 2b interacts strongly with AGO1 on the surface of the PAZ module that contains the RNA-binding groove, and also partly with the domain containing the PIWI-box (Zhang *et al.*, 2006).

A second example of interference between a viral suppressor and RISC function is found in poleroviruses. When first described, the P0 protein of polerovirus was postulated to act similarly to the HCPro from potyvirus. The P0 was shown to interfere with siRNA activity without being able to inhibit silencing signal spread (Pfeffer *et al.*, 2002). The accumulation of several miRNAs in plants overexpressing P0 also pointed the RISC complex as the suppressors target in the interference with silencing pathways. Further analysis of this protein has revealed the presence of an F-box motif in the P0 sequence, presumably responsible of mediating interaction between P0 and AGO1. The F-box proteins are components of E3 ubiquitin ligase complexes that mediate the labelling of selected lysine residues for being guided to proteasome degradation. Nevertheless, the proteasome does not seem to be implicated in P0 activity. The proposed mechanism for this viral suppressor is the destabilization of AGO1 by targeting the AGO-PAZ domain motifs by protein–protein interaction, modulating gene silencing at the RISC level (Pazhouhandeh *et al.*, 2006; Baumberger *et al.*, 2007; Bortolamiol *et al.*, 2007).

### Interference with amplification of the signal

The tymovirus Turnip yellow mosaic virus (TYMV) codes for the protein p69 initially described as an essential component for virus spread, and which affects strongly symptoms severity in infected plants (Bozarth *et al*, 1992; Tsai and Dreher, 1993), most likely by alteration of the endogenous miRNA levels. The protein is able to suppress RNA silencing induced by sense RNA, but not by an inverted repeat construct, and this fact points towards its activity as a repressor of dsRNA formation, and indeed the virus seems to be able to inhibit RDRP activity (Chen *et al.*, 2004).

Another case of suppressor that might act similarly is the NSs of TSWV. This protein is the first identified silencing suppressors in negative-stranded RNA viruses (Takeda *et al.*, 2002), and it has been also implicated in the interference with the generation of dsRNA required for synthesis of secondary siRNAs (Ding and Voinnet, 2007).

The transitivity loop of the RNA silencing pathway is also involved in the amplification of the signal. This phenomenon demands target RNA cleavage by primary siRNAs, thus cleaved ssRNA can be double-stranded by RDRP. DsRNA are then processed by DCL enzymes resulting in secondary siRNAs. Thus, removal of the primary siRNAs by siRNA binding silencing suppressor such as p19 or HCPro will inhibit amplification of the RNA silencing signal as part of their action (Lakatos *et al.*, 2006).

### Other mechanisms

As mentioned above, for most viral suppressors there are uncertainties regarding how they exert their action. In fact, as we have seen it is not uncommon that more than one step in the different silencing pathways is targeted, what can generate contradictory interpretation of results obtained in different tests. Furthermore, other way in which viruses can raise their capacity to surpass the RNA silencing mechanisms is by incorporating in their genomes more than one suppressor of silencing. Extreme cases of this strategy could be found in citrus tristeza virus and tomato chorosis virus (Lu *et al.*, 2004; Canizares *et al.*, 2008). Once again, this can result in difficulties to interpret conflicting data.

Nevertheless, it is also possible that new mechanisms of action, still not well characterized, wait to be identified. For instance, interaction with and the eventual inactivation of new essential elements of the silencing pathways are likely to provide explanation of the activity of some suppressors. For instance, studies performed with two different geminiviruses, the begomovirus tomato golden mosaic virus (TGMV) and the curtovirus beet curly top virus (BCTV), lead to the identification of two new suppressors in the proteins AL2 and L2 respectively (Wang *et al.*, 2005). Interestingly, these suppressors were known to inactivate a cellular adenosine kinase (ADK) (Wang *et al.*, 2003), a fact that suggests the involvement of this kinase in RNA silencing (Wang *et al.*, 2005).

## Influence of viral suppressors of RNA silencing in symptoms development

### Phenotype alterations by mimicking miRNAs abnormalities

Several viral suppressors were first identified as determinants of virulence, responsible for virus induced symptoms. As shown, this peculiarity was in some cases decisive for finding out its ability to interfere with host defences, such as with the 2b protein of CMV (Ding *et al.*, 1995). However, the majority of the viral silencing suppressors have been studied using transient expression assays, which provided information about the mechanism of suppression in a molecular level, but leaving in most cases their real implication in viral symptomatology as a matter of speculation.

As indicated, the CMV 2b protein is a good case of study to analyse the role of silencing suppression in causation of phenotype alterations. This protein is known to be a pathogenic determinant that controls the symptoms development of infected plants. Previous studies have classified CMV strains in two subgroups, I (also further classified as IA and IB) and II. Strains from group I are more virulent than those of subgroup II, and this difference was used to identify the virulence determinants, finding that replacements of the 2b ORF of the FNY CMV strain (subgroup IA) into different strains belonging to subgroup IB altered both the symptoms and even the host range and the levels of RNA accumulation, demonstrating the important role of the suppressor of silencing in symptom expression (Du *et al.*, 2007b). Since 2b has been shown to localize in the nucleus, thanks to the presence of an arginine-rich nuclear localization signal, and its direct interaction with AGO1 was proven to inhibit the cleavage activity (Zhang *et al.*, 2006), it is tempting to speculate that the high accumulations level of viral RNA in the most virulent strains of CMV might be correlated with the high alteration of the miRNA pathways. Notably, transgenic lines overexpressing the 2b protein of FNY CMV share characteristic phenotypes with *ago1* mutant alleles, such as altered shoot meristems, spoon-shaped cotyledons, narrow leaves, and alteration in flowers with narrow sepals (Bohmert *et al.*, 1998; Morel *et al.*, 2002; Baumberger and Baulcombe, 2005; Kidner and Martienssen, 2005b).

The correlation between siRNA-mediated antiviral silencing and interference with miRNA pathways with the induction of developmental abnormalities in plants has been also established in several studies. The better characterized viral suppressors HCPro, P19 and p21 have been postulated to interfere with miRNA/miRNA* duplex stability, preventing them from being loaded into RISC. This fact could explain aberrant accumulation of miRNA* in plants, which might contribute to produce phenotypes mimicking miRNA abnormalities (Chapman *et al.*, 2004; Dunoyer *et al.*, 2004; Lakatos *et al.*, 2006).

A good example is provided by the transgenic expression in *Arabidopsis* plants of the HCPro of turnip mosaic potyvirus (TuMV), where developmental effects mimicked mutations of DCL1 affecting the miRNA pathway (Kasschau *et al.*, 2003). As mentioned before, many miRNAs negatively regulate genes involved in developmental processes, and thus it is reasonable to consider that viral symptoms could be, at least partially, a direct consequence of the interference of silencing suppressors with endogenous miRNAs. Indeed, infection of *Arabidopsis* plants with TuMV provokes developmental defects resembling those observed in miRNA-deficient dcl1 mutants. TuMV infected plants showed an increase in the level of several miRNAs with essential functions, such as miR171, involved in the regulation of Scarecrow-like transcription factors (Llave *et al.*, 2002), indicating that the presence of the viral suppressor can cause a wide range of phenotypic abnormalities by deregulation of miRNA functions.

Interestingly, transgenic plant expressing HCPro has been shown to be more susceptible to diverse viral pathogens, as expected by the frequent synergistic interactions between potyviruses and other viruses (Vance, 1991; Pruss *et al.*, 1997), but it was a surprise to find at the same time that they seemed to be more resistant to unrelated heterologous pathogens such us the oomycete *Peronospora tabacina* (Pruss *et al.*, 2004). This effect may occur as a consequence of the interaction of silencing suppressors with the host defensive systems, resulting in enhancement of resistance. For instance, the Salicylic Acid (SA)-dependent and independent pathways are good candidates to account for this type of intriguing observations, since recent evidences point towards a deep connection between virus pathogenicity and SA signalling pathways (Alamillo *et al.*, 2006; Atsumi *et al.*, 2009). In the case of CMV, studies with deletion mutants deprived of the 2b suppressor also suggest a link between SA-dependent virus resistance and RNA silencing (Diaz-Pendon *et al.*, 2007).

In other viruses, mutations in the coding region for the silencing suppressors have been used to determine their influence in symptom severity. Thus, plants inoculated with a Cym19Stop mutant, a mutant virus unable to express the viral p19 protein, are able to recover from infections that otherwise would be fulminant, pointing the imperative role of this protein in silencing suppression and spread of symptoms (Szittya *et al.*, 2002, 2003). Other studies have been performed also with the potyviral protein HCPro, where point mutations are able to clearly decrease the rates of virus accumulation, associating decrease of silencing suppression activity with symptom expression (Atreya *et al.*, 1992; Yambao *et al.*, 2008). However, when a large number of mutants were analysed, both reduction and enhancement of severity were observed (Torres-Barcelo *et al.*, 2008), which clearly shows that the natural infections are evolutionarily balanced to reach an equilibrium between enough virus replication and excessive damage to the host.

## DI RNAs, Sat RNAs and viroids: influence on RNA silencing and symptoms modulation

As a consequence of the activation of RNA silencing mechanisms in some viral infections, plants are able to recover from systemic infection in the upper leaves (Ratcliff *et al.*, 1997). This phenomenon might be also associated to the modulation of symptoms observed when Defective Interfering (DI), satellites or other RNA elements are present during a viral infection.

Small molecules of RNA known as DI RNAs appear in certain viral infections as a consequence of deletions generated during replication of the parental viral genome. The presence of the helper virus is necessary to multiply and accumulate a DI. Several studies on tombusvirus DIs have probed their influence in modulation of symptom expression under infection, by impeding viral RNA accumulation, probably caused by competition with the helper viruses for the replication components, and therefore attenuating the symptoms (Russo *et al.*, 1994). RNA silencing seems to play an important role on the symptom modulation caused by DI RNAs (Havelda *et al.*, 1998). Short DI RNAs are considered poorly accessible for processing as precursors of silencing, while long DI RNAs are targeted more efficiently, and this difference determines that they can act as strong inducers of silencing. Theoretically, natural selection would favour those non-processed small DI RNAs (poor targets of RNA silencing) to last longer and accumulate at higher levels in infected plant cells, and consequently compete efficiently for the replicase complex, thus reducing titre of the helper virus. On the other hand, long DI RNAs, as well as high amounts of small DI RNAs, may induce RNA silencing against the related helper virus, therefore reducing the final accumulation on virus in the cells (Szittya *et al.*, 2002). The appearance and fate of DI RNAs during natural virus infection will be the outcome of these opposite balancing forces.

A similar case can be modelled to explain the presence on virus-infected cells of small RNAs also dependent on helper viruses for replication, but with little or no sequence homology with the related virus. These molecules are known as satellite RNAs, and they also appear alongside certain viral infections, modulating the accumulation of virus and being mostly responsible of virulence enhancement or other effects in symptoms development (Dalmay and Rubino, 1994). RNA silencing plays also a role on viral satellite pathogenicity, as shown in transgenic plants expressing the TEV P1/HCPro region that includes the RNA silencing suppressor: when these plants were infected with CMV and Y-SatRNA, a drastic reduction of the characteristic symptoms was observed despite the high accumulation levels of the satellite RNA (Wang *et al.*, 2004). These results suggest that RNA silencing mechanisms are required for the development of symptoms caused by the satellite.

In the turnip crinkle carmovirus (TCV), the presence of the SatC interferes with virion formation. In this example, a model has been proposed to explain the influence of RNA silencing, the silencing suppressor P38 of TCV and the levels of SatC in the final pathogenicity. The protein P38 is the Coat Protein of TCV when associated to virions, but it also acts as silencing suppressor when it is present as a free protein in the cytoplasm (Thomas *et al.*, 2003; Zhang & Simon, 2003). The presence of SatC results in an increase of free CP in the cytoplasm, which means more silencing suppressor activity, and consequently an enhanced spread of virus and an increased virulence.

It is worth mentioning the existence of other causative agents of plant molecular diseases that may influence the plant responses mediated by RNA silencing. Viroids represent the smallest single-stranded RNA molecules able to replicate in susceptible cells without the requirement of helper viruses (Tabler and Tsagris, 2004; Flores *et al.*, 2005). They are formed by circular genomes in which secondary structure presents extensive double-stranded regions, and recent evidences suggest that this genomic folding might favour inaccessibility of DCL enzymes, minimizing siRNA formation, and therefore, evading silencing (Wang *et al.*, 2004).

DI RNAs, Sat RNAs and viroids are examples of small molecules that may have strong influence in the activation of silencing responses without coding for proteins, not only by acting as precursors of RNA silencing, but also by playing an active role in interference with the molecular system in viruses carrying their own silencing suppressor proteins, and causing strong influence in symptoms development. An explanation for the existence, maintenance in evolution and biological significance of these elements could be that silencing suppressors ultimately affect various endogenous pathways of the plant, being overly harsh with their hosts. To avoid extreme situations, viruses would have developed strategies to control and modulate the expression of their genetic material, and thus the viral suppressors might be helped by these molecules in this task of control.

## Concluding remarks

As has been reviewed in the previous sections, a great deal of substantially important data have been produced in the last decade by many groups working with plant viruses in relation to RNA silencing and the activity of viral suppressors. From this abundant literature on the topic, a better focused picture is starting to emerge, with insights that clarify formerly dark areas of the research on plant viruses. As mentioned, the phenomenon of RNA silencing is serving to better explain pathogen derived resistance, recovery, synergism, pathogenicity, symptom modulation, and many other aspects of the plant–virus interaction. Besides this, the characterization of viral silencing suppressor is providing tools to better understand the mechanisms of the different silencing pathways operating in plants. New strategies for efficient virus control, not only in plants but in any other host susceptible to suffer virus infection, is an obvious outcome that is being pursued actively. Finally, even biotechnological applications can be envisaged, and the potential use of the viral suppressors of silencing to modulate and boost the production of foreign high-value molecules in plant-based biofactories is already a reality (Mallory *et al.*, 2002).

However, we are still far from a complete understanding of how all the players are acting during an RNA silencing process. The interactions among the already known actors certainly will deserve further studies, and also new essential molecules participating in the different pathways wait to be discovered and characterized. The catalogue of modes of action of viral suppressors is still incomplete, and much work will undoubtedly continue in this direction in the coming years. Far from being exhaustive, the present chapter briefly updates the current situation of the plant virology discipline in these exciting years after the discovery of RNA silencing, and illustrates how deeply it has impacted in the research efforts that are being undertaken. We have no doubt that this influence will continue in the coming years.

## References


Alamillo, J.M., Saenz, P., and Garcia, J.A. (2006). Salicylic acid-mediated and RNA-silencing defence mechanisms cooperate in the restriction of systemic spread of plum pox virus in tobacco. Plant J. *48*, 217–227.

Allen, E., Xie, Z., Gustafson, A.M., and Carrington, J.C. (2005). microRNA-directed phasing during trans-acting siRNA biogenesis in plants. Cell *121*, 207–221.

Anandalakshmi, R., Marathe, R., Ge, X., Herr, J.M., Jr., Mau, C., Mallory, A., Pruss, G., Bowman, L., and Vance, V.B. (2000). A calmodulin-related protein that suppresses posttranscriptional gene silencing in plants. Science *290*, 142–144.

Anandalakshmi, R., Pruss, G.J., Ge, X., Marathe, R., Mallory, A.C., Smith, T.H., and Vance, V.B. (1998). A viral suppressor of gene silencing in plants. Proc. Natl. Acad. Sci. U.S.A. *95*, 13079–13084.

Atreya, C.D., Atreya, P.L., Thornbury, D.W., and Pirone, T.P. (1992). Site-directed mutations in the potyvirus HC-Pro gene affect helper component activity, virus accumulation, and symptom expression in infected tobacco plants. Virology *191*, 106–111.

Atsumi, G., Kagaya, U., Kitazawa, H., Nakahara, K.S., and Uyeda, I. (2009). Activation of the Salicylic Acid Signaling Pathway Enhances Clover yellow vein virus Virulence in Susceptible Pea Cultivars. Mol. Plant Microbe Interact. *22*, 166–175.

Bartel, D.P. (2004). MicroRNAs: genomics, biogenesis, mechanism, and function. Cell *116*, 281–297.

Baulcombe, D. (2002). Viral suppression of systemic silencing. Trends Microbiol. *10*, 306–308.

Baumberger, N., and Baulcombe, D.C. (2005). *Arabidopsis* ARGONAUTE1 is an RNA Slicer that selectively recruits microRNAs and short interfering RNAs. Proc. Natl. Acad. Sci. U.S.A. *102*, 11928–11933.

Baumberger, N., Tsai, C.H., Lie, M., Havecker, E., and Baulcombe, D.C. (2007). The Polerovirus silencing suppressor P0 targets ARGONAUTE proteins for degradation. Curr. Biol. *17*, 1609–1614.

Bayne, E.H., Rakitina, D.V., Morozov, S.Y., and Baulcombe, D.C. (2005). Cell-to-cell movement of potato potexvirus X is dependent on suppression of RNA silencing. Plant J. *44*, 471–482.

Bernstein, E., Caudy, A.A., Hammond, S.M., and Hannon, G.J. (2001). Role for a bidentate ribonuclease in the initiation step of RNA interference. Nature *409*, 363–366.

Bisaro, D.M. (2006). Silencing suppression by geminivirus proteins. Virology *344*, 158–168.

Blevins, T., Rajeswaran, R., Shivaprasad, P.V., Beknazariants, D., Si-Ammour, A., Park, H.S., Vazquez, F., Robertson, D., Meins, F., Jr., Hohn, T., *et al.* (2006).

Four plant Dicers mediate viral small RNA biogenesis and DNA virus induced silencing. Nucleic Acids Res. *34*, 6233–6246.

Bohmert, K., Camus, I., Bellini, C., Bouchez, D., Caboche, M., and Benning, C. (1998). AGO1 defines a novel locus of *Arabidopsis* controlling leaf development. EMBO J. *17*, 170–180.

Borsani, O., Zhu, J., Verslues, P.E., Sunkar, R., and Zhu, J.K. (2005). Endogenous siRNAs derived from a pair of natural cis-antisense transcripts regulate salt tolerance in *Arabidopsis*. Cell *123*, 1279–1291.

Bortolamiol, D., Pazhouhandeh, M., Marrocco, K., Genschik, P., and Ziegler-Graff, V. (2007). The Polerovirus F box protein P0 targets ARGONAUTE1 to suppress RNA silencing. Curr. Biol. *17*, 1615–1621.

Bouche, N., Lauressergues, D., Gasciolli, V., and Vaucheret, H. (2006). An antagonistic function for *Arabidopsis* DCL2 in development and a new function for DCL4 in generating viral siRNAs. EMBO J. *25*, 3347–3356.

Bozarth, C.S., Weiland, J.J., and Dreher, T.W. (1992). Expression of ORF-69 of turnip yellow mosaic virus is necessary for viral spread in plants. Virology *187*, 124–130.

Brigneti, G., Voinnet, O., Li, W.X., Ji, L.H., Ding, S.W., and Baulcombe, D.C. (1998). Viral pathogenicity determinants are suppressors of transgene silencing in *Nicotiana benthamiana*. EMBO J. *17*, 6739–6746.

Brodersen, P., Sakvarelidze-Achard, L., Bruun-Rasmussen, M., Dunoyer, P., Yamamoto, Y.Y., Sieburth, L., and Voinnet, O. (2008). Widespread translational inhibition by plant miRNAs and siRNAs. Science *320*, 1185–1190.

Brodersen, P., and Voinnet, O. (2006). The diversity of RNA silencing pathways in plants. Trends Genet. *22*, 268–280.

Bucher, E., Hemmes, H., de Haan, P., Goldbach, R., and Prins, M. (2004). The influenza A virus NS1 protein binds small interfering RNAs and suppresses RNA silencing in plants. J. Gen. Virol. *85*, 983–991.

Burgyan, J. (2008). Role of silencing suppressor proteins. Methods Mol. Biol. *451*, 69–79.

Callaway, A., Giesman-Cookmeyer, D., Gillock, E.T., Sit, T.L., and Lommel, S.A. (2001). The multifunctional capsid proteins of plant RNA viruses. Annu. Rev. Phytopathol. *39*, 419–460.

Canizares, M.C., Navas-Castillo, J., and Moriones, E. (2008). Multiple suppressors of RNA silencing encoded by both genomic RNAs of the crinivirus, Tomato chlorosis virus. Virology *379*, 168–174.

Carmell, M.A., and Hannon, G.J. (2004). RNase III enzymes and the initiation of gene silencing. Nat. Struc. Mol. Biol. *11*, 214–218.

Carrington, J.C., Kasschau, K.D., and Johansen, L.K. (2001). Activation and suppression of RNA silencing by plant viruses. Virology *281*, 1–5.

Chapman, E.J., and Carrington, J.C. (2007). Specialization and evolution of endogenous small RNA pathways. Nat. Rev. Genet. *8*, 884–896.

Chapman, E.J., Prokhnevsky, A.I., Gopinath, K., Dolja, V.V., and Carrington, J.C. (2004). Viral RNA silencing suppressors inhibit the microRNA pathway at an intermediate step. Genes Dev. *18*, 1179–1186.

Chen, J., Li, W.X., Xie, D., Peng, J.R., and Ding, S.W. (2004). Viral virulence protein suppresses RNA silencing-mediated defense but up-regulates the role of microrna in host gene expression. Plant Cell *16*, 1302–1313.

Dalmay, T., Hamilton, A., Rudd, S., Angell, S., and Baulcombe, D.C. (2000). An RNA-dependent RNA polymerase gene in *Arabidopsis* is required for post-transcriptional gene silencing mediated by a transgene but not by a virus. Cell *101*, 543–553.

Dalmay, T., and Rubino, L. (1994). The nature of multimeric forms of cymbidium ringspot tombusvirus satellite RNA. Arch. Virol. *138*, 161–167.

Dalmay, T., Rubino, L., Burgyan, J., Kollar, A., and Russo, M. (1993). Functional analysis of cymbidium ringspot virus genome. Virology *194*, 697–704.

Deleris, A., Gallego-Bartolome, J., Bao, J., Kasschau, K.D., Carrington, J.C., and Voinnet, O. (2006). Hierarchical action and inhibition of plant Dicer-like proteins in antiviral defense. Science *313*, 68–71.

Delgadillo, M.O., Saenz, P., Salvador, B., Garcia, J.A., and Simon-Mateo, C. (2004). Human influenza virus NS1 protein enhances viral pathogenicity and acts as an RNA silencing suppressor in plants. J. Gen. Virol. *85*, 993–999.

Diaz-Pendon, J.A., Li, F., Li, W.X., and Ding, S.W. (2007). Suppression of antiviral silencing by cucumber mosaic virus 2b protein in *Arabidopsis* is associated with drastically reduced accumulation of three classes of viral small interfering RNAs. Plant Cell *19*, 2053–2063.

Ding, S.W., Anderson, B.J., Haase, H.R., and Symons, R.H. (1994). New overlapping gene encoded by the cucumber mosaic virus genome. Virology *198*, 593–601.

Ding, S.W., Li, W.X., and Symons, R.H. (1995). A novel naturally occurring hybrid gene encoded by a plant RNA virus facilitates long distance virus movement. EMBO J. *14*, 5762–5772.

Ding, S.W., and Voinnet, O. (2007). Antiviral immunity directed by small RNAs. Cell *130*, 413–426.

Donaire, L., Barajas, D., Martinez-Garcia, B., Martinez-Priego, L., Pagan, I., and Llave, C. (2008). Structural and genetic requirements for the biogenesis of tobacco rattle virus-derived small interfering RNAs. J. Virol. *82*, 5167–5177.

Du, Q.S., Duan, C.G., Zhang, Z.H., Fang, Y.Y., Fang, R.X., Xie, Q., and Guo, H.S. (2007a). DCL4 targets Cucumber mosaic virus satellite RNA at novel secondary structures. J. Virol. *81*, 9142–9151.

Du, Z.Y., Chen, F.F., Liao, Q.S., Zhang, H.R., Chen, Y.F., and Chen, J.S. (2007b). 2b ORFs encoded by subgroup IB strains of cucumber mosaic virus induce differential virulence on Nicotiana species. J. Gen. Virol. *88*, 2596–2604.

Dunoyer, P., Lecellier, C.H., Parizotto, E.A., Himber, C., and Voinnet, O. (2004). Probing the microRNA and small interfering RNA pathways with virus-encoded suppressors of RNA silencing. Plant Cell *16*, 1235–1250.

Dunoyer, P., Pfeffer, S., Fritsch, C., Hemmer, O., Voinnet, O., and Richards, K.E. (2002). Identification, subcellular localization and some properties of a cysteine-rich suppressor of gene silencing encoded by peanut clump virus. Plant J. *29*, 555–567.

Ebhardt, H.A., Thi, E.P., Wang, M.B., and Unrau, P.J. (2005). Extensive 3′ modification of plant small RNAs is modulated by helper component-proteinase expression. Proc. Natl. Acad. Sci. U.S.A. *102*, 13398–13403.

Ekwall, K. (2004). The RITS complex-A direct link between small RNA and heterochromatin. Mol. Cell *13*, 304–305.

Elbashir, S.M., Martinez, J., Patkaniowska, A., Lendeckel, W., and Tuschl, T. (2001). Functional anatomy of siRNAs for mediating efficient RNAi in *Drosophila melanogaster* embryo lysate. EMBO J. *20*, 6877–6888.

Fagard, M., Boutet, S., Morel, J.B., Bellini, C., and Vaucheret, H. (2000). AGO1, QDE-2, and RDE-1 are related proteins required for post-transcriptional gene silencing in plants, quelling in fungi, and RNA interference in animals. Proc. Natl. Acad. Sci. U.S.A. *97*, 11650–11654.

Falk, B.W., and Tsai, J.H. (1998). Biology and molecular biology of viruses in the genus Tenuivirus. Annu. Rev. Phytopathol. *36*, 139–163.

Flores, R., Hernandez, C., Martinez de Alba, A.E., Daros, J.A., and Di Serio, F. (2005). Viroids and viroid–host interactions. Annu. Rev. Phytopathol. *43*, 117–139.

Fusaro, A.F., Matthew, L., Smith, N.A., Curtin, S.J., Dedic-Hagan, J., Ellacott, G.A., Watson, J.M., Wang, M.B., Brosnan, C., Carroll, B.J., *et al.* (2006). RNA interference-inducing hairpin RNAs in plants act through the viral defence pathway. EMBO Rep. *7*, 1168–1175.

Girard, A., and Hannon, G.J. (2008). Conserved themes in small-RNA-mediated transposon control. Trends Cell Biol. *18*, 136–148.

Goodin, M.M., Zaitlin, D., Naidu, R.A., and Lommel, S.A. (2008). *Nicotiana benthamiana*: its history and future as a model for plant–pathogen interactions. Mol. Plant Microbe Interact. *21*, 1015–1026.

Goto, K., Kobori, T., Kosaka, Y., Natsuaki, T., and Masuta, C. (2007). Characterization of silencing suppressor 2b of cucumber mosaic virus based on examination of its small RNA-binding abilities. Plant Cell Physiol. *48*, 1050–1060.

Guo, H.S., and Ding, S.W. (2002). A viral protein inhibits the long range signaling activity of the gene silencing signal. EMBO J. *21*, 398–407.

Gutierrez, C. (2000). DNA replication and cell cycle in plants: learning from geminiviruses. EMBO J. *19*, 792–799.

Haas, G., Azevedo, J., Moissiard, G., Geldreich, A., Himber, C., Bureau, M., Fukuhara, T., Keller, M., and Voinnet, O. (2008). Nuclear import of CaMV P6 is required for infection and suppression of the RNA silencing factor DRB4. EMBO J. *27*, 2102–2112.

Haasnoot, J., de Vries, W., Geutjes, E.J., Prins, M., de Haan, P., and Berkhout, B. (2007). The Ebola virus VP35 protein is a suppressor of RNA silencing. PLoS Pathog. *3*, e86.

Hamilton, A.J., and Baulcombe, D.C. (1999). A species of small antisense RNA in posttranscriptional gene silencing in plants. Science *286*, 950–952.

Hammond, S.M., Bernstein, E., Beach, D., and Hannon, G.J. (2000). An RNA-directed nuclease mediates post-transcriptional gene silencing in *Drosophila* cells. Nature *404*, 293–296.

Han, M.H., Goud, S., Song, L., and Fedoroff, N. (2004). The *Arabidopsis* double-stranded RNA-binding protein HYL1 plays a role in microRNA-mediated gene regulation. Proc. Natl. Acad. Sci. U.S.A. *101*, 1093–1098.

Havelda, Z., Szittya, G., and Burgyan, J. (1998). Characterization of the molecular mechanism of defective interfering RNA-mediated symptom attenuation in tombusvirus-infected plants. J. Virol. *72*, 6251–6256.

Hemmes, H., Lakatos, L., Goldbach, R., Burgyan, J., and Prins, M. (2007). The NS3 protein of Rice hoja blanca tenuivirus suppresses RNA silencing in plant and insect hosts by efficiently binding both siRNAs and miRNAs. RNA *13*, 1079–1089.

Hiraguri, A., Itoh, R., Kondo, N., Nomura, Y., Aizawa, D., Murai, Y., Koiwa, H., Seki, M., Shinozaki, K., and Fukuhara, T. (2005). Specific interactions between Dicer-like proteins and HYL1/DRB-family dsRNA-binding proteins in *Arabidopsis thaliana*. Plant Mol. Biol. *57*, 173–188.

Hock, J., and Meister, G. (2008). The Argonaute protein family. Genome biology *9*, 210.

Hutvagner, G., and Simard, M.J. (2008). Argonaute proteins: key players in RNA silencing. Nat. Rev. Mol. Cell Biol. *9*, 22–32.

Hull, R. (2002). Matthew's Plant Virology, Fourth edition. (San Diego: Academic Press).

Janssen, D., Martin, G., Velasco, L., Gomez, P., Segundo, E., Ruiz, L., and Cuadrado, I.M. (2005). Absence of a coding region for the helper component-proteinase in the genome of cucumber vein yellowing virus, a whitefly-transmitted member of the Potyviridae. Arch. Virol. *150*, 1439–1447.

Johansen, L.K., and Carrington, J.C. (2001). Silencing on the spot. Induction and suppression of RNA silencing in the Agrobacterium-mediated transient expression system. Plant Physiol. *126*, 930–938.

Jones-Rhoades, M.W., Bartel, D.P., and Bartel, B. (2006). MicroRNAS and their regulatory roles in plants. Annu. Rev. Plant Biol. *57*, 19–53.

Joshua-Tor, L. (2006). The Argonautes. Cold Spring Harb. Symp. Quant. Biol. *71*, 67–72.

Karjee, S., Islam, M.N., and Mukherjee, S.K. (2008). Screening and identification of virus-encoded RNA silencing suppressors. Methods Mol. Biol. *442*, 187–203.

Kasschau, K.D., and Carrington, J.C. (1998). A counterdefensive strategy of plant viruses: suppression of posttranscriptional gene silencing. Cell *95*, 461–470.

Kasschau, K.D., Cronin, S., and Carrington, J.C. (1997). Genome amplification and long-distance movement functions associated with the central domain of tobacco etch potyvirus helper component-proteinase. Virology *228*, 251–262.

Kasschau, K.D., Xie, Z., Allen, E., Llave, C., Chapman, E.J., Krizan, K.A., and Carrington, J.C. (2003). P1/HC-Pro, a viral suppressor of RNA silencing, interferes with *Arabidopsis* development and miRNA function. Dev. Cell *4*, 205–217.

Kidner, C.A., and Martienssen, R.A. (2005a). The developmental role of microRNA in plants. Curr. Opin. Plant Biol. *8*, 38–44.

Kidner, C.A., and Martienssen, R.A. (2005b). The role of ARGONAUTE1 (AGO1) in meristem formation and identity. Dev. Biol. *280*, 504–517.

Lakatos, L., Csorba, T., Pantaleo, V., Chapman, E.J., Carrington, J.C., Liu, Y.P., Dolja, V.V., Calvino, L.F., Lopez-Moya, J.J., and Burgyan, J. (2006). Small RNA binding is a common strategy to suppress RNA silencing by several viral suppressors. EMBO J. *25*, 2768–2780.

Lakatos, L., Szittya, G., Silhavy, D., and Burgyan, J. (2004). Molecular mechanism of RNA silencing suppression mediated by p19 protein of tombusviruses. EMBO J. *23*, 876–884.

Li, W., Hilf, M.E., Webb, S.E., Baker, C.A., and Adkins, S. (2008). Presence of P1b and absence of HC-Pro in Squash vein yellowing virus suggests a general feature of the genus Ipomovirus in the family Potyviridae. Virus Res. *135*, 213–219.

Li, W.X., and Ding, S.W. (2001). Viral suppressors of RNA silencing. Curr. Opin. Biotechnol. *12*, 150–154.

Lindbo, J.A., and Dougherty, W.G. (2005). Plant pathology and RNAi: a brief history. Annu. Rev. Phytopathol. *43*, 191–204.

Lingel, A., Simon, B., Izaurralde, E., and Sattler, M. (2004). Nucleic acid 3'-end recognition by the Argonaute2 PAZ domain. Nat. Struc. Mol. Biol. *11*, 576–577.

Lingel, A., Simon, B., Izaurralde, E., and Sattler, M. (2005). The structure of the flock house virus B2 protein, a viral suppressor of RNA interference, shows a novel mode of double-stranded RNA recognition. EMBO Rep. *6*, 1149–1155.

Llave, C., Kasschau, K.D., and Carrington, J.C. (2000). Virus-encoded suppressor of posttranscriptional gene silencing targets a maintenance step in the silencing pathway. Proc. Natl. Acad. Sci. U.S.A. *97*, 13401–13406.

Llave, C., Xie, Z., Kasschau, K.D., and Carrington, J.C. (2002). Cleavage of Scarecrow-like mRNA targets directed by a class of *Arabidopsis* miRNA. Science *297*, 2053–2056.

Lozsa, R., Csorba, T., Lakatos, L., and Burgyan, J. (2008). Inhibition of 3' modification of small RNAs in virus-infected plants require spatial and temporal co-expression of small RNAs and viral silencing-suppressor proteins. Nucleic Acids Res. *36*, 4099–4107.

Lu, R., Folimonov, A., Shintaku, M., Li, W.X., Falk, B.W., Dawson, W.O., and Ding, S.W. (2004). Three distinct suppressors of RNA silencing encoded by a 20-kb viral RNA genome. Proc. Natl. Acad. Sci. U.S.A. *101*, 15742–15747.

Ma, J.B., Ye, K., and Patel, D.J. (2004). Structural basis for overhang-specific small interfering RNA recognition by the PAZ domain. Nature *429*, 318–322.

Maia, I.G., Haenni, A., and Bernardi, F. (1996). Potyviral HC-Pro: a multifunctional protein. J. Gen. Virol. *77*, 1335–1341.

Mallory, A.C., Ely, L., Smith, T.H., Marathe, R., Anandalakshmi, R., Fagard, M., Vaucheret, H., Pruss, G., Bowman, L., and Vance, V.B. (2001). HC-Pro suppression of transgene silencing eliminates the small RNAs but not transgene methylation or the mobile signal. Plant Cell *13*, 571–583.

Mallory, A.C., Mlotshwa, S., Bowman, L.H., and Vance, V.B. (2003). The capacity of transgenic tobacco to send a systemic RNA silencing signal depends on the nature of the inducing transgene locus. Plant J. *35*, 82–92.

Mallory, A.C., Parks, G., Endres, M.W., Baulcombe, D., Bowman, L.H., Pruss, G.J., and Vance, V.B. (2002). The amplicon-plus system for high-level expression of transgenes in plants. Nat. Biotechnol. *20*, 622–625.

Mallory, A.C., Reinhart, B.J., Jones-Rhoades, M.W., Tang, G., Zamore, P.D., Barton, M.K., and Bartel, D.P. (2004). MicroRNA control of PHABULOSA in leaf development: importance of pairing to the microRNA 5' region. EMBO J. *23*, 3356–3364.

Mallory, A.C., and Vaucheret, H. (2006). Functions of microRNAs and related small RNAs in plants. Nat. Genet. *38 Suppl.*, S31–36.

Martinez-Turino, S., and Hernandez, C. (2009). Inhibition of RNA silencing by the coat protein of Pelargonium flower break virus: distinctions from closely related suppressors. J. Gen. Virol. *90*, 519–525.

Matranga, C., Tomari, Y., Shin, C., Bartel, D.P., and Zamore, P.D. (2005). Passenger-strand cleavage facilitates assembly of siRNA into Ago2-containing RNAi enzyme complexes. Cell *123*, 607–620.

Merai, Z., Kerenyi, Z., Molnar, A., Barta, E., Valoczi, A., Bisztray, G., Havelda, Z., Burgyan, J., and Silhavy, D. (2005). Aureusvirus P14 is an efficient RNA silencing suppressor that binds double-stranded RNAs without size specificity. J. Virol. *79*, 7217–7226.

Meyers, B.C., Axtell, M.J., Bartel, B., Bartel, D.P., Baulcombe, D., Bowman, J.L., Cao, X., Carrington, J.C., Chen, X., Green, P.J., *et al.* (2008). Criteria for annotation of plant microRNAs. Plant Cell *20*, 3186–3190.

Millar, A.A., and Waterhouse, P.M. (2005). Plant and animal microRNAs: similarities and differences. Func. Integr. Genomics *5*, 129–135.

Mlotshwa, S., Voinnet, O., Mette, M.F., Matzke, M., Vaucheret, H., Ding, S.W., Pruss, G., and Vance, V.B. (2002). RNA silencing and the mobile silencing signal. Plant Cell *14 Suppl*, S289–301.

Moissiard, G., and Voinnet, O. (2006). RNA silencing of host transcripts by cauliflower mosaic virus requires coordinated action of the four *Arabidopsis* Dicer-like proteins. Proc. Natl. Acad. Sci. U.S.A. *103*, 19593–19598.

Morel, J.B., Godon, C., Mourrain, P., Beclin, C., Boutet, S., Feuerbach, F., Proux, F., and Vaucheret, H. (2002). Fertile hypomorphic ARGONAUTE (ago1) mutants impaired in post-transcriptional gene silencing and virus resistance. Plant Cell *14*, 629–639.

Mourrain, P., Beclin, C., Elmayan, T., Feuerbach, F., Godon, C., Morel, J.B., Jouette, D., Lacombe, A.M., Nikic, S., Picault, N., *et al.* (2000). *Arabidopsis* SGS2 and SGS3 genes are required for posttranscriptional gene silencing and natural virus resistance. Cell *101*, 533–542.

Nagy, P.D., and Pogany, J. (2008). Multiple roles of viral replication proteins in plant RNA virus replication. Methods Mol. Biol. *451*, 55–68.

Palukaitis, P., Carr, J.P., and Schoelz, J.E. (2008). Plant–virus interactions. Methods Mol. Biol. *451*, 3–19.

Park, W., Li, J., Song, R., Messing, J., and Chen, X. (2002). CARPEL FACTORY, a Dicer homolog, and HEN1, a novel protein, act in microRNA metabolism in *Arabidopsis thaliana*. Curr. Biol. *12*, 1484–1495.

Pazhouhandeh, M., Dieterle, M., Marrocco, K., Lechner, E., Berry, B., Brault, V., Hemmer, O., Kretsch, T., Richards, K.E., Genschik, P., *et al.* (2006). F-box-like domain in the polerovirus protein P0 is required for silencing suppressor function. Proc. Natl. Acad. Sci. U.S.A. *103*, 1994–1999.

Peragine, A., Yoshikawa, M., Wu, G., Albrecht, H.L., and Poethig, R.S. (2004). SGS3 and SGS2/SDE1/RDR6 are required for juvenile development and the production of trans-acting siRNAs in *Arabidopsis*. Genes Dev. *18*, 2368–2379.

Pfeffer, S., Dunoyer, P., Heim, F., Richards, K.E., Jonard, G., and Ziegler-Graff, V. (2002). P0 of beet Western yellows virus is a suppressor of posttranscriptional gene silencing. J. Virol. *76*, 6815–6824.

Pruss, G., Ge, X., Shi, X.M., Carrington, J.C., and Bowman Vance, V. (1997). Plant viral synergism: the potyviral genome encodes a broad-range pathogenicity enhancer that transactivates replication of heterologous viruses. Plant Cell *9*, 859–868.

Pruss, G.J., Lawrence, C.B., Bass, T., Li, Q.Q., Bowman, L.H., and Vance, V. (2004). The potyviral suppressor of RNA silencing confers enhanced resistance to multiple pathogens. Virology *320*, 107–120.

Qian, S., Zhong, X., Yu, L., Ding, B., de Haan, P., and Boris-Lawrie, K. (2009). HIV-1 Tat RNA silencing suppressor activity is conserved across kingdoms and counteracts translational repression of HIV-1. Proc. Natl. Acad. Sci. U.S.A. *106*, 605–610.

Qu, F., and Morris, T.J. (2005). Suppressors of RNA silencing encoded by plant viruses and their role in viral infections. FEBS Lett. *579*, 5958–5964.

Qu, F., Ren, T., and Morris, T.J. (2003). The coat protein of turnip crinkle virus suppresses posttranscriptional gene silencing at an early initiation step. J. Virol. *77*, 511–522.

Qu, F., Ye, X., and Morris, T.J. (2008). *Arabidopsis* DRB4, AGO1, AGO7, and RDR6 participate in a DCL4-initiated antiviral RNA silencing pathway negatively regulated by DCL1. Proc. Natl. Acad. Sci. U.S.A. *105*, 14732–14737.

Ramachandran, V., and Chen, X. (2008). Small RNA metabolism in *Arabidopsis*. Trends Plant Sci. *13*, 368–374.

Ratcliff, F., Harrison, B.D., and Baulcombe, D.C. (1997). A similarity between viral defense and gene silencing in plants. Science *276*, 1558–1560.

Reed, J.C., Kasschau, K.D., Prokhnevsky, A.I., Gopinath, K., Pogue, G.P., Carrington, J.C., and Dolja, V.V. (2003). Suppressor of RNA silencing encoded by Beet yellows virus. Virology *306*, 203–209.

Roth, B.M., Pruss, G.J., and Vance, V.B. (2004). Plant viral suppressors of RNA silencing. Virus Res. *102*, 97–108.

Rubino, L., and Russo, M. (1997). Molecular analysis of the pothos latent virus genome. J. Gen. Virol. *78*, 1219–1226.

Ruiz, M.T., Voinnet, O., and Baulcombe, D.C. (1998). Initiation and maintenance of virus-induced gene silencing. Plant Cell *10*, 937–946.

Russo, M., Burgyan, J., and Martelli, G.P. (1994). Molecular biology of tombusviridae. Adv. Virus Res. *44*, 381–428.

Schnettler, E., Hemmes, H., Goldbach, R., and Prins, M. (2008). The NS3 protein of rice hoja blanca virus suppresses RNA silencing in mammalian cells. J. Gen. Virol. *89*, 336–340.

Scholthof, H.B., Scholthof, K.B., and Jackson, A.O. (1995). Identification of tomato bushy stunt virus host-specific symptom determinants by expression of individual genes from a potato virus X vector. Plant Cell 7, 1157–1172.

Silhavy, D., Molnar, A., Lucioli, A., Szittya, G., Hornyik, C., Tavazza, M., and Burgyan, J. (2002). A viral protein suppresses RNA silencing and binds silencing-generated, 21- to 25-nucleotide double-stranded RNAs. EMBO J. *21*, 3070–3080.

Soldan, S.S., Plassmeyer, M.L., Matukonis, M.K., and Gonzalez-Scarano, F. (2005). La Crosse virus nonstructural protein NSs counteracts the effects of short interfering RNA. J. Virol. *79*, 234–244.

Song, J.J., Smith, S.K., Hannon, G.J., and Joshua-Tor, L. (2004). Crystal structure of Argonaute and its implications for RISC slicer activity. Science *305*, 1434–1437.

Stenger, D.C., French, R., and Gildow, F.E. (2005). Complete deletion of Wheat streak mosaic virus HC-Pro: a null mutant is viable for systemic infection. J. Virol. *79*, 12077–12080.

Stenger, D.C., Young, B.A., Qu, F., Morris, T.J., and French, R. (2007). Wheat streak mosaic virus lacking helper component-proteinase is competent to produce disease synergism in double infections with maize chlorotic mottle virus. Phytopathology *97*, 1213–1221.

Szittya, G., Molnar, A., Silhavy, D., Hornyik, C., and Burgyan, J. (2002). Short defective interfering RNAs of tombusviruses are not targeted but trigger post-transcriptional gene silencing against their helper virus. Plant Cell *14*, 359–372.

Szittya, G., Silhavy, D., Molnar, A., Havelda, Z., Lovas, A., Lakatos, L., Banfalvi, Z., and Burgyan, J. (2003). Low temperature inhibits RNA silencing-mediated defence by the control of siRNA generation. EMBO J. *22*, 633–640.

Tabler, M., and Tsagris, M. (2004). Viroids: petite RNA pathogens with distinguished talents. Trends Plant Sci. *9*, 339–348.

Takeda, A., Sugiyama, K., Nagano, H., Mori, M., Kaido, M., Mise, K., Tsuda, S., and Okuno, T. (2002). Identification of a novel RNA silencing suppressor, NSs protein of Tomato spotted wilt virus. FEBS Lett. *532*, 75–79.

Taliansky, M., Torrance, L., and Kalinina, N.O. (2008). Role of plant virus movement proteins. Methods Mol. Biol. *451*, 33–54.

Thomas, C.L., Leh, V., Lederer, C., and Maule, A.J. (2003). Turnip crinkle virus coat protein mediates suppression of RNA silencing in *Nicotiana benthamiana*. Virology *306*, 33–41.

Tolia, N.H., and Joshua-Tor, L. (2007). Slicer and the argonautes. Nat. Chem. Biol. *3*, 36–43.

Tomari, Y., and Zamore, P.D. (2005). Perspective: machines for RNAi. Genes Dev. *19*, 517–529.

Torres-Barcelo, C., Martin, S., Daros, J.A., and Elena, S.F. (2008). From hypo- to hypersuppression: effect of amino acid substitutions on the RNA-silencing suppressor activity of the Tobacco etch potyvirus HC-Pro. Genetics *180*, 1039–1049.

Tsai, C.H., and Dreher, T.W. (1993). Increased viral yield and symptom severity result from a single amino acid substitution in the turnip yellow mosaic virus movement protein. Mol. Plant Microbe Interact. *6*, 268–273.

Valli, A., Dujovny, G., and Garcia, J.A. (2008). Protease activity, self interaction, and small interfering RNA binding of the silencing suppressor p1b from cucumber vein yellowing ipomovirus. J. Virol. *82*, 974–986.

Valli, A., Lopez-Moya, J.J., and Garcia, J.A. (2007). Recombination and gene duplication in the evolutionary diversification of P1 proteins in the family Potyviridae. J. Gen. Virol. *88*, 1016–1028.

Valli, A., Martin-Hernandez, A.M., Lopez-Moya, J.J., and Garcia, J.A. (2006). RNA silencing suppression by a second copy of the P1 serine protease of Cucumber vein yellowing ipomovirus, a member of the family Potyviridae that lacks the cysteine protease HCPro. J. Virol. *80*, 10055–10063.

Vance, V.B. (1991). Replication of potato virus X RNA is altered in coinfections with potato virus Y. Virology *182*, 486–494.

Vanitharani, R., Chellappan, P., and Fauquet, C.M. (2005). Geminiviruses and RNA silencing. Trends Plant Sci. *10*, 144–151.

Vargason, J.M., Szittya, G., Burgyan, J., and Hall, T.M. (2003). Size selective recognition of siRNA by an RNA silencing suppressor. Cell *115*, 799–811.

Vaucheret, H. (2006). Post-transcriptional small RNA pathways in plants: mechanisms and regulations. Genes Dev. *20*, 759–771.

Vaucheret, H., Vazquez, F., Crete, P., and Bartel, D.P. (2004). The action of ARGONAUTE1 in the miRNA pathway and its regulation by the miRNA pathway are crucial for plant development. Genes Dev. *18*, 1187–1197.

Vazquez, F., Gasciolli, V., Crete, P., and Vaucheret, H. (2004a). The nuclear dsRNA binding protein HYL1 is required for microRNA accumulation and plant development, but not posttranscriptional transgene silencing. Curr. Biol. *14*, 346–351.

Vazquez, F., Vaucheret, H., Rajagopalan, R., Lepers, C., Gasciolli, V., Mallory, A.C., Hilbert, J.L., Bartel, D.P., and Crete, P. (2004b). Endogenous trans-acting siRNAs regulate the accumulation of *Arabidopsis* mRNAs. Mol. Cell *16*, 69–79.

Verdel, A., Jia, S., Gerber, S., Sugiyama, T., Gygi, S., Grewal, S.I., and Moazed, D. (2004). RNAi-mediated targeting of heterochromatin by the RITS complex. Science *303*, 672–676.

Voinnet, O. (2005). Induction and suppression of RNA silencing: insights from viral infections. Nat. Rev. Genet. *6*, 206–220.

Voinnet, O., and Baulcombe, D.C. (1997). Systemic signalling in gene silencing. Nature *389*, 553.

Voinnet, O., Lederer, C., and Baulcombe, D.C. (2000). A viral movement protein prevents spread of the gene silencing signal in *Nicotiana benthamiana*. Cell *103*, 157–167.

Voinnet, O., Pinto, Y.M., and Baulcombe, D.C. (1999). Suppression of gene silencing: a general strategy used by diverse DNA and RNA viruses of plants. Proc. Natl. Acad. Sci. U.S.A. *96*, 14147–14152.

Voinnet, O., Rivas, S., Mestre, P., and Baulcombe, D. (2003). An enhanced transient expression system in plants based on suppression of gene silencing by the p19 protein of tomato bushy stunt virus. Plant J. *33*, 949–956.

Wang, H., Buckley, K.J., Yang, X., Buchmann, R.C., and Bisaro, D.M. (2005). Adenosine kinase inhibition and suppression of RNA silencing by geminivirus AL2 and L2 proteins. J. Virol. *79*, 7410–7418.

Wang, H., Hao, L., Shung, C.Y., Sunter, G., and Bisaro, D.M. (2003). Adenosine kinase is inactivated by geminivirus AL2 and L2 proteins. Plant Cell *15*, 3020–3032.

Wang, M.B., Bian, X.Y., Wu, L.M., Liu, L.X., Smith, N.A., Isenegger, D., Wu, R.M., Masuta, C., Vance, V.B., Watson, J.M., *et al.* (2004). On the role of RNA silencing in the pathogenicity and evolution of viroids and viral satellites. Proc. Natl. Acad. Sci. U.S.A. *101*, 3275–3280.

Wang, M.B., and Metzlaff, M. (2005). RNA silencing and antiviral defense in plants. Curr. Opin. Plant Biol. *8*, 216–222.

Wassenegger, M., and Krczal, G. (2006). Nomenclature and functions of RNA-directed RNA polymerases. Trends Plant Sci. *11*, 142–151.

Wijkamp, I., van Lent, J., Kormelink, R., Goldbach, R., and Peters, D. (1993). Multiplication of tomato spotted wilt virus in its insect vector, *Frankliniella occidentalis*. J. Gen. Virol. *74 (Pt 3)*, 341–349.

Xie, Z., Kasschau, K.D., and Carrington, J.C. (2003). Negative feedback regulation of Dicer-Like1 in *Arabidopsis* by microRNA-guided mRNA degradation. Curr. Biol. *13*, 784–789.

Xie, Z., and Qi, X. (2008). Diverse small RNA-directed silencing pathways in plants. Biochim. Biophys. Acta *1779*, 720–724.

Yambao, M.L., Yagihashi, H., Sekiguchi, H., Sekiguchi, T., Sasaki, T., Sato, M., Atsumi, G., Tacahashi, Y., Nakahara, K.S., and Uyeda, I. (2008). Point mutations in helper component protease of clover yellow vein virus are associated with the attenuation of RNA-silencing suppression activity and symptom expression in broad bean. Arch. Virol. *153*, 105–115.

Yang, S.J., Carter, S.A., Cole, A.B., Cheng, N.H., and Nelson, R.S. (2004). A natural variant of a host RNA-dependent RNA polymerase is associated with increased susceptibility to viruses by *Nicotiana benthamiana*. Proc. Natl. Acad. Sci. U.S.A. *101*, 6297–6302.

Ye, K., Malinina, L., and Patel, D.J. (2003). Recognition of small interfering RNA by a viral suppressor of RNA silencing. Nature *426*, 874–878.

Ye, K., and Patel, D.J. (2005). RNA silencing suppressor p21 of Beet yellows virus forms an RNA binding octameric ring structure. Structure *13*, 1375–1384.

Yelina, N.E., Savenkov, E.I., Solovyev, A.G., Morozov, S.Y., and Valkonen, J.P. (2002). Long-distance movement, virulence, and RNA silencing suppression controlled by a single protein in hordei- and potyviruses: complementary functions between virus families. J. Virol. *76*, 12981–12991.

Yu, B., Yang, Z., Li, J., Minakhina, S., Yang, M., Padgett, R.W., Steward, R., and Chen, X. (2005). Methylation as a crucial step in plant microRNA biogenesis. Science *307*, 932–935.

Yuan, Y.R., Pei, Y., Ma, J.B., Kuryavyi, V., Zhadina, M., Meister, G., Chen, H.Y., Dauter, Z., Tuschl, T., and Patel, D.J. (2005). Crystal structure of A. aeolicus argonaute, a site-specific DNA-guided endoribonuclease, provides insights into RISC-mediated mRNA cleavage. Mol. Cell *19*, 405–419.

Zhang, B., Wang, Q., and Pan, X. (2007). MicroRNAs and their regulatory roles in animals and plants. J. Cell Physiol. *210*, 279–289.

Zhang, F., and Simon, A.E. (2003). Enhanced viral pathogenesis associated with a virulent mutant virus or a virulent satellite RNA correlates with reduced virion accumulation and abundance of free coat protein. Virology *312*, 8–13.

Zhang, X., Yuan, Y.R., Pei, Y., Lin, S.S., Tuschl, T., Patel, D.J., and Chua, N.H. (2006). Cucumber mosaic virus-encoded 2b suppressor inhibits *Arabidopsis* Argonaute1 cleavage activity to counter plant defense. Genes Dev. *20*, 3255–3268.

Ziegler-Graff, V., and Brault, V. (2008). Role of vector-transmission proteins. Methods Mol. Biol. *451*, 81–96.

# The Properties and Roles of Virus-encoded MicroRNAs

3

Mélanie Tanguy and Sébastien Pfeffer

Abstract

The discovery that viruses could encode micro (mi)RNAs, similarly to the eukaryotic organisms they infect, has opened new perspectives in the study of host–virus interactions. These small regulatory RNAs, which are critically involved in an ever-increasing number of biological processes, have revolutionized the way we used to see gene regulation. Some mammalian viruses, mainly from the herpesvirus and polyomavirus families, have hijacked this mechanism in order to help them achieve the infection of their host. In this chapter, we will present the diversity of known viral miRNAs, their specific properties, their viral and cellular targets and the roles they play during the course of infection. We will see that more and more it appears that virally encoded miRNAs seem to be critically involved in every step of the virus life cycle.

## Introduction

The prime role of RNA silencing during a viral infection seems to act as a defence system to protect the organism. This is commonly admitted in plant and insect organisms, where viruses are both effectors and targets of the silencing machinery (see Chapter 2). However, the situation is much less clear in mammals where it remains to be formally demonstrated that there really is an antiviral role of RNAi in the context of a physiologically relevant infection. The missing indisputable proof for this peculiar protection mechanism would be the identification of siRNAs of viral origin in infected cells or tissue. Yet, in the search of these elusive viral siRNAs, what has been surprisingly found instead was that some viruses do indeed express small RNAs, but these rather resemble micro (mi)RNAs and they appear to be beneficial for the virus rather than part of a degradation or protection scheme.

Among the different classes of small regulatory RNAs, miRNAs are one of the most represented and studied in vertebrates. These 21–24 nucleotides (nt) long molecules are key managers of gene expression and are involved in a very large number of biological processes. Although they were discovered in the nematode *Caenorhabditis elegans* in 1993 (Lee *et al.*, 1993; Wightman *et al.*, 1993), they were only shown to be present in other eukaryotes in 2001 (Lagos-Quintana *et al.*, 2001; Lau *et al.*, 2001). Their biogenesis has been under intense scrutiny since, and has been reviewed extensively elsewhere (for example see (Bartel, 2004; Du and Zamore, 2005; Kim, 2005; Meister and Tuschl, 2004; Okamura *et al.*, 2008)]. Fig. 3.1 depicts the critical steps and enzymes involved in this pathway. Briefly, miRNAs are typically transcribed by RNA polymerase II, and in rare occasions by RNA pol III, in the nucleus of the cell. The large pri-miRNA transcript is then processed by the microprocessor complex, where the type III ribonuclease Drosha is the active component, to give rise to the stem–loop precursor miRNA (pre-miRNA). In few cases, pre-miRNAs named mirtrons can be excised in a Drosha-independent manner by direct splicing. The pre-miRNA is then exported to the cytoplasm via the exportin 5 factor. Once in the cytoplasm, the pre-miRNA is processed into an RNA duplex, very similar to an siRNA duplex, by a second RNase named Dicer.

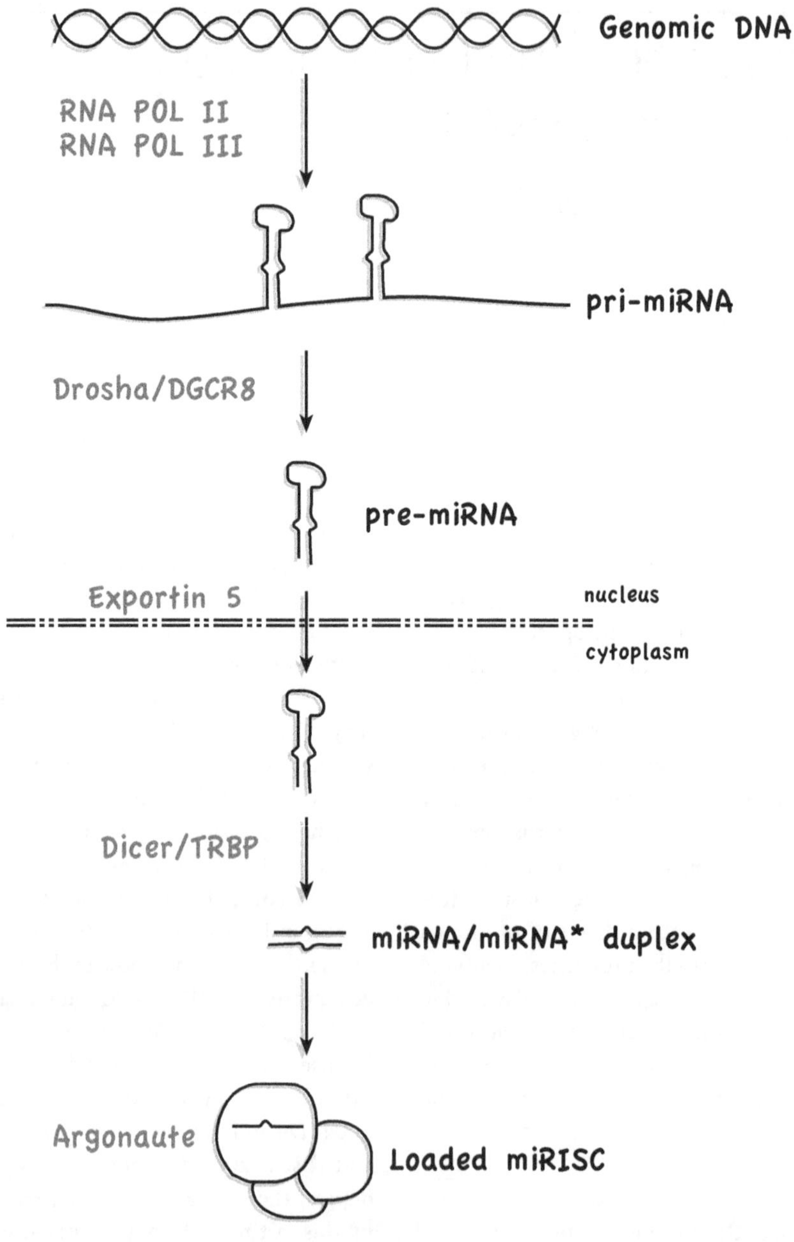


**Figure 3.1** Key factors involved in the biogenesis of cellular and viral miRNAs. DGCR8, DiGeorge critical syndrome region 8; TRBP, Tat-responsive element RNA-binding protein.

After this step, either the mature strand or both strands are incorporated in the RNA-induced silencing complex (RISC), which invariably contains a member of the Piwi/Argonaute family of proteins. The active RISC will be directed to target sequences by the loaded mature miRNA. The targeted transcript will afterward undergo two possible fates that, at least partially, depend on its degree of complementarity with the small RNA. In case of a perfectly matched binding site and only when RISC contains Argonaute 2, the targeted transcript will be cleaved and degraded by exonucleases. When the pairing is only incomplete, with stretches of paired nucleotides and bulges, the translation of the target mRNA will be inhibited. This translation block can be achieved in different ways from the inhibition of initiation to the destabilization of the RNA by de-adenylation and decapping (reviewed in Eulalio *et al.*, 2008; Pillai *et al.*, 2007). Often, the miRNA target will be translocated to so-called cytoplasmic processing or GW182 bodies where it will not be available to the translation machinery. Interestingly, this sequestration is reversible in certain stress conditions (Bhattacharyya *et al.*, 2006). Most targets that have been identified up to known are recognized by miRNAs through binding sites in their 3′-UTR (Lim *et al.*, 2005), although it appears that miRNAs can also attach to other parts of the mRNA (Orom *et al.*, 2008; Tay *et al.*, 2008). Some of the miRNA nucleotides appear to be more important than others in the recognition of the target. In particular, several publications have pointed to the importance of the so-called seed of the miRNA, which is composed of nucleotides 2–8, with some variations depending of the reports (Lewis *et al.*, 2005).

Viral miRNAs that have been described up to now mostly belong to three distinct virus families: Herpesviridae, Polyomaviridae and Adenoviridae. We will review here the current knowledge on virus-encoded miRNAs from the extensive description of the great diversity of known viral miRNAs to their possible functions during the infectious cycle.

## The diversity of viral miRNAs

To date, more than 140 miRNAs of viral origin have been identified (Table 3.1). In this section, we will present the viruses in which they have been found, their genomic localization and their mode of expression. The vast majority of viral miRNAs are not evolutionary conserved between distant virus relatives or with their host. However, there are a few notable exceptions where some limited sequence homologies can be found between a viral miRNA and a cellular miRNA, but this rather seems to be due to a co-evolution in function. As we will see below, there also are a number of cases where the role of the miRNA has been conserved between distantly related viruses with no conservation at the primary sequence level.

### Gammaherpesviruses

Epstein–Barr virus (EBV) was the first human virus shown to encode miRNAs. Pfeffer *et al.* initially identified 5 miRNAs in the BL41 cell line latently infected with the B95.8 strain of EBV (Pfeffer *et al.*, 2004), before an additional 18 were found, among which 13 were localized in a region of 12 kb that is deleted in this particular strain (Cai *et al.*, 2006; Grundhoff *et al.*, 2006). EBV primo-infection causes infectious mononucleosis in young patients before it goes for a lifelong latent infection – up to 95% of people are latently infected with this virus in the Western world. EBV was the first oncogenic virus identified in humans, and was primarily reported to be the aetiological agent of Burkitt's lymphoma. Additionally, EBV is known to be associated with a number of other neoplastic disorders such as B and T cells lymphomas, as well as nasopharyngeal and gastric carcinomas (Kieff and Rickinson, 2007). The miRNAs in EBV are located in three different clusters: two within the BART (BamHI A rightward transcripts) locus, which coding potential is uncertain and another one within the BHRF1 (Bam H1 rightward open reading frame 1) transcript, an early lytic transcript associated with the replicative cycle and a distant homologue of Bcl2 (Oudejans *et al.*, 1995). This transcript is also involved in the translation of EBNA proteins (EBV nuclear antigens). MiR-BHRF1-1 to 3 are potentially located within the 5′ and 3′-UTR of BHRF1 transcript, but some have reported that they could be expressed from the large EBNA1 transcript, and not from the shorter BHRF1 mRNA. The BART clusters include miR-BART-1 to miR-BART-20, which map to intronic regions of the BART transcript. A recent study has shown that the BART miRNAs are produced prior to splicing and that a specific form of the spliced precursor transcript is favouring miRNAs production (Edwards *et al.*, 2008). Interestingly, EBV miRNAs have been shown to accumulate differently in different types of carcinomas. Thus, BART miRNAs, but not BHRF1 miRNAs, are highly expressed in gastric carcinoma cell lines and tumour samples (Kim do *et al.*, 2007), which has an important implication for their putative role in the development of this disease.

Cai and coworkers cloned small RNAs from another, distant member of the gammaherpesvirus genus, the rhesus lymphocryptovirus (rLCV) (Cai *et al.*, 2006). Notably, the 16 identified miRNAs precursors mapped to regions similar to those found in EBV, and eight mature miRNAs

**Table 3.1** Known virus-encoded miRNAs

| Virus family, genus | Name | No. of pre-miRNAs | Names | References |
|---|---|---|---|---|
| Herpesviruses | | | | |
| α/Simplexvirus | HSV1 | 6 | miR-H1 to H6 | Cui *et al.* (2006), Umbach *et al.* (2008) |
| | HSV2 | 3 | miR-I to III | Tang *et al.* (2008, 2009) |
| α/Mardivirus | MDV1 | 14 | miR-M1 to M13, miR-M31 | Burnside *et al.* (2006, 2008), Yao *et al.* (2008) |
| | MDV2 | 17 | miR-M14 to M30 | Yao *et al.* (2007) |
| β/Cytomegalovirus | HCMV | 11 | miR-UL22A, UL36, UL70, UL112, UL148D, US4, US5-1, US5-2, US25-1, US25-2, US33 | Dunn *et al.* (2005), Grey *et al.* (2005), Pfeffer *et al.* (2005) |
| | MCMV | 18 | miR-m01-1 to m01-4, m21-1, m22-1, M23-1, M23-2, M44-1, M55-1, m59-1, m59-2, M87-1, m88-1, M95-1, m107-1, m108-1, m108-2 | Buck *et al.* (2007), Dolken *et al.* (2007) |
| γ/Lymphocryptovirus | EBV | 23 | miR-BART1 to 20, BHRF1-1 to 3 | Cai *et al.* (2006), Grundhoff *et al.* (2006), Pfeffer *et al.* (2004) |
| | rLCV | 16 | miR-rL1-1 to 16 | Cai *et al.* (2006) |
| γ/rhadinovirus | KSHV | 12 | miR-K12-1 to 12 | Cai *et al.* (2005), Grundhoff *et al.* (2006), Pfeffer *et al.*, (2005), Samols *et al.* (2005) |
| | RRV | 7 | miR-rR1-1 to 7 | Schafer *et al.* (2007) |
| | MHV68 | 9 | miR-M1-1 to 9 | Pfeffer *et al.* (2005) |
| Polyomaviruses | SV40 | 1 | miR-S1 | Sullivan *et al.* (2005) |
| | SA12 | 1 | No name | Cantalupo *et al.* (2005) |
| | BKV | 1 | miR-B1 | Seo *et al.* (2008b) |
| | JCV | 1 | miR-J1 | Seo *et al.* (2008b) |
| | MCV | 1 | miR-M1 | Seo *et al.* (2008a) |
| Adenoviruses | hAV | 1 | No name | Andersson *et al.* (2005), Aparicio *et al.* (2006), Sano *et al.* (2006), Xu *et al.* (2007) |
| Ascoviruses | HvAv | 1 | miR-1 | Hussain *et al.* (2008) |

were conserved in sequence suggesting that an evolutionary pressure maintained these sequences.

Kaposi's sarcoma-associated herpesvirus (KSHV), another gammaherpesvirus, is the primary aetiological agent of Kaposi's sarcoma, an endothelial tumour developing in immunocompromised individuals. KSHV is also associated with aggressive lymphomas such as body-cavity-based lymphoma, pleural effusion lymphoma and multicentric Castleman's disease (Ganem, 2007). It encodes 12 miRNAs (Cai *et al.*, 2005; Grundhoff *et al.*, 2006; Pfeffer *et al.*,

2005; Samols *et al.*, 2005). These miRNAs are clustered in one segment located at the vicinity of the major KSHV latency transcript K12. The K12 locus encodes three isoforms of Kaposin proteins, which role is not well known yet, although the A isoform of Kaposin has been shown to have transformation properties (Kliche *et al.*, 2001). KSHV-miR-K12-1 to miR-K12-9 and miR-K12-11 are located in the intron of the larger kaposin transcript, while miR-K12-10 maps to the coding region and miR-K12-12 resides within the 3′-UTR of the K12 coding sequence. One of KSHV miRNA, miR-K12-10, is edited at position 2 of the miRNA sequence. The two variants presenting either an adenosine or a guanine at this position seem to accumulate to similar level in the latently infected BCBL1 cell line (Pfeffer *et al.*, 2005). The role of this editing is not known, but its location at a critical residue within the miRNA seed region suggests that it could be important during target recognition.

Similarly than rLCV and EBV, miRNAs have been identified in a monkey model of KSHV, the rhesus rhadinovirus (RRV), and seem to have been maintained when these viruses diverged million years ago. However, in this case, the miRNA sequences are not conserved, but their genomic location is in a locus similar to the KSHV one. The RRV pri-miRNA could be expressed from the ORF73 gene as a unique transcript. Eleven RRV-derived miRNAs were discovered in latently infected epithelial cells. These 11 functional miRNAs derive from 7 precursors, and, as for KSHV, seem to be mostly latent genes. However, at least three of them are strongly induced during the virus lytic cycle, indicating that the pre-miRNAs could be transcribed from two different promoters depending on the latent vs. lytic state of infection (Schafer *et al.*, 2007).

The last gammaherpesvirus in which miRNAs have been identified is the mouse gammaherpesvirus 68 (MHV68). This virus is evolutionary related to the human gammaherpesviruses EBV and KSHV and is often referred to as the small animal model for the biology of these viruses (Speck and Virgin, 1999). In 2005, Pfeffer *et al.* identified nine miRNAs (miR-M1-1 to M1-9) encoded by MHV68 in a latently infected mouse B lymphoma cell line. Interestingly, all of them mapped to a 6-kb region near the M1 ORF, and were located downstream of previously identified tRNA genes. The function of these tRNAs had remained an enigma since it had been shown that none of them was aminoacylated. Upon discovery of the miRNAs, it became clear that these tRNAs acted as RNA pol III promoters for their expression in a non conventional manner (Pfeffer *et al.*, 2005).

## Betaherpesviruses

The betaherpesvirus human cytomegalovirus (HCMV) has the largest genome of all herpesviruses. Acute disease occurs only in a small fraction of infected individuals, and only when the ability to mount an efficient immune response is jeopardized, i.e. transplacental transmission or primary infection of immunocompromised patients (Mocarski *et al.*, 2007). HCMV miRNAs were first described by Pfeffer *et al.* who cloned and sequenced small RNAs from lytically infected primary fibroblasts and identified nine miRNAs (Pfeffer *et al.*, 2005). Dunn *et al.* also described in 2005 2 HCMV-encoded miRNAs, miR UL23, leading to two mature miRNAs (hcmv-miR-UL23-5p and UL23-3p, now reported as miR-UL22A and UL22A*) and hcmv-miR-US24 (now reported as miR-US25-1). They are expressed, respectively, at early and immediate-early stage of infection (Dunn *et al.*, 2005). Finally, Grey *et al.* described 5 HCMV miRNAs among which were two novel miRNAs (miR-US4-1 and miR-UL70-1) (Grey *et al.*, 2005). The later were discovered by looking for stem–loop structures conservation between the HCMV and the chimpanzee cytomegalovirus (CCMV). Out of 110 HCMV stem loop sequences conserved in CCMV, Grey *et al.* predicted 13 potential miRNAs. Again, these miRNAs are expressed with an immediate early (UL70-1) or early kinetics (US4-1, US5-1, US5-2, UL36-1). HCMV miRNAs are spread throughout the viral genome and are located either in non-coding regions, intronic regions or complementary strand of known ORF of viral genes. As an example, miR-UL-112–1 is located antisense to the viral uracil DNA glycosylase UL114. One of these miRNAs, miR-US36-1, is located in the intron of the UL36 gene, but does not seem to follow the immediate-early expression pattern of this gene. This indicates that there could be an alternate promoter driving its expression or that there is a stabilization of the intron after UL36 transcription.

The murine cytomegalovirus (MCMV) encodes 18 pre-miRNAs that are both early and late gene products (Buck *et al.*, 2007; Dolken *et al.*, 2007). No virally encoded miRNAs has been described during latency, probably due to the lack of an amenable system to study MCMV latency phase. Similar to HCMV these miRNAs are spread along the genome and expressed either individually or in cluster, but none of them is conserved at the primary sequence level. Also, their genomic localization differs between the two viruses, with the notable exception of miRNAs in the M23 region of MCMV corresponding to the UL23 region of HCMV. Notably, it has been shown that MCMV miRNAs are expressed from both strands of the viral DNA genome. Thus, the star sequence of miR-M23-2 is complementary to 16 nucleotides of the miR-m21-1 sequence. The same sequence complementarity exists for miR-M23-1–5p and miR-m22-1, and for miR-m107-1 and miR-m108-2-3p. These miRNAs are all detectable by Northern blot so we should admit that these genomic localizations do not interfere with the miRNAs accumulation. In addition, some MCMV miRNAs were described to be polyuridylated (Dolken *et al.*, 2007), although the physiological relevance of this modification is not yet understood. One hypothesis is that it could be a signal for degradation, as plant miRNAs have been shown to be degraded following polyuridylation when their 3′ end is not protected by a methyl group (Yu *et al.*, 2005). However, another recent report claims that a uridine stretch at the 3′ end of a miRNA could have a protective role against an *Arabidopsis* exonuclease (Ramachandran and Chen, 2008), which indicates that the exact role of this post-transcriptional modification is far from being elucidated.

## Alphaherpesviruses

The Alphaherpesvirinae subfamily includes several miRNAs coding viruses. Two simplexviruses, HSV1 and HSV2, and two mardiviruses, MDV1 and MDV2, have been shown to use this way of modulating gene expression.

Herpes simplex virus type 1 (HSV1), the aetiological agent of oral herpes encodes six microRNAs, miR-H1 to miR-H6 (Cui *et al.*, 2006; Umbach *et al.*, 2008). Except for miR-H1 that was detected primarily in lytically infected cells (Cui *et al.*, 2006), the other HSV1 miRNAs are expressed in latently infected cells. MiR-H1 is a late gene product transcribed upstream of the latency-associated transcript (LAT) transcription start site. Interestingly, miR-H6 was found to be expressed from the same genomic location as miR-H1, but on the complementary strand (Umbach *et al.*, 2008). The LAT transcript is the source of the remaining four miRNAs, which are all located in the second exon of the 6.3kb spliced LAT transcript. When cloned by the Cullen lab from a vector expressing the LAT transcript, the most abundant miRNAs were mir-H2-3p and miR-H4-5p. The star sequences of these miRNAs were also detected (miR-H2-5p and miR-H4-5p). Among these latently expressed miRNAs, miR-H2 is transcribed in an antisense orientation to ICP0, a viral regulator important for productive HSV-1 replication and probably in reactivation from latency (see below). HSV1-miR-H3 and miR-H4-5p are located antisense to ICP34.5.

The close relative herpes virus simplex 2 (HSV2), the aetiological agent of genital herpes, is also a source of miRNAs. This virus encodes a LAT associated miRNA named miR-I. This miRNA is expressed *in vivo* under the control of LAT promoter in the latency model used, i.e. the guinea pig, and it is located on the strand antisense to the ICP34.5 transcript, a key viral neurovirulence factor (Tang *et al.*, 2008). It has been shown recently that HSV-2 LAT exon 2 encodes two additional miRNAs, miR-II and miR-III, which are located in an antisense orientation to, respectively, the 5′-UTR of ICP34.5 and the coding region of ICP0-exon3 (Tang *et al.*, 2009). These miRNAs are less abundant than miR-I. Interestingly, this novel study reinforces the LAT promoter independent transcription of these miRNAs. Indeed several other promoters active during productive infection could also be involved in miRNAs expression. Thus, miR-I and miR-II were detected early after inoculation, implying that the LAT promoter was not involved in their initial expression. Consistent with this postulate, the deletion of the LAT promoter did not totally abolish the presence of miR-I but strongly reduced its level of expression (Tang *et al.*, 2008). MiR-I has also been shown to be expressed during acute infection in sensory ganglia. By luciferase assay and Northern blotting experiments, Tang and coworkers showed that ICP4 is down-regulating the LAT and also the ORF O promoters activity (Tang *et al.*, 2009).

Two other alphaherpesviruses, Marek's visease virus (MDV) 1 and 2, express miRNAs. Marek's disease (MD) is a chicken lymphoproliferative disorder associated with aggressive T-cell lymphomas. The serotype 1 of MDV has been reported to be oncogenic but the serotype 2 is not. By high-throughput sequencing of small RNA cloned from infected chicken embryotic fibroblasts, Burnside *et al.* first found that MDV1 encode eight different miRNAs (Burnside *et al.*, 2006). Yao *et al.* who worked with an MDV-transformed CD4 T-cell line (MSB-1) identified an additional six miRNAs in MDV1 genome (Yao *et al.*, 2008). Some arguments led to think that the MDV miRNAs were transcribed from three independent primary clusters: one in the LAT region and two upstream and downstream of the Meq transcript. Meq, which is indeed transcribed in tumours and transformed cells, is a protein with oncogenic potential and one of his spliced variant has been reported to be the major MDV oncoprotein (Liu and Kung, 2000). Eight of the miRNA precursors are clustered in the Meq region and show the same expression pattern. While miR-M2 to M5 and M11 are located upstream of the Meq promoter and lies antisense to RLORF8, miR-M1 is found within the intron of Meqsp, the spliced variant of Meq. The remaining miRNAs map to the LAT region of the MDV genome. LAT in MDV is expressed in an antisense orientation to ICP4 and, up to now, no translation product of the corresponding transcript has been found.

The other Marek's disease virus, MDV2, encodes 17 miRNAs, 16 of which are clustered in a 4.2-kb-long repeat region that encodes R-LORF2 to R-LORF5 putative open reading frames. The single miRNA outside of this cluster is located in the short repeat region, within the C-terminal region of a major transcriptional regulatory protein, an ICP4 homologue (Yao *et al.*, 2007).

## Polyomaviruses

The simian virus 40 (SV40) encodes one precursor of miRNA that was discovered by Sullivan *et al.* in 2005. The authors used a computational approach to predict likely precursors of miRNAs, and validated their prediction by Northern blotting. Using probes complementary to each strand of the predicted precursor, they showed that both arms of the pre-miRNA were expressed as functional miRNAs, miR-S1-3p and miR-S1-5p. SV40 miRNAs accumulate at late times of infection. This is totally consistent with the late expression of the viral transcript that encloses the precursor. These miRNAs are perfectly complementary to early T antigen mRNAs and are involved in their cleavage (see below) (Sullivan *et al.*, 2005).

Interestingly, the function of miRNAs in polymaviruses seems to have been conserved, although not at the primary sequence level. Using an approach similar to the one they used for SV40, Sullivan and co-workers showed that the human BK and JC polyomaviruses do each express a single miRNA in a genomic localization similar to SV40. SV40, JC and BK viruses' miRNAs show some limited nucleotides identities notably in their seed sequences. The most abundant miRNA of JCV (jcv-miR-J1-5p) is detectable *in vivo* in tissues of infected patients. Despite these sequences divergence, all three miRNAs are expressed in the late phase and target their complementary early mRNAs, thus allowing these viruses to shut down the expression of early viral genes in late infection (Seo *et al.*, 2008). Similarly, a miRNA with conserved functions has been found in Merkel cell polymavirus, a recently identified human virus associated with cancer (Seo *et al.*, 2009). Finally, a report from Cantalupo *et al.* in 2005 argues that the simian agent virus 12 (SA12) also encodes miRNAs. SA12 is closely related in sequence to the human BK virus, raising the possibility that both viruses circulate between baboon and human. Bioinformatics analyses showed that the pre-miRNA described in the other viruses of this family is also conserved in SA12 genome. Northern blotting experiments confirmed the presence of miRNAs derived from both strand of the stem loop precursor but the functional activity of SA12 miRNAs has not been investigated (Cantalupo *et al.*, 2005).

## Other viruses

The genomes of adenovirus type 2 and 5 encode two non-coding virus-associated (VA) RNAs, VAI RNA and VAII RNA. VA RNAs are non-coding RNAs transcribed by RNA polymerase III that accumulate in very large amounts in the late phase of infection. The VAI RNA, via its highly structured stems, blocks the PKR-dependent anti-viral pathway by binding PKR and preventing the inhibition of protein synthesis (Berk, 2007). These VA RNAs are exported to the cytoplasm via exportin 5, the same export factor as for pre-miRNAs, and have been shown to function as inhibitors of siRNA and miRNA function (Andersson *et al.*, 2005; Lu and Cullen, 2004). Additionally, it was recently reported that VA RNAs could be processed into small RNAs (Aparicio *et al.*, 2006; Sano *et al.*, 2006). Although their processing by Dicer is quite inefficient (only 2–5%), given the very large amount of VA RNA molecules this results in a significant quantity of svaRNAs. In fact, these small RNA are efficiently assembled in the RISC, and even compete with cellular miRNAs for Argonaute 2 binding (see below) (Aparicio *et al.*, 2006; Xu *et al.*, 2007). One could speculate that a complete processing of VA RNA would be deleterious for its other activities. It is not totally clear whether adenovirus small RNA have any cellular targets, but given their characteristics and their assembly into RISC, they can be seen as a form of viral miRNAs.

The insect virus *Heliothis virescens* ascovirus (HvAV) also encodes one single miRNA that was identified by bioinformatics prediction, and validated by Northern blot analysis. HvAv-miR-1 is located in the major capsid protein gene. It was reported to be present at late point after infection (Hussain *et al.*, 2008).

The presence of miRNAs in retroviruses such as HIV remains a matter of debate that will not be developed here. Although some have reported that regions such as Nef (Omoto and Fujii, 2005) or Tar (Ouellet *et al.*, 2008) could be a potential source of miRNAs, others failed to clone miRNAs from HIV-infected cells (Pfeffer *et al.*, 2005).

## Targets of viral miRNAs: finding the needle in a haystack

The precise functions of individual viral miRNAs remain to be determined, but an increasing number of studies are starting to unveil some of the roles that these small RNAs might be playing. Intuitively, one would imagine that the involvement of miRNAs in the viral life cycle could be at two main levels: to modify the cellular environment so as to provide an optimized setting for the infection to take place, or to act directly on the virus to fine-tune the expression of its genes. In the first case scenario, viral miRNAs would for example help the virus escape the immune response, or they could participate in the oncogenesis induction to immortalize the cells producing the virus. In the second hypothesis, the miRNAs would regulate the expression of viral genes at key steps of the virus like cycle (for example, the transition from lytic to latent stage). The regulation of their targets by miRNAs can be done in several manners that depend on the level of complementarity of the small RNAs and the targeted transcripts. Like cellular miRNAs, viral miRNAs can engage in different possible interactions with their targets, from a fully complementary binding, to a more canonical binding involving the seed of the miRNA (Fig. 3.2).

### Viral targets and modes of recognition

The first targets of viral miRNAs that have been identified are of viral origin, and an increasing number of reports confirms that it is a widespread phenomenon (Table 3.2). In the original paper describing the identification of virus-encoded miRNAs, it was proposed that one of EBV miRNAs, namely miR-BART2, was potentially targeting the viral DNA polymerase BALF5. This hypothesis was inferred from the fact that miR-BART2 gene was expressed from a position opposite to the genomic location of BALF5 transcript (Pfeffer *et al.*, 2004). There was also some support that miR-BART2 was involved in the cleavage of BALF5 transcript from an earlier study that showed this RNA to accumulate in two forms of distinct lengths, and that mapped precisely the site of divergence between the two to a locus that latter turned out to correspond exactly to the position opposite to the middle of miR-BART2 (Furnari *et al.*, 1993). A recent report has confirmed the involvement of miR-BART2 in that process (Barth *et al.*, 2008). This mode of regulation by cleavage, which is not predominant among animal miRNAs, is also found in the polyomavirus SV40. Indeed, SV40 miRNA miR-S1 is also localized antisense to the early T antigen (Ag) transcript,

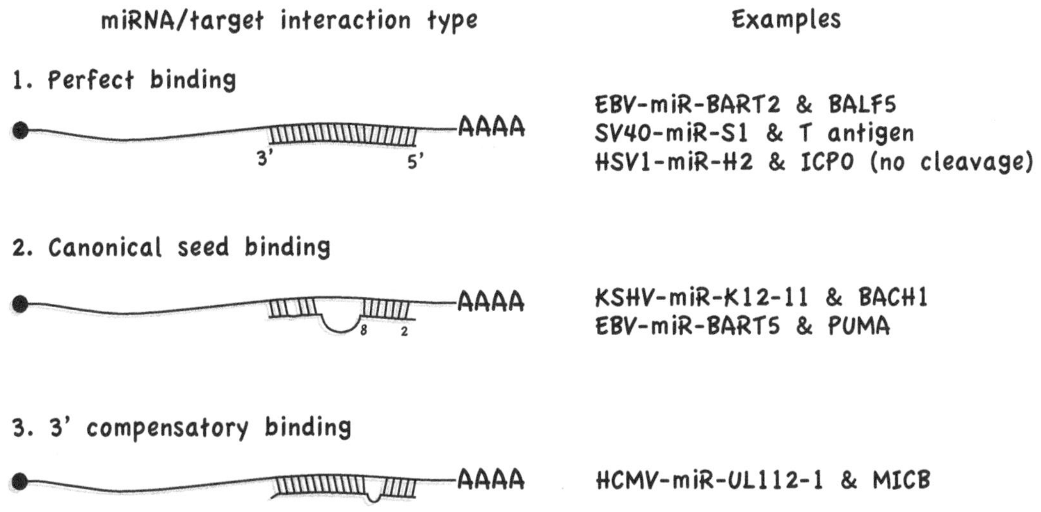


**Figure 3.2** Modes of target recognition by viral miRNAs. The targeted transcript is represented with the bound miRNA in the different configurations possible. See text for details.

and engages its cleavage (Sullivan *et al.*, 2005). Interestingly, the regulation of T Ag expression by a viral miRNA has been conserved between the simian virus and the human BK and JC viruses (Seo *et al.*, 2009), as well as in the Merkel cell polyomavirus (Seo *et al.*, 2009). The conservation is not at the level of the sequence of the miRNA itself, but in all cases, the miRNA gene is located antisense to the large T-antigen transcript and is responsible for its cleavage by RISC.

It is noteworthy that the recognition of these viral transcripts occurs via cleavage after a perfect binding of the target (Fig. 3.2). However, it should be noted that at least in one case, the

**Table 3.2** Viral targets of viral miRNAs

| Virus | miRNA | Target | Function | References |
|---|---|---|---|---|
| SV40 | miR-S1 | T antigen | Early protein | Sullivan *et al.* (2005) |
| BKV | miR-B1 | T antigen | | Seo *et al.* (2008b) |
| JCV | miR-J1 | T antigen | | Seo *et al.* (2008b) |
| MCV | miR-M1 | T antigen | | Seo *et al.* (2008a) |
| EBV | miR-BART2 | BALF5 | DNA polymerase | Barth *et al.* (2008), Pfeffer *et al.* (2004) |
| | miR-BART1, 16 | LMP1 | Signalling molecule | Lo *et al.* (2007) |
| HCMV | miR-UL112-1 | UL112/113, UL120/121 | Immediate early protein | Grey *et al.* (2007) |
| | | UL123 (IE72, IE1) | Immediate early protein | Grey *et al.* (2007), Murphy *et al.* (2008) |
| HSV-1 | miR-H2-3p | ICP0 | Immediate early protein | Umbach *et al.* (2008) |
| | miR-H6 | ICP4 | Immediate early protein | Umbach *et al.* (2008) |
| HSV-2 | miR-I | ICP34.5 | Pathogenicity factor | Tang *et al.* (2008) |
| | miR-II, miR-III | ICP34.5, ICP0 | | Tang *et al.* (2009) |
| HvAv | miR-1 | ORF1 | DNA polymerase | Hussain *et al.* (2008) |

perfect complementarity between a viral miRNAs and its target does not lead to regulation, and in another example, the regulation exists but does not involve cleavage and degradation of the transcript. In the first case, Grey *et al.* used a computational approach to identify the putative viral target of HCMV miR-UL112-1. By restricting their search to targets that were conserved between the human and the chimpanzee cytomegaloviruses, they came up with a list of 14 candidate 3′ UTRs, of which they could validate three (UL112/113, UL120/121 and UL123) by luciferase assay (see below). Surprisingly, UL114, which is perfectly complementary to miR-UL112-1 because it lies antisense to it on the genome, could not be validated as a target (Grey *et al.*, 2007). Although this was only measured by reporter assay, this finding could indicate that the degree of complementarity between a miRNA and its target cannot explain everything. It could be that in this particular case, the binding site is not accessible due to a strong local secondary structure, or to the interaction with an RNA-binding protein. The second scenario features a miRNA encoded by HSV1, Cullen and colleagues identified the miRNAs of HSV1 by small RNA cloning and deep sequencing (see above), and showed that three of them lay antisense to genes of known function. Thus, miR-H2 could perfectly basepair with the transcript coding for ICP0, and both miR-H3 and H4 are antisense to the ICP34.5 gene. The targeting of ICP0 by miR-H2 was confirmed by Western blot analysis, but interestingly, although there was a very strong effect of miR-H2 at the protein level, the level of ICP0's messenger RNA was not affected (Umbach *et al.*, 2008). This observation is in agreement with a recent report by Brodersen *et al.* who demonstrated that the predominant mode of regulation by small RNAs in plants was at the level of translation inhibition, even in the cases of a perfectly matched miRNA/transcript pair (Brodersen *et al.*, 2008). Umbach *et al.* did not examine the regulation of ICP34.5 by miR-H2 and H3, but Tang *et al.* from the laboratory of P. Krause showed that HSV2 was also encoding a miRNA, named miR-I, which genomic localization was on the strand opposite to the one of ICP34.5. They confirmed both by reporter assay and Western blot analysis that ICP34.5 was indeed targeted by this miRNA, however they did not measure the effect at the RNA level (Tang *et al.*, 2008a). In another report, this group latter found that two other miRNAs, miR-II and miR-III, expressed at a lower level by HSV2, were respectively targeting ICP34.5 and ICP0 with a perfect complementarity similarly to HSV1 miRNAs (Tang *et al.*, 2009). Together with the data obtained with polyomaviruses (Seo *et al.*, 2008), this is further evidence that some function of viral miRNAs are conserved between distantly related viruses without a conservation at the sequence level.

Other targets of viral origin that do not require perfect binding of the viral miRNA have been identified. In these examples, the interaction is more canonical and occurs via a classical 'seed-type' binding (Fig. 3.2). We have already mentioned the recognition of three viral transcripts coding for immediate early proteins by HCMV miR-UL112-1 (Grey *et al.*, 2007). Murphy *et al.* also used a bioinformatics approach to predict viral targets of viral miRNAs. They validated the regulation of HCMV UL123 (IE1) by the same miRNA, miR-UL112-1, using luciferase reporter assays and Western blot analysis (Murphy *et al.*, 2008). Among HSV1 miRNAs, miR-H6 has also been shown to be involved in the regulation of ICP4 via a canonical interaction. Co-expression of a plasmid encoding ICP4 and of a synthetic oligonucleotide mimicking miR-H6 results in the down-regulation of ICP4 expression from the plasmid, whereas a mutant version of ICP4 in which the miRNA binding site has been mutated is not affected (Umbach *et al.*, 2008). Besides targeting BALF5 for cleavage, some of EBV miRNAs have been shown to deregulate the expression of LMP1 through imperfect binding. Lo *et al.* have reported that miR-BART1-5p, miR-BART16 and miR-BART-17-5p all have potential binding sites within LMP1 3′-UTR, and a luciferase reporter fused to LMP1 3′-UTR is down-regulated in the presence of these miRNAs. Additionally, the level of LMP1 protein could be lowered following overexpression of either miRNAs, or up-regulated following blockage of the miRNAs by antisense oligonucleotides (Lo *et al.*, 2007). Interestingly, among the putative binding sites of these BART miRNAs, some displayed some mismatches within the seed region, an observation that has also been made for cellular targets of viral miRNAs (see below). This should attract our attention to the fact that in the search of miRNA targets we should not restrict ourselves to candidates that strictly follow pre-established rules. In fact, even cellular miRNAs can in some cases act through non-perfect seed/seed-matches pairing, including the very first identified miRNA, Lin-4 that binds to its target lin-14 with a bulge in its seed. Amazingly, this bulge even seems to be required for the regulation mediated by the small RNA (Ha *et al.*, 1996).

Finally, the last virus for which a viral target has been identified is the insect DNA virus *Heliothis virescens ascovirus*. HvAv miRNA miR-1 can partially basepair with the viral DNA polymerase transcript. Interestingly, it seems that the effect on the targeted transcript is at the RNA level, although some more experiments are needed to confirm the initial observation (Hussain *et al.*, 2008).

### Cellular targets

The number of validated cellular targets of viral miRNAs is still lower than for viral targets (Table 3.3), although we should expect this tendency to be reversed in a near future. The identification of cellular targets remains a challenge due to the lack of amenable high-throughput validation method. What holds true for cellular miRNAs is true for viral miRNAs as well. Actually, finding targets for viral miRNAs can prove even more difficult than for host miRNAs. Indeed, a number of people have relied on bioinformatics prediction to predefine candidates to be followed on. In order to increase the signal-to-noise ratio in these predictions, people have used the propensity of miRNAs to be conserved between species to look for binding sites in regions that were also conserved cross-species (Rajewsky, 2006). This approach cannot be directly transposed to viral miRNAs, which are, for a very large proportion, not conserved with the host miRNAs or between distantly related viral species. Nevertheless, there are notable exceptions to this observation. The most striking of these exceptions is the KSHV miRNA, miR-K12-11, which has partial homology with the cellular miRNA, miR-155. Moreover, this conservation is somewhat restricted to the seed region of the miRNA. This observation inspired two groups to postulate that the functions of miR-155 and of miR-K12-11 were conserved, and that the two miRNAs must share a common set of targets. To challenge their hypothesis, Skalsky *et al.* and Gottwein *et al.* turned to a global approach for target identification that relies on the use of transcriptomic analyses to determine which transcripts are down-regulated upon a given miRNA overexpression in a transgenic cell line (Gottwein *et al.*, 2007; Skalsky *et al.*, 2007). This method has initially been used to identify the targets of tissue-specific miRNAs after their misexpression in HeLa cells. Lim and co-workers showed that overexpression of the neuron specific miR-124, or muscle specific miR-1, led to the down-regulation of a large number of target mRNAs, and that among the deregulated transcripts, there was a significant enrichment in seed match motifs (Lim *et al.*, 2005). The down-regulation of transcripts accumulation upon miRNA recognition can be a consequence of their de-adenylation and/or decapping, although this does not occur

**Table 3.3** Cellular targets of viral miRNAs

| Virus | miRNA | Target gene | Function | References |
|---|---|---|---|---|
| KSHV | miR-K12-6–3P and others | THBS1 | Angiogenesis regulator | Samols *et al.* (2007) |
| | miR-K12-11 | BACH1, Fos | Transcription factors | Gottwein *et al.* (2007), Skalsky *et al.* (2007) |
| EBV | miR-BHRF1-3 | CXCL-11 | Chemokine | Xia *et al.* (2008) |
| | miR-BART5 | PUMA | Pro-apoptotic protein | Choy *et al.* (2008) |
| HCMV | miR-UL112-1 | MICB | NK cell ligand | Stern-Ginossar *et al.* (2007) |
| MDV | miR-M4 | PU.1 | Transcription factor | Zhao *et al.* (2009) |

systematically (Eulalio *et al.*, 2008). In the case of KSHV miR-K12-11, the two reports showed that Affymetrix analyses of cell lines expressing either miR-K12-11 or miR-155 could identify a common set of targets that had a repeated occurrence of the conserved seed match motif. Among the targets that were validated by luciferase reporter assays, and by Western blots, were the transcription factors Fos and Bach-1 (Gottwein *et al.*, 2007; Skalsky *et al.*, 2007). Very interestingly, the conservation of the seed region between a cellular and a viral miRNA has been predicted in a few other cases (Gottwein and Cullen, 2008), and the functionality of this conservation has been confirmed for another viral orthologue of miR-155. Indeed, MDV1 miR-M4 also has the same seed that the chicken miR-155, and has been shown to target PU.1, a known target of miR-155 (Zhao *et al.*, 2009). The microarray analysis approach has also been used in a broader attempt at finding cellular targets for ten out of the known twelve miRNAs encoded by KSHV. The fact that all of KSHV miRNAs are clustered on the same genomic locus helped Samols *et al.* to express them in HEK293 cells. After transcriptomic analysis of these cells, they identify a number of putative targets of which they could validate THBS1, a protein playing a role in angiogenesis (Samols *et al.*, 2007). Intriguingly, all 10 miRNAs were predicted to target THBS1 3′ UTR, although this observation was not truly validated.

Despite the underlying problem of the lack of viral miRNAs conservation, computational target predictions have been used successfully in some instances. To bypass the problem, Stern-Ginossar *et al.* designed a novel algorithm for target prediction that did not rely on cross-species conservation. Rather, they looked for the occurrence of repeated motifs within 3′ UTRs and searched for their potential interactions with HCMV miRNAs. This approach enabled them to predict and validate the NK cell ligand MICB as a target of HCMV miR-UL112-1 (Stern-Ginossar *et al.*, 2007). Interestingly, the binding of miR-UL112-1 to MICB 3′-UTR occurs in a non-canonical manner with a two-nucleotide bulge within the seed, and a stronger binding of the 3′ end of the miRNA (Fig. 3.2). Intriguingly, a closely related isoform of MICB, MICA, which differs only by one nucleotide in the putative miR-UL112-1 binding site, is not a target of this miRNA.

The use of algorithms initially designed to predict cellular miRNAs targets has also proved useful. Thus, Choy *et al.* used two different programs, namely MiRanda and RNAhybrid, to predict targets for each EBV BART miRNA. By using luciferase assays, they could validate one cellular target of miR-BART5, namely PUMA (P53 up-regulated modulator of apoptosis), which is involved in the control of apoptosis (Choy *et al.*, 2008). It should be noted however, that this was the only target validated out of nine short listed predicted targets for this miRNA, which is not too bad but also shows the limits of this approach. Another cellular target has been validated for EBV miR-BHRF1-3. Xia *et al.* took on to validate a prediction done with the program MiRanda and previously published for this miRNA (Pfeffer *et al.*, 2004), the chemokine CXCL-11. The authors showed that transfection of miR-BHRF1-3 resulted in a decrease in CXCL-11 protein level, and that conversely, blocking miR-BHRF1-3 in infected cells resulted in an increase of protein level (Xia *et al.*, 2008). One word of caution though, the authors state in their article that miR-BHRF1-3 is 100% complementary to its target, but actually the alignment is only imperfect with some mismatches and GU wobbles. However, there are three predicted binding sites for this miRNA in CXCL-11 3′-UTR (Pfeffer *et al.*, 2004).

## Roles and implications of viral and cellular targets regulation

We portrayed before the viral and cellular targets that have been identified so far, and the different means by which a miRNA can recognize and regulate its targets. We will now focus more precisely on the functions that the regulation of these targets play, and the implications in context of the viral life cycle. Although there is a large diversity of both viral and cellular targets, some unifying themes are emerging.

### Regulation of latency

A key step in all herpesviruses life cycle is the transition from a productive lytic infection to a latent infection where the virus will remain dormant in the infected host. A number of evidences now point to a strong involvement of viral miRNAs in this transition through the regulation of lytic viral proteins. Thus, the regulation of EBV DNA polymerase BALF5 by the miRNA miR-BART2 was shown to play a role in the control of virus accumulation. Through cleavage of the BALF5 transcript, this miRNA might degrade aberrantly transcribed BALF5 messenger RNA to keep the virus replication in check. However, the control is not entirely mediated by miRNA because blocking their expression by chemically modified antisense oligonucleotides does not induce lytic replication in latently infected cells (Barth *et al.*, 2008). A striking parallel with the role of miR-BART2 in regulating EBV DNA polymerase was reported recently in the insect virus HvAv. Here also, the viral DNA polymerase expression seems to be controlled in a temporal manner by a virus-encoded miRNA. However in the latter case, the miRNA expression alone seems sufficient to control virus replication (Hussain *et al.*, 2008).

The viral DNA polymerase is not the only essential gene during lytic infection to be targeted by miRNAs. The human cytomegalovirus is another fascinating illustration of how critical small RNAs can be for the virus replication. Two reports have showed that one of HCMV miRNAs, miR-UL112-1, can target immediate-early viral genes (Grey *et al.*, 2007; Murphy *et al.*, 2008). Grey *et al.* first identified UL112/113, UL120/121 and UL123 (encoding the IE72 or IE1 protein) as targets of miR-UL112-1, and both UL112/113 and IE72 encode products important for the virus replication. They also showed that miR-UL112-1 overexpression during infection could inhibit immediate-early gene expression and viral DNA replication. Since IE72 is necessary for an efficient replication of the virus through trans-activation of early and late viral genes, one could imagine that its repression by a miRNA could play an important role for the establishment of a latent or persistent infection. These results were confirmed by Murphy *et al.* who independently predicted the same binding site for miR-UL112-1 in the 3′-UTR of IE72 (IE1). They also confirmed the involvement of this miRNA in IE1 down-regulation, and constructed a mutated virus in which miR-UL112-1 expression was impaired. Although the mutant virus replication was not perturbed in fibroblasts, a significant higher level of IE1 accumulated in cells infected by this virus compared to the wild type. Very interestingly, Murphy *et al.* also predicted that other viral miRNAs could potentially target IE genes in four herpesviruses, for example KSHV miR-K12-6-3p was predicted to target Zta and Rta, which share a 3′-UTR and are strong trans-activators of lytic gene expression (Murphy *et al.*, 2008).

The alphaherpesviruses also use miRNAs to regulate the expression of immediate-early genes. Strikingly, all the miRNAs that have been identified in this family of viruses are localized in latency-associated transcripts. For both HSV genera, a miRNA has been found to be expressed antisense to the ICP0 transcript, a key immediate-early transcriptional activator that might be important for the entry into the productive lytic cycle. For the two viruses, down-modulation of ICP0 expression by their respective miRNAs (i.e. miR-H2-3p and miR-III) was confirmed (Tang *et al.*, 2009; Umbach *et al.*, 2008). Interestingly, Tang *et al.* showed that HSV2 miRNAs expression was controlled by the viral factor ICP4. Since in HSV1, ICP4 is itself regulated by the viral miRNA miR-H6, it appears that the switch to latency is very tightly controlled at multiple levels where miRNAs play a central role. Even though there are some notable differences between HSV1 and HSV2 regarding the level of expression of each viral miRNAs, and the sequences of the miRNAs themselves, their genomic localization, and mode of action seem to have been partially conserved between the two viruses. The implication of MDV1 and 2 miRNAs in the control of latency has not been explored yet, however given their localization in the latency associated transcript of this virus, and their high abundance in latently infected cells, it is likely that they will also play a prominent role in this process (Burnside *et al.*, 2006; Yao *et al.*, 2008).

## Controlling the immune response

An important role of virus-encoded miRNAs is also to help the virus escape the immune response and remain unnoticed in the infected host for long period of time. The use of small RNAs for that purpose is a perfect example of how viruses causing persistent infection have co-evolved with their hosts. It is indeed remarkable that some viruses have been able to hijack the cellular miRNA biogenesis machinery, which enables them to express short non-immunogenic RNA molecules with potential deleterious effects to the host. As an example, the polyomavirus SV40 miRNA miR-S1 regulates the expression of the large T antigen transcript by cleavage, but this regulation does not seem to impact on the virus replication. Rather, it has been shown by using a mutant virus not expressing the miRNA that the role of this control was to lower the antigen load in infected cells to prevent recognition by cytotoxic T cells. Host cells infected with the miRNA mutant virus were much more efficiently killed by cytotoxic T lymphocytes (CTL) than cells infected with the wild-type virus (Sullivan *et al.*, 2005). As we noted previously, the regulation of T Ag by miRNA has been conserved in the human polyomaviruses BK, JC and Merkell cell (Seo *et al.*, 2008, 2009), and most probably in the baboon polyomavirus SA12 (Cantalupo *et al.*, 2005). Even though the direct effect on CTL recognition has not been measured for these viruses, it is very likely that the T Ag regulation plays the same role than during SV40 infection.

The same HCMV miRNA that was shown to play important roles in the control of viral replication is coincidently also involved in the regulation of a cellular gene very important for keeping HCMV infected cells in check. Indeed, Stern-Ginossar *et al.* showed that the major histocompatibility complex class I-related chain B (MICB) gene was targeted by HCMV-miR-UL112-1. MICB is a stress-induced ligand of a natural killer (NK) cell activating receptor, and is known to be critical for NK cell killing of virus-infected cells. The regulation of MICB by miR-UL112-1 also takes place in the real context of infection, and cells expressing miR-UL112-1 ectopically are less sensitive to specific killing by NK cells. Moreover, a mutant virus not expressing miR-UL112-1 results in cells being more effectively killed by NK cells after infection, very much like what has been previously observed with SV40 virus and CTL (Stern-Ginossar *et al.*, 2007). It is worth noting that the virus UL16 protein already lowers MICB expression at the cell surface (Cosman *et al.*, 2001), thus there seems to be a cumulative effect due to miRNA expression that helps the virus achieve the difficult task to escape the immune response of the host.

The last example of immune surveillance regulation by viral miRNAs is illustrated by one of EBV miRNA, miR-BHRF1-3, which regulates the expression of the chemokine CXCL-11, also known as the IFN-inducible T-cell attracting chemokine (I-TAC). The level of this viral miRNA is high in type III latency cell line, and in primary EBV-associated AIDS-related diffuse large B-cell lymphomas (DLBCL). Conversely, primary Burkitt lymphoma and primary effusion lymphoma cell lines express high level of BART miRNAs, but no BHRF1-3 miRNA. The level of CXCL-11 in these different cell types is inversely correlated with the level of miR-BHRF1-3, and blocking this miRNA expression results in an increase in CXCL-11 level (Xia *et al.*, 2008). These results indicate that the immunosuppressive effect via regulation of the chemokine CXCL-11 is highly dependent of the immune status of the infected person and imply that the involvement of viral miRNAs in these processes is not a simple task to address.

## Roles in oncogenesis

EBV emerged as the first candidate human tumour virus in the sixties and is now the prototype for the oncogenic gamma-herpesviruses subfamily, which is associated with a number of lymphomas, carcinomas and sarcomas (Kieff and Rickinson, 2007). KSHV is also tightly associated with a number of oncogenic disorders. For both of these viruses, recent evidences pointed to an implication of viral miRNAs in the development of cancer linked with the viral infection. The involvement of viral miRNAs in oncogenesis can be seen both due to the targeting of viral genes and of cellular genes. Hence, some of EBV BART miRNAs down-regulate the expression of the viral LMP1, a well-described viral oncogene, which explains the lack of correlation between LMP1 mRNA and protein expression in different EBV infected epithelial cells, and nasopharyngeal carcinomas (NPC). Indeed, there is an inverse correlation of BART miRNAs and LMP1 protein levels in NPC tissues (Lo *et al.*, 2007). LMP1, a constitutively active tumour necrosis factor receptor, is a multifaceted protein that is both capable of inducing cell growth and transformation, and, when overexpressed, growth inhibition and sensitization to apoptosis (Eliopoulos *et al.*,

1996; Liu *et al.*, 2002). Therefore, BART miRNAs might act to fine-tune the expression of this protein during different steps of the infection, and as a consequence, drive LMP1 in inducing cell transformation rather than apoptosis. It is also intriguing to note that LMP1 is known to induce, through the action of NF-κB, the expression of a cellular miRNA, miR-146a, which is an important factor in lymphocyte signalling pathways (Cameron *et al.*, 2008). The latest observation adds yet another layer of complexity in the already intricate interplay between viruses and the miRNA machinery.

Control of apoptosis is an efficient way to drive cell into proliferation, and it seems that it is being used extensively by viral miRNAs. Accordingly, EBV miRNA miR-BART5 targets PUMA, both a P53-dependent and independent pro-apoptotic gene, and blocking miR-BART5 activity results in an increase in etoposide-induced apoptosis (Choy *et al.*, 2008). Upon induction of its expression by P53, PUMA can bind to Bcl-2, localizes to the mitochondria to induce cytochrome c release, and activate the rapid induction of programmed cell death (Nakano and Vousden, 2001). The regulation of a downstream target of P53 by a viral miRNA also adds to the growing complexity of interactions between P53 and miRNAs, as P53 has been shown to be a transcription factor for the cellular miRNAs of the miR-34 family (He *et al.*, 2007). The anti-apoptotic effect mediated by the repression of PUMA by miR-BART5 completes a panel of tools that EBV can employ to inhibit cell death, such as the viral Bcl-2 homologue BHRF1 (Henderson *et al.*, 1993).

KSHV miRNAs could also play a role in the development of cancer. Their genomic localization in one of KSHV most transforming region (Muralidhar *et al.*, 2000) is already an indication that they could be directly involved in oncogenesis. More concrete evidence came with the finding that KSHV miRNAs were targeting THBS1 in 293 cells (Samols *et al.*, 2007). A gene involved in the regulation of angiogenesis, THBS1 codes for a matricellular protein playing important roles in cell–cell and cell–matrix adhesion, and is down-regulated in numerous cancers. It possesses both anti-proliferative and anti-angiogenic activity (de Fraipont *et al.*, 2001). Although some have reported THBS1 to be poorly expressed in Kaposi's sarcoma lesions (Taraboletti *et al.*, 1999), to date there is no direct evidence about the involvement of KSHV miRNAs in this regulation in context of a real infection and not only in an artificial system using 293 cell lines. Other cellular genes, which are targeted by KSHV miR-155 orthologue, miR-K12-11, have been identified, such as the transcription factors BACH1 and Fos (Gottwein *et al.*, 2007; Skalsky *et al.*, 2007). Although there is no direct connection to oncogenesis, it is interesting to note that miR-155 has been implicated in B-lymphoproliferative disorders (Kluiver *et al.*, 2005) and in normal B cell development (Thai *et al.*, 2007), and it could be that there are additional cellular targets for miR-K12-11 that will explain its role in KSHV-induced pleural effusion lymphomas (PEL). In fact, PEL do not express miR-155, but they do express miR-K12-11 at high levels. Therefore, the viral miRNA could complement the lack of miR-155 and contributes to lymphomagenesis. Intriguingly, EBV does not express an orthologue of miR-155, but EBV infection has been shown to induce the expression of this cellular miRNA (Gatto *et al.*, 2008). Thus, these two gammaherpesviruses use different strategies to attain a similar objective. Finally, the chicken Marek's disease virus also expresses an orthologue or miR-155 that was shown to target PU.1 (Zhao *et al.*, 2009) and could also be involved in the virus-induced lymphomagenesis.

## Other functions

Besides the evident functions that we described in the previous paragraphs, there are a number of other ways viral miRNAs can use to perturb the normal functioning of the cell or subtly modifying its environment. One should also consider the interactions that viruses engage with cellular miRNAs, as we saw above for a few cases. This can take place either through modulation of cellular miRNAs expression following infection, or through direct interactions of cellular miRNAs with the viral genome or viral transcripts. Although there have been a number of reports describing the latter interactions, we will not describe those into details. Rather, we refer the reader to Chapter 5. It is also possible that virus-encoded miRNAs can act indirectly by saturating, or modifying the miRNA biogenesis machinery. Hence, the adenovirus non-coding VA RNAs are expressed at very high levels and have been shown to saturate the exportin 5 factor (Lu and Cullen, 2004), and to compete for Dicer with endogenous miRNAs or transfected hairpin RNAs (Andersson *et al.*, 2005) thus acting as a suppressor of RNA silencing. Therefore, these non coding RNAs could very well act by preventing the function of cellular miRNAs, and not by directly targeting viral or cellular transcripts. Nevertheless, although Dicer processes VA RNAs very inefficiently, a non-negligible number of small RNAs can be assembled into RISC, and the possibility that they do act as true miRNAs remains open. Although no target has been identified to date, two groups showed that VA-derived small RNAs are bound to Argonaute 2, and that they can regulate the expression of artificial reporter genes (Aparicio *et al.*, 2006; Xu *et al.*, 2007). The work by Aparicio *et al.* also reported that blocking VA small RNAs affects efficient adenovirus production, indicating that they play important roles during the virus infectious cycle.

Similarly, miRNAs identified for the mouse CMV during lytic infection of fibroblasts accumulate in very high amount compared to cellular miRNAs, and it could be that they do partially function by saturating the machinery (Dolken *et al.*, 2007). However, the situation encountered upon *in vitro* infection of cells in culture with a high dose of virus, and where each individual cell will be infected is very remote from what happens during infection of the natural host. Nevertheless, it could very well be that during *in vivo* infection, the virus will express high levels of viral miRNAs in some cell types, and lower amount in others, thereby specifying a distinct mode of action in each situation.

## Concluding remarks

Viral miRNAs have been discovered only recently, but their importance in the study of host–virus interactions has risen exponentially since. We cannot overlook the fact that these tiny RNA molecules might be central in a number of processes during the different steps of the infection. Actually, some observations that could not be readily explained upon their discovery turned out to involve miRNAs. Thus, the cleavage of EBV BALF5 transcripts in two forms of different lengths, the precise role of the large EBV BART RNA, or the stabilization of the LAT intron in HSV1 and 2, are a few examples of data that became clearer once miRNAs were taken into account. The existence of virally derived miRNAs also points to fundamental differences in the relationship between viruses and RNA silencing in mammals. Although the debate whether RNA silencing can be seen as an antiviral defence in mammals like in plants and insects remains controversial but open, it is evident that some viruses have evolved to utilize this cellular machinery for their own good. The positive regulation of HCV replication by a cellular miRNA is another instance of viral hijacking of the miRNA pathway. What clearly lacks to formally prove that RNA silencing is (or is not) a true antiviral system in mammalian organisms would be the detection of siRNAs of viral origin in infected tissues. So far, there have been only non-conclusive or negative data (Lin and Cullen, 2007; Pfeffer *et al.*, 2005). However, recent evidence has emerged that endogenous siRNAs deriving from repeats can be cloned from germline specific tissues, such as oocytes and testes (Tam *et al.*, 2008; Watanabe *et al.*, 2006), or from ES cells (Babiarz *et al.*, 2008). These are clearly distinct from the previously described piwi-associated small RNAs (piRNAs) (Aravin *et al.*, 2006; Girard *et al.*, 2006), and point to a possible involvement of siRNAs in some processes during mammalian development. Thus, we cannot completely exclude that in some particular conditions, siRNAs of viral origin could arise, although it remains to be shown. Nevertheless, there is an involvement of small RNAs in antiviral defence, as it has also been shown that miRNAs of cellular origin can have deleterious effects on viruses. Hence, miR-32 negatively regulates the primate foamy virus (Lecellier *et al.*, 2005), and Dicer-deficient mice are hypersusceptible to vesicular stomatitis virus due to the lack of peculiar miRNAs accumulation (Otsuka *et al.*, 2007), and interferon-induced miRNAs have a negative impact on HCV (Pedersen *et al.*, 2007). The latest case nicely exemplifies that miRNAs can have both a positive and a negative impact on the very same virus, and it could be that for some DNA viruses encoding miRNAs that are beneficial for the virus, there will be cellular miRNAs, that will have direct or indirect detrimental effects to the infection. Actually, this hypothesis is not only fictional, as Pellett and coworkers nicely showed that two cellular miRNAs, miR-100 and miR-101, regulate the mTOR pathway, which is important for HCMV infection. In turn, HCMV down-regulates the expression of these two miRNAs, indicating that the virus can counteract the potential problem (Wang *et al.*, 2008).

Finally, one can wonder whether viral miRNAs have positive effects only for the virus they derive from. It is puzzling to note that most of the viral miRNAs have been found in viruses that have co-evolved with their host, and that cause persistent infection most of the time without the development of any symptom. A provocative view could be that these viruses are in some respect beneficial to their host, at least when the host is immunocompetent. This vision gained some support recently in a report that showed that herpesvirus latency confers a surprising benefit to the host. Indeed, Barton *et al.* (2007) reported that mice latently infected by either MHV68 or MCMV were resistant to infection with the bacterial pathogens *Listeria monocytogenes* and *Yersinia pestis*. These results suggest that latency could be seen as a symbiotic relationship rather than a potentially harmful ticking bomb. In this respect, viral miRNAs have clearly a potential to be beneficial for the host through the modulation of genes important in, for example, innate immunity, or other yet unknown processes.

## References


Andersson, M.G., Haasnoot, P.C., Xu, N., Berenjian, S., Berkhout, B., and Akusjarvi, G. (2005). Suppression of RNA interference by adenovirus virus-associated RNA. J. Virol. *79*, 9556–9565.

Aparicio, O., Razquin, N., Zaratiegui, M., Narvaiza, I., and Fortes, P. (2006). Adenovirus virus-associated RNA is processed to functional interfering RNAs involved in virus production. J. Virol. *80*, 1376–1384.

Aravin, A., Gaidatzis, D., Pfeffer, S., Lagos-Quintana, M., Landgraf, P., Iovino, N., Morris, P., Brownstein, M.J., Kuramochi-Miyagawa, S., Nakano, T., *et al.* (2006). A novel class of small RNAs bind to MILI protein in mouse testes. Nature *442*, 203–207.

Babiarz, J.E., Ruby, J.G., Wang, Y., Bartel, D.P., and Blelloch, R. (2008). Mouse ES cells express endogenous shRNAs, siRNAs, and other Microprocessor-independent, Dicer-dependent small RNAs. Genes Dev. 22, 2773–2785.

Bartel, D.P. (2004). MicroRNAs: genomics, biogenesis, mechanism, and function. Cell *116*, 281–297.

Barth, S., Pfuhl, T., Mamiani, A., Ehses, C., Roemer, K., Kremmer, E., Jaker, C., Hock, J., Meister, G., and Grasser, F.A. (2008). Epstein-Barr virus-encoded microRNA miR-BART2 down-regulates the viral DNA polymerase BALF5. Nucleic Acids Res. *36*, 666–675.

Barton, E.S., White, D.W., Cathelyn, J.S., Brett-McClellan, K.A., Engle, M., Diamond, M.S., Miller, V.L., and Virgin, H.W. t. (2007). Herpesvirus latency confers symbiotic protection from bacterial infection. Nature *447*, 326–329.

Berk, A.J. (2007). Adenoviridae: the viruses and their replication, In Fields Virology, Knipe, D.M., Howley, P.M., Griffin, D.E., Lamb, R.A., Martin, M.A., Roizman, B., and Straus, S.E., eds. (Philadelphia: Lippincott, Williams & Wilkins), pp. 2355–2394.

Bhattacharyya, S.N., Habermacher, R., Martine, U., Closs, E.I., and Filipowicz, W. (2006). Relief of microRNA-mediated translational repression in human cells subjected to stress. Cell *125*, 1111–1124.

Brodersen, P., Sakvarelidze-Achard, L., Bruun-Rasmussen, M., Dunoyer, P., Yamamoto, Y.Y., Sieburth, L., and Voinnet, O. (2008). Widespread translational inhibition by plant miRNAs and siRNAs. Science *320*, 1185–1190.

Buck, A.H., Santoyo-Lopez, J., Robertson, K.A., Kumar, D.S., Reczko, M., and Ghazal, P. (2007). Discrete clusters of virus-encoded micrornas are associated with complementary strands of the genome and the 7.2-kilobase stable intron in murine cytomegalovirus. J. Virol. *81*, 13761–13770.

Burnside, J., Bernberg, E., Anderson, A., Lu, C., Meyers, B.C., Green, P.J., Jain, N., Isaacs, G., and Morgan, R.W. (2006). Marek's disease virus encodes MicroRNAs that map to meq and the latency-associated transcript. J. Virol. *80*, 8778–8786.

Burnside, J., Ouyang, M., Anderson, A., Bernberg, E., Lu, C., Meyers, B.C., Green, P.J., Markis, M., Isaacs, G., Huang, E., and Morgan, R.W. (2008). Deep sequencing of chicken microRNAs. BMC Genomics *9*, 185.

Cai, X., Lu, S., Zhang, Z., Gonzalez, C.M., Damania, B., and Cullen, B.R. (2005). Kaposi's sarcoma-associated herpesvirus expresses an array of viral microRNAs in latently infected cells. Proc. Natl. Acad. Sci. U.S.A. *102*, 5570–5575.

Cai, X., Schafer, A., Lu, S., Bilello, J.P., Desrosiers, R.C., Edwards, R., Raab-Traub, N., and Cullen, B.R. (2006). Epstein–Barr Virus MicroRNAs Are Evolutionarily Conserved and Differentially Expressed. PLoS Pathog. *2*, e23.

Cameron, J.E., Yin, Q., Fewell, C., Lacey, M., McBride, J., Wang, X., Lin, Z., Schaefer, B.C., and Flemington, E.K. (2008). Epstein–Barr virus latent membrane protein 1 induces cellular MicroRNA miR-146a, a modulator of lymphocyte signaling pathways. J. Virol. *82*, 1946–1958.

Cantalupo, P., Doering, A., Sullivan, C.S., Pal, A., Peden, K.W., Lewis, A.M., and Pipas, J.M. (2005). Complete nucleotide sequence of polyomavirus SA12. J. Virol. *79*, 13094–13104.

Choy, E.Y., Siu, K.L., Kok, K.H., Lung, R.W., Tsang, C.M., To, K.F., Kwong, D.L., Tsao, S.W., and Jin, D.Y. (2008). An Epstein–Barr virus-encoded microRNA targets

PUMA to promote host cell survival. J. Exp. Med. *205*, 2551–2560.

Cosman, D., Mullberg, J., Sutherland, C.L., Chin, W., Armitage, R., Fanslow, W., Kubin, M., and Chalupny, N.J. (2001). ULBPs, novel MHC class I-related molecules, bind to CMV glycoprotein UL16 and stimulate NK cytotoxicity through the NKG2D receptor. Immunity *14*, 123–133.

Cui, C., Griffiths, A., Li, G., Silva, L.M., Kramer, M.F., Gaasterland, T., Wang, X.J., and Coen, D.M. (2006). Prediction and identification of herpes simplex virus 1-encoded microRNAs. J. Virol. *80*, 5499–5508.

de Fraipont, F., Nicholson, A.C., Feige, J.J., and Van Meir, E.G. (2001). Thrombospondins and tumor angiogenesis. Trends Mol Med 7, 401–407.

Dolken, L., Perot, J., Cognat, V., Alioua, A., John, M., Soutschek, J., Ruzsics, Z., Koszinowski, U., Voinnet, O., and Pfeffer, S. (2007). Mouse cytomegalovirus microRNAs dominate the cellular small RNA profile during lytic infection and show features of posttranscriptional regulation. J. Virol. *81*, 13771–13782.

Du, T., and Zamore, P.D. (2005). microPrimer: the biogenesis and function of microRNA. Development *132*, 4645–4652.

Dunn, W., Trang, P., Zhong, Q., Yang, E., van Belle, C., and Liu, F. (2005). Human cytomegalovirus expresses novel microRNAs during productive viral infection. Cell. Microbiol. 7, 1684–1695.

Edwards, R.H., Marquitz, A.R., and Raab-Traub, N. (2008). Epstein–Barr virus BART microRNAs are produced from a large intron prior to splicing. J. Virol. *82*, 9094–9106.

Eliopoulos, A.G., Dawson, C.W., Mosialos, G., Floettmann, J.E., Rowe, M., Armitage, R.J., Dawson, J., Zapata, J.M., Kerr, D.J., Wakelam, M.J., *et al.* (1996). CD40-induced growth inhibition in epithelial cells is mimicked by Epstein–Barr Virus-encoded LMP1: involvement of TRAF3 as a common mediator. Oncogene *13*, 2243–2254.

Eulalio, A., Huntzinger, E., and Izaurralde, E. (2008). Getting to the root of miRNA-mediated gene silencing. Cell *132*, 9–14.

Furnari, F.B., Adams, M.D., and Pagano, J.S. (1993). Unconventional processing of the 3′ termini of the Epstein–Barr virus DNA polymerase mRNA. Proc. Natl. Acad. Sci. U.S.A. *90*, 378–382.

Ganem, D. (2007). Kaposi's sarcoma-associated herpesvirus, In Fields Virology, Knipe, D.M., Howley, P.M., Griffin, D.E., Lamb, R.A., Martin, M.A., Roizman, B., and Straus, S.E., eds. (Philadelphia: Lippincott, Williams & Wilkins), pp. 2847–2888.

Gatto, G., Rossi, A., Rossi, D., Kroening, S., Bonatti, S., and Mallardo, M. (2008). Epstein–Barr virus latent membrane protein 1 trans-activates miR-155 transcription through the NF-kappaB pathway. Nucleic Acids Res. *36*, 6608–6619.

Girard, A., Sachidanandam, R., Hannon, G.J., and Carmell, M.A. (2006). A germline-specific class of small RNAs binds mammalian Piwi proteins. Nature *442*, 199–202.

Gottwein, E., and Cullen, B.R. (2008). Viral and cellular microRNAs as determinants of viral pathogenesis and immunity. Cell Host Microbe *3*, 375–387.

Gottwein, E., Mukherjee, N., Sachse, C., Frenzel, C., Majoros, W.H., Chi, J.T., Braich, R., Manoharan, M., Soutschek, J., Ohler, U., and Cullen, B.R. (2007). A viral microRNA functions as an orthologue of cellular miR-155. Nature *450*, 1096–1099.

Grey, F., Antoniewicz, A., Allen, E., Saugstad, J., McShea, A., Carrington, J.C., and Nelson, J. (2005). Identification and characterization of human cytomegalovirus-encoded microRNAs. J. Virol. *79*, 12095–12099.

Grey, F., Meyers, H., White, E.A., Spector, D.H., and Nelson, J. (2007). A human cytomegalovirus-encoded microRNA regulates expression of multiple viral genes involved in replication. PLoS Pathog. *3*, e163.

Grundhoff, A., Sullivan, C.S., and Ganem, D. (2006). A combined computational and microarray-based approach identifies novel microRNAs encoded by human gamma-herpesviruses. RNA *12*, 733–750.

Ha, I., Wightman, B., and Ruvkun, G. (1996). A bulged lin-4/lin-14 RNA duplex is sufficient for *Caenorhabditis elegans* lin-14 temporal gradient formation. Genes Dev. *10*, 3041–3050.

He, L., He, X., Lowe, S.W., and Hannon, G.J. (2007). microRNAs join the p53 network--another piece in the tumour-suppression puzzle. Nat Rev Cancer 7, 819–822.

Henderson, S., Huen, D., Rowe, M., Dawson, C., Johnson, G., and Rickinson, A. (1993). Epstein–Barr virus-coded BHRF1 protein, a viral homologue of Bcl-2, protects human B cells from programmed cell death. Proc. Natl. Acad. Sci. U.S.A. *90*, 8479–8483.

Hussain, M., Taft, R.J., and Asgari, S. (2008). An insect virus-encoded microRNA regulates viral replication. J. Virol. *82*, 9164–9170.

Kieff, E.D., and Rickinson, A.B. (2007). Epstein–Barr virus and its replication, In Field's Virology, Knipe, D.M., Howley, P.M., Griffin, D.E., Lamb, R.A., Martin, M.A., Roizman, B., and Straus, S.E., eds. (Philadelphia: Lippincott, Williams & Wilkins), pp. 2603–2654.

Kim do, N., Chae, H.S., Oh, S.T., Kang, J.H., Park, C.H., Park, W.S., Takada, K., Lee, J.M., Lee, W.K., and Lee, S.K. (2007). Expression of viral microRNAs in Epstein–Barr virus-associated gastric carcinoma. J. Virol. *81*, 1033–1036.

Kim, V.N. (2005). MicroRNA biogenesis: coordinated cropping and dicing. Nat. Rev. Mol. Cell Biol. *6*, 376–385.

Kliche, S., Nagel, W., Kremmer, E., Atzler, C., Ege, A., Knorr, T., Koszinowski, U., Kolanus, W., and Haas, J. (2001). Signaling by human herpesvirus 8 kaposin A through direct membrane recruitment of cytohesin-1. Mol. Cell 7, 833–843.

Kluiver, J., Poppema, S., de Jong, D., Blokzijl, T., Harms, G., Jacobs, S., Kroesen, B.J., and van den Berg, A. (2005). BIC and miR-155 are highly expressed in Hodgkin, primary mediastinal and diffuse large B cell lymphomas. J. Pathol. *207*, 243–249.

Lagos-Quintana, M., Rauhut, R., Lendeckel, W., and Tuschl, T. (2001). Identification of novel genes coding for small expressed RNAs. Science *294*, 853–858.

Lau, N.C., Lim, L.P., Weinstein, E.G., and Bartel, D.P. (2001). An abundant class of tiny RNAs with probable regulatory roles in *Caenorhabditis elegans*. Science *294*, 858–862.

Lecellier, C.H., Dunoyer, P., Arar, K., Lehmann-Che, J., Eyquem, S., Himber, C., Saib, A., and Voinnet, O. (2005). A cellular microRNA mediates antiviral defense in human cells. Science *308*, 557–560.

Lee, R.C., Feinbaum, R.L., and Ambros, V. (1993). The C. elegans heterochronic gene lin-4 encodes small RNAs with antisense complementarity to lin-14. Cell *75*, 843–854.

Lewis, B.P., Burge, C.B., and Bartel, D.P. (2005). Conserved seed pairing, often flanked by adenosines, indicates that thousands of human genes are microRNA targets. Cell *120*, 15–20.

Lim, L.P., Lau, N.C., Garrett-Engele, P., Grimson, A., Schelter, J.M., Castle, J., Bartel, D.P., Linsley, P.S., and Johnson, J.M. (2005). Microarray analysis shows that some microRNAs down-regulate large numbers of target mRNAs. Nature *433*, 769–773.

Lin, J., and Cullen, B.R. (2007). Analysis of the interaction of primate retroviruses with the human RNA interference machinery. J. Virol. *81*, 12218–12226.

Liu, J.L., and Kung, H.J. (2000). Marek's disease herpesvirus transforming protein MEQ: a c-Jun analogue with an alternative life style. Virus Genes *21*, 51–64.

Liu, Y., Wang, X., Lo, A.K., Wong, Y.C., Cheung, A.L., and Tsao, S.W. (2002). Latent membrane protein-1 of Epstein–Barr virus inhibits cell growth and induces sensitivity to cisplatin in nasopharyngeal carcinoma cells. J. Med. Virol. *66*, 63–69.

Lo, A.K., To, K.F., Lo, K.W., Lung, R.W., Hui, J.W., Liao, G., and Hayward, S.D. (2007). Modulation of LMP1 protein expression by EBV-encoded microRNAs. Proc. Natl. Acad. Sci. U.S.A. *104*, 16164–16169.

Lu, S., and Cullen, B.R. (2004). Adenovirus VA1 noncoding RNA can inhibit small interfering RNA and MicroRNA biogenesis. J. Virol. *78*, 12868–12876.

Meister, G., and Tuschl, T. (2004). Mechanisms of gene silencing by double-stranded RNA. Nature *431*, 343–349.

Mocarski, E.S., Shenk, T., and Pass, R.F. (2007). Cytomegaloviruses, In Fields Virology, Knipe, D.M., Howley, P.M., Griffin, D.E., Lamb, R.A., Martin, M.A., Roizman, B., and Straus, S.E., eds. (Philadelphia: Lippincott, Williams & Wilkins), pp. 2701–2772.

Muralidhar, S., Veytsmann, G., Chandran, B., Ablashi, D., Doniger, J., and Rosenthal, L.J. (2000). Characterization of the human herpesvirus 8 (Kaposi's sarcoma-associated herpesvirus) oncogene, kaposin (ORF K12). J Clin Virol *16*, 203–213.

Murphy, E., Vanicek, J., Robins, H., Shenk, T., and Levine, A.J. (2008). Suppression of immediate-early viral gene expression by herpesvirus-coded microRNAs: implications for latency. Proc. Natl. Acad. Sci. U.S.A. *105*, 5453–5458.

Nakano, K., and Vousden, K.H. (2001). PUMA, a novel proapoptotic gene, is induced by p53. Mol. Cell *7*, 683–694.

Okamura, K., Chung, W.J., and Lai, E.C. (2008). The long and short of inverted repeat genes in animals: microRNAs, mirtrons and hairpin RNAs. Cell Cycle *7*, 2840–2845.

Omoto, S., and Fujii, Y.R. (2005). Regulation of human immunodeficiency virus 1 transcription by nef microRNA. J. Gen. Virol. *86*, 751–755.

Orom, U.A., Nielsen, F.C., and Lund, A.H. (2008). MicroRNA-10a binds the 5′-UTR of ribosomal protein mRNAs and enhances their translation. Mol. Cell *30*, 460–471.

Otsuka, M., Jing, Q., Georgel, P., New, L., Chen, J., Mols, J., Kang, Y.J., Jiang, Z., Du, X., Cook, R., *et al.* (2007). Hypersusceptibility to vesicular stomatitis virus infection in Dicer1-deficient mice is due to impaired miR24 and miR93 expression. Immunity *27*, 123–134.

Oudejans, J.J., van den Brule, A.J., Jiwa, N.M., de Bruin, P.C., Ossenkoppele, G.J., van der Valk, P., Walboomers, J.M., and Meijer, C.J. (1995). BHRF1, the Epstein–Barr virus (EBV) homologue of the BCL-2 protooncogene, is transcribed in EBV-associated B-cell lymphomas and in reactive lymphocytes. Blood *86*, 1893–1902.

Ouellet, D.L., Plante, I., Landry, P., Barat, C., Janelle, M.E., Flamand, L., Tremblay, M.J., and Provost, P. (2008). Identification of functional microRNAs released through asymmetrical processing of HIV-1 TAR element. Nucleic Acids Res. *36*, 2353–2365.

Pedersen, I.M., Cheng, G., Wieland, S., Volinia, S., Croce, C.M., Chisari, F.V., and David, M. (2007). Interferon modulation of cellular microRNAs as an antiviral mechanism. Nature *449*, 919–922.

Pfeffer, S., Sewer, A., Lagos-Quintana, M., Sheridan, R., Sander, C., Grasser, F.A., van Dyk, L.F., Ho, C.K., Shuman, S., Chien, M., *et al.* (2005). Identification of microRNAs of the herpesvirus family. Nat Methods *2*, 269–276.

Pfeffer, S., Zavolan, M., Grasser, F.A., Chien, M., Russo, J.J., Ju, J., John, B., Enright, A.J., Marks, D., Sander, C., and Tuschl, T. (2004). Identification of virus-encoded microRNAs. Science *304*, 734–736.

Pillai, R.S., Bhattacharyya, S.N., and Filipowicz, W. (2007). Repression of protein synthesis by miRNAs: how many mechanisms? Trends Cell Biol. *17*, 118–126.

Rajewsky, N. (2006). microRNA target predictions in animals. Nat. Genet. *38 (Suppl.)*, S8–13.

Ramachandran, V., and Chen, X. (2008). Degradation of microRNAs by a family of exoribonucleases in *Arabidopsis*. Science *321*, 1490–1492.

Samols, M.A., Hu, J., Skalsky, R.L., and Renne, R. (2005). Cloning and identification of a microRNA cluster within the latency-associated region of Kaposi's sarcoma-associated herpesvirus. J. Virol. *79*, 9301–9305.

Samols, M.A., Skalsky, R.L., Maldonado, A.M., Riva, A., Lopez, M.C., Baker, H.V., and Renne, R. (2007). Identification of cellular genes targeted by KSHV-encoded microRNAs. PLoS Pathog. *3*, e65.

Sano, M., Kato, Y., and Taira, K. (2006). Sequence-specific interference by small RNAs derived from adenovirus VAI RNA. FEBS Lett. *580*, 1553–1564.

Schafer, A., Cai, X., Bilello, J.P., Desrosiers, R.C., and Cullen, B.R. (2007). Cloning and analysis of microRNAs encoded by the primate gamma-herpesvirus rhesus monkey rhadinovirus. Virology *364*, 21–27.

Seo, G.J., Chen, C.J., and Sullivan, C.S. (2009). Merkel cell polyomavirus encodes a microRNA with the ability to autoregulate viral gene expression. Virology *383*, 183–187.

Seo, G.J., Fink, L.H., O'Hara, B., Atwood, W.J., and Sullivan, C.S. (2008). Evolutionarily conserved function of a viral microRNA. J. Virol. *82*, 9823–9828.

Skalsky, R.L., Samols, M.A., Plaisance, K.B., Boss, I.W., Riva, A., Lopez, M.C., Baker, H.V., and Renne, R. (2007). Kaposi's sarcoma-associated herpesvirus encodes an ortholog of miR-155. J. Virol. *81*, 12836–12845.

Speck, S.H., and Virgin, H.W. (1999). Host and viral genetics of chronic infection: a mouse model of gamma-herpesvirus pathogenesis. Curr Opin Microbiol *2*, 403–409.

Stern-Ginossar, N., Elefant, N., Zimmermann, A., Wolf, D.G., Saleh, N., Biton, M., Horwitz, E., Prokocimer, Z., Prichard, M., Hahn, G., *et al.* (2007). Host immune system gene targeting by a viral miRNA. Science *317*, 376–381.

Sullivan, C.S., Grundhoff, A.T., Tevethia, S., Pipas, J.M., and Ganem, D. (2005). SV40-encoded microRNAs regulate viral gene expression and reduce susceptibility to cytotoxic T cells. Nature *435*, 682–686.

Tam, O.H., Aravin, A.A., Stein, P., Girard, A., Murchison, E.P., Cheloufi, S., Hodges, E., Anger, M., Sachidanandam, R., Schultz, R.M., and Hannon, G.J. (2008). Pseudogene-derived small interfering RNAs regulate gene expression in mouse oocytes. Nature *453*, 534–538.

Tang, S., Bertke, A.S., Patel, A., Wang, K., Cohen, J.I., and Krause, P.R. (2008). An acutely and latently expressed herpes simplex virus 2 viral microRNA inhibits expression of ICP34.5, a viral neurovirulence factor. Proc. Natl. Acad. Sci. U.S.A. *105*, 10931–10936.

Tang, S., Patel, A., and Krause, P.R. (2009). Novel less-abundant viral miRNAs encoded by herpes simplex virus 2 Latency-associated transcript and their roles in regulating ICP34.5 and ICP0 mRNAs. J. Virol. *83*, 1433–1442.

Taraboletti, G., Benelli, R., Borsotti, P., Rusnati, M., Presta, M., Giavazzi, R., Ruco, L., and Albini, A. (1999). Thrombospondin-1 inhibits Kaposi's sarcoma (KS) cell and HIV-1 Tat-induced angiogenesis and is poorly expressed in KS lesions. J. Pathol. *188*, 76–81.

Tay, Y., Zhang, J., Thomson, A.M., Lim, B., and Rigoutsos, I. (2008). MicroRNAs to Nanog, Oct4 and Sox2 coding regions modulate embryonic stem cell differentiation. Nature *455*, 1124–1128.

Thai, T.H., Calado, D.P., Casola, S., Ansel, K.M., Xiao, C., Xue, Y., Murphy, A., Frendewey, D., Valenzuela, D., Kutok, J.L., *et al.* (2007). Regulation of the germinal center response by microRNA-155. Science *316*, 604–608.

Umbach, J.L., Kramer, M.F., Jurak, I., Karnowski, H.W., Coen, D.M., and Cullen, B.R. (2008). MicroRNAs expressed by herpes simplex virus 1 during latent infection regulate viral mRNAs. Nature *454*, 780–783.

Wang, F.Z., Weber, F., Croce, C., Liu, C.G., Liao, X., and Pellett, P.E. (2008). Human cytomegalovirus infection alters the expression of cellular microRNA species that affect its replication. J. Virol. *82*, 9065–9074.

Watanabe, T., Takeda, A., Tsukiyama, T., Mise, K., Okuno, T., Sasaki, H., Minami, N., and Imai, H. (2006). Identification and characterization of two novel classes of small RNAs in the mouse germline: retrotransposon-derived siRNAs in oocytes and germline small RNAs in testes. Genes Dev. *20*, 1732–1743.

Wightman, B., Ha, I., and Ruvkun, G. (1993). Posttranscriptional regulation of the heterochronic gene lin-14 by lin-4 mediates temporal pattern formation in C. elegans. Cell *75*, 855–862.

Xia, T., O'Hara, A., Araujo, I., Barreto, J., Carvalho, E., Sapucaia, J.B., Ramos, J.C., Luz, E., Pedroso, C., Manrique, M., *et al.* (2008). EBV microRNAs in primary lymphomas and targeting of CXCL-11 by ebv-mir-BHRF1-3. Cancer Res. *68*, 1436–1442.

Xu, N., Segerman, B., Zhou, X., and Akusjarvi, G. (2007). Adenovirus virus-associated RNAII-derived small RNAs are efficiently incorporated into the rna-induced silencing complex and associate with polyribosomes. J. Virol. *81*, 10540–10549.

Yao, Y., Zhao, Y., Xu, H., Smith, L.P., Lawrie, C.H., Sewer, A., Zavolan, M., and Nair, V. (2007). Marek's disease virus type 2 (MDV-2)-encoded microRNAs show no sequence conservation with those encoded by MDV-1. J. Virol. *81*, 7164–7170.

Yao, Y., Zhao, Y., Xu, H., Smith, L.P., Lawrie, C.H., Watson, M., and Nair, V. (2008). MicroRNA profile of Marek's disease virus-transformed T-cell line MSB-1: predominance of virus-encoded microRNAs. J. Virol. *82*, 4007–4015.

Yu, B., Yang, Z., Li, J., Minakhina, S., Yang, M., Padgett, R.W., Steward, R., and Chen, X. (2005). Methylation as a crucial step in plant microRNA biogenesis. Science *307*, 932–935.

Zhao, Y., Yao, Y., Xu, H., Lambeth, L., Smith, L.P., Kgosana, L., Wang, X., and Nair, V. (2009). A functional micro-RNA-155 ortholog encoded by the oncogenic Marek's disease virus. J. Virol. 83, 489–492.

# Virus-encoded Suppressors of RNA Silencing and the Role of Cellular miRNAs in Mammalian Antiviral Immune Responses

4

Joost Haasnoot and Ben Berkhout

Abstract

Small RNA-directed silencing mechanisms play important roles in the regulation of eukaryotic gene expression. In plants, insects, nematodes and fungi RNA silencing mechanisms are also involved in innate antiviral defence responses. To counter antiviral RNA silencing, viruses from plants, insects and fungi encode RNA silencing suppressors (RSSs). Recent studies suggest that RNA silencing in mammals, or RNA interference (RNAi), is also involved in antiviral responses. In particular, there is increasing evidence that cellular regulatory microRNAs (miRNAs) have a function in restricting virus replication in mammalian cells. Similar to plant and insect viruses, several mammalian viruses encode RSS factors that inhibit the RNAi mechanism. Several of these suppressors are multifunctional proteins that were previously shown to block innate antiviral immune responses involving the interferon (IFN) pathway. Here we summarize the current data on mammalian virus-encoded RSS factors. In addition, different aspects of antiviral RNAi and the role of cellular miRNAs in restricting virus replication in mammalian cells are discussed.

## Introduction

Innate antiviral immune responses are the first line of defence against invading pathogenic viruses. In mammals, these antiviral responses involve the induction of expression of the type I IFNs, IFN-α and IFN-β, and pro-inflammatory cytokines. The IFNs trigger an antiviral programme that involves the expression of many IFN-stimulated genes, which ultimately results in inhibition of protein synthesis and instalment of both innate and adaptive immune responses (Bowie and Unterholzner, 2008). Recent studies suggest that the RNAi pathway, in addition to the IFN pathway, plays a role in antiviral responses in mammals.

RNAi is a conserved sequence-specific gene silencing mechanism in eukaryotes that, similar to the IFN pathway, is triggered by double-stranded RNA (dsRNA). It plays a pivotal role in post-transcriptional regulation of cellular gene expression via the endogenously encoded small non-coding miRNAs. miRNAs function as guides for the RNA-induced silencing complex (RISC) that triggers translational inhibition via imperfect basepairing to sequences in the 3′ untranslated region (3′-UTR) of target mRNAs (Bartel, 2004). Hundreds of human miRNAs have been cloned and it is estimated that these miRNAs regulate the expression of at least 30% of human genes (Krek *et al.*, 2005). In plants, insects, nematodes and fungi RNAi also plays an essential role in innate antiviral immune responses (Waterhouse *et al.*, 2001; Voinnet, 2001; Wilkins *et al.*, 2005; Wang *et al.*, 2006a; Segers *et al.*, 2007). In these organisms, virus infection results in the production of small interfering RNA (siRNAs) duplexes of 21 nucleotides that correspond to the viral RNA. These virus-derived siRNAs (sometimes referred to as viRNAs) guide RISC to the complementary viral transcripts which, after binding, results in sequence-specific cleavage of the target RNA. To overcome the inhibitory effects of antiviral RNA silencing, plant and insect viruses encode specific RSSs (Ding and Voinnet, 2007; Li and

Ding, 2001). Each RSS blocks antiviral RNA silencing at a distinct step in the pathway, enabling the virus to replicate to higher titres. Because RNAi and related RNA silencing mechanisms are conserved within eukaryotes, it seems likely that the mammalian RNAi pathway also plays a role in innate antiviral responses. However, the typical viRNAs, which are characteristic for antiviral RNAi responses in plants and insects, could thus far not be detected in virus-infected mammalian cells (Pfeffer *et al.*, 2004). It has been argued that mammalian cells do not need RNAi-based antiviral responses because they have acquired the 'more sophisticated' IFN-mediated antiviral response (Cullen, 2006). The antiviral role of RNAi in mammals has therefore remained controversial (Cullen, 2006; Berkhout and Jeang, 2007).

Despite the controversy on the role of RNAi in mammalian antiviral responses, it is clear that mammalian viruses interact in multiple ways with the cellular RNAi pathway. For example, virus infection affects the expression of cellular miRNAs (Houzet *et al.*, 2008; Yeung *et al.*, 2005). This in turn will change the expression of many cellular genes, which may have important implications in terms of virus-induced disease development. Virus-induced changes in miRNA expression may also affect the level of cellular protein cofactors that are required for viral replication. For example, human immunodeficiency virus type 1 (HIV-1) inhibits the expression of miR-20 and miR-17-5p, resulting in increased expression of the histone acetylase PCAF that is required for optimal HIV-1 transcription (Triboulet *et al.*, 2007). In addition, several virus species, in particular from the family of herpesviruses, have co-opted the cellular RNAi pathway and encode miRNAs that regulate the expression of host or viral genes (Gottwein and Cullen, 2008; Cullen, 2009). Finally, it has also been shown that cellular miRNAs, besides regulating host gene expression, can also target and inhibit viral gene expression in mammalian cells (Berkhout and Jeang, 2007). This finding supports the idea that the cellular RNAi mechanism has a function in restricting virus replication in mammals despite the fact that in most cases viRNAs cannot be detected.

If RNAi acts as an antiviral mechanism in mammalian cells, infecting viruses are expected to encode RSS factors similar to their plant and insect counterparts. Suppressors of RNAi have been identified for a number of mammalian viruses (Table 4.1) (de Vries and Berkhout, 2008). In fact, the identification of these RSSs formed one of the first indications for an antiviral RNAi role in mammals. Here, we summarize the available data on the RSS functions of mammalian viruses. For clarity, we will first present the various branches of the mammalian RNAi-pathway that may contribute to RNAi-based immune responses.

## The RNAi mechanism in mammals

Only in recent years it has become clear that RNA plays a major role in cellular gene regulation. The role of small non-coding RNAs appears to be more important than anyone could have ever imagined (Bartel, 2004). Currently, several species of small RNAs have been identified that associate with Argonaute or Piwi proteins and serve as guides in the sequence-specific regulation of gene expression, transposon silencing, and genome organization (Rana, 2007). Small RNAs of 21–22 nucleotides that associate with a member of the Argonaute protein family are involved in RNAi and miRNA-mediated regulation of gene expression (Kim *et al.*, 2009). In mammals, the biogenesis of cellular-encoded miRNAs and the regulation of gene expression via miRNAs have received a lot of attention. In contrast, only a limited number of studies addressed the production of siRNA or viRNAs in virus-infected mammalian cells.

miRNAs are derived from stem–loop structures of about 80 nucleotides that are initially expressed as part of long primary transcripts (pri-miRNAs, see Fig. 4.1) (Lee *et al.*, 2004). The stem–loop structures are processed in the nucleus by the RNAse III enzyme Drosha and its cofactor DGCR8, leaving a characteristic two-nucleotide 3′ overhang (Han *et al.*, 2004). The resulting stem–loop structure, the precursor miRNA (pre-miRNA) of approximately 60 nucleotides, is exported from the nucleus to the cytoplasm by the nuclear export factor Exportin 5 (Exp5) and Ran-GTP. The pre-miRNAs are further processed in the cytoplasm by the RNAse III enzyme Dicer and its cofactor TAR RNA binding protein (TRBP) (Grishok *et al.*, 2001). The Dicer–TRBP

**Table 4.1** RNAi/silencing suppressor factors encoded by mammalian viruses

| Genome structure | Virus | RSS | Action | Cell type | Other functions | References |
|---|---|---|---|---|---|---|
| Positive-strand RNA | HCV | Core | Direct interaction with Dicer | Human | Capsid formation, transcription, cell proliferation, apoptosis | Wang *et al.* (2006b), Chen *et al.* (2008) |
| | | E2 | Direct interaction with Ago2 | Human | Envelope protein, PKR inhibitor | Ji *et al.* (2008) |
| | NoV | B2 | dsRNA binding, preventing Dicer cleavage | Insect, human | No other function described | Sullivan and Ganem (2005) |
| Negative-strand RNA | Influenza | NS1 | dsRNA binding | Plants, insect, human | PKR inhibitor, RIG-I antagonist | Li *et al.* (2004), Delgadillo *et al.* (2004), Bucher *et al.* (2004) |
| | Ebola virus | VP35 | dsRNA binding | Human | Structural protein, RIG-I antagonist | Haasnoot (2007a) |
| | LACV | NSs | siRNA inhibition | Human | Suppression type I IFN pathway | Soldan *et al.* (2005) |
| dsRNA | Reovirus | σ3 | dsRNA binding | Plants, human | Structural protein, PKR inhibitor | Lichner *et al.* (2003) |
| DNA | Vaccinia virus | E3L | dsRNA binding | Insects, human | PKR inhibitor | Li *et al.* (2004) |
| | adenovirus | VA RNA | Saturation RNAi pathway (Dicer, Ago2, Exp5) | Human | PKR inhibitor | Andersson *et al.* (2005), Lu and Cullen (2005), Aparicio *et al.* (2006), Xu *et al.* (2007) |
| retrovirus | HIV-1 | Tat | Inhibition Dicer, binding si/miRNAs | Human, plants | Transcriptional activator, PKR inhibitor | Bennasser *et al.* (2005, 2006), Haasnoot (2007a), Qian (2009) |
| | PFV-1 | Tas | Inhibition miR-32 | Human, plants | Transcriptional activator | Lecellier *et al.* (2005) |

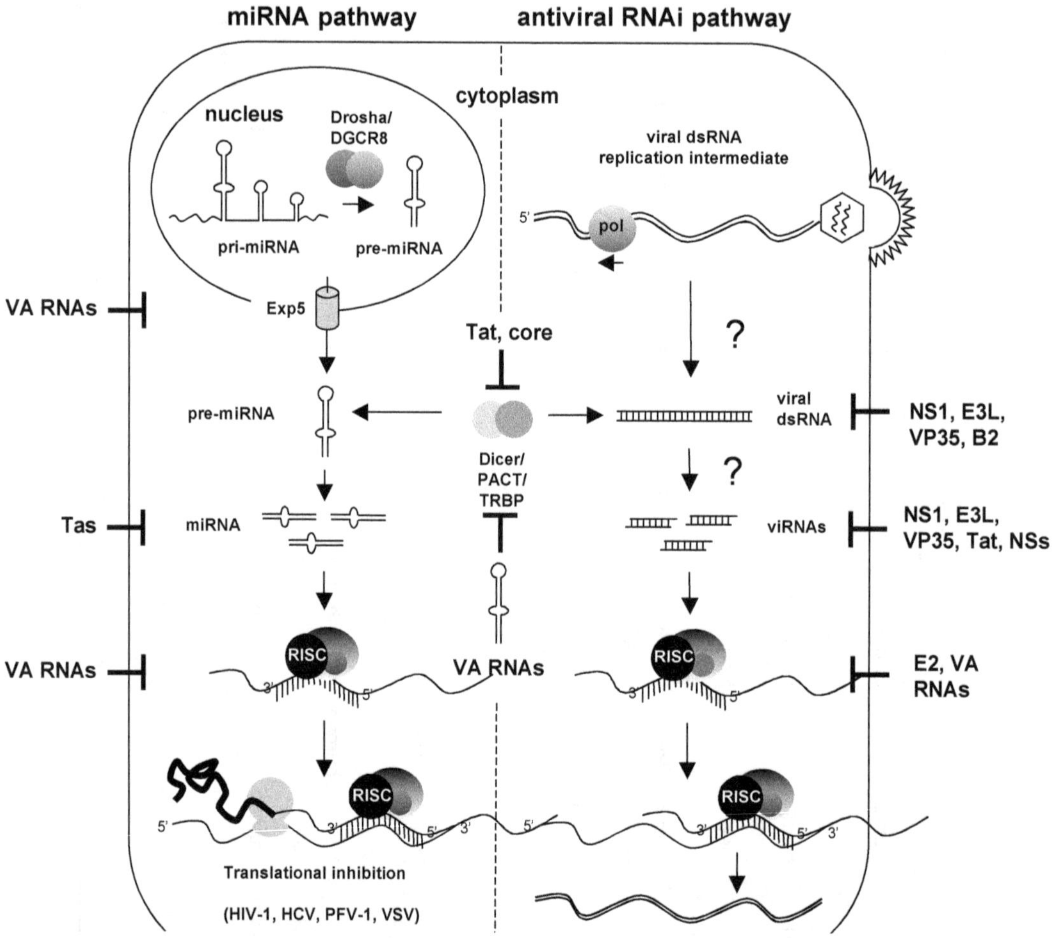


**Figure 4.1** Schematic of the mammalian RNAi pathway and the possible interactions with viral RSS factors. The left panel depicts miRNA-mediated translational inhibition of both cellular and viral transcripts. The right panel shows the putative antiviral RNAi pathway, resulting in the production of viRNAs that induce cleavage of the viral target RNAs. Shown are the RSS factors HIV-1 Tat, PFV-1 Tas, vaccinia virus E3L, influenza A virus NS1, EBOV VP35, adenovirus VA RNAs, HCV core and E2, NoV B2 and LACV NSs. These RSS factors can block RNAi at the indicated steps in the pathways.

complex cuts off the terminal loop, resulting in the miRNA duplex with a two-nucleotide 3′ overhang on either side. This miRNA duplex is subsequently loaded into RISC which involves Dicer, the RNA-binding proteins TRBP and protein kinase R-activating protein (PACT), and one of the four Argonaute proteins (Ago1–4) (Kok *et al.*, 2007). miRNAs with central mismatches preferentially interact with Ago1, whereas perfectly matching siRNA duplexes are incorporated by Ago2, the slicer Argonaute protein that can catalyse cleavage of the homologous target mRNA (Kim *et al.*, 2009). Once loaded into RISC the miRNA duplex is unwound into the mature miRNA or guide strand and the complementary passenger strand, which is degraded. RISC containing the mature miRNA strand, sometimes referred to as miRISC, can target mRNAs bearing complementary sequences within the 3′-UTR resulting in translational repression and in some cases RNA cleavage (Hutvagner and Zamore, 2002). The exact criteria for target recognition and the mechanism behind translational repression are still not clear. However, pairing of the 5′ 7–8 nucleotides of the miRNA (seed region) to the 3′-UTR of a target mRNA is of critical importance. Furthermore, translocation of the miRISC-targeted mRNA to cellular processing (P)-bodies is important. Targeted mRNAs are de-adenylated, de-capped, degraded, or stored within the P-body (Nilsen, 2007).

Unlike miRNAs, siRNAs or viRNAs are fully complementary to the mRNA target sequence and, when loaded in Ago2 containing RISC, they trigger cleavage of the target RNA rather than translational repression (Fig. 4.1). In plants and insects, viRNAs arise from long virus-derived dsRNAs that are cleaved by Dicer into the viRNA duplexes. In case of RNA viruses, the precursor dsRNA can represent replication intermediates (Fig. 4.2) (Ding and Voinnet, 2007). For plant viruses, viRNAs have also been shown to originate from sequences that fold into extensive secondary structures (Molnar *et al.*, 2005). For DNA viruses and retroviruses, convergent transcriptional activity can result in the production of long viral dsRNAs as precursors for viRNAs (Fig. 4.2). Unlike in plants and insects, viRNAs are not readily detectable in virus-infected mammalian cells (Pfeffer *et al.*, 2004). However, recent findings indicate the accumulation of small virus-derived RNAs in cells infected with HIV-1, poliovirus,

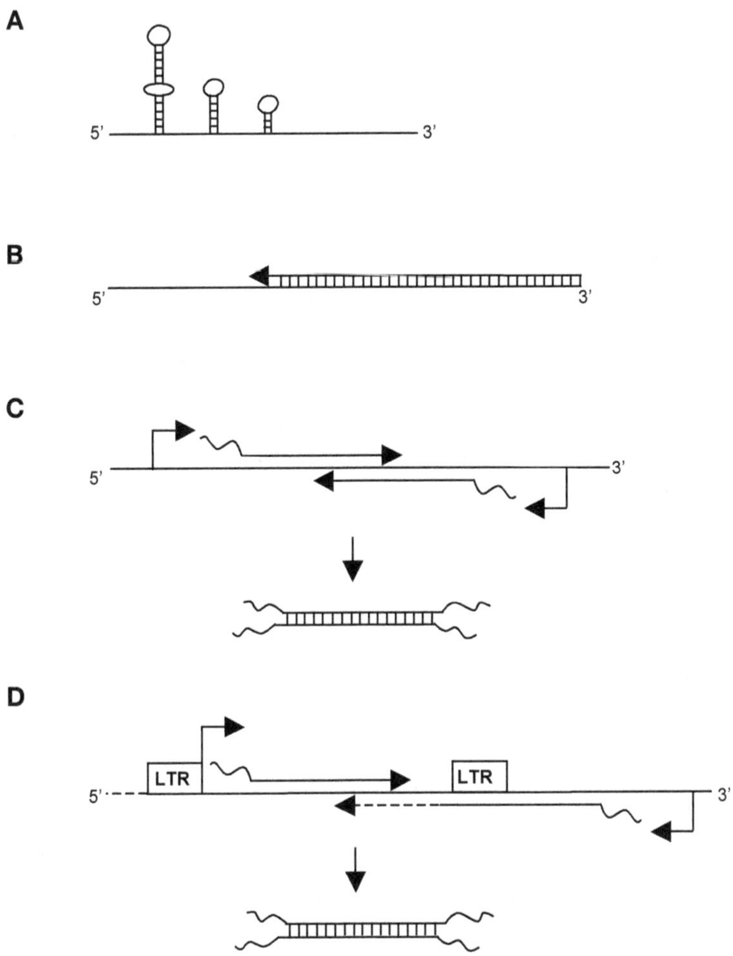


**Figure 4.2** The origin of virus-derived dsRNA. During virus infection, virus-specific dsRNA as a source for viRNAs can be generated in various ways. (A) Extensive RNA structures in the viral RNA genome (+ or – strand) may be recognized by Dicer and processed into viRNAs. (B) RNA viruses can generate dsRNA replication intermediates. (C) Some viruses, such as SV40 and hepatitis B virus, encode genes that are transcribed in opposite directions, which can result in the formation of dsRNA. (D) In cells containing integrated retroviruses, cellular transcription opposite to that of the provirus can yield dsRNA. Annealing the complementary viral transcripts produces dsRNA that can be processed into viRNAs.

hepatitis C virus (HCV), vesicular stomatitis virus (VSV), flock house virus (FHV), *Drosophila* C virus (DCV), and Sindbis virus (Parameswaran *et al* 2008; K.-T. Jeang, personal communication). These data were obtained using sensitive deep-sequencing technology. However, it is currently unclear whether these small RNAs indeed function as viRNAs or whether they represent inactive viral RNA degradation products. In any case, the accumulation of these small RNAs in mammalian cells is relatively modest when compared to the accumulation of viRNAs in plants and insects.

In addition to its natural functions, the use of RNAi as a tool to specifically silence genes has revolutionized molecular biological research and it provides new possibilities in drug design (Haasnoot *et al.*, 2007b). Common strategies for activation of artificial RNAi in mammalian cells include transfection of synthetic siRNAs and endogenous expression of short hairpin RNAs (shRNAs). These shRNAs are modelled after pre-miRNA stem–loop structures and need to be processed by Dicer into siRNAs in order to become active. Both synthetic siRNAs and shRNA expression strategies are currently being used in the development of new antiviral therapeutics, from a transient therapy for respiratory syncytial virus (RSV) to a durable gene therapy for HIV-1. However, error-prone viruses relatively easily select escape variants that are resistant to the expressed siRNA inhibitor. The resistant virus variants have acquired point mutations or a deletion within the siRNA-target sequence (Boden *et al.*, 2003; Das *et al.*, 2004). Therefore, the combined expression of multiple siRNAs in a combinatorial RNAi approach is required to persistently block virus replication (Ter Brake *et al.*, 2006; Grimm and Kay, 2007; Liu and Berkhout, 2008).

## Cellular miRNAs targeting viral transcripts

Besides regulating the expression of cellular genes, cellular miRNAs can also target and inhibit the expression of viral genes (Fig. 4.1, Table 4.2). These 'antiviral' miRNAs may represent the RNAi-based immune response in mammals as an equivalent of the RNAi function in plants, insects, nematodes and fungi. The first reported example of a cellular miRNA that targets a virus concerns miR-32. This miRNA targets the retrovirus primate foamy virus type 1 (PFV-1) in a genome region that is present in the 3′-UTR of all viral transcripts and that encodes both the viral Bt and Env Bet proteins (Lecellier *et al.*, 2005). Although basepairing between miR-32 and its target is not very extensive, the miRNA triggers specific translational inhibition of reporter genes with the target sequence. Targeting of the viral RNA by miR-32 results in reduced virus replication, and PFV-1 requires the RSS activity of the viral Tas protein to overcome this inhibition (Lecellier *et al.*, 2005).

Another example of a virus that is restricted by cellular miRNAs is VSV (Otsuka *et al.*, 2007).

**Table 4.2** Cellular miRNAs targeting viral genes

| Virus | miRNA | Target | Effect | References |
|---|---|---|---|---|
| HIV-1 | miR-28, miR-125b, miR-382, miR-150, miR-223 | TAR, Env, NF-kb, Env, Nef | Negative (induces viral latency) | Huang *et al.* (2007), Wang *et al.* (2008) |
| | miR-29a | Nef | Negative | Ahluwalia *et al.* (2008) |
| PFV-1 | miR-32 | Bt and EnvBet proteins, 3′-UTR | Negative (Tas protein counters this effect) | Lecellier *et al.* (2005) |
| VSV | miR-24, miR-93 | L and P protein | Negative | Otsuka *et al.* (2007) |
| HCV | miR-122 | 5′-UTR (IRES) | Stimulates translation (liver specific) | Jopling *et al.* (2005) |
| | miR-1, miR-30, miR-128, miR-196, miR-296, miR-351, miR-431, miR-448 | C, NS5A (others not specified) | Negative (induced by IFN signalling) | Pedersen *et al.* (2007) |
| | miR-199a* | 5′-UTR (IRES) | Negative | Murakami (2008) |

Ostsuka and co-workers showed that VSV transcripts encoding the viral large protein (L protein) and the phosphoprotein (P protein) are targeted by miR-24 and miR-93, respectively. Reduced expression of miR-24 and miR-93 in Dicer-deficient mice caused a strong increase of VSV replication leading to virus-induced lethality. No viRNAs could be detected in VSV-infected cells, and no change was observed in the IFN-mediated antiviral responses, suggesting that host miRNAs play a role in virus/host interactions. Besides VSV, the authors also tested the susceptibility of Dicer-deficient macrophages for infection by other viruses such as encephalo-myocarditis virus, lymphocytic choriomeningitis virus, Coxsackievirus group B serotype 3, influenza A virus, herpes simplex virus type 1 (HSV-1) and vaccinia virus. Interestingly, only HSV-1 displayed increased virus production, suggesting that Dicer-mediated antiviral RNAi responses do not represent a general antiviral mechanism in mammals. However, miRNAs are differentially expressed among different cell types, and viruses likewise infect a specific subset of cells. Therefore, the lack of an effect in the Dicer-deficient macrophages does not exclude the possibility that the virus is affected by Dicer deficiency in the biologically more relevant cell types or tissues (Otsuka *et al.*, 2007).

Cellular miRNAs also target and potently repress HIV-1 gene expression (Huang *et al.*, 2007). It was shown that miR-28, miR-125b, miR-150, miR-223 and miR-382 can target sequences near the 3′ end of HIV-1 transcripts and thus suppress the expression of viral mRNAs in resting $CD4^+$ T cells. Interestingly, the expression of this set of miRNAs is decreased in activated $CD4^+$ T cells, thus allowing virus replication. The results suggest that miRNAs play a role in the establishment of viral latency (Huang *et al.*, 2007). HIV-1 latency occurs in resting $CD4^+$ T cells when the HIV-1 provirus is stably integrated into the host genome without producing new viral transcripts and proteins. HIV-1 latency may allow the virus to escape from the immune system. Thus, miRNA-mediated inhibition appears to be beneficial for the virus. This idea is supported by the observation that HIV-1 retains these miRNA targets and no escape variants with mutated target sequences emerge. HIV-1 latency poses a serious problem for the treatment of HIV-1 infected individuals with antiviral drugs. After diminishing the bulk of replicating HIV-1 and a subsequent therapy stop, the virus will rapidly re-emerge from this latent viral reservoir. Huang and co-workers showed that specific 2′-O-methyl-oligoribonucleotide antisense inhibitors of the antiviral miRNAs can counter latency induction, thus forcing virus production. Manipulation of the anti-HIV miRNAs could possibly be used as a new approach to purge the latent HIV-1 reservoir. Several other miRNAs including miR-29a were also predicted to target the HIV-1 genome (Ahluwalia *et al.*, 2008; Hariharan *et al.*, 2005). It was recently shown that miR-29a indeed recognizes a target within the HIV-1 Nef gene to restrict expression of Nef protein and virus replication (Ahluwalia *et al.*, 2008). Finally, the latency inducing miRNAs appear to determine the susceptibility of peripheral blood monocytes for HIV-1 infection. These cells express the essential receptors for HIV-1 entry, but are infrequently infected. Such miRNA-mediated virus suppression may be part of the innate immune system that could have a role in determining a particular cell tropism (Wang *et al.*, 2008). Restriction of HIV-1 replication by miRNAs correlates with increased HIV-1 production in Drosha and Dicer knockdown cells (Triboulet *et al.*, 2007). In turn, this is in agreement with the observation that the cellular miRNA expression pattern is significantly affected by virus infection (Houzet *et al.*, 2008; Yeung *et al.*, 2005).

Researchers have also recognized the use of miRNA-mediated repression as a tool in the design of novel viral vectors. This strategy can be employed to alter vector tropism or to generate attenuated virus variants that may be used for future vaccine development. For example, insertion of mi-142–3p targets in a lentiviral vector restricted transgene expression in the haematopoietic cell lineage, whereas expression was maintained in non-haematopoietic cells (Brown *et al.*, 2006). Insertion of the liver-specific miR-122 target in adenovirus severely attenuated replication in hepatocytes but allowed normal replication in other cells (Ylosmaki *et al.*, 2008). Insertion of targets for the muscle specific miR-133 and miR-206 in the 3′-UTR of the oncolytic picornavirus Coxsackie virus A21 strongly attenuated viral pathogenicity (Kelly *et al.*, 2008). In general, the inserted targets are fully complementary to the cellular miRNAs and therefore targeting will probably result in cleavage of the viral/vector RNA. Interestingly, viral escape variants that contain mutations in the target site have been selected (Kelly *et al.*, 2008; Vignuzzi *et al.*, 2008). This phenomenon is very similar to the escape mutants that emerge during prolonged culturing of viruses in the presence of shRNA inhibitors.

In conclusion, these combined results indicate that cellular miRNAs can potently restrict virus replication in mammalian cells and thus could represent the RNAi-based branch of the antiviral immune response in mammals. Besides inhibiting viral gene expression, the total set of miRNAs expressed in a certain cell type will determine the level of evolutionary freedom of the virus. Certain nucleotide changes in viral targets will not be tolerated because they create more optimal miRNA target sequences that are more potently repressed by the miRNA machinery. This will restrict the virus in its ability to adjust to changing environments, such as the adaptive immune response.

## Cellular miRNAs targeting HCV

Not all cellular miRNAs that target viral mRNAs have a negative effect on gene expression. It was convincingly demonstrated that miR-122 is required for HCV replication (Table 4.2) (Jopling *et al.*, 2005; Jopling *et al.*, 2008; Chang *et al.*, 2008; Randall *et al.*, 2007). This miRNA is highly expressed in human liver cells and was recently demonstrated to stimulate translation of HCV. This stimulation is caused by direct interaction of miR-122 with two target sites in the 5′-UTR, which harbours the internal ribosomal entry site (IRES) that drives HCV translation (Henke *et al.*, 2008). Possibly, this miR-122 co-factor could be targeted with so-called antagomirs, antisense oligonucleotides that specifically inhibit miRNA function. Indeed, inhibition of miR-122 resulted in a strong decrease of HCV RNA replication. Besides the positive role of miR-122, a specific set of cellular miRNAs seem involved in restricting HCV gene expression. These miRNAs (miR-1, miR-30, miR-128, miR-196, miR-296, miR-351, miR-431 and miR-448) were found to be up-regulated in response to IFN signalling that is triggered during HCV replication (Pedersen *et al.*, 2007). The transfection of synthetic miRNA-mimics reproduced the antiviral effect of IFN-β on HCV replication and infection. In addition, IFN treatment leads to a significant reduction in the expression of miR-122 (Pedersen *et al.*, 2007; Ye *et al.*, 2008). Therefore, most antiviral activity of IFN appears to be caused by respectively up or down regulation of cellular miRNAs. However, a recent study in which liver biopsies from patients with chronic HCV infection undergoing IFN therapy were analysed revealed no correlation between miR-122 expression and the viral load (Sarasin-Filipowicz 2009).

## RNAi-mediated transposon silencing

Small RNAs also play an important role in blocking transposon activity in mammals including humans, insects, nematodes and plants (Czech *et al.*, 2008; Watanabe *et al.*, 2008; Yang and Kazazian, Jr., 2006; Sijen and Plasterk, 2003; Girard and Hannon, 2008). Transposons are endogenous nucleic acid parasites that are capable of movement and propagation within host genomes (Kazazian, Jr., 2004). It was recently shown that endogenous transposon-specific siRNAs, so-called endo-siRNAs, are produced in mouse oocytes to block transposon activity. In addition, these endo-siRNAs down-regulate the expression of certain protein-coding transcripts. The endo-siRNAs are processed from naturally occurring dsRNAs precursors that are formed by the annealing of protein-coding transcripts with antisense transcripts from a homologous pseudogene. Interestingly, knock-down of Dicer or Ago2 resulted in decreased levels of endo-siRNAs and increased levels of the corresponding retrotransposon and protein-coding transcripts (Tam *et al.*, 2008; Watanabe *et al.*, 2008).

Other small RNAs involved in transposon silencing are the 25- to 27-nucleotide long Piwi-interacting RNAs (piRNAs). These small RNAs are incorporated into Piwi-containing effector complexes, which target transposon transcripts for post-transcriptional silencing (Girard and Hannon, 2008). In contrast to siRNAs, piRNAs are generated and active in a Dicer and Ago2 independent manner. Because retroviruses such as HIV-1 are related to retrotransposons it is not unlikely that, in a similar way as endo-siRNAs,

HIV-1-specific siRNAs can arise in HIV-1 infected cells and inhibit virus replication.

## RNAi suppressors and methods used for their identification

Viral RSS factors were first identified for the plant viruses (Ding and Voinnet, 2007; Li and Ding, 2001). These RSS factors were initially characterized as multifunctional virulence or pathogenicity factors that contribute to high virus production and efficient viral spread (Cronin *et al.*, 1995; Ding *et al.*, 1995). As we have discussed, there are now strong indications that RNAi imposes an inhibitory effect on virus replication in mammals (Fig. 4.1 and Table 4.1). Therefore, mammalian viruses will probably have evolved similar RSS factors and/or strategies that block, or at least attenuate, such antiviral RNAi effects in order to replicate efficiently. A total of 11 mammalian virus-encoded RSS factors have been described so far, compared to over 40 plant virus RSS proteins. To complete the survey, 4 and 1 RSSs have been described for insect and fungus infecting viruses, respectively (Sullivan and Ganem, 2005; de Vries and Berkhout, 2008; Li *et al.*, 2002; Soldan *et al.*, 2005; van Rij *et al.*, 2006; Segers *et al.*, 2006).

Most RSS proteins, even the ones from related viruses, show a striking lack of sequence similarity. This may illustrate the multifunctional role played by most of these factors. One important characteristic that is shared by many RSSs is the dsRNA-binding capacity (Lakatos *et al.*, 2006). Via dsRNA-binding, RSS proteins can block the processing of precursor dsRNA into effective siRNAs by Dicer. Alternatively, viRNAs can be inactivated by being sequestered via the dsRNA-binding domain (dsRBD). For example, the P19 protein of the plant tombus virus suppresses RNAi both in plants and mammalian cells by binding and sequestering siRNAs from the RNAi pathway (Scholthof, 2006). The potyviral HC-Pro protein interacts with siRNAs and suppresses RNA slicing by preventing RISC assembly (Shiboleth *et al.*, 2007; Diaz-Pendon and Ding, 2008). Another way to block RNAi is via inhibition of Dicer activity. The turnip crinkle virus P83 protein was shown to specifically block Dicer-like 4 activity in plants (Deleris *et al.*, 2006).

Most RSS factors have been identified using RNAi reporter systems. In these experiments reporter genes are silenced by co-expression of a specific RNAi inducer. To test for RSS activity, the RSS candidate is co-expressed with the -to be silenced- reporter. Reporter expression is maintained in the presence of silencing suppression activity. The relevance of RSS activity measured in reporter assays has been questioned because non-specific binding of siRNAs by overexpression of any dsRNA-binding protein may also result in RNAi suppression (Cullen, 2006; Lichner *et al.*, 2003). RSS factors from plant and insect viruses have been tested for their impact on the accumulation of viRNAs in infected cells. However, not all suppressors will score an effect in this type of assay because some RSSs affect the RNAi pathway after the viRNA processing step. So far, inhibition of viRNA accumulation *in vivo* has not been described for mammalian virus-encoded RSS factors, although inhibition of siRNA production has been reported *in vitro* (Bennasser *et al.*, 2005; Bennasser and Jeang, 2006; Andersson *et al.*, 2005; Wang *et al.*, 2006b).

Viruses with an inactivated RSS generally exhibit a severe replication defect. Knockdown of the RNAi pathway rescues virus replication in case the sole defect is the lack of RSS activity. These defective viruses can sometimes be rescued by co-expression of a heterologous RSS protein from another virus. Such trans-complementation assays have been widely used to identify new RSS factors. Trans-complementation experiments have also been used for the identification of mammalian virus-encoded RSS factors (Li *et al.*, 2004). Interestingly, there are a number of examples in which the complementation experiment has been used successfully across kingdoms. RSS factors from plant viruses are active in mammalian cells and vice versa (Li *et al.*, 2004; Schnettler *et al.*, 2009; Triboulet *et al.*, 2007).

An important difference between studies on plant virus-encoded RSS factors versus those from mammalian viruses is that the latter are primarily studied using established cell lines. It is known that such cell lines frequently show disturbances in cellular pathways. It is therefore quite possible that RNAi pathways in these cells do not represent the physiological situation. In contrast, most studies on plant viral RSS factors have been performed *in vivo* in intact plants. Such studies revealed that antiviral RNA silencing is a tissue specific phenomenon (Hirai *et al.*, 2008). Certain cells display antiviral RNA silencing activity whereas others do not have this activity. The same may be true for mammalian cells. Therefore, it is of critical importance to study mammalian viral RSS activity in the most natural test system available.

## RNAi suppressors of mosquito-transmitted viruses

RNA silencing acts as an adaptive antiviral defence mechanism in *Drosophila* (Li *et al.*, 2002). Li *et al.* showed that insect cells accumulate viRNAs when infected with the positive-stranded RNA virus FHV. In addition, they reported that knock-down of Ago2 results in increased virus replication. In subsequent *in vivo* studies using *Drosophila* embryos and adult flies, Dicer-2 was shown to process newly synthesized virus-specific dsRNA into viRNAs that guide Ago-2 containing RISC to target and degrade the viral RNA (Galiana-Arnoux *et al.*, 2006). Similar results were reported for flies infected with DCV (van Rij *et al.*, 2006).

The FHV B2 protein was the first animal virus-encoded RSS protein that was identified (Li *et al.*, 2002). This protein suppresses RNAi in plant and insects cells and certain types of mammalian cells. The B2 protein RSS activity depends on sequestration of viRNAs. In addition, B2 has been reported to directly prevent loading of siRNAs into RISC, and to protect viral dsRNA from processing by Dicer (Chao *et al.*, 2005). Related to FHV is the Nodamura virus (NoV), which causes lethal infections in insects and mammals. The NoV B2 protein shares only 29% sequence identity with the FHV B2 protein, but has an overall similarity in size and distribution of charged and hydrophilic amino acids. The NoV B2 protein was shown to inhibit miRNA generation both *in vitro* and *in vivo* (Sullivan and Ganem, 2005). In addition, NoV B2 was shown to inhibit Dicer-mediated cleavage of long precursor dsRNA *in vitro* and RISC activity in cells. The activity involves binding to Dicer, precursor RNAs and siRNAs, similar to the properties of the FHV B2 protein.

Another mosquito-transmitted virus for which an RSS was identified is La Cross encephalitis virus (LACV). This virus is a member of the Bunyaviridae family that causes severe encephalitis in children. Soldan and co-workers (2005) showed that the viral NSs protein is able to suppress siRNA-induced RNAi in human HEK293T cells (Soldan *et al.*, 2005). Further studies indicated that NSs also had a strong inhibitory effect on IFN (Blakqori *et al.*, 2007). Deletion of the NSs gene produced no phenotype in cells in which the IFN system was inactivated. This observation could indicate that anti-LACV RNAi acts downstream of the IFN response, similar to what has been reported for HCV (Pedersen *et al.*, 2007).

## RNAi suppression by the IFN antagonistic proteins NS1 (influenza virus) and E3L (vaccinia virus)

Mammalian viruses have to overcome a broad spectrum of IFN-mediated antiviral responses in order to replicate. Therefore, many mammalian viruses encode specific IFN-antagonistic factors (Bowie and Unterholzner, 2008). The IFN pathway is activated upon virus infection when viral nucleic acids are sensed by host pattern-recognition receptors. These receptors, such as Toll-like receptors (TLRs), can detect foreign DNA, single-stranded RNA, dsRNA and glycoproteins. For example, TLR3 is localized on the endosomal membrane and senses viral dsRNA after viral entry via endocytosis. The cellular RNA helicases retinoic acid-inducible gene I (RIG-I) and melanoma differentiation-associated gene 5 (MDA5) are cytoplasmic dsRNA sensors (Kato *et al.*, 2006; Marques *et al.*, 2006). Activation of these sensors results in phosphorylation of interferon-regulatory factor-3 (IRF-3) and IRF-7, which triggers production of type 1 IFNs-$\alpha/\beta$ and induction of the antiviral state of cells. This leads to inhibition of viral replication and cell proliferation, apoptosis, and destruction of virus infected cells by natural killer cells. Other cellular dsRNA sensors include 2′–5′ oligo adenylate synthetase enzymes (OAS), cellular endoribonuclease L (RNase L) and dsRNA-dependent protein kinase (PKR). The latter protein detects dsRNA in the cytoplasm (Bowie and Unterholzner, 2008).

Activation of PKR by dsRNA binding leads to phosphorylation of the α subunit of the translation initiation factor (eIF)–2α, which abolishes translation of both viral and cellular proteins. OAS catalyses the conversion of ATP to 2′–5′ oligoadenylates, which activate the RNAse L enzyme that cleaves both cellular and viral mRNAs.

The influenza virus NS1 protein and vaccinia virus E3L protein are two well-known viral factors that are involved in suppression of several of the above mentioned antiviral immune responses. The influenza A virus NS1 protein is 230 amino acids long and harbours a dsRNA binding domain (dsRBD) within the N-terminal 73 amino acids. NS1 has been reported to block OAS-mediated activation of RNAse L (Min and Krug, 2006) and limits the induction of IFN-β (Ludwig *et al.*, 2002; Wang *et al.*, 2000). Furthermore, NS1 interacts with and inhibits the cytoplasmic dsRNA sensor RIG-I (Guo *et al.*, 2007; Mibayashi *et al.*, 2007) and blocks PKR-mediated inhibition of protein synthesis (Bergmann *et al.*, 2000; Min and Krug, 2006). Similar to NS1, vaccinia virus E3L protein is able to inhibit intracellular IFN-induced pathways and is required for viral pathogenesis. E3L has a C-terminal dsRBD that sequesters viral dsRNAs and thereby prevents activation of PKR and OAS (Chang *et al.*, 1992; Rivas *et al.*, 1998; Shors *et al.*, 1997). Interestingly, E3L is the only RSS identified so far of which the dsRBD has sequence similarity with the cellular Dicer protein (Li and Ding 2006).

Although influenza virus pathogenicity is a multigenic trait, the NS1 protein is an important virulence determinant in avian, swine and human influenza viruses (Jackson *et al.*, 2008; Li *et al.*, 2006; Seo *et al.*, 2002). Similarly, E3L is important for vaccinia virus replication and pathogenicity. NS1 and E3L share this characteristic with plant virus-encoded RSS factors. In addition, both NS1 and E3L contain dsRBDs similar to plant virus-encoded RSS factors. Li and co-workers used FHV to study the RNA silencing suppression activity of NS1 and E3L (Li *et al.*, 2004). The FHV B2 protein inhibits RNAi by sequestering long viral dsRNAs and siRNAs (Fenner *et al.*, 2006). Both E3L and NS1 were able to rescue replication of a FHV variant in which the B2 gene had been replaced by GFP, indicating that both E3L and NS1 are functionally equivalent to the B2 protein. In addition, NS1 was shown to suppress RNA silencing in plants and to increase the virulence of potato virus X in *Nicotiana benthamiana* (Bucher *et al.*, 2004; Delgadillo *et al.*, 2004). Furthermore, NS1 was shown to bind siRNAs *in vitro*.

In a similar trans-complementation assay, we analysed whether NS1 and E3L could functionally replace the RSS function of the HIV-1 Tat protein. Indeed, both NS1 and E3L could rescue the production of an HIV-1 variant that lacks Tat RSS activity (Haasnoot *et al.*, 2007a). In follow-up experiments we compared the IFN antagonistic activity and RSS activity of NS1 for several influenza strains (de Vries *et al.*, 2009). Interestingly, the various NS1 proteins were all capable of inhibiting the activation of an IFN-responsive promoter, but we measured differential RSS activity among the NS1 variants. The variable ability of NS1 to suppress RNAi may contribute to differences in viral replication capacity and pathogenicity. For both NS1 and E3L other groups have reported that these proteins do not display RSS activity in mammalian cells (Kok and Jin, 2006; Lantermann *et al.*, 2007). However, both studies used an excess of RNAi inducer, which may have prevented detection of RSS activity.

## RNAi suppressor encoded by Ebola virus

Ebola virus (EBOV) is a negative-stranded RNA virus from the family of the Filoviridae. This virus is well known for its severe pathogenicity. During viral outbreaks, the mortality rate can be as high as 90% as a result of virus-induced severe haemorrhagic fever. An important determinant of EBOV pathogenicity is the VP35 protein. VP35 is a 340 amino acid protein with a dsRBD that shows similarities to the dsRBD of influenza virus A NS1 protein (Hartman *et al.*, 2004). Similar to NS1 and E3L, the EBOV VP35 protein has been shown to inhibit RIG-I-induced activation of type I IFN responses and activation of IRF-3 (Cardenas *et al.*, 2006; Hartman *et al.*, 2008). This inhibition is dependent, at least in part, on the dsRNA-binding capacity of VP35. Using a luciferase-based, shRNA-induced RNAi reporter system we demonstrated that this dsRBD is also important for VP35-mediated RSS activity (Haasnoot *et al.*, 2007a). In complementation studies, VP35 was able to replace the Tat RSS function and support production of an HIV-1 variant lacking Tat. These data suggest that VP35-mediated RNAi suppression and the VP35-dsRNA-dependent IFN antagonistic properties are linked. Possibly, VP35 sequesters viRNAs or dsRNA precursors of viRNA, resulting in suppression of an antiviral RNAi response that acts upstream of the OAS/RNAse L/PKR and IFN pathways.

## RNAi suppression by the adenovirus VA RNAs

Most RSS factors identified so far are proteins that act via sequestration of viRNAs or dsRNA precursors of viRNAs. The human adenovirus encodes an RSS function that follows a different strategy. Adenovirus is a lytic virus of which the genome consists of 39 kbp dsDNA. During the late phase of infection the virus expresses high levels of the two structured non-coding virus-associated (VA) RNAs, RNAI and RNAII. These RNAs block PKR activity. The highly structured nature of the VA molecules suggests some similarity to pre-miRNAs, raising the possibility that the VA RNAs are substrates for Dicer. Several studies have now convincingly demonstrated that the VA-RNAs are recognized and processed by Dicer into virus-derived si/miRNAs that are functionally incorporated into RISC (Andersson *et al.*, 2005; Aparicio *et al.*, 2006; Lu and Cullen, 2004; Sano *et al.*, 2006; Xu *et al.*, 2007). Their extremely high copy number of up to $1 \times 10^8$ copies per cell causes saturation of the RNAi pathway. Thus, the VA RNAs act as suppressor by saturation of the RNAi pathway. This saturation takes place at multiple levels. *In vitro* studies indicated that the VA RNAs can saturate Dicer. In addition, it has been shown that 80% of the RISC complexes in virus-infected cells contain VA RNA-derived siRNAs. VA RNAs also compete with cellular miRNAs for transport from the nucleus to the cytoplasm by Exp5 (Lu and Cullen, 2004). A detailed study of VA RNA processing by Dicer revealed that VA RNAII, rather than VA RNAI, is the better substrate for Dicer cleavage (Xu *et al.*, 2007). This was surprising, since VA RNAII is expressed at a much lower level than VA RNAI. Because VA RNAI was previously shown to be the decisive factor in suppressing PKR activity, VA RNAII may be dedicated to saturate the RNAi pathway. The top part of the VA RNA hairpin structures appear to be essential for interaction with and inhibition of PKR. The lower part of the structures, in particular the basepaired stem, is a Dicer substrate that enters the RNAi silencing pathway, thus competing with cellular miRNAs. Adenovirus mutants that lack the VA RNAs are severely replication-impaired. However, it is unclear whether this is caused by the loss of the viral ability to block PKR or because the virus can no longer suppress antiviral RNAi responses, or both.

Viruses from the DNA virus family of herpesviruses encode several miRNAs that are believed to reshape the cellular environment for their own benefit (Cullen, 2009). Similarly, the VA RNAs have been suggested to function as virus-encoded miRNAs instead of or in addition to their ability to block cellular RNAi processes (Xu *et al.*, 2007). These miRNAs may target cellular mRNAs that are involved in antiviral defence mechanisms. In agreement with this notion is the observation that VA RNA-derived miRNAs can be found on polyribosomes, suggesting that they are involved in the regulation of translation (Xu *et al.*, 2007). However, the target mRNAs have as yet not been identified. The fact that adenovirus is a lytic virus may argue against an miRNA role of the VA RNAs as regulation of gene expression appears to be less important in the short period of infection. As said, the VA RNAs may have multiple functions. Interestingly, early studies on the VA RNAI function revealed that it can be replaced by two non-coding structured RNAs encoded by Epstein–Barr virus, the EBERs, which may have a similar function (Bhat and Thimmappaya, 1985).

## Retroviral transcription activator proteins as RNAi suppressors

Transposons are controlled by RNA silencing in plants, nematodes, insects and mammals (Czech *et al.*, 2008; Watanabe *et al.*, 2008; Yang and Kazazian, Jr., 2006; Sijen and Plasterk, 2003; Girard and Hannon, 2008; Schramke and Allshire, 2004). Suppression of RNA silencing triggers mobilization of the transposons. This observation forms an early indication that mammalian exogenous retroviruses could perhaps also be controlled by RNAi (Lecellier *et al.*, 2005). As we have discussed above, two viruses from the family of retroviruses, PFV-1 and HIV-1, are indeed controlled by RNAi via cellular miRNAs (Huang *et al.*, 2007; Lecellier *et al.*, 2005). Retroviruses have a single-stranded RNA genome that, upon infection of a new cell, is reversed transcribed into double-stranded DNA that integrates into the host genome. This stable provirus forms the template for expression of new viral RNA and proteins. Important in this process are the transcription activation proteins Tat (HIV-1) and Tas (PFV-1). Tat protein stimulates transcription from the viral long terminal repeat (LTR) promoter by interacting with the transactivation response (TAR) element, a stable RNA stem–loop structure present at the 5′ terminus of each viral transcript (Berkhout *et al.*, 1989). Lecellier *et al.* showed that the PFV-1 Tas protein could also act as a suppressor of RNAi (Lecellier *et al.*, 2005). They initially showed that co-expression of the plant virus RSS P19 protein increased PFV-1 production, indicating that PFV-1 is restricted by RNAi. Via computational analyses they identified miR-32 as the effector of this inhibition. The Tas protein was subsequently identified as an RSS that represses miR-32 accumulation, thus allowing efficient replication (Lecellier *et al.*, 2005). Interestingly, Tas also showed considerable RSS activity in *Arabidopsis* plants, where it was able to decrease siRNA accumulation.

Jeang and co-workers demonstrated that the HIV-1 Tat protein has RSS activity (Bennasser *et al.*, 2005; Bennasser and Jeang, 2006). They first analysed the accumulation of HIV-1 specific siRNAs during virus infection. A single siRNA was identified that is processed from a dsRNA duplex which is formed by basepairing of complementary sequences in the Pol and Env open reading frames. Tat-mediated RSS activity was required to overcome this inhibition. Tat RSS activity was shown to involve a direct interaction with Dicer and requires dsRNA binding activity. Band shift assays indicated that specific residues in Tat are required for the interaction with Dicer. Tat RSS activity appears to be independent of the transcriptional trans-activation properties of this HIV-1 protein. We confirmed these results using an HIV-1 variant that is independent of Tat for viral transcription (Haasnoot *et al.*, 2007a). Despite this independence, a frame-shift mutation in Tat caused a dramatic decrease in virus production, which could only be restored by heterologous viral factors (influenza virus A NS1 protein, vaccinia virus E3L and EBOV VP35) that showed RSS activity in a luciferase RNAi reporter system.

We recently showed that the plant virus RSS NS3 of rice hoja blanca virus (RHBV) is also able to complement the RSS function of HIV-1 Tat (Schnettler *et al.*, 2009). RHBV NS3 is an RSS that exclusively binds small dsRNA molecules (miRNAs and siRNAs), and which is also active in mammalian cells (Hemmes *et al.*, 2007; Schnettler *et al.*, 2008). An NS3 mutant that is deficient in dsRNA binding and RSS activity is not able to rescue production of the Tat-negative HIV-1 variant. This result suggests that HIV-1 replication is inhibited either by siRNAs or miRNAs and that Tat RSS activity is required to counter this inhibition.

Another study reported that the Tat RSS function can be replaced by the P19 protein of tomato bushy stunt virus (Qian *et al.*, 2009). The HIV-1 Tat protein exhibited RSS activity in *Nicotiana benthamiama* protoplasts. Unlike the earlier reported Tat-mediated inhibition of processing of precursor dsRNAs by Dicer, Qian *et al.* showed that the Tat RSS activity is positioned downstream of siRNA production (Bennasser *et al.*, 2005). The cross-kingdom RNAi suppression of both Tas and Tat is a strong indication of the potency of these RSSs and indicates the significance of the antiviral RNAi response in mammalian cells (Lecellier *et al.*, 2005). These studies also demonstrate the usefulness of well-defined RSS proteins.

Other groups have challenged the importance of Tas and Tat as RSS factors (Lin and Cullen, 2007). These authors did not score any suppression activity, but only promoter activation. These studies were performed with an excess of RNAi inducer, which is not the correct setting for identifying RSS activity. The relevance of the HIV-1 derived siRNA described by Bennasser *et al.* has also been questioned (Cullen, 2006; Bennasser *et al.*, 2005). The proposed RNA structure was suggested not to be able to form because the sequence is already part of the Rev-responsive element (RRE).

## HIV-1 TAR and TRBP

The processing of long dsRNA precursors into siRNAs by Dicer is assisted by the TRBP protein. Intriguingly, TRBP was originally identified as HIV-1 TAR RNA binding protein. TRBP is a co-factor that is required for HIV-1 transcription (Gatignol *et al.*, 1991). During the initial phase of HIV-1 transcription, cooperative binding of HIV-1 Tat, TRBP and cyclin-T1 to TAR activates elongation and initiation of RNA polymerase II-mediated transcription. Thus, HIV-1 employs a cellular co-factor that is now implicated as part of the RNAi machinery. Several studies have focussed on the interactions of TRBP with TAR and the RNAi machinery. Bennasser *et al.*, 2006 were the first to show that overexpression of TAR can sequester TRBP and thus inhibit RNAi (Bennasser *et al.*, 2006). The authors suggested that this may provide the virus with an additional strategy to block antiviral RNAi responses. Besides sequestering TRBP, the TAR hairpin may also serve as a substrate for Dicer to be processed into mi/siRNAs that target cellular or viral transcripts. Klase *et al.*, showed that the HIV-1 TAR hairpin is indeed a substrate for Dicer (Klase *et al.*, 2007). Furthermore, it was shown that TAR-derived miRNAs are present in HIV-1 infected CD4+ T lymphocytes. Both the miRNAs from the 5′ and 3′ arm of TAR could be detected, although miR-TAR-3′ was processed more efficiently, yielding better knock-down of a man-made reporter gene with TAR-target (Ouellet *et al.*, 2008). Recently, Klase and co-workers demonstrated that the TAR miRNA protects infected cells from apoptosis by down-regulation of the cellular genes ERCC1 and IER3 (Klase *et al.*, 2009). The fact that TAR is a substrate for Dicer is in agreement with its ability to interact with TRBP. This, in turn, correlates with Tat-mediated inhibition of Dicer. Currently, it is unclear how these multiple TRBP interactions are coordinated to control viral transcription and/or antiviral RNA silencing. TRBP is required for efficient transcription and replication of HIV-1 and knockdown of TRBP decreases HIV-1 replication. This finding does not support a role of TRBP in antiviral RNAi processes, but the situation is rather complex with both positive and negative TRBP effects occurring during HIV-1 replication.

## RSS factors encoded by HCV

HCV is the prototype member of the hepacivirus genus within the family of Flaviviridae, and causes chronic infection of the liver. HCV has a 9.6-kb single-stranded RNA genome of positive polarity. HCV has an interesting interaction with cellular RNAi pathways. The virus is both stimulated and repressed by cellular miRNAs (Jopling *et al.*, 2005; Jopling *et al.*, 2008; Chang *et al.*, 2008; Randall *et al.*, 2007). In addition, both HCV core protein and the envelope protein 2 show RSS activity *in vitro*. The core protein has been shown to block shRNA-induced RNAi, but not siRNA-induced gene silencing (Wang *et al.*, 2006b). Furthermore, HCV core protects long dsRNA from being processed by Dicer into siRNAs (Chen *et al.*, 2008). For plant viruses, coat proteins regularly function as RNA silencing suppressor proteins. For example, turnip crinkle virus coat protein (CP p38) suppresses Dicer activity. Immunoprecipitation studies revealed a direct interaction between HCV core and Dicer. Deletion studies suggested that the N-terminal 62 amino acids are required for this interaction. Besides HCV core, the viral E2 protein was demonstrated to interact with Ago-2 in co-immunoprecipitation experiments, thus suppressing shRNA-induced RNAi against a reporter gene (Ji *et al.*, 2008). The HCV E2 protein was earlier shown to have PKR-antagonistic properties (He and Katze, 2002).

## Interaction between RNAi and antiviral IFN-mediated immune responses

Most mammalian virus-encoded RSS factors described to date have also been linked to the IFN-mediated antiviral immune responses. HIV-1 TAR RNA and the RSS protein Tat have been reported to act as PKR antagonist (Cai *et al.*, 2000; Park *et al.*, 1994; SenGupta *et al.*, 1990). TRBP, which interacts with HIV-1 TAR, is involved in siRNA production and is also an inhibitory factor for PKR. The cellular protein PACT, a TRBP co-factor in siRNA production, also acts as a cellular activator of PKR (Daher *et al.*, 2009). Similarly, EBOV VP35, influenza virus NS1, and vacccinia virus E3L have RSS activity and act as antagonists of RIG-I, PKR and OAS (Katze *et al.*, 2002; Xiang *et al.*, 2002; Lu *et al.*, 1995; Min and Krug, 2006;

Rivas *et al.*, 1998; Mibayashi *et al.*, 2007; Chang *et al.*, 1992). The double function of many viral RSS/IFN antagonistic proteins is perhaps not that surprising considering that both the RNAi and IFN pathways are induced by virus-derived dsRNAs. These findings suggest that the RNAi and IFN pathways cooperate in the innate defence responses against viruses. The IFN-induced expression of anti-HCV miRNAs is a good example of how IFN and RNAi co-operate to combat virus infections (Pedersen *et al.*, 2007). In this case antiviral RNAi appears to act downstream of the IFN response. Other scenarios may include a sequence-specific antiviral RNAi response that acts upstream of the IFN-response. The antiviral RNAi response would specifically block virus replication as a first attempt to clear the virus without triggering the apoptosis-inducing IFN pathway.

## Effect of viral RNAi suppressors on regulatory pathways

Because the miRNA pathway and antiviral RNAi responses overlap, the activity of RNAi suppressors will not only block the antiviral effect, but may also interfere with processing, maturation and function of host miRNAs that regulate cellular transcripts. Such an effect has been demonstrated for plant virus RNAi suppressors. For example, turnip mosaic virus HC-Pro blocks miR-171 activity, resulting in abnormal plant development (Chapman *et al.*, 2004; Chellappan *et al.*, 2005; Kasschau *et al.*, 2003). For the geminivirus-encoded RSS AC4 protein from African cassava mosaic virus, Chellapan and co-workers reported that transgenic plants that express this protein exhibit decreased miRNA expression and abnormal development. In human cells, adenovirus VA RNAI was shown to interfere with miRNA biogenesis (Andersson *et al.*, 2005; Lu and Cullen, 2004). Furthermore, HIV-1 production affects the cellular miRNA expression pattern, possibly via RSS activity of the Tat protein (Yeung *et al.*, 2005). Recently, Qian and co-workers showed that HIV-1 Tat blocks miRNA activity in an experimental setting (Qian *et al.*, 2009). The impact of these changes in miRNA levels on cellular gene expression is not known. However, it remains an important question because altered cellular miRNA expression has been associated with disease, e.g. certain forms of cancer. Viral pathogenicity could therefore be affected by these changes in cellular miRNAs (Lu *et al.*, 2005; He *et al.*, 2005; Calin *et al.*, 2002). In case the virus-induced changes in miRNA expression are specific, miRNA mimics or inhibitors could be used to restore the normal miRNA levels and thus act as antiviral therapeutics.

## RNAi suppressor activity and antiviral therapeutics

RNAi technology has been developed into a potentially powerful therapeutic tool to combat a variety of viruses. Profound inhibition has been reported for many important human pathogens (Haasnoot *et al.*, 2007b). Because some of these viruses are known to encode an RSS factor, an obvious question is whether or not viral RSSs can interfere with an antiviral RNAi-based therapy. So far, no studies have addressed this question directly. However, the fact that viruses can efficiently be inhibited by induced-RNAi illustrates that viral RNAi suppression does not pose a serious problem for therapeutic RNAi. For example, efficient RNAi-mediated inhibition has been reported for RSS-encoding viruses such as HIV-1, influenza A virus and EBOV (Ter Brake *et al.*, 2006; Geisbert *et al.*, 2006; Ge *et al.*, 2004; Tompkins *et al.*, 2004). Most likely, the therapeutic siRNAs will saturate the RNAi suppressor factor, rendering it ineffective. On the other hand, it is also known that viruses can overcome the inhibitory effect at low concentrations of RNAi inducers (Bitko *et al.*, 2005; Liu *et al.*, 2007). RNAi suppressor activity could play a role in these cases. We recently described the phenomenon of viral pseudo-escape from RNAi (Ter Brake *et al.*, 2008; von Eije *et al.*, 2008). This phenomenon describes that viruses show break-through replication in cells that express inhibitory siRNAs, but without mutation of the target sequence. We hypothesised that ongoing low level virus replication in the presence of sub-optimal shRNA inhibitors will eventually lead to signs of viral break-through in the culture (von Eije *et al.*, 2009). In this scenario, RSS factors may play a role in supporting a certain level of virus replication in the presence of antiviral RNAi-inducers. Another observation may point to RSS activity. A single point mutation in the siRNA target sequence can result in HIV-1 escape (Das *et al.*, 2004). However, luciferase reporters bearing these escape targets can still be efficiently repressed by siRNAs (Westerhout *et al.*, 2005). This suggests that the imposed RNAi pressure is somehow less effective in the viral context. RNAi suppressor activity might play a role, although this needs to be verified experimentally.

Analyses of virus-encoded immune suppressor genes, either IFN antagonists and/or RSS factors, may be important in the development of live-attenuated virus vaccines. Viral RSS factors are important, but usually not essential for virus replication. Mutation or removal of RSS factors therefore often results in attenuation of the viral replication capacity. Such variants may represent ideal vaccine strains. For example, a swine influenza A strain vaccine lacking NS1 protected swine against infection with the wild type virus (Richt *et al.*, 2006). From another therapeutic point of view, the mammalian virus-encoded RSSs may also represent interesting targets for antiviral drugs. Inhibition of a viral RSS would trigger the recruitment of the cellular antiviral RNAi responses. For example, the cellular miRNAs that were reported to target HIV-1, PFV-1 or HCV may become much more effective in repressing viral gene expression in case the RSS activity of Tat (HIV-1), Tas (PFV-1) or core (HCV) is inhibited.

Finally, viral vector and vaccine production may be restricted by RNAi, in particular for constructs that carry RNAi-inducer cassettes. We reported that the level of Sindbis and adenovirus replication can be increased significantly by suppressing RNAi in cells via stable expression of a virus-encoded RSS such as VP35, NS1, E3L protein (de Vries *et al.*, 2008). Similarly, NoV B2 protein and adenovirus VA RNAI were shown to increase the production of vectors harbouring shRNA-expression cassettes (Poluri and Sutton, 2007). Viral RSSs may therefore become favourable tools for boosting the titres of these vectors (de Vries *et al.*, 2008). This would also be important for the production of live-attenuated vaccines that lack their own RSS gene function. However, expression of heterologous viral RSS factors may be undesirable during vaccine production. Therefore, artificial RSS factors may be constructed that are based on minimal dsRBDs.

## Concluding remarks

Mammalian virus-encoded RSS factors have been studied for several years. Although there is some insight in the way that RSS factors counter RNAi in experimental systems, so far little is known about their function during natural virus replication. These studies on RSSs have been hampered by the fact that the trigger for antiviral RNAi in mammalian cells has long remained illusive. The current data suggest that miRNA-mediated inhibition of viral gene expression represents an RNAi-based antiviral immune response in mammals. It will be interesting to test whether viral RSS factors can counter the antiviral effects of these cellular miRNAs, similar to what has been reported for the PFV-1 RSS protein Tas. It remains unclear whether the fortuitous targeting of viral transcripts by cellular miRNAs can be considered an antiviral response in mammals. The miRNA targets in viral genes may represent the balanced end product of co-evolution of the virus with its host, and the virus may somehow benefit from miRNA-mediated repression. For example, miRNA-mediated repression may enable the virus to replicate at a lower level in certain cell types so that it can prevent activation of other (IFN-mediated) innate immune responses. The establishment of HIV-1 latency under miRNA-control forms another example. The role of viral RSS factors may therefore lie in counteracting miRNA-induced restriction during specific phases of virus replication in certain cell types. It is also evident that there are close interactions between the mammalian RNAi pathway and IFN-mediated immune responses. Both pathways can be affected by the same virus-encoded RSS factors. We would argue that this field of research is only just beginning. Future research will provide a better understanding of viral RSS functions, the role of antiviral RNAi and its interaction with other innate immune responses in mammals.

### Acknowledgements

We thank W. de Vries and C. Tiepel for critically reading the manuscript. RNAi research in the Berkhout lab is sponsored by ZonMw (Vici grant and Translational Gene Therapy program), NWO-CW (Top grant), the European Union (LSHP-CT-2006-037301) and the Technology Foundation STW (grant AGT.7708).

## References


Ahluwalia, J.K., Khan, S.Z., Soni, K., Rawat, P., Gupta, A., Hariharan, M., Scaria, V., Lalwani, M., Pillai, B., Mitra, D., and Brahmachari, S.K. (2008). Human cellular microRNA hsa-miR-29a interferes with viral nef protein expression and HIV-1 replication. Retrovirology 5, 117.

Andersson, M.G., Haasnoot, P.C.J., Xu, N., Berenjian, S., Berkhout, B., and Akusjarvi, G. (2005). Suppression of RNA interference by adenovirus virus-associated RNA. J. Virol. *79*, 9556–9565.

Aparicio, O., Razquin, N., Zaratiegui, M., Narvaiza, I., and Fortes, P. (2006). Adenovirus virus-associated RNA is processed to functional interfering RNAs involved in virus production. J. Virol. *80*, 1376–1384.

Bartel, D.P. (2004). MicroRNAs: genomics, biogenesis, mechanism, and function. Cell *116*, 281–297.

Bennasser, Y., and Jeang, K.T. (2006). HIV-1 Tat interaction with Dicer: requirement for RNA. Retrovirology 3, 95.

Bennasser, Y., Le, S.Y., Benkirane, M., and Jeang, K.T. (2005). Evidence that HIV-1 encodes an siRNA and a suppressor of RNA silencing. Immunity *22*, 607–619.

Bennasser, Y., Yeung, M.L., and Jeang, K.T. (2006). HIV-1 TAR RNA subverts RNA interference in transfected cells through sequestration of TAR RNA-binding protein, TRBP. J. Biol. Chem. *281*, 27674–27678.

Bergmann, M., Garcia-Sastre, A., Carnero, E., Pehamberger, H., Wolff, K., Palese, P., and Muster, T. (2000). Influenza virus NS1 protein counteracts PKR-mediated inhibition of replication. J. Virol. *74*, 6203–6206.

Berkhout, B., and Jeang, K.T. (2007). RISCy business: microRNAs, pathogenesis, and viruses. J. Biol. Chem. *282*, 26641–26645.

Berkhout, B., Silverman, R.H., and Jeang, K.T. (1989). Tat trans-activates the human immunodeficiency virus through a nascent RNA target. Cell 59, 273–282.

Bhat, R.A., and Thimmappaya, B. (1985). Construction and analysis of additional adenovirus substitution mutants confirm the complementation of VAI RNA function by two small RNAs encoded by Epstein–Barr virus. J. Virol. *56*, 750–756.

Bitko, V., Musiyenko, A., Shulyayeva, O., and Barik, S. (2005). Inhibition of respiratory viruses by nasally administered siRNA. Nat. Med. *11*, 50–55.

Blakqori, G., Delhaye, S., Habjan, M., Blair, C.D., Sanchez-Vargas, I., Olson, K.E., ttarzadeh-Yazdi, G., Fragkoudis, R., Kohl, A., Kalinke, U., Weiss, S., Michiels, T., Staeheli, P., and Weber, F. (2007). La Crosse bunyavirus nonstructural protein NSs serves to suppress the type I interferon system of mammalian hosts. J. Virol. *81*, 4991–4999.

Boden, D., Pusch, O., Lee, F., Tucker, L., and Ramratnam, B. (2003). Human immunodeficiency virus type 1 escape from RNA interference. J. Virol. *77*, 11531–11535.

Bowie, A.G., and Unterholzner, L. (2008). Viral evasion and subversion of pattern-recognition receptor signalling. Nat. Rev. Immunol. *8*, 911–922.

Brown, B.D., Venneri, M.A., Zingale, A., Sergi, S.L., and Naldini, L. (2006). Endogenous microRNA regulation suppresses transgene expression in hematopoietic lineages and enables stable gene transfer. Nat. Med. *12*, 585–591.

Bucher, E., Hemmes, H., de Haan, P., Goldbach, R., and Prins, M. (2004). The influenza A virus NS1 protein binds small interfering RNAs and suppresses RNA silencing in plants. J. Gen. Virol. *85*, 983–991.

Cai, R., Carpick, B., Chun, R.F., Jeang, K.T., and Williams, B.R. (2000). HIV-I TAT inhibits PKR activity by both RNA-dependent and RNA-independent mechanisms. Arch. Biochem Biophys. *373*, 361–367.

Calin, G.A., Dumitru, C.D., Shimizu, M., Bichi, R., Zupo, S., Noch, E., Aldler, H., Rattan, S., Keating, M., Rai, K., Rassenti, L., Kipps, T., Negrini, M., Bullrich, F., and Croce, C.M. (2002). Frequent deletions and down-regulation of micro- RNA genes miR15 and miR16 at 13q14 in chronic lymphocytic leukemia. Proc. Natl. Acad. Sci. U.S.A. *99*, 15524–15529.

Cardenas, W.B., Loo, Y.M., Gale, M., Jr., Hartman, A.L., Kimberlin, C.R., Martinez-Sobrido, L., Saphire, E.O., and Basler, C.F. (2006). Ebola virus VP35 protein binds double-stranded RNA and inhibits alpha/beta interferon production induced by RIG-I signaling. J. Virol. *80*, 5168–5178.

Chang, H.W., Watson, J.C., and Jacobs, B.L. (1992). The E3L gene of vaccinia virus encodes an inhibitor of the interferon-induced, double-stranded RNA-dependent protein kinase. Proc. Natl. Acad. Sci. U.S.A. *89*, 4825–4829.

Chang, J., Guo, J.T., Jiang, D., Guo, H., Taylor, J.M., and Block, T.M. (2008). Liver-specific microRNA miR-122 enhances the replication of hepatitis C virus in nonhepatic cells. J. Virol. *82*, 8215–8223.

Chao, J.A., Lee, J.H., Chapados, B.R., Debler, E.W., Schneemann, A., and Williamson, J.R. (2005). Dual modes of RNA-silencing suppression by Flock House virus protein B2. Nat. Struct. Mol. Biol. *12*, 952–957.

Chapman, E.J., Prokhnevsky, A.I., Gopinath, K., Dolja, V.V., and Carrington, J.C. (2004). Viral RNA silencing suppressors inhibit the microRNA pathway at an intermediate step. Genes Dev. *18*, 1179–1186.

Chellappan, P., Vanitharani, R., and Fauquet, C.M. (2005). MicroRNA-binding viral protein interferes with *Arabidopsis* development. Proc. Natl. Acad. Sci. U.S.A. *102*, 10381–10386.

Chen, W., Zhang, Z., Chen, J., Zhang, J., Zhang, J., Wu, Y., Huang, Y., Cai, X., and Huang, A. (2008). HCV core protein interacts with Dicer to antagonize RNA silencing. Virus Res. *133*, 250–258.

Cronin, S., Verchot, J., Haldeman-Cahill, R., Schaad, M.C., and Carrington, J.C. (1995). Long-distance movement factor: a transport function of the potyvirus helper component proteinase. Plant Cell 7, 549–559.

Cullen, B.R. (2006). Is RNA interference involved in intrinsic antiviral immunity in mammals? Nat. Immunol. *7*, 563–567.

Cullen, B.R. (2009). Viral and cellular messenger RNA targets of viral microRNAs. Nature *457*, 421–425.

Czech, B., Malone, C.D., Zhou, R., Stark, A., Schlingeheyde, C., Dus, M., Perrimon, N., Kellis, M., Wohlschlegel, J.A., Sachidanandam, R., Hannon, G.J., and Brennecke, J. (2008). An endogenous small interfering RNA pathway in *Drosophila*. Nature. *453*, 729–731.

Daher, A., Laraki, G., Singh, M., Melendez-Pena, C.E., Bannwarth, S., Peters, A.H., Meurs, E.F., Braun, R.E., Patel, R.C., and Gatignol, A. (2009). TRBP control of PACT-induced phosphorylation of protein kinase R is reversed by stress. Mol. Cell Biol. *29*, 254–265.

Das, A.T., Brummelkamp, T.R., Westerhout, E.M., Vink, M., Madiredjo, M., Bernards, R., and Berkhout, B. (2004). Human immunodeficiency virus type 1 escapes from RNA interference-mediated inhibition. J. Virol. *78*, 2601–2605.

de Vries, W., and Berkhout, B. (2008). RNAi suppressors encoded by pathogenic human viruses. Int. J. Biochem. Cell Biol. *40*, 2007–2012.

de Vries, W., Haasnoot, J., van der Velden, J., van Montfort, T., Zorgdrager, F., Paxton, W., Cornelissen, M., van Kuppeveld, F., de Haan, P., and Berkhout, B. (2008). Increased virus replication in mammalian cells by blocking intracellular innate defense responses. Gene Ther. *15*, 545–552.

de Vries, W., Haasnoot, J., Fouchier, R., de Haan, P., and Berkhout, B. (2009). Differential RNA silencing suppression activity of NS1 proteins from different influenza A virus strains. J. Gen. Virol. *90*, 1916–1922.

Deleris, A., Gallego-Bartolome, J., Bao, J., Kasschau, K.D., Carrington, J.C., and Voinnet, O. (2006). Hierarchical action and inhibition of plant Dicer-like proteins in antiviral defense. Science *313*, 68–71.

Delgadillo, M.O., Saenz, P., Salvador, B., Garcia, J.A., and Simon-Mateo, C. (2004). Human influenza virus NS1 protein enhances viral pathogenicity and acts as an RNA silencing suppressor in plants. J. Gen. Virol. *85*, 993–999.

Diaz-Pendon, J.A., and Ding, S.W. (2008). Direct and indirect roles of viral suppressors of RNA silencing in pathogenesis. Annu. Rev. Phytopathol. *46*, 303–326.

Ding, S.W., Li, W.X., and Symons, R.H. (1995). A novel naturally occurring hybrid gene encoded by a plant RNA virus facilitates long distance virus movement. EMBO J. *14*, 5762–5772.

Ding, S.W., and Voinnet, O. (2007). Antiviral immunity directed by small RNAs. Cell *130*, 413–426.

Fenner, B.J., Goh, W., and Kwang, J. (2006). Sequestration and protection of double-stranded RNA by the betanodavirus b2 protein. J. Virol. *80*, 6822–6833.

Galiana-Arnoux, D., Dostert, C., Schneemann, A., Hoffmann, J.A., and Imler, J.L. (2006). Essential function *in vivo* for Dicer-2 in host defense against RNA viruses in drosophila. Nat. Immunol. 7, 590–597.

Gatignol, A., Buckler-White, A., Berkhout, B., and Jeang, K.T. (1991). Characterization of a human TAR RNA-binding protein that activates the HIV-1 LTR. Science *251*, 1597–1600.

Ge, Q., Filip, L., Bai, A., Nguyen, T., Eisen, H.N., and Chen, J. (2004). Inhibition of influenza virus production in virus-infected mice by RNA interference. Proc. Natl. Acad. Sci. U.S.A. *101*, 8676–8681.

Geisbert, T.W., Hensley, L.E., Kagan, E., Yu, E.Z., Geisbert, J.B., Daddario-DiCaprio, K., Fritz, E.A., Jahrling, P.B., McClintock, K., Phelps, J.R., Lee, A.C., Judge, A., Jeffs, L.B., and Maclachlan, I. (2006). Postexposure protection of guinea pigs against a lethal ebola virus challenge is conferred by RNA interference. J Infect. Dis. *193*, 1650–1657.

Girard, A., and Hannon, G.J. (2008). Conserved themes in small-RNA-mediated transposon control. Trends Cell Biol. *18*, 136–148.

Gottwein, E., and Cullen, B.R. (2008). Viral and cellular microRNAs as determinants of viral pathogenesis and immunity. Cell Host Microbe 3, 375–387.

Grimm, D., and Kay, M.A. (2007). Combinatorial RNAi: a winning strategy for the race against evolving targets? Mol Ther. *15*, 878–888.

Grishok, A., Pasquinelli, A.E., Conte, D., Li, N., Parrish, S., Ha, I., Baillie, D.L., Fire, A., Ruvkun, G., and Mello, C.C. (2001). Genes and mechanisms related to RNA interference regulate expression of the small temporal RNAs that control C. elegans developmental timing. Cell *106*, 23–34.

Guo, Z., Chen, L.M., Zeng, H., Gomez, J.A., Plowden, J., Fujita, T., Katz, J.M., Donis, R.O., and Sambhara, S. (2007). NS1 protein of influenza A virus inhibits the function of intracytoplasmic pathogen sensor, RIG-I. Am J Respir. Cell Mol Biol *36*, 263–269.

Haasnoot, J., de Vries, W., Geutjes, E.J., Prins, M., de Haan, P., and Berkhout, B. (2007a). The Ebola virus VP35 protein is a suppressor of RNA silencing. PLoS Pathog. *3*, e86.

Haasnoot, J., Westerhout, E.M., and Berkhout, B. (2007b). RNA interference against viruses: strike and counterstrike. Nat. Biotechnol. *25*, 1435–1443.

Han, J., Lee, Y., Yeom, K.H., Kim, Y.K., Jin, H., and Kim, V.N. (2004). The Drosha–DGCR8 complex in primary microRNA processing. Genes Dev. *18*, 3016–3027.

Hariharan, M., Scaria, V., Pillai, B., and Brahmachari, S.K. (2005). Targets for human encoded microRNAs in HIV genes. Biochem Biophys. Res. Commun. *337*, 1214–1218.

Hartman, A.L., Bird, B.H., Towner, J.S., Antoniadou, Z.A., Zaki, S.R., and Nichol, S.T. (2008). Inhibition of IRF-3 activation by VP35 is critical for the high level of virulence of ebola virus. J. Virol. *82*, 2699–2704.

Hartman, A.L., Towner, J.S., and Nichol, S.T. (2004). A C-terminal basic amino acid motif of Zaire ebolavirus VP35 is essential for type I interferon antagonism and displays high identity with the RNA-binding domain of another interferon antagonist, the NS1 protein of influenza A virus. Virol. *328*, 177–184.

He, L., Thomson, J.M., Hemann, M.T., Hernando-Monge, E., Mu, D., Goodson, S., Powers, S., Cordon-Cardo, C., Lowe, S.W., Hannon, G.J., and Hammond, S.M. (2005). A microRNA polycistron as a potential human oncogene. Nature *435*, 828–833.

He, Y., and Katze, M.G. (2002). To interfere and to anti-interfere: the interplay between hepatitis C virus and interferon. Viral Immunol. *15*, 95–119.

Hemmes, H., Lakatos, L., Goldbach, R., Burgyan, J., and Prins, M. (2007). The NS3 protein of Rice hoja blanca tenuivirus suppresses RNA silencing in plant and insect hosts by efficiently binding both siRNAs and miRNAs. RNA *13*, 1079–1089.

Henke, J.I., Goergen, D., Zheng, J., Song, Y., Schuttler, C.G., Fehr, C., Junemann, C., and Niepmann, M. (2008). microRNA-122 stimulates translation of hepatitis C virus RNA. EMBO J. *27*, 3300–3310.

Hirai, K., Kubota, K., Mochizuki, T., Tsuda, S., and Meshi, T. (2008). Antiviral RNA silencing is restricted to the marginal region of the dark green tissue in the mosaic leaves of tomato mosaic virus-infected tobacco plants. J. Virol. *82*, 3250–3260.

Houzet, L., Yeung, M.L., de Lame, V., Desai, D., Smith, S.M., and Jeang, K.T. (2008). MicroRNA profile changes in human immunodeficiency virus type 1 (HIV-1) seropositive individuals. Retrovirology *5*, 118.

Huang, J., Wang, F., Argyris, E., Chen, K., Liang, Z., Tian, H., Huang, W., Squires, K., Verlinghieri, G., and Zhang, H. (2007). Cellular microRNAs contribute to HIV-1 latency in resting primary CD4(+) T lymphocytes. Nat. Med. *13*, 1241–1247.

Hutvagner, G., and Zamore, P.D. (2002). A microRNA in a multiple-turnover RNAi enzyme complex. Science *297*, 2056–2060.

Jackson, D., Hossain, M.J., Hickman, D., Perez, D.R., and Lamb, R.A. (2008). A new influenza virus virulence determinant: the NS1 protein four C-terminal residues modulate pathogenicity. Proc. Natl. Acad. Sci. U.S.A. *105*, 4381–4386.

Ji, J., Glaser, A., Wernli, M., Berke, J.M., Moradpour, D., and Erb, P. (2008). Suppression of short interfering RNA-mediated gene silencing by the structural proteins of hepatitis C virus. J. Gen. Virol. *89*, 2761–2766.

Jopling, C.L., Schütz, S., and Sarnow, P. (2008). Position-dependent function for a tandem microRNA miR-122-binding site located in the hepatitis C virus RNA genome. Cell Host. Microbe *4*, 77–85.

Jopling, C.L., Yi, M., Lancaster, A.M., Lemon, S.M., and Sarnow, P. (2005). Modulation of hepatitis C virus RNA abundance by a liver-specific microRNA. Science *309*, 1577–1581.

Kasschau, K.D., Xie, Z., Allen, E., Llave, C., Chapman, E.J., Krizan, K.A., and Carrington, J.C. (2003). P1/HC-Pro, a viral suppressor of RNA silencing, interferes with *Arabidopsis* development and miRNA function. Dev. Cell *4*, 205–217.

Kato, H., Takeuchi, O., Sato, S., Yoneyama, M., Yamamoto, M., Matsui, K., Uematsu, S., Jung, A., Kawai, T., Ishii, K.J., Yamaguchi, O., Otsu, K., Tsujimura, T., Koh, C.S., Reis e Sousa, Matsuura, Y., Fujita, T., and Akira, S. (2006). Differential roles of MDA5 and RIG-I helicases in the recognition of RNA viruses. Nature *441*, 101–105.

Katze, M.G., He, Y., and Gale, M., Jr. (2002). Viruses and interferon: a fight for supremacy. Nat. Rev. Immunol. *2*, 675–687.

Kazazian, H.H., Jr. (2004). Mobile elements: drivers of genome evolution. Science *303*, 1626–1632.

Kelly, E.J., Hadac, E.M., Greiner, S., and Russell, S.J. (2008). Engineering microRNA responsiveness to decrease virus pathogenicity. Nat. Med. *14*, 1278–1283.

Kim, V.N., Han, J., and Siomi, M.C. (2009). Biogenesis of small RNAs in animals. Nat. Rev. Mol. Celliol. *10*, 126–139.

Klase, Z., Kale, P., Winograd, R., Gupta, M.V., Heydarian, M., Berro, R., McCaffrey, T., and Kashanchi, F. (2007). HIV-1 TAR element is processed by Dicer to yield a viral micro-RNA involved in chromatin remodeling of the viral LTR. BMC. Mol. Biol. *8*, 63.

Klase, Z., Winograd, R., Davis, J., Carpio, L., Hildreth, R., Heydarian, M., Fu, S., McCaffrey, T., Meiri, E., yash-Rashkovsky, M., Gilad, S., Bentwich, Z., and Kashanchi, F. (2009). HIV-1 TAR miRNA protects against apoptosis by altering cellular gene expression. Retrovirology *6*, 18.

Kok, K.H., and Jin, D.Y. (2006). Influenza A virus NS1 protein does not suppress RNA interference in mammalian cells. J. Gen. Virol. *87*, 2639–2644.

Kok, K.H., Ng, M.H., Ching, Y.P., and Jin, D.Y. (2007). Human TRBP and PACT directly interact with each other and associate with dicer to facilitate the production of small interfering RNA. J. Biol. Chem. *282*, 17649–17657.

Krek, A., Grun, D., Poy, M.N., Wolf, R., Rosenberg, L., Epstein, E.J., MacMenamin, P., da Piedade, I., Gunsalus, K.C., Stoffel, M., and Rajewsky, N. (2005). Combinatorial microRNA target predictions. Nat. Genet. *37*, 495–500.

Lakatos, L., Csorba, T., Pantaleo, V., Chapman, E.J., Carrington, J.C., Liu, Y.P., Dolja, V.V., Calvino, L.F., Lopez-Moya, J.J., and Burgyan, J. (2006). Small RNA binding is a common strategy to suppress RNA silencing by several viral suppressors. EMBO J. *25*, 2768–2780.

Lantermann, M., Schwantes, A., Sliva, K., Sutter, G., and Schnierle, B.S. (2007). Vaccinia virus double-stranded RNA-binding protein E3 does not interfere with siRNA-mediated gene silencing in mammalian cells. Virus Res. *126*, 1–8.

Lecellier, C.H., Dunoyer, P., Arar, K., Lehmann-Che, J., Eyquem, S., Himber, C., Saib, A., and Voinnet, O. (2005). A cellular microRNA mediates antiviral defense in human cells. Science *308*, 557–560.

Lee, Y., Kim, M., Han, J., Yeom, K.H., Lee, S., Baek, S.H., and Kim, V.N. (2004). MicroRNA genes are transcribed by RNA polymerase II. EMBO J. *23*, 4051–4060.

Li, H., Li, W.X., and Ding, S.W. (2002). Induction and suppression of RNA silencing by an animal virus. Science *296*, 1319–1321.

Li, W.X., and Ding, S.W. (2001). Viral suppressors of RNA silencing. Curr. Opin. Biotechnol. *12*, 150–154.

Li, W.X., Li, H., Lu, R., Li, F., Dus, M., Atkinson, P., Brydon, E.W., Johnson, K.L., Garcia-Sastre, A., Ball, L.A., Palese, P., and Ding, S.W. (2004). Interferon antagonist proteins of influenza and vaccinia viruses are suppressors of RNA silencing. Proc. Natl. Acad. Sci. U.S.A. *101*, 1350–1355.

Li, Z., Jiang, Y., Jiao, P., Wang, A., Zhao, F., Tian, G., Wang, X., Yu, K., Bu, Z., and Chen, H. (2006). The NS1 gene contributes to the virulence of H5N1 avian influenza viruses. J. Virol. *80*, 11115–11123.

Lichner, Z., Silhavy, D., and Burgyan, J. (2003). Double-stranded RNA-binding proteins could suppress RNA interference-mediated antiviral defences. J. Gen. Virol. *84*, 975–980.

Lin, J., and Cullen, B.R. (2007). Analysis of the interaction of primate retroviruses with the human RNA interference machinery. J. Virol. *81*, 12218–12226.

Liu, Y.P., and Berkhout, B. (2008). Combinatorial RNAi strategies against HIV-1 and other escape-prone viruses. Int. J. BioSci. Technol. *1*, 1–10.

Liu, Y.P., Haasnoot, J., and Berkhout, B. (2007). Design of extended short hairpin RNAs for HIV-1 inhibition. Nucleic Acids Res. *35*, 5683–5693.

Lu, J., Getz, G., Miska, E.A., Alvarez-Saavedra, E., Lamb, J., Peck, D., Sweet-Cordero, A., Ebert, B.L., Mak, R.H., Ferrando, A.A., Downing, J.R., Jacks, T., Horvitz, H.R., and Golub, T.R. (2005). MicroRNA expression profiles classify human cancers. Nature *435*, 834–838.

Lu, S., and Cullen, B.R. (2004). Adenovirus VA1 non-coding RNA can inhibit small interfering RNA and microRNA biogenesis. J. Virol. *78*, 12868–12876.

Lu, Y., Wambach, M., Katze, M.G., and Krug, R.M. (1995). Binding of the influenza virus NS1 protein to double-stranded RNA inhibits the activation of the protein kinase that phosphorylates the elF-2 translation initiation factor. Virology *214*, 222–228.

Ludwig, S., Wang, X., Ehrhardt, C., Zheng, H., Donelan, N., Planz, O., Pleschka, S., Garcia-Sastre, A., Heins, G., and Wolff, T. (2002). The influenza A virus NS1 protein inhibits activation of Jun N-terminal kinase and AP-1 transcription factors. J. Virol. *76*, 11166–11171.

Marques, J.T., Devosse, T., Wang, D., Zamanian-Daryoush, M., Serbinowski, P., Hartmann, R., Fujita, T., Behlke, M.A., and Williams, B.R. (2006). A structural basis for discriminating between self and nonself double-stranded RNAs in mammalian cells. Nat. Biotechnol. *24*, 559–565.

Mibayashi, M., Martinez-Sobrido, L., Loo, Y.M., Cardenas, W.B., Gale, M., Jr., and Garcia-Sastre, A. (2007). Inhibition of retinoic acid-inducible gene I-mediated induction of beta interferon by the NS1 protein of influenza A virus. J. Virol. *81*, 514–524.

Min, J.Y., and Krug, R.M. (2006). The primary function of RNA binding by the influenza A virus NS1 protein in infected cells: Inhibiting the 2′–5′ oligo(A) synthetase/RNase L pathway. Proc. Natl. Acad. Sci. U.S.A. *103*, 7100–7105.

Molnar, A., Csorba, T., Lakatos, L., Varallyay, E., Lacomme, C., and Burgyan, J. (2005). Plant virus-derived small interfering RNAs originate predominantly from highly structured single-stranded viral RNAs. J. Virol. *79*, 7812–7818.

Murakami, Y., Aly, H.H., Tajima, A., Inoue, I., and Shimotohno, K. (2008). Regulation of the hepatitis C virus genome replication by miR-199a. J. Hepat. *50*, 453–460.

Nilsen, T.W. (2007). Mechanisms of microRNA-mediated gene regulation in animal cells. Trends Genet. *23*, 243–249.

Otsuka, M., Jing, Q., Georgel, P., New, L., Chen, J., Mols, J., Kang, Y.J., Jiang, Z., Du, X., Cook, R., Das, S.C., Pattnaik, A.K., Beutler, B., and Han, J. (2007). Hypersusceptibility to vesicular stomatitis virus infection in Dicer1-deficient mice is due to impaired miR24 and miR93 expression. Immunity *27*, 123–134.

Ouellet, D.L., Plante, I., Landry, P., Barat, C., Janelle, M.E., Flamand, L., Tremblay, M.J., and Provost, P. (2008). Identification of functional microRNAs released through asymmetrical processing of HIV-1 TAR element. Nucleic Acids Res. *36*, 2353–2365.

Parameswaran, P., Fire, A., Jalili, R., Gharizadeh, B., Ronaghi, M., Lu, R., Ding, S.W., Burgon, T., Jackson, W., Kirkegaard, K., *et al.* (2008). Investigating the interplay between viral replication and the RNAi macinery in vertebrates & invertebrates. In the abstract book of the Keystone Symposium on RNAi, MicroRNA, and Non-coding RNA (Silverstone, CO: Keystone Symposia), , poster abstract number 358, page 127.

Park, H., Davies, M.V., Langland, J.O., Chang, H.W., Nam, Y.S., Tartaglia, J., Paoletti, E., Jacobs, B.L., Kaufman, R.J., and Venkatesan, S. (1994). TAR RNA-binding protein is an inhibitor of the interferon-induced protein kinase PKR. Proc. Natl. Acad. Sci. U.S.A. *91*, 4713–4717.

Pedersen, I.M., Cheng, G., Wieland, S., Volinia, S., Croce, C.M., Chisari, F.V., and David, M. (2007). Interferon modulation of cellular microRNAs as an antiviral mechanism. Nature *449*, 919–922.

Pfeffer, S., Zavolan, M., Grasser, F.A., Chien, M., Russo, J.J., Ju, J., John, B., Enright, A.J., Marks, D., Sander, C., and Tuschl, T. (2004). Identification of virus-encoded microRNAs. Science *304*, 734–736.

Poluri, A., and Sutton, R.E. (2007). Titers of HIV-based vectors encoding shRNAs are reduced by a Dicer-dependent mechanism. Mol Ther. *16*, 378–386.

Qian, S., Zhong, X., Yu, L., Ding, B., de Haan, P., and Boris-Lawrie, K. (2009). HIV-1 Tat RNA silencing suppressor activity is conserved across kingdoms and counteracts translational repression of HIV-1. Proc. Natl. Acad. Sci. U.S.A. *106*, 605–610.

Rana, T.M. (2007). Illuminating the silence: understanding the structure and function of small RNAs. Nat. Rev. Mol. Cell Biol. *8*, 23–36.

Randall, G., Panis, M., Cooper, J.D., Tellinghuisen, T.L., Sukhodolets, K.E., Pfeffer, S., Landthaler, M., Landgraf, P., Kan, S., Lindenbach, B.D., Chien, M., Weir, D.B., Russo, J.J., Ju, J., Brownstein, M.J., Sheridan, R., Sander, C., Zavolan, M., Tuschl, T., and Rice, C.M. (2007). Cellular cofactors affecting hepatitis C virus infection and replication. Proc. Natl. Acad. Sci. U.S.A. *104*, 12884–12889.

Richt, J.A., Lekcharoensuk, P., Lager, K.M., Vincent, A.L., Loiacono, C.M., Janke, B.H., Wu, W.H., Yoon, K.J., Webby, R.J., Solorzano, A., and Garcia-Sastre, A. (2006). Vaccination of pigs against swine influenza viruses by using an NS1-truncated modified live-virus vaccine. J. Virol. *80*, 11009–11018.

Rivas, C., Gil, J., Melkova, Z., Esteban, M., and Diaz-Guerra, M. (1998). Vaccinia virus E3L protein is an inhibitor of the interferon (IFN)-induced 2-5A synthetase enzyme. Virol. *243*, 406–414.

Sano, M., Kato, Y., and Taira, K. (2006). Sequence-specific interference by small RNAs derived from adenovirus VAI RNA. FEBS Lett. *580*, 1553–1564.

Schnettler, E., de Vries, W., Hemmes, H., Haasnoot, J., Kormelink, R., Goldbach, R., and Berkhout, B. (2009). The NS3 protein of rice hoja blanca virus complements the RNAi suppressor function of HIV-1 Tat. EMBO Rep. *10*, 258–263.

Schnettler, E., Hemmes, H., Goldbach, R., and Prins, M. (2008). The NS3 protein of rice hoja blanca virus suppresses RNA silencing in mammalian cells. J. Gen. Virol. *89*, 336–340.

Scholthof, H.B. (2006). The Tombusvirus-encoded P19: from irrelevance to elegance. Nat. Rev. Microbiol. *4*, 405–411.

Schramke, V., and Allshire, R. (2004). Those interfering little RNAs! Silencing and eliminating chromatin. Curr. Opin. Genet. Dev. *14*, 174–180.

Segers, G.C., van, W.R., Zhang, X., Hong, Y., and Nuss, D.L. (2006). Hypovirus papain-like protease p29 suppresses RNA silencing in the natural fungal host and in a heterologous plant system. Eukaryot. Cell *5*, 896–904.

Segers, G.C., Zhang, X., Deng, F., Sun, Q., and Nuss, D.L. (2007). Evidence that RNA silencing functions as an antiviral defense mechanism in fungi. Proc. Natl. Acad. Sci. U.S.A. *104*, 12902–12906.

SenGupta, D.N., Berkhout, B., Gatignol, A., Zhou, A.M., and Silverman, R.H. (1990). Direct evidence for translational regulation by leader RNA and Tat protein of human immunodeficiency virus type 1. Proc. Natl. Acad. Sci. U.S.A. *87*, 7492–7496.

Seo, S.H., Hoffmann, E., and Webster, R.G. (2002). Lethal H5N1 influenza viruses escape host anti-viral cytokine responses. Nat. Med *8*, 950–954.

Shiboleth, Y.M., Haronsky, E., Leibman, D., Arazi, T., Wassenegger, M., Whitham, S.A., Gaba, V., and Gal-On, A. (2007). The conserved FRNK box in HC-Pro, a plant viral suppressor of gene silencing, is required for small RNA binding and mediates symptom development. J. Virol. *81*, 13135–13148.

Shors, T., Kibler, K.V., Perkins, K.B., Seidler-Wulff, R., Banaszak, M.P., and Jacobs, B.L. (1997). Complementation of vaccinia virus deleted of the E3L gene by mutants of E3L. Virol. *239*, 269–276.

Sijen, T., and Plasterk, R.H. (2003). Transposon silencing in the Caenorhabditis elegans germ line by natural RNAi. Nature *426*, 310–314.

Soldan, S.S., Plassmeyer, M.L., Matukonis, M.K., and Gonzalez-Scarano, F. (2005). La Crosse virus nonstructural protein NSs counteracts the effects of short interfering RNA. J. Virol. *79*, 234–244.

Sullivan, C.S., and Ganem, D. (2005). A virus-encoded inhibitor that blocks RNA interference in mammalian cells. J. Virol. *79*, 7371–7379.

Tam, O.H., Aravin, A.A., Stein, P., Girard, A., Murchison, E.P., Cheloufi, S., Hodges, E., Anger, M., Sachidanandam, R., Schultz, R.M., and Hannon, G.J. (2008). Pseudogene-derived small interfering RNAs regulate gene expression in mouse oocytes. Nature *453*, 534–538.

Ter Brake, O., 't Hooft, K., Liu, Y.P., Centlivre, M., von Eije, K.J., and Berkhout, B. (2008). Lentiviral vector design for multiple shRNA expression and durable HIV-1 inhibition. Mol. Ther. *16*, 557–564.

Ter Brake, O., Konstantinova, P., Ceylan, M., and Berkhout, B. (2006). Silencing of HIV-1 with RNA interference: a multiple shRNA approach. Mol. Ther. *14*, 883–892.

Tompkins, S.M., Lo, C.Y., Tumpey, T.M., and Epstein, S.L. (2004). Protection against lethal influenza virus challenge by RNA interference *in vivo*. Proc. Natl. Acad. Sci. U.S.A. *101*, 8682–8686.

Triboulet, R., Mari, B., Lin, Y.L., Chable-Bessia, C., Bennasser, Y., Lebrigand, K., Cardinaud, B., Maurin, T., Barbry, P., Baillat, V., Reynes, J., Corbeau, P., Jeang, K.T., and Benkirane, M. (2007). Suppression of microRNA-silencing pathway by HIV-1 during virus replication. Science *315*, 1579–1582.

van Rij, R.P., Saleh, M.C., Berry, B., Foo, C., Houk, A., Antoniewski, C., and Andino, R. (2006). The RNA silencing endonuclease Argonaute 2 mediates specific antiviral immunity in *Drosophila melanogaster*. Genes Dev. *20*, 2985–2995.

Vignuzzi, M., Wendt, E., and Andino, R. (2008). Engineering attenuated virus vaccines by controlling replication fidelity. Nat. Med. *14*, 154–161.

Voinnet, O. (2001). RNA silencing as a plant immune system against viruses. Trends Genet. *17*, 449–459.

von Eije, K.J., Ter Brake, O., and Berkhout, B. (2008). Human immunodeficiency virus type 1 escape is restricted when conserved genome sequences are targeted by RNA interference. J. Virol. *82*, 2895–2903.

von Eije, K.J., ter Brake, O., and Berkhout, B. (2009). Stringent testing identifies highly potent and escape-proof anti-HIV short hairpin RNAs. J. Gene Med. *11*, 459–467.

Wang, X., Li, M., Zheng, H., Muster, T., Palese, P., Beg, A.A., and Garcia-Sastre, A. (2000). Influenza A virus NS1 protein prevents activation of NF-kappaB and induction of alpha/beta interferon. J. Virol. *74*, 11566–11573.

Wang, X.H., Aliyari, R., Li, W.X., Li, H.W., Kim, K., Carthew, R., Atkinson, P., and Ding, W. (2006a). RNA interference directs innate immunity against viruses in adult *Drosophila*. Science *312*, 452–454.

Wang, Y., Kato, N., Jazag, A., Dharel, N., Otsuka, M., Taniguchi, H., Kawabe, T., and Omata, M. (2006b). Hepatitis C virus core protein is a potent inhibitor of RNA silencing-based antiviral response. Gastroenterology *130*, 883–892.

Wang, X., Ye, L., Hou, W., Zhou, Y., Wang, Y.J., Metzger, D.S., and Ho, W.Z. (2008). Cellular microRNA expression correlates with susceptibility of monocytes/macrophages to HIV-1 infection. Blood *113*, 671–674.

Watanabe, T., Totoki, Y., Toyoda, A., Kaneda, M., Kuramochi-Miyagawa, S., Obata, Y., Chiba, H., Kohara, Y., Kono, T., Nakano, T., Surani, M.A., Sakaki, Y., and Sasaki, H. (2008). Endogenous siRNAs from naturally formed dsRNAs regulate transcripts in mouse oocytes. Nature. *453*, 539–543.

Waterhouse, P.M., Wang, M.B., and Lough, T. (2001). Gene silencing as an adaptive defence against viruses. Nature *411*, 834–842.

Westerhout, E.M., Ooms, M., Vink, M., Das, A.T., and Berkhout, B. (2005). HIV-1 can escape from RNA interference by evolving an alternative structure in its RNA genome. Nucleic Acids Res. 33, 796–804.

Wilkins, C., Dishongh, R., Moore, S.C., Whitt, M.A., Chow, M., and Machaca, K. (2005). RNA interference is an antiviral defence mechanism in Caenorhabditis elegans. Nature *436*, 1044–1047.

Xiang, Y., Condit, R.C., Vijaysri, S., Jacobs, B., Williams, B.R., and Silverman, R.H. (2002). Blockade of interferon induction and action by the E3L double-stranded RNA binding proteins of vaccinia virus. J. Virol. *76*, 5251–5259.

Xu, N., Segerman, B., Zhou, X., and Akusjarvi, G. (2007). Adenovirus virus-associated RNAII-derived small RNAs are efficiently incorporated into the rna-induced silencing complex and associate with polyribosomes. J. Virol. *81*, 10540–10549.

Yang, N., and Kazazian, H.H., Jr. (2006). L1 retrotransposition is suppressed by endogenously encoded small interfering RNAs in human cultured cells. Nat. Struct. Mol Biol *13*, 763–771.

Ye, L., Wang, X., Wang, S., Wang, Y., Song, L., Hou, W., Zhou, L., Li, H., and Ho, W. (2008). CD56(+) T cells inhibit hepatitis C virus replication in human hepatocytes. Hepatology *49*, 753–762.

Yeung, M.L., Bennasser, Y., Myers, T.G., Jiang, G., Benkirane, M., and Jeang, K.T. (2005). Changes in microRNA expression profiles in HIV-1-transfected human cells. Retrovirology *2*, 81.

Ylosmaki, E., Hakkarainen, T., Hemminki, A., Visakorpi, T., Andino, R., and Saksela, K. (2008). Generation of a conditionally replicating adenovirus based on targeted destruction of E1A mRNA by a cell type-specific MicroRNA. J. Virol. *82*, 11009–11015.

# HCV and the Interaction with miR-122 in the Liver

5

Gabriele Fuchs and Cara T. Pager


## Abstract

MicroRNAs (miRNA) are endogenous non-coding RNA molecules, typically 21 nucleotides in length that can inhibit target gene expression when bound to 3′ untranslated regions in target mRNAs. In this review we summarize how miRNAs are generated, how miRNA–target interactions occur and the different outcomes of these interactions. We describe in detail the unusual interaction between the liver-specific miRNA miR-122 and hepatitis C virus (HCV) RNA, and review what is known about the mechanism by which miR-122 modulates HCV gene expression. Furthermore, we discuss the role of miR-122 in the liver, and how therapeutic approaches that alter miR-122 levels can influence HCV replication and hepatocarcinogenesis.


## Introduction

MicroRNAs (miRNAs) are a fast-growing class of small, non-coding RNAs (Kim *et al.*, 2009) that have been shown to be expressed in eukaryotes from fungi to mammals (Berezikov and Plasterk, 2005). Since their first discovery in *Caenorhabditis elegans* (Lee *et al.*, 1993; Olsen and Ambros, 1999; Wightman *et al.*, 1993), miRNAs have been shown to play a crucial regulatory role in gene expression and development (Alvarez-Garcia and Miska, 2005). To date, approximately 500 miRNA genes have been identified in the human genome, and these are predicted to regulate between 30% and 60% of protein-encoding genes in mammalian cells (Friedman *et al.*, 2009; Lewis *et al.*, 2003). Generally, miRNAs target complementary sequences in the 3′ untranslated region (UTR) of messenger RNAs (mRNAs) affecting mRNA stability, or causing translational repression of mRNA by a mechanism that is not yet fully understood (reviewed in Filipowicz *et al.*, 2008; Wu and Belasco, 2008).

## MicroRNA biogenesis

Mammalian miRNAs are encoded as single or clustered transcription units. They are transcribed as long primary transcripts by RNA polymerase II either from independent miRNA genes or as miRNA introns ('Mirtrons') within protein-coding or non-protein-coding transcripts (Rodriguez *et al.*, 2004; and reviewed in Filipowicz *et al.*, 2008, and Kim, 2005). These long primary transcripts (pri-miRNAs) (Lee *et al.*, 2002) fold into hairpin structures that are recognized and further processed by the nuclear endoribonuclease Drosha into 70 nucleotide (nt) precursor (pre-) miRNAs (Fig. 5.1) (Lee *et al.*, 2003). A first link between miRNA function and human diseases came from the identification of patients suffering from DiGeorge syndrome who presented a deletion on chromosome 22 encoding the protein DGCR8 (DiGeorge syndrome critical region protein 8) (Shiohama *et al.*, 2003). It has been shown that DGCR8 stabilizes the pri-miRNA and aids Drosha in the cleavage event (Gregory *et al.*, 2004; Han *et al.*, 2004; Landthaler *et al.*, 2004). Lack of DGCR8 may result in dysfunction of miRNA biogenesis, accumulation of unprocessed pri-miRNAs and presentation of the disease phenotype in patients carrying this deletion. Indeed, mice with a haploinsufficiency of the DGCR8 gene display an altered expression of miRNAs,

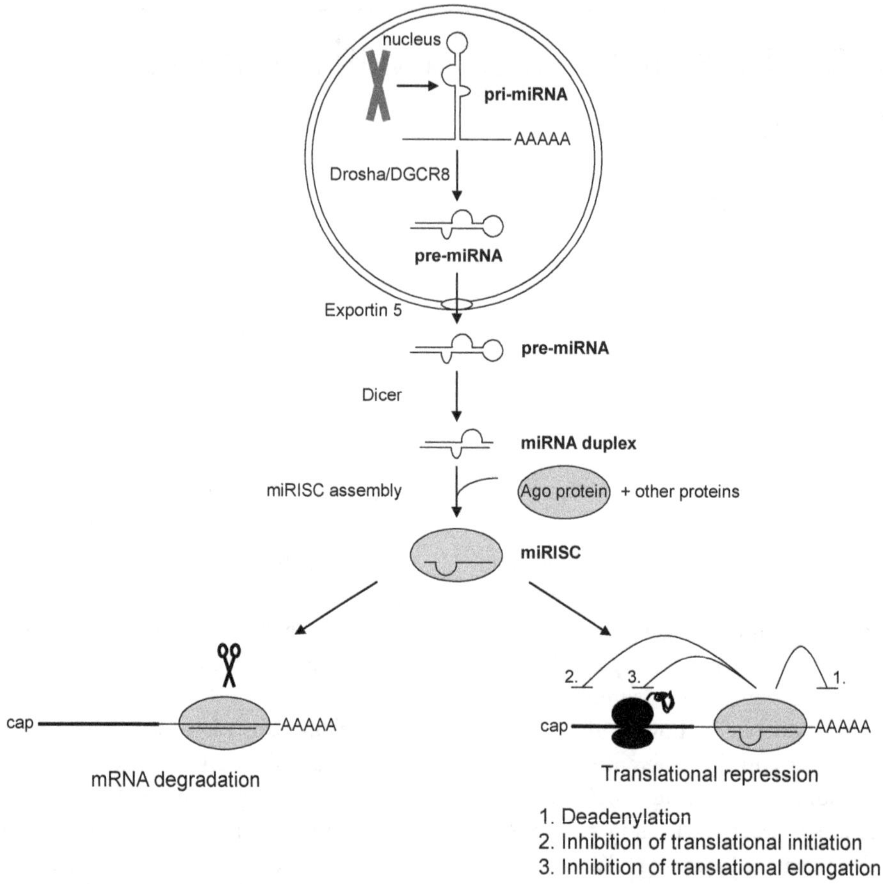


**Figure 5.1** Schematic of microRNA biogenesis. RNA polymerase II transcribes microRNA (miRNA) genes to generate a primary miRNA (pri-miRNA). In the nucleus, Drosha/DGCR cleaves the pri-miRNA to a 70 nt pre-miRNA. Exportin5 and its co-factor RanGTP facilitate transport of the pre-miRNA into the cytoplasm. In the cytoplasm Dicer, a RNase III enzyme, processes the pre-miRNA to a miRNA duplex. The miRNA duplex is incorporated into a ribonucleoprotein particle (RNP) containing numerous proteins including the Argonaute (AGO) proteins. This miRNA–RNP complex is known as the miRNA-containing RNA-induced silencing complex (miRISC). The miRISC separates the duplex, and retains the mature miRNA. Gene expression is modulated by the miRNA–miRISC complex by mRNA degradation and/or translational repression.

and behavioural and neuronal deficits (Stark *et al.*, 2008); possibly linking some of the symptoms found in human patients to the dysfunction of the miRNA biogenesis pathway.

After nuclear processing, pre-miRNAs are exported via exportin5 (Bohnsack *et al.*, 2004; Lund *et al.*, 2004; Yi *et al.*, 2003) and further processed in the cytoplasm. The cytoplasmic complex, containing the RNase III endonuclease Dicer and the TAR RNA binding protein (TRBP), removes the terminal loop from the pre-miRNA hairpin to generate 21-nt miRNA duplexes that contain 2-nt 3′-overhangs (Grishok *et al.*, 2001; Hutvagner *et al.*, 2001) (Fig. 5.1). One strand of the miRNA duplex (the guiding strand) is incorporated into a micro ribonucleoprotein particle (miRNP), while the other strand (the passenger strand) is degraded (Matranga *et al.*, 2005; Rand *et al.*, 2005). The miRNP complex, also known as the RNA-induced silencing complex (RISC), is composed of the miRNA and numerous proteins including Dicer and Argonaute (AGO) proteins (Sasaki *et al.*, 2003). Although all four AGO (AGO1–4) proteins are expressed in mammalian cells, only the PIWI domain of AGO2 has been shown to have endonucleolytic cleavage activity when mRNA and the miRNA are perfectly complementary (Liu *et al.*, 2004; Meister *et al.*,

2004b). Since this so-called 'slicer'-activity has not been found in AGO1, AGO3 and AGO4, it has been proposed that AGO1, AGO3 and AGO4 proteins might mediate alternative effects, such as translational repression in the presence of imperfect miRNA-target basepairing. Recent results in *Drosophila melanogaster* provide supporting evidence for this hypothesis (Iwasaki *et al.*, 2009; Okamura *et al.*, 2004). While AGO1 has been shown to be crucial for miRNA translational repression (Eulalio *et al.*, 2008), AGO2 mainly functions in RNAi (Peters and Meister, 2007).

Although perfect complementarity between an mRNA and a miRNA results in endonucleolytic cleavage and mRNA decay, most mRNA–miRNA interactions display imperfect complementarity. Nevertheless, perfect basepairing between the mRNA with nucleotides 2 to 8 of the miRNA, the so-called seed sequence (Lewis *et al.*, 2005) is required for miRNA-mediated translational repression. While the remaining nucleotides towards the 3'-end of the miRNAs are less conserved, insufficient basepairing within the seed region requires strong compensatory 3' basepairing (Brennecke *et al.*, 2005) for miRNA action.

Lastly, the originally described, exclusive binding of miRNPs to the 3'-UTR of mRNAs is probably obsolete, as more recent data suggests possible interactions with other regions of the mRNA such as the 5'-UTR and coding regions (Lytle *et al.*, 2007).

## Mechanism of control by miRNAs

mRNA–miRNA interactions have been shown to mostly facilitate translational repression and degradation of target mRNAs in cycling cells (Behm-Ansmant *et al.*, 2006). However positive effects, such as miRNA-mediated enhanced gene expression in growth-arrested cells (Vasudevan *et al.*, 2007), and a miRNA requirement for replication of Hepatitis C viral RNA have also been described (Jopling *et al.*, 2005). Therefore, the main challenge is to understand the molecular mechanism by which miRNAs can execute these diverse tasks.

Translation of mRNAs can be divided into three different stages namely: initiation, elongation and termination. It is a matter of dispute whether miRNAs interfere with the initiation process, at a post-initiation step, or both. Experiments using reporter assays to study miRNA function revealed that miRNAs only repress mRNAs containing a functional mRNA cap, but not mRNAs containing a non-functional cap or an internal ribosome entry site (IRES) (Humphreys *et al.*, 2005; Pillai *et al.*, 2005). This suggested that miRNAs might repress translation initiation (Humphreys *et al.*, 2005; Pillai *et al.*, 2005). In contrast, in the original experiments in *Caenorhabditis elegans* it was shown that the lin-4 miRNA targets (lin-14 and lin-28) remain associated with ribosomes, suggesting that miRNAs interfere with a step post initiation (Olsen and Ambros, 1999). This model in which miRNAs block elongation is further supported by two more recent studies using reporter mRNA assays in mammalian cells, which showed that the repressed mRNA is associated with active polysomes (Nottrott *et al.*, 2006; Petersen *et al.*, 2006). Furthermore, both Petersen *et al.* (2006) and Lytle *et al.* (2007) have reported that IRES-mediated translation of a downstream open reading frame (ORF) in a bicistronic reporter can be inhibited by miRNAs, strengthening the model of miRNA inhibition post-initiation. Lastly, it has been shown that miRNAs can trigger mRNA decay by shortening the poly(A) tail in a reaction that is independent from an endonucleolytic cleavage by AGO2 (Wu *et al.*, 2006). Initial rapid deadenylation of the mRNA targets by the CCR4:NOT deadenylase complex is followed by mRNA degradation that requires the decapping complex DCP1:DCP2 and GW182 (Behm-Ansmant *et al.*, 2006). Interestingly, very recent evidence from *Drosophila melanogaster* suggests that the specific AGO protein assembled into RISC defines the mechanism of translational repression. While AGO1-RISC has been shown to promote deadenylation of the mRNA target and to require the protein GW182 (Behm-Ansmant *et al.*, 2006; Eulalio *et al.*, 2008), translational repression by AGO2-RISC is independent of deadenylation and GW182 (Iwasaki *et al.*, 2009). Moreover, Iwasaki *et al.* (2009) showed that AGO1-RISC repressed translation of a $m^7$G-capped target as well as an A-capped, IRES-containing target, independent of a poly(A) tail. These results suggested that AGO1-RISC primarily induces deadenylation, but can also interfere with a step after cap recognition. In contrast, AGO2-RISC only repressed the $m^7G$-capped target, but not the A-capped target, which is consistent with a role in cap recognition (Iwasaki *et al.*, 2009). Indeed, mutational analysis revealed that AGO2-RISC and eIF4G have overlapping binding sites on eIF4E resulting in competition for eIF4E binding (Iwasaki *et al.*, 2009).

Despite the elaborate research that has been undertaken to understand the mechanism of posttranslational repression, a coherent picture has not yet emerged. Nonetheless, progress has been made suggesting that more than one mechanism of translational repression might exist, and that composition of RISC determines which mechanism is preferentially employed.

## MicroRNAs and viruses

In addition to regulating cellular gene targets, miRNAs also function to regulate viral infection. While viral encoded miRNAs and their control of viral and cellular genes is extensively discussed in Chapter 3, mammalian viruses such as primate foamy virus (Lecellier *et al.*, 2005), human immunodeficiency virus (Huang *et al.*, 2007; Triboulet *et al.*, 2007), vesicular stomatitis virus (Otsuka *et al.*, 2007), and hepatitis C virus (HCV) (Jopling *et al.*, 2005) are targets of cellular microRNAs. The interaction of HCV with the liver-specific miRNA miR-122 is however unusual, as miR-122 up-regulates viral gene expression. The identification and role of miR-122 in viral infection is described here in detail.

### Hepatitis C virus (HCV)

HCV is the leading cause of human liver disease, globally infecting an estimated 170 million individuals. Well known consequences of HCV infections are hepatitis, fibrosis, cirrhosis of the liver and hepatocellular carcinoma. Additionally, chronic HCV infections contribute most frequently to liver transplantations in the United States. Current interferon (IFN)-$\alpha$ and ribavirin therapies for chronic HCV infections are often ineffective, and neither alternative inhibitor treatments nor vaccines are currently available (Major *et al.*, 2001). Thus, the development of novel antiviral strategies is critical.

HCV is an enveloped RNA virus classified within the *Flaviviridae* family (Lindenbach and Rice, 2001). The ~9,600 nt single-stranded, positive-sense genome encodes a single open reading frame. Highly conserved untranslated regions at the 5′ and 3′ ends of the genome are essential for expression of the long ORF and genome replication. An early obligatory step of HCV infection is translation of the viral genome into a polyprotein. An IRES within the 5′-UTR directs cap-independent translation of the polyprotein (Wang *et al.*, 1993). The HCV polyprotein is subsequently cleaved by both cellular and viral proteases to release structural (core, E1 and E2) and non-structural (p7, NS2, NS3, NS4 and NS5) proteins (reviewed in (Lindenbach and Rice, 2005)). The non-structural proteins coordinate the virus life cycle. HCV genome replication occurs on cytoplasmic membranes (Egger *et al.*, 2002), although the exact mechanism is unknown. The replication complex consisting of the NS3 RNA helicase, the phosphoprotein NS5A and the RNA-dependent RNA polymerase NS5B, mediates RNA synthesis via a negative-strand RNA intermediate that is initiated by copying the 3′ end of the positive strand. The 3′-UTR in the HCV genome lacks a poly(A) tail but consists of a variable region, a poly(U/UC) tract, and a highly conserved three stem–loop 3′ X domain (Blight and Rice, 1997; Kolykhalov *et al.*, 1996; Tanaka *et al.*, 1995; Tanaka *et al.*, 1996). The poly(U/UC) tract and the 3′ X domain are essential for HCV replication (Friebe and Bartenschlager, 2002; Friebe *et al.*, 2005; Yi and Lemon, 2003a,b). Although the IRES is the prominent feature of the 5′ UTR, the first 125 nucleotides are also required for HCV replication (Friebe *et al.*, 2001). Recently, HCV gene expression was shown to require a specific interaction between the 5′-UTR and a liver-specific miRNA, miR-122 (Jopling *et al.*, 2005). The discovery and proposed outcomes of this novel cellular miRNA–viral RNA interaction are discussed in detail.

### HCV RNA is modulated by miR-122

Tissue specific cloning of mouse miRNAs identified miR-122 as a liver-specific miRNA, accounting for 72% of all cloned miRNAs (Lagos-Quintana *et al.*, 2001). Indeed levels of miR-122 quantified by RNase protection assay identified 66,000 miR-122 copies per average liver cell. Interestingly, 135,000 miR-122 copies/average cell are found in human primary hepatocytes; and liver-derived cell lines express ten-fold or fewer miR-122 copies per average cell. For example Huh7, a human hepatocellular carcinoma cell line, only expresses 16,000 copies/average cell, while human hepatoblastoma HepG2 cells express fewer than 140 copies/average cell (Chang *et al.*, 2004). That miR-122 is the predominant miRNA in the liver, and HCV replication is restricted to liver and liver-derived cell lines, raised the possibility that miR-122 may play a role in HCV infection and contribute to tissue tropism.

To determine whether HCV could indeed be targeted by miR-122, the HCV genome was inspected for potential miR-122 binding sites. Two predicted miR-122 seed match sequences were initially identified in the HCV RNA genome (Jopling *et al.*, 2005). These binding sites were located in the 5′ and 3′ UTRs of the viral genome (Fig. 5.2), and were found to be conserved across all six HCV genotypes. To establish whether miR-122 modulates HCV gene expression, Huh7 cells, stably expressing dicistronic HCV genotype 1b replicon RNA (HCV NNeo/C-5B replicon cells), were transfected with a miR-122 antisense 2′-O-methylated oligonucleotides (122–2′OMe). 2′-OMe oligonucleotides have a ribose ring, in which the 2′ oxygen of the hydroxyl group has been modified with a methyl group. This 2′-O-methyl modification renders the oligonucleotides less susceptible to alkaline hydrolysis and degradation by nucleases. As such, 2′-OMe oligonucleotides efficiently sequester targeted-miRNAs and render them non-functional. Sequestration and inactivation of miR-122 was confirmed by co-transfecting 5B replicon cells with a GFP reporter plasmid, harbouring a complementary miR-122 target sequence in the 3′ UTR. As expected, GFP reporter mRNA was degraded following transfection into 5B replicon cells, and levels of GFP reporter mRNA were rescued following co-transfection of plasmid and 122–2′OMe. Unexpectedly, the opposite was true for HCV replicon RNA. In the presence of miR-122, HCV RNA and viral core protein were abundant; while sequestration and inactivation of miR-122 significantly decreased the amount of HCV RNA. A similar result was observed following transfection of Huh7 cells with a replication

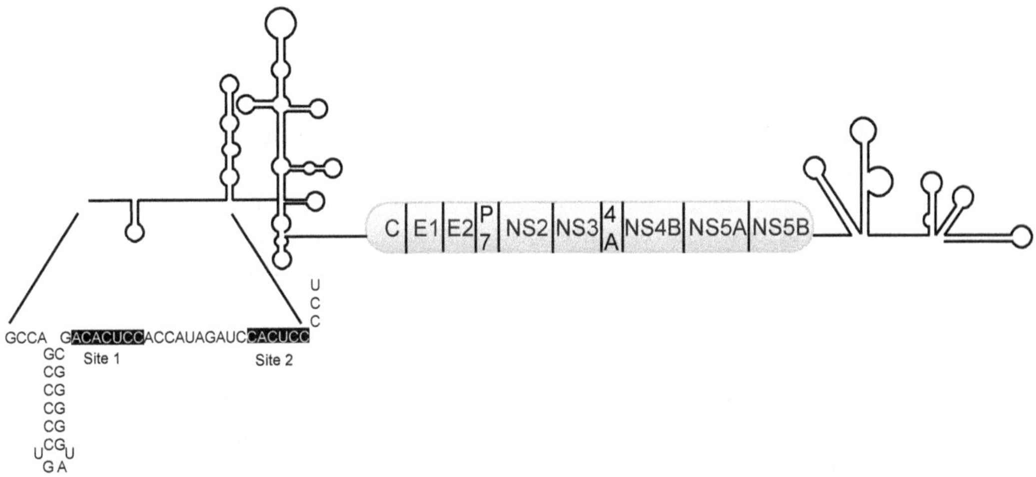


**Figure 5.2** Hepatitis C virus (HCV) genome and location of the miR-122 binding sites. The viral genome encodes a long polyprotein (indicated as a box) from one open reading frame. Both cellular and viral proteases process the polyprotein to structural proteins (Core [C], and envelope proteins E1 and E2) and non-structural (NS) proteins (p7, NS2, NS3, NS4 and NS5). Untranslated regions (UTR) abut the open reading frame at the 5′ and 3′ ends. An enlarged region, within the schematic, highlights the two miR-122 seed match sequences within the 5′-UTR of the viral genome.

competent HCV genotype 1a (H77c) RNA and 122–2′OMe, suggesting that maintenance of HCV RNA required functional miR-122. The dependence of HCV RNA on miR-122 may be via a direct interaction between miR-122 and the HCV RNA, or by miR-122 modulating a cellular mRNA. To establish whether miR-122 directly interacted with the HCV RNA, the miR-122 seed match sequences in the 5′ and 3′ UTRs of H77c were mutated, and mutant H77c RNA electroporated into Huh7 cells. A four nucleotide mutation within the 3′-UTR seed match sequence did not decrease of the level of HCV RNA. In contrast, single and double nucleotide mutations within the 5′-UTR seed match sequence significantly diminished the levels of HCV RNA. Yet 5′ UTR-mutated HCV RNA levels could be restored by transfection of ectopic miR-122 harbouring complementary mutations, indicating that miR-122 modulation of HCV RNA was the result of a direct interaction with the viral RNA.

Cellular miRNAs are thought to down-regulate target mRNAs either by modulating translation or promoting RNA degradation (Filipowicz *et al.*, 2008). Although miR-122 stimulated HCV RNA levels, the mechanism by which this occurs may similarly be by modulating translation of the viral RNA, stabilizing viral RNA or affecting viral replication. To establish whether miR-122 regulated HCV IRES-directed translation, viral replication and polyprotein translation needed to be uncoupled. HCV RNA is duplicated by a virally encoded RNA-dependent RNA polymerase. While deletion of GDD residues within the polymerase catalytic domain inhibits viral replication, translation of the polyprotein persists. Jopling *et al.* (2005) therefore transfected Huh7 cells with replication-competent or replication-deficient viral RNA, containing either a functional or non-functional miR-122 seed match sequence (Jopling *et al.*, 2005). Twenty-hours post-transfection no significant difference in the levels of core protein was detected, suggesting that viral RNA (containing or lacking a miR-122 binding site) was efficiently translated. Although miR-122 did not modulate translation of the HCV polyprotein, the possibility remained that mutations in the 5′-UTR miR-122 seed match sequence altered HCV RNA stability. However, an assay specifically measuring the intracellular stability of HCV RNA established that viral RNA containing a mutated miR-122 seed match sequence exhibited a similar stability to a replication incompetent replicon RNA, confirming that disruption of miR–122 interaction with the HCV 5′-UTR did not destabilize the viral RNA, resulting in RNA decay and decreased viral expression. Since miR-122 did not enable translation, nor contribute to viral RNA stability; Jopling *et al.* (2005) proposed that miR-122 probably affected replication of the viral RNA (Jopling *et al.*, 2005).

Recently, Henke *et al.* (2008) used reporter plasmids, *in vitro* translational silencing assays and profile analysis of translating ribosomes to show that miR-122 enhanced translation of HCV RNA (Henke *et al.*, 2008). To determine whether miR-122 modulated translation of a full-length HCV RNA, miR-122 and the brain-specific miR-124 duplex RNAs were transfected into Huh7 cells. These cells were then electroporated with replication-competent and replication-deficient HCV RNAs. Translation efficiency was measured by assaying HCV core protein and viral RNA levels 4h post electroporation. HCV RNA levels were not affected by ectopic expression of miR-122 or control miR-124. In contrast, miR-122 stimulated translation of core protein from replication-competent viral RNA 1.5-fold. From these results Henke *et al.* (2008) concluded that miR-122 modulated translation of the HCV genome (Henke *et al.*, 2008). While the manipulation of viral RNA levels might be a result of miR-122 stimulating viral translation, the moderate increase in core antigen from input viral RNA incapable of viral replication, suggests such an effect occurs early in HCV infection as Jopling *et al.* (Jopling *et al.*, 2005) did not detect changes in core protein 20 h post-transfection.

The effect of miR-122 on HCV translation was further explored by examining translation of reporter plasmids containing the HCV UTRs at the respective 5′ and 3′ ends of the reporter gene (Henke *et al.*, 2008). Northern analysis of miR-122 levels demonstrated that human cervical carcinoma-derived HeLa cells do not express endogenous miR-122. However transfection of miR-122 duplex increased translation the luciferase reporter plasmid, while ectopic expression of control miR-124 duplex did not promote translation. The miR-122 effect could additionally be quenched by 122–2′OMe sequestration of transfected miR-122, suggesting that even in a non-hepatoma cell line, miR-122 could stimulate translation of reporter RNA (Henke *et al.*, 2008).

*In vitro* translation assays performed in rabbit reticulocyte lysate (RRL) faithfully recapitulate translation in the cell, and have been used to dissect different stages of translation. To decipher at which step in translation miR-122 might act to promote HCV IRES-directed translation of a luciferase plasmid, Henke *et al.* (2008) applied an *in vitro* miRNA-directed translational silencing assay first described by Wang *et al.* (2006). Although miR-122 in RRL was undetectable by Northern blot, increasing amounts of mature single-stranded miR-122 promoted translation of the HCV luciferase reporter plasmid, while miR-122 duplex had no significant effect. As with HCV luciferase reporter translation in HeLa cells, specificity was assumed from assays with non-specific miR-124 mature and duplexed miRNA, and 2′-OMe sequestration of miR-122. Although Wang *et al.* (2006) report that the RRL *in vitro* system lacks the ability to load siRNA duplex into miRNPs (Wang *et al.*, 2006), the increased translation efficiency Henke *et al.* (2008) observed, may be the result of the mature miR-122 oligonucleotide interacting with the seed match sequence to stabilize the RNA structure and promote efficient ribosome binding (Henke *et al.*, 2008). As such, addition of a single RNA strand complementary to another region within the 5′-UTR of HCV, possibly to nucleotides between the two seed match sites, would address this caveat. An mRNA encoding a short 96 nt ORF, and flanked by 5′ and 3′ HCV UTRs, was used to determine whether miR-122 influences translation initiation of this HCV mRNA mimic in RRL. When $^{32}$P-labelled mRNA and initiation complexes were examined by sucrose gradient analysis, in the presence and absence of miR-122, an enhanced 48S peak was observed in the presence of miR-122 (Henke *et al.*, 2008). Furthermore this 48S peak formed within 2 min, suggesting the mature miR-122 strand facilitated the association of the small ribosomal subunit with the short mRNA, and subsequent formation of polysomes. Although the elevated 48S peak was less obvious when the full length HCV genome was incubated in RRL, Henke *et al.* (2008) concluded that miR-122 enhances the association of ribosomes with the viral RNA to stimulate translation of the polyprotein.

## Interaction of miR-122 with HCV RNA

MiRNPs typically recognize sequences within the 3′-UTR of target mRNAs. The finding by Jopling *et al.* (2005) that miR-122 recognized seed match sequences in the 5′-UTR, and promoted gene up-regulation was unprecedented. Numerous questions therefore arise about the nature of interaction of miR-122 with the HCV genome. For example: does miR-122 modulate HCV RNA via interaction of a specialized miRNP complex with the miR-122 binding site? Are the viral sequences surrounding the miR-122 binding site, and the location of the miR-122 seed match sequence in the 5′-UTR important? Or does miR-122 exhibit a unique ability to stimulate gene expression?

To characterize the position of the miR-122 binding site within the HCV RNA genome, Jopling *et al.* (2008) created two luciferase reporter plasmids containing HCV RNA 5′-UTR sequences, encompassing the miR-122 binding site, either upstream of a luciferase gene or within the 3′-UTR of the reporter plasmid (Jopling *et al.*, 2008). While the miR-122 binding site positioned at the 5′-end of the reporter RNA did not influence luciferase translation directed from the HCV IRES, introduction of the miR-122 binding site into the 3′-UTR of the reporter mRNA significantly down-regulated luciferase gene expression. Thus, the position of the miR-122 binding site guides miR-122 directed regulation of reporter gene expression.

The number of miRNA binding sites within the mRNA 3′-UTR has been shown to influence the outcome of miRNA-directed gene expression. In addition to the miR-122 seed match sequence at nt 22–28, a putative miR-122 seed match sequence at nt 38–43 was identified (Fig. 5.2). In the context of the reporter plasmid this second miR-122 seed match sequence within the HCV nt 1–45 was found to be functional (Jopling *et al.*, 2008). To investigate the significance of this second miR-122 binding site within the HCV RNA genome, two nucleotides within the seed match sequence of site 1, site 2 or both were mutated, and RNA abundance examined following electroporation of wild-type (wt) or mutant viral RNA (Jopling *et al.*, 2008). HCV RNA levels were dramatically reduced when seed match site 2 was mutated, and could be partially rescued upon ectopic expression of the compensatory mutant miR-122. When both seed match sites were mutated, the amount of HCV RNA could be restored to wt levels with ectopic expression of compensatory mutant miR-122 (Jopling *et al.*, 2008). Thus, miR-122 interacts at both seed match sites to modulate HCV RNA levels. Additionally, introduction of unique changes within each seed match sequence and complementation with mutant miR-122, either individually or in combination, demonstrated that within one HCV RNA molecule miR-122 interacts at both sites to modulate viral RNA levels. Whether such an interaction occurs concurrently or sequentially remains to be determined.

Fourteen nucleotides separate the start of seed match site 1 with the start of site 2. Spacing between adjacent miRNA binding sites contributes to efficient regulation of gene expression by miRNAs (Saetrom *et al.*, 2007). Substitution or deletion of two nucleotides immediately upstream of the first seed match site did not alter miRNA-directed repression of the luciferase reporter mRNA. In contrast, HCV RNA levels were dramatically reduced by nucleotide substitution, deletion and insertion mutations within the spacer sequences separating the miR-122 binding sites within the 5′-UTR (Jopling *et al.*, 2008). The first 125 nucleotides within the 5′-UTR regulate viral replication (Friebe *et al.*, 2001). Thus, nucleotide mutagenesis between sites 1 and 2 may alter RNA structure and/or signals essential for HCV RNA replication. While neither miRNA seed match sequences localized within 5′ UTRs nor stimulation of gene expression by a miRNA are unique to miR-122, this study shows that the ability of miR-122 to manipulate viral RNA levels does depend on two seed match sequences that are position and sequence-dependent within the HCV 5′-UTR (Jopling *et al.*, 2008).

## Oligonucleotide approaches to study miRNA targets

The discovery of miRNAs, and more specifically their role in regulation of gene expression, has directed the exploration of miRNAs as possible novel therapeutic components. Similar to gene therapy, one approach uses viral or non-viral vectors to express short hairpin RNA (shRNA). These shRNAs share features of miRNAs and have been shown to be functional in the RNAi pathway (McManus *et al.*, 2002; Paddison *et al.*, 2002; Yu *et al.*, 2002). Viral vectors have been shown to have undesirable side-effects such as strong immune and inflammatory responses, as well as the insertion of the viral genome into the host genome causing cell transformation and malignancy. Although non-viral vectors do not cause these side-effects, the vectors do display a general low toxicity. However, the main caveat of using shRNA is dosage control. It has been shown that intravenously administered AAV/shRNA vectors in mice strongly competed with cellular small RNAs for limiting cellular processing factors (Grimm *et al.*, 2006). Additionally, an activation of the IFN response was observed suggesting that this approach might cause a strong immune response with unclear consequences in long-term treatment (Witting *et al.*, 2008).

The second strategy explores the possibility of a direct administration of complementary nucleic acid oligonucleotides that are not as dependent on the RNase III enzymes, Drosha and Dicer. Although the potential of nucleic acid oligonucleotides as a tool in cell biology had been shown more than 30 years ago (Zamecnik and Stephenson, 1978), unreliable results have mostly restricted their use in biomedical applications. Progress has, however, been made by using modified oligonucleotides that display higher, and more specific binding to cellular nucleic acid targets (Meister *et al.*, 2004a).

In addition to the previously described and widely used 2′-OMe oligonucleotides, locked nucleic acids (LNAs), containing an O-methylene bridge between the 2′ and 4′ position of the ribose, have recently been synthesized (Koshkin *et al.*, 1998; Kumar *et al.*, 1998; Singh *et al.*, 1998). This bicyclic structure locks the ribose sugar in its 3′-*endo* structure, and increases the thermal stability of oligonucleotides containing LNAs. Lastly, peptide nucleic acids (PNAs), another oligonucleotide analogue with an uncharged peptide-like backbone, can easily be synthesized and display some properties that differ from those of LNAs. Specifically, the solubility of PNAs strongly depends on their sequence composition.

Also, although degradation of miRNAs targeted by PNAs has been observed, the mechanism of this degradation is unclear because PNAs have not been shown to activate RNase H (Knudsen and Nielsen, 1996).

Taken together, 2′-OMe, locked nucleic acid, and peptide nucleic acid oligonucleotides have the potential to be used in biomedical applications. In the following section we will summarize how different oligonucleotides have been used to identify cellular miRNA targets in mammalian cells. Furthermore, recent results from animal experiments strongly encourage using oligonucleotides for therapeutic approaches.

## Roles for miR-122 in the liver

LNA and PNA oligonucleotides have also been used to sequester the liver-specific miRNA miR-122 to identify its cellular targets. These studies have confirmed some miR-122 target genes predicted by computational analysis. Additionally, these results suggest roles for miR-122 in liver metabolism and hepatocarcinogenesis. Lastly, recent promising results using 2′-OMe and LNAs in rodents and non-human primates imply that oligonucleotides could potentially be used as a new therapeutic approach.

### Amino acid starvation and stress response

To identify putative miRNA target genes, computational tools based on the known rules for miRNA-mRNA recognition were developed (Lewis *et al.*, 2003). Chang *et al.* (2004) used these Lewis-based predictions and identified a putative miR-122 binding site in the 3′-UTR of the cationic amino acid transporter (CAT-1) mRNA (Chang *et al.*, 2004). Cationic amino acid transporters function in the bidirectional transport of positively charged amino acids such as lysine and arginine (Hatzoglou *et al.*, 2004). Since nutritional stress such as amino acid starvation results in increased CAT-1 mRNA stabilization and translation, it is likely that these events are regulated via a posttranscriptional mechanism (Fernandez *et al.*, 2003). To test if the CAT-1 mRNA is indeed a miR-122 target, Chang *et al.* (2004) fused 400 nt of the CAT-1 3′-UTR to the hepatitis delta small antigen (δAg-S) open reading frame. In Huh7 cells that express miR-122 at high levels, expression of the δAg-S was decreased. However no change in δAg-S levels was observed in 293T cells that do not significantly express miR-122. Yet, when miR-122 was expressed exogenously in 293T cells, repression of δAg-S was observed. Moreover, sequestering miR-122 with complementary 2′-OMe oligonucleotides resulted in derepression of δAg-S in Huh7 cells (Chang *et al.*, 2004). These results, together with the observation that CAT-1 and miR-122 expression are reciprocal during the liver development in mice, strongly suggest that expression of CAT-1 protein is regulated by miR-122 (Chang *et al.*, 2004). More evidence supporting that translation of the CAT-1 mRNA is indeed inhibited by miR-122 comes from experiments in which Huh7 cells were subjected to different stress conditions. Amino acid starvation resulted in the release of CAT-1 mRNA from cytoplasmic processing bodies into translationally active polysomes. This shift requires the AU-rich element containing protein HuR, and confirms not only CAT-1 as a miR-122 target *in vivo*, but also suggests a role for miR-122 in the cellular stress response (Bhattacharyya *et al.*, 2006).

### Tumorigenesis

Based on the early description of a genetic interaction between the proto-oncogene c-myc and miR-122 (at that time named hcr), and its importance for development of liver tumours in woodchucks, it was suggested that alterations in the expression pattern of miRNAs might contribute to cancer (Etiemble *et al.*, 1989; Moroy *et al.*, 1986, 1989). To test this hypothesis, rats were fed a diet deficient in methionine and devoid of folic acid and choline (FMD diet), which is know to cause hepatocellular carcinomas (HCC) (Ghoshal and Farber, 1984; Lombardi *et al.*, 1991; Pogribny *et al.*, 2004). Indeed, when miRNA expression was compared in hepatocellular carcinomas (HCC) from rats on a FMD diet with age-matched rats on a normal diet, miR-122 was found to be significantly down-regulated in HCCs (Kutay *et al.*, 2006). This trend was confirmed in human primary HCCs when compared to control tissues (Kutay *et al.*, 2006). An independent study using genome-wide miRNA microarray, Northern blot and real-time PCR analysis confirmed the down-regulation of miR-122a in HCCs and

HCC-derived cell lines (Gramantieri *et al.*, 2007). Using computational algorithms to predict miR-122 targets, the gene for cyclin G1 was identified as a putative miR-122 target (Gramantieri *et al.*, 2007). HEP3B and SNU449 cells that normally do not express miR-122 showed a reduction of cyclin G1 protein levels following transfection with miR-122 oligonucleotides. When a part of the 3′-UTR of the cyclin G1 gene was fused to the firefly luciferase ORF and cotransfected with miR-122, luciferase expression was found to be repressed. These results suggest that cyclin G1 is a likely miR-122 target. Interestingly, the inverse correlation between the expression of miR-122 and cyclin G1 protein in primary liver carcinomas found by Gramantieri *et al.* (2007) emphasize the crucial part of miR-122 in hepatocarcinogenesis (Gramantieri *et al.*, 2007).

### Cholesterol pathway

Another approach to identify miR-122 targets was employed by Krutzfeldt *et al.* (2005) and Esau *et al.* (2006) who used antisense oligonucleotides to silence miR-122. Krutzfeldt *et al.* (2005) used an antagomir (a cholesterol-conjugated 2′-OMe oligoribonucleotide complementary to miR-122) to identify novel miR-122 targets not predicted by computer analysis. Total liver RNA from mice treated with miR-122 or a mismatch control antagomir was analysed by cDNA microarray. Of the several hundred transcripts that were up-regulated at least 1.4-fold, some were confirmed by reverse-transcriptase-mediated polymerase chain reaction (RT-PCR) (Krutzfeldt *et al.*, 2005). Krutzfeldt *et al.* (2005) therefore identified a large number of new *in vivo* miRNA targets not predicted by computer algorithms, by a silencing approach combined with a cDNA microarray. Similar results were obtained by Esau *et al.* (2006) who used the computational analysis software TargetScan to predict six putative miR-122 targets. Following transfection of primary hepatocytes with 2′-O-methoxyethyl (2′-MOE) phosphorothioate-modified antisense oligonucleotides, mRNA levels of these putative miR-122 targets were measured. Five out of the six predicted targets were indeed found to be de-repressed after sequestration of miR-122 by antisense oligonucleotides. Regulation of CAT-1, glycogen synthase and aldolase A by miR-122 was confirmed in an independent study (Fabani and Gait, 2008). Additionally, GTF2b (transcription initiation factor IIB) was identified as a novel miR-122 targets in primary rat hepatocytes (Fabani and Gait, 2008). When mice were treated intraperitoneally with different doses of miR-122 antisense oligonucleotides four of the five mRNAs, identified as miR-122 targets in the previous cell culture experiments, were up-regulated in these mice (Esau *et al.*, 2006). Strikingly, both studies showed that hepatocytes from mice treated with miR-122 antisense oligonucleotides exhibited a significant decrease of total plasma cholesterol and triglyceride levels, while the rate of fatty acid oxidation was increased. Moreover, mice with a high-fat diet-induced liver steatosis showed improvement in histological analysis of liver samples following miR-122 inhibition (Esau *et al.*, 2006). Taken together, inhibition of miR-122 may be a novel therapeutic approach to reduce steatosis and obesity in humans.

Specifically, the pharmacodynamic study by Krutzfeldt *et al.* (2005) using a miR-122 antagomir was a step towards the utilization of antisense oligonucleotides as a therapeutic approach *in vivo*. Within 24 h after injection of the antagomir into the tail veins of mice, levels of miR-122 in the liver were undetectable by Northern blot analysis. This effect was neither due to the specific miRNA nor the organ, since the ubiquitously expressed miR-16 was effectively silenced with a miR-16 antagomir in all tissues tested with the exception of the brain, possibly due to the restricted access of substances through the blood–brain barrier (Krutzfeldt *et al.*, 2005). Up to 23 days after injection of the miR-122 antagomir, levels of miR-122 remained undetectable by Northern blot analysis, and mice treated with miR-122 antagomir showed low cholesterol levels for at least two weeks (Krutzfeldt *et al.*, 2005). Taken together, these results provided the first evidence that the issue of dosage control of shRNAs can be overcome by using miRNA antagomirs, and that miRNAs could serve as therapeutic targets.

## MicroRNA-based therapeutic approaches in the liver using small nucleic acids

Due to their high affinity binding and compatibility with the intracellular RNAi machinery,

LNAs have been proposed to have reduced off-target RNAi silencing effects (Braasch *et al.*, 2003; Elmen *et al.*, 2005; Wahlestedt *et al.*, 2000). Additionally, LNAs are substantially stable in blood serum and do not cause detectable toxic reactions in rat brain (Braasch *et al.*, 2003; Wahlestedt *et al.*, 2000). As such, LNAs have been considered advantageous to be used for *in vivo* RNAi therapy. Indeed, locked-nucleic acid-modified DNA oligonucleotides (LNA anti-miRs) tested in a luciferase reporter assay for miR-122 in cultured Huh7 liver cells, showed nearly 1.5 fold de-repression of the luciferase reporter. Moreover, high affinity LNA anti-miRs, with more than 50% LNA content and with melting temperatures above 80°C, proved to be highly efficient in inhibition of HCV replication in Huh7 cells containing the HCV-N replicon NNeo/C-5B when compared to a 2′-O-methyl oligonucleotide (Elmen *et al.*, 2008). However, major challenges that have limited the use of oligonucleotides as therapeutic agents are efficient *in vivo* delivery and toxicity of oligonucleotides. In contrast to the study by Krutzfeld *et al.* (2005), who used a cholesterol-conjugated 2′-OMe oligoribonucleotide, Elmen *et al.* (2008) used unconjugated, high-affinity miR-122 LNA-anti-miRs. When LNAs with different backbone modifications were tested in diet-induced obese mice, complete sequestration and heteroduplex formation of mature miR-122 with the high affinity miR-122 LNA-anti-miR containing an entire phosphorothioate backbone was observed (Elmen *et al.*, 2008). This suggests that the miR-122 LNA-anti-miR was taken up efficiently by the murine liver after three intraperitoneal injection. In agreement with previous results showing that sequestration of miR-122 results in a decrease in cholesterol levels (Esau *et al.*, 2006; Krutzfeldt *et al.*, 2005), total cholesterol levels in mice treated with the miR-122 LNA-anti-miR were decreased by approximately 30% without any signs of hepatotoxicity (Elmen *et al.*, 2008). These encouraging results, of miR-122 silencing experiments in rodents, led the way to explore the effect of a miR-122 LNA anti-miRs in non-human primates. Elmen *et al.* (2008) administered African green monkeys unconjugated, PBS-formulated miR-122 LNA anti-miR intravenously and examined its uptake into the primate hepatocytes, and monitored total cholesterol levels as a readout for efficient sequestration of miR-122 by the LNA-anti-miR. As previously observed in mice, total cholesterol levels were decreased in the primates in a dose-dependent manner. More importantly, neither acute nor subchronic toxicity to the liver, the kidneys, or the pancreas was observed by liver biopsies or clinical chemistry.

PNAs that have been proposed to act through a steric block (Knudsen and Nielsen, 1996) have also been shown to release miR-122 repressed targets in Huh7 cells and primary rat hepatocytes (Fabani and Gait, 2008). This was the first time that a non-negatively charged oligonucleotide has been shown to interfere with miRNA function. Both an unconjugated PNA 23-mer complementary to miR-122 with one additional Lys at the N-terminus and three Lys residues at the C-terminus as well as the PNA with the cell-penetrating peptide R6-Penetratin (Abes *et al.*, 2007; Turner *et al.*, 2005) were both efficiently taken up by cells without any transfection reagent (Fabani and Gait, 2008). Remarkably, while the peptide-conjugated PNA resulted in some loss of mature miR-122, the unconjugated PNA resulted in complete disappearance of the mature miR-122 band (Fabani and Gait, 2008). It is currently unclear how PNAs act on a miRNA and the pathway used for their degradation. However, since it is unlikely that PNAs are cleaved by the AGO2 protein of the RNAi RISC complex (Krutzfeldt *et al.*, 2007) and PNAs do not activate RNase H cleavage (Knudsen and Nielsen, 1996), they might become useful therapeutics for certain applications when these degradation pathways are impaired. With the advantage of low toxicity of antisense oligonucleotides over vector-based gene therapy approaches, all these results strongly encourage the exploration of antisense oligonucleotides as a therapeutic approach in humans in the near future.

## Anti-HCV therapy

At present, the pegylated IFN and ribavirin combination therapy is the only effective treatment available for individuals with chronic HCV (Feld and Hoofnagle, 2005). This therapy is however expensive, has numerous side-effects and sub-optimal patient response rates. While IFN treatment in cell culture modulates distinct miRNAs, including miR-122 (Pedersen *et al.*,

2007), the impact of IFN/ribavirin treatment on patient miRNA levels is unknown. Sarasin-Filipowicz *et al.* (2009) examined miR-122 and other miRNA levels in liver biopsies from patients with chronic HCV (Sarasin-Filipowicz *et al.*, 2009). Liver biopsies were acquired from 42 patients undergoing IFN treatment, and from six non-chronic HCV individuals. A second biopsy four hours after the first injection of pegylated IFN was obtained from eleven of the 42 patients. Serum HCV RNA levels were determined prior, and 4 and 12 weeks post-treatment. Surprisingly, miR-122 levels were low in patients who did not respond to treatment, compared to patients with undetectable HCV RNA levels at week 12. The unexpected correlation between miR-122 levels and response to treatment could not be attributed to virus genotype, levels of HCV RNA in serum or liver, or the extent of liver fibrosis. Although nonresponsive patients had low levels of miR-122, the investigators suggest that these levels may not be limiting for HCV replication (Sarasin-Filipowicz *et al.*, 2009). Alternatively, the number of miR-122 copies in the biopsies might not accurately reflect the levels of miR-122 in infected cells. The different levels of miR-122, and their correlation to the treatment outcome using pegIFN-α and ribavirin may suggest that miR-122 levels in non-responsive patients are not limiting, and that HCV might adjust to establish a persistent infection depending on the immune response. It is feasible that a prolonged activation of the IFN system in nonresponsive patients prior to HCV infection resulted in a persistent HCV infection independent of miR-122.

## Cellular miRNAs negatively regulate HCV

MiR-199a was recently described to target nt 62–84 within the 5′-UTR of the HCV genome (Murakami *et al.*, 2009). MiR-199a was cloned into pSilencerH1-puro, and expression of this plasmid in replicon and HCV-infected Huh7 cells lowered HCV RNA levels. This decrease could be reversed with a miR-199a antisense oligonucleotide. The authors also investigated the association of miR-199a with HCV RNA by immunoprecipitation of an AGO2-containing RNP, and by mutational analysis. A direct interaction between miR-199a and HCV RNA was however ambiguous. Brown *et al.* (2007) determined a threshold of miRNA copies, required to efficiently modulate gene expression (Brown *et al.*, 2007). In liver and Huh7 cells miR-199a is present in low copy number. Thus, the biological repercussions of miR-199a and HCV remain inconclusive.

Combination pegylated IFN and ribavirin treatment is frequently applied to individuals with chronic HCV (Feld and Hoofnagle, 2005). While IFN is known to stimulate gene transcription, the impact of IFN on the miRNA landscape had not previously been investigated. Pederson *et al.* (2007) demonstrated that IFN-β can modulate the levels of cellular miRNAs (Pedersen *et al.*, 2007). Microarray analysis of IFN-β/α-treated cells identified 30 miRNAs either up- or down-regulated. Although eight of the 30 miRNAs identified have seed sequences within conserved areas of the HCV genome, only five miRNAs were up-regulated in Huh7 cells upon IFN-β stimulation (Pedersen *et al.*, 2007). IFN-β was shown to rapidly induce the transcription of these miRNAs in a dose-dependent manner. Interestingly, levels of miR-122 decreased within 30 minutes of IFN-β treatment, and returned to normal levels within 24 h. The impact of these IFN-β-induced miRNAs on HCV was examined by transfecting JFH-1 replicon and virion infected cells with synthetic miRNA duplexes (mimicking miR-1, miR-30, miR-128, miR-196, miR-296, miR-351, miR-431, and miR-448), and viral RNA levels determined by quantitative PCR (qPCR). MiR-196, miR-296, miR-351, miR-431 and miR-448 individually and in combination, decreased viral RNA levels. IFN is known to up-regulate many factors that contribute to the antiviral state. The decrease of viral RNA following transfection of these anti-HCV miRNAs might however be a direct or indirect effect. To address this, Pedersen *et al.* (2007) designed a complementation assay (Pedersen *et al.*, 2007). Seed-match sequences for miR-196 and miR-448 differ by one nucleotide between JFH-1 and J6CF HCV strains. Although J6CF HCV is fastidious in cell culture, the chimeric J6/JFH-1 strain does infect Huh7 cells. With generation of this chimeric virus, the miR-196 and miR-448 seed match sequences were also exchanged. JFH-1 viral RNA levels decreased following transfection of JFH-1 specific miR-196 and miR-448. In contrast, synthetic miR-196 and miR-448 recognizing J6/JFH-1 seed match sequences did not decrease JFH-1 RNA levels. The reverse infection and specific miRNA transfection strategy produced a similar result, suggesting that miR-196 and miR-448 recognize and genetically interact with the viral genome (Pedersen *et al.*, 2007). That miR-196 and miR-448 recognize sequences within NS5A and core regions of the viral genome is interesting, and raises many questions as to how these miRNAs negatively affect HCV. Does IFN treatment modify (either or both) miRNA and viral RNPs to permit access and down-regulation by miR-196 and miR-448? Do IFN-induced miRNAs preferentially target mRNA coding versus non-coding regions? Although JFH-1 replicon cells treated with IFN-β show dramatically reduced viral RNA levels, these levels are partially rescued by exogenous addition of miR-122 and/or sequestering IFN-β-induced miRNAs, suggesting the IFN-β-induced miRNAs contribute to the antiviral response.

IFN treatment of Huh7 cells modulates levels of miRNAs including significantly decreasing miR-122 (Pedersen *et al.*, 2007). Recently Sarasin-Filipowicz *et al.* (2009) similarly examined the levels of miR-1, miR-196b, miR-296–5p, miR-351 and miR-448 in liver biopsies from patients chronically infected with HCV, and undergoing IFN treatment. Only miR-296–5p was found to be abundant, while the remaining miRNAs were either absent or present at low levels (Sarasin-Filipowicz *et al.*, 2009). This study highlights the variability and cautions the applicability of cell culture studies to patient treatment.

## Overexpressed HCV proteins interact with cellular RNAi machinery

HCV, similar to many RNA viruses, produces double-stranded (ds) RNA intermediates during viral replication. Recognition of HCV dsRNA by RIG-I activates IFN-α/β expression and induces an antiviral state (Saito *et al.*, 2007). It has been proposed that HCV dsRNA could potentially be a substrate for the RNAi machinery. Indeed, Wang *et al.* (2006) show that HCV RNA is susceptible to Dicer digestion (Wang *et al.*, 2006). The ability of overexpressed HCV core and non-structural proteins to rescue luciferase expression in the presence of a luciferase-specific shRNA was examined. HCV core protein rescued shRNA-targeted but not siRNA-directed gene silencing, suggesting that the when core protein is overexpressed, Dicer function is abrogated (Chen *et al.*, 2008; Wang *et al.*, 2006). Overexpression of core protein did not decrease levels of endogenous Dicer, implying that core protein interacts with, and inhibits Dicer activity (Chen *et al.*, 2008). Immunofluorescence and co-immunoprecipitation studies show Dicer and core protein to co-localize and interact (Chen *et al.*, 2008; Wang *et al.*, 2006). However, the nature of the Dicer–core interaction was not further explored. Chen *et al.* (2008) applied an *in vitro* assay to demonstrate that core protein blocked the ability of Dicer to process a long dsRNA target to a siRNA-like molecule (Chen *et al.*, 2008). The HCV structural protein E2 has also been proposed to restrict the RNAi machinery (Ji *et al.*, 2008). The HCV polyprotein as well as individual HCV proteins were expressed in the U-2 OS tetracycline-controlled cells. The HCV E1/E2 envelope complex was shown to interact with the AGO2 protein and decrease siRNA-directed gene silencing. While overexpression of individual HCV proteins may interact with the RNAi machinery to exert antiviral activity, the nature and mechanism of these interactions in the context of a viral infection remain unknown. Additionally, miR-122 (Jopling *et al.*, 2005), and the components of the RNAi machinery such as Dicer, Drosha, DGCR8 and AGO proteins are essential for HCV infection (Randall *et al.*, 2007). It thus, seems likely HCV has usurped the RNAi machinery to facilitate, rather than hinder, viral replication and establish a persistent infection.

## Concluding remarks

Many physiological processes are regulated by miRNAs, where dysfunction can result in diseases such as cancer. It is therefore critical to uncover the many functions of miRNAs. The highly abundant liver miRNA miR-122 has been shown to have physiological roles in cholesterol biosynthesis, lipid metabolism, and hepatocarcinogenesis, and is required for efficient HCV replication. Recently, important advances using antisense oligonucleotides as potential therapeutics *in vitro* and *in vivo* have been made. Specifically, unconjugated high affinity LNAs with low toxicity have potential to be approved for a clinical trial.

The consistent decline in cholesterol levels and the reduced hepatic steatosis in mice on a high-fat diet are particularly encouraging to explore miR-122 antagomirs in humans. However, of the several hundred genes found to be possibly regulated by miR-122 only a fraction have been confirmed and described in more detail. Given the numbers of potential targets, most roles for miR-122 in the normal or pathological liver have yet to be ascribed.

## References


Abes, S., Turner, J.J., Ivanova, G.D., Owen, D., Williams, D., Arzumanov, A., Clair, P., Gait, M.J., and Lebleu, B. (2007). Efficient splicing correction by PNA conjugation to an R6-Penetratin delivery peptide. Nucleic Acids Res. 35, 4495–4502.

Alvarez-Garcia, I., and Miska, E.A. (2005). MicroRNA functions in animal development and human disease. Development *132*, 4653–4662.

Behm-Ansmant, I., Rehwinkel, J., Doerks, T., Stark, A., Bork, P., and Izaurralde, E. (2006). mRNA degradation by miRNAs and GW182 requires both CCR4:NOT deadenylase and DCP1:DCP2 decapping complexes. Genes Dev. *20*, 1885–1898.

Berezikov, E., and Plasterk, R.H. (2005). Camels and zebrafish, viruses and cancer: a microRNA update. Hum. Mol. Genet *14 Spec. No. 2*, R183–190.

Bhattacharyya, S.N., Habermacher, R., Martine, U., Closs, E.I., and Filipowicz, W. (2006). Relief of microRNA-mediated translational repression in human cells subjected to stress. Cell *125*, 1111–1124.

Blight, K.J., and Rice, C.M. (1997). Secondary structure determination of the conserved 98-base sequence at the 3′ terminus of hepatitis C virus genome RNA. J. Virol. *71*, 7345–7352.

Bohnsack, M.T., Czaplinski, K., and Gorlich, D. (2004). Exportin 5 is a RanGTP-dependent dsRNA-binding protein that mediates nuclear export of pre-miRNAs. Rna *10*, 185–191.

Braasch, D.A., Jensen, S., Liu, Y., Kaur, K., Arar, K., White, M.A., and Corey, D.R. (2003). RNA interference in mammalian cells by chemically-modified RNA. Biochemistry *42*, 7967–7975.

Brennecke, J., Stark, A., Russell, R.B., and Cohen, S.M. (2005). Principles of microRNA-target recognition. PLoS Biol. 3, e85.

Brown, B.D., Gentner, B., Cantore, A., Colleoni, S., Amendola, M., Zingale, A., Baccarini, A., Lazzari, G., Galli, C., and Naldini, L. (2007). Endogenous microRNA can be broadly exploited to regulate transgene expression according to tissue, lineage and differentiation state. Nat. Biotechnol. 25, 1457–1467.

Chang, J., Nicolas, E., Marks, D., Sander, C., Lerro, A., Buendia, M.A., Xu, C., Mason, W.S., Moloshok, T., Bort, R., *et al.* (2004). miR-122, a mammalian liver-specific microRNA, is processed from hcr mRNA and may down-regulate the high affinity cationic amino acid transporter CAT-1. RNA Biol *1*, 106–113.

Chen, W., Zhang, Z., Chen, J., Zhang, J., Zhang, J., Wu, Y., Huang, Y., Cai, X., and Huang, A. (2008). HCV core protein interacts with Dicer to antagonize RNA silencing. Virus Res. *133*, 250–258.

Egger, D., Wolk, B., Gosert, R., Bianchi, L., Blum, H.E., Moradpour, D., and Bienz, K. (2002). Expression of hepatitis C virus proteins induces distinct membrane alterations including a candidate viral replication complex. J. Virol. *76*, 5974–5984.

Elmen, J., Lindow, M., Schutz, S., Lawrence, M., Petri, A., Obad, S., Lindholm, M., Hedtjarn, M., Hansen, H.F., Berger, U., *et al.* (2008). LNA-mediated microRNA silencing in non-human primates. Nature *452*, 896–899.

Elmen, J., Thonberg, H., Ljungberg, K., Frieden, M., Westergaard, M., Xu, Y., Wahren, B., Liang, Z., Orum, H., Koch, T., *et al.* (2005). Locked nucleic acid (LNA) mediated improvements in siRNA stability and functionality. Nucleic Acids Res. *33*, 439–447.

Esau, C., Davis, S., Murray, S.F., Yu, X.X., Pandey, S.K., Pear, M., Watts, L., Booten, S.L., Graham, M., McKay, R., *et al.* (2006). miR-122 regulation of lipid metabolism revealed by *in vivo* antisense targeting. Cell Metabolism 3, 87–98.

Etiemble, J., Moroy, T., Jacquemin, E., Tiollais, P., and Buendia, M.A. (1989). Fused transcripts of c-myc and a new cellular locus, hcr in a primary liver tumor. Oncogene *4*, 51–57.

Eulalio, A., Huntzinger, E., and Izaurralde, E. (2008). GW182 interaction with Argonaute is essential for miRNA-mediated translational repression and mRNA decay. Nat. Struct. Mol. Biol. *15*, 346–353.

Fabani, M.M., and Gait, M.J. (2008). miR-122 targeting with LNA/2′-O-methyl oligonucleotide mixmers, peptide nucleic acids (PNA), and PNA–peptide conjugates. Rna *14*, 336–346.

Feld, J.J., and Hoofnagle, J.H. (2005). Mechanism of action of interferon and ribavirin in treatment of hepatitis C. Nature *436*, 967–972.

Fernandez, J., Lopez, A.B., Wang, C., Mishra, R., Zhou, L., Yaman, I., Snider, M.D., and Hatzoglou, M. (2003). Transcriptional control of the arginine/lysine transporter, cat-1, by physiological stress. J. Biol. Chem. *278*, 50000–50009.

Filipowicz, W., Bhattacharyya, S.N., and Sonenberg, N. (2008). Mechanisms of post-transcriptional regulation by microRNAs: are the answers in sight? Nat Rev Genet 9, 102–114.

Friebe, P., and Bartenschlager, R. (2002). Genetic analysis of sequences in the 3′ nontranslated region of hepatitis C virus that are important for RNA replication. J. Virol. *76*, 5326–5338.

Friebe, P., Boudet, J., Simorre, J.-P., and Bartenschlager, R. (2005). Kissing-loop interaction in the 3′ end of the hepatitis C virus genome essential for RNA replication. J. Virol. *79*, 380–392.

Friebe, P., Lohmann, V., Krieger, N., and Bartenschlager, R. (2001). Sequences in the 5′ nontranslated region of hepatitis C virus required for RNA replication. J. Virol. *75*, 12047–12057.

Friedman, R.C., Farh, K.K., Burge, C.B., and Bartel, D.P. (2009). Most mammalian mRNAs are conserved targets of microRNAs. Genome Res. *19*, 92–105.

Ghoshal, A.K., and Farber, E. (1984). The induction of liver cancer by dietary deficiency of choline and methionine without added carcinogens. Carcinogenesis *5*, 1367–1370.

Gramantieri, L., Ferracin, M., Fornari, F., Veronese, A., Sabbioni, S., Liu, C.G., Calin, G.A., Giovannini, C., Ferrazzi, E., Grazi, G.L., *et al.* (2007). Cyclin G1 is a target of miR-122a, a microRNA frequently downregulated in human hepatocellular carcinoma. Cancer Res. *67*, 6092–6099.

Gregory, R.I., Yan, K.P., Amuthan, G., Chendrimada, T., Doratotaj, B., Cooch, N., and Shiekhattar, R. (2004). The Microprocessor complex mediates the genesis of microRNAs. Nature *432*, 235–240.

Grimm, D., Streetz, K.L., Jopling, C.L., Storm, T.A., Pandey, K., Davis, C.R., Marion, P., Salazar, F., and Kay, M.A. (2006). Fatality in mice due to oversaturation of cellular microRNA/short hairpin RNA pathways. Nature *441*, 537–541.

Grishok, A., Pasquinelli, A.E., Conte, D., Li, N., Parrish, S., Ha, I., Baillie, D.L., Fire, A., Ruvkun, G., and Mello, C.C. (2001). Genes and mechanisms related to RNA interference regulate expression of the small temporal RNAs that control C. elegans developmental timing. Cell *106*, 23–34.

Han, J., Lee, Y., Yeom, K.H., Kim, Y.K., Jin, H., and Kim, V.N. (2004). The Drosha–DGCR8 complex in primary microRNA processing. Genes Dev. *18*, 3016–3027.

Hatzoglou, M., Fernandez, J., Yaman, I., and Closs, E. (2004). Regulation of cationic amino acid transport: the story of the CAT-1 transporter. Annu Rev Nutr *24*, 377–399.

Henke, J.I., Goergen, D., Zheng, J., Song, Y., Schuttler, C.G., Fehr, C., Junemann, C., and Niepmann, M. (2008). microRNA-122 stimulates translation of hepatitis C virus RNA. EMBO J. *27*, 3300–3310.

Huang, J., Wang, F., Argyris, E., Chen, K., Liang, Z., Tian, H., Huang, W., Squires, K., Verlinghieri, G., and Zhang, H. (2007). Cellular microRNAs contribute to HIV-1 latency in resting primary CD4+ T lymphocytes. Nat. Med. *13*, 1241–1247.

Humphreys, D.T., Westman, B.J., Martin, D.I., and Preiss, T. (2005). MicroRNAs control translation initiation by inhibiting eukaryotic initiation factor 4E/cap and poly(A) tail function. Proc. Natl. Acad. Sci. U.S.A. *102*, 16961–16966.

Hutvagner, G., McLachlan, J., Pasquinelli, A.E., Balint, E., Tuschl, T., and Zamore, P.D. (2001). A cellular function for the RNA-interference enzyme Dicer in the maturation of the let-7 small temporal RNA. Science *293*, 834–838.

Iwasaki, S., Kawamata, T., and Tomari, Y. (2009). *Drosophila* argonaute1 and argonaute2 employ distinct mechanisms for translational repression. Mol. Cell *34*, 58–67.

Ji, J., Glaser, A., Wernli, M., Berke, J.M., Moradpour, D., and Erb, P. (2008). Suppression of short interfering RNA-mediated gene silencing by the structural proteins of hepatitis C virus. J. Gen. Virol. *89*, 2761–2766.

Jopling, C.L., Schutz, S., and Sarnow, P. (2008). Position-dependent function for a tandem microRNA miR-122-binding site located in the hepatitis C virus RNA genome. Cell Host Microbe *4*, 77–85.

Jopling, C.L., Yi, M., Lancaster, A.M., Lemon, S.M., and Sarnow, P. (2005). Modulation of hepatitis C virus RNA abundance by a liver-specific MicroRNA. Science *309*, 1577–1581.

Kim, V.N. (2005). MicroRNA biogenesis: coordinated cropping and dicing. Nat. Rev. Mol. Cell Biol. *6*, 376–385.

Kim, V.N., Han, J., and Siomi, M.C. (2009). Biogenesis of small RNAs in animals. Nat. Rev. Mol. Cell Biol. *10*, 126–139.

Knudsen, H., and Nielsen, P.E. (1996). Antisense properties of duplex- and triplex-forming PNAs. Nucleic Acids Res. *24*, 494–500.

Kolykhalov, A.A., Feinstone, S.M., and Rice, C.M. (1996). Identification of a highly conserved sequence element at the 3′ terminus of hepatitis C virus genome RNA. J. Virol. *70*, 3363–3371.

Koshkin, A.A., Singh, S.K., Nielsen, P., Rajwanshi, V.K., Kumar, R., Meldgaard, M., Olsen, C.E., and Wengel, J. (1998). LNA (locked nucleic acids): synthesis of the adenine, cytosine, guanine, 5-methylcytosine, thymine and uracil bicylonucloside monomers, oligomerisation and unprecedented nucleic acid recognition. Tetrahedron *54*, 3607–3630.

Krutzfeldt, J., Kuwajima, S., Braich, R., Rajeev, K.G., Pena, J., Tuschl, T., Manoharan, M., and Stoffel, M. (2007). Specificity, duplex degradation and subcellular localization of antagomirs. Nucleic Acids Res. *35*, 2885–2892.

Krutzfeldt, J., Rajewsky, N., Braich, R., Rajeev, K.G., Tuschl, T., Manoharan, M., and Stoffel, M. (2005). Silencing of microRNAs *in vivo* with 'antagomirs'. Nature *438*, 685–689.

Kumar, R., Singh, S.K., Koshkin, A.A., Rajwanshi, V.K., Meldgaard, M., and Wengel, J. (1998). The first analogues of LNA (locked nucleic acids): phosphorothioate-LNA and 2′-thio-LNA. Bioorg. Med. Chem. Lett. *8*, 2219–2222.

Kutay, H., Bai, S., Datta, J., Motiwala, T., Pogribny, I., Frankel, W., Jacob, S.T., and Ghoshal, K. (2006). Down-regulation of miR-122 in the rodent and human hepatocellular carcinomas. J. Cell Biochem. *99*, 671–678.

Lagos-Quintana, M., Rauhut, R., Lendeckel, W., and Tuschl, T. (2001). Identification of novel genes coding for small expressed RNAs. Science *294*, 853–858.

Landthaler, M., Yalcin, A., and Tuschl, T. (2004). The human DiGeorge syndrome critical region gene 8 and Its *D. melanogaster* homolog are required for miRNA biogenesis. Curr. Biol. *14*, 2162–2167.

Lecellier, C.H., Dunoyer, P., Arar, K., Lehmann-Che, J., Eyquem, S., Himber, C., Saib, A., and Voinnet, O. (2005). A cellular microRNA mediates antiviral defense in human cells. Science *308*, 557–560.

Lee, R.C., Feinbaum, R.L., and Ambros, V. (1993). The *C. elegans* heterochronic gene lin-4 encodes small RNAs with antisense complementarity to lin-14. Cell *75*, 843–854.

Lee, Y., Ahn, C., Han, J., Choi, H., Kim, J., Yim, J., Lee, J., Provost, P., Radmark, O., Kim, S., *et al.* (2003). The nuclear RNase III Drosha initiates microRNA processing. Nature *425*, 415–419.

Lee, Y., Jeon, K., Lee, J.T., Kim, S., and Kim, V.N. (2002). MicroRNA maturation: stepwise processing and subcellular localization. EMBO J. *21*, 4663–4670.

Lewis, B.P., Burge, C.B., and Bartel, D.P. (2005). Conserved seed pairing, often flanked by adenosines, indicates that thousands of human genes are microRNA targets. Cell *120*, 15–20.

Lewis, B.P., Shih, I.H., Jones-Rhoades, M.W., Bartel, D.P., and Burge, C.B. (2003). Prediction of mammalian microRNA targets. Cell *115*, 787–798.

Lindenbach, B.D., and Rice, C.M. (2001). *Flaviviridae*: The Viruses and Their Replication, Vol. 1, 3rd edn (Philadelphia: Lippincott-Raven Publishers).

Lindenbach, B.D., and Rice, C.M. (2005). Unravelling hepatitis C virus replication from genome to function. Nature *436*, 933–938.

Liu, J., Carmell, M.A., Rivas, F.V., Marsden, C.G., Thomson, J.M., Song, J.J., Hammond, S.M., Joshua-Tor, L., and Hannon, G.J. (2004). Argonaute2 is the catalytic engine of mammalian RNAi. Science *305*, 1437–1441.

Lombardi, B., Chandar, N., and Locker, J. (1991). Nutritional model of hepatocarcinogenesis. Rats fed choline-devoid diet. Dig. Dis. Sci. *36*, 979–984.

Lund, E., Guttinger, S., Calado, A., Dahlberg, J.E., and Kutay, U. (2004). Nuclear export of microRNA precursors. Science *303*, 95–98.

Lytle, J.R., Yario, T.A., and Steitz, J.A. (2007). Target mRNAs are repressed as efficiently by microRNA-binding sites in the 5′-UTR as in the 3′ UTR. Proc. Natl. Acad. Sci. U.S.A. *104*, 9667–9672.

Major, M., Rehermann, B., and Feinstone, S.M. (2001). Hepatitis C Viruses, Vol. 2, 4th edn (Philadelphia: Lippincott-Raven Publishers).

Matranga, C., Tomari, Y., Shin, C., Bartel, D.P., and Zamore, P.D. (2005). Passenger-strand cleavage facilitates assembly of siRNA into Ago2-containing RNAi enzyme complexes. Cell *123*, 607–620.

McManus, M.T., Petersen, C.P., Haines, B.B., Chen, J., and Sharp, P.A. (2002). Gene silencing using micro-RNA designed hairpins. Rna *8*, 842–850.

Meister, G., Landthaler, M., Dorsett, Y., and Tuschl, T. (2004a). Sequence-specific inhibition of microRNA- and siRNA-induced RNA silencing. Rna *10*, 544–550.

Meister, G., Landthaler, M., Patkaniowska, A., Dorsett, Y., Teng, G., and Tuschl, T. (2004b). Human Argonaute2 mediates RNA cleavage targeted by miRNAs and siRNAs. Mol. Cell *15*, 185–197.

Moroy, T., Etiemble, J., Bougueleret, L., Hadchouel, M., Tiollais, P., and Buendia, M.A. (1989). Structure and expression of hcr, a locus rearranged with c-myc in a woodchuck hepatocellular carcinoma. Oncogene *4*, 59–65.

Moroy, T., Marchio, A., Etiemble, J., Trepo, C., Tiollais, P., and Buendia, M.A. (1986). Rearrangement and enhanced expression of c-myc in hepatocellular carcinoma of hepatitis virus infected woodchucks. Nature *324*, 276–279.

Murakami, Y., Aly, H.H., Tajima, A., Inoue, I., and Shimotohno, K. (2009). Regulation of the hepatitis C virus genome replication by miR-199a(*). J Hepatol *50*, 453–460.

Nottrott, S., Simard, M.J., and Richter, J.D. (2006). Human let-7a miRNA blocks protein production on actively translating polyribosomes. Nat. Struct. Mol. Biol. *13*, 1108–1114.

Okamura, K., Ishizuka, A., Siomi, H., and Siomi, M.C. (2004). Distinct roles for Argonaute proteins in small RNA-directed RNA cleavage pathways. Genes Dev. *18*, 1655–1666.

Olsen, P.H., and Ambros, V. (1999). The lin-4 regulatory RNA controls developmental timing in Caenorhabditis elegans by blocking LIN-14 protein synthesis after the initiation of translation. Dev. Biol. *216*, 671–680.

Otsuka, M., Jing, Q., Georgel, P., New, L., Chen, J., Mols, J., Kang, Y.J., Jiang, Z., Du, X., Cook, R., *et al.* (2007). Hypersusceptibility to vesicular stomatitis virus infection in Dicer1-deficient mice is due to impaired miR24 and miR93 expression. Immunity *27*, 123–134.

Paddison, P.J., Caudy, A.A., Bernstein, E., Hannon, G.J., and Conklin, D.S. (2002). Short hairpin RNAs (shRNAs) induce sequence-specific silencing in mammalian cells. Genes Dev. *16*, 948–958.

Pedersen, I.M., Cheng, G., Wieland, S., Volinia, S., Croce, C.M., Chisari, F.V., and David, M. (2007). Interferon modulation of cellular microRNAs as an antiviral mechanism. Nature *449*, 919–922.

Peters, L., and Meister, G. (2007). Argonaute proteins: mediators of RNA silencing. Mol. Cell *26*, 611–623.

Petersen, C.P., Bordeleau, M.E., Pelletier, J., and Sharp, P.A. (2006). Short RNAs repress translation after initiation in mammalian cells. Mol. Cell *21*, 533–542.

Pillai, R.S., Bhattacharyya, S.N., Artus, C.G., Zoller, T., Cougot, N., Basyuk, E., Bertrand, E., and Filipowicz, W. (2005). Inhibition of translational initiation by Let-7 MicroRNA in human cells. Science *309*, 1573–1576.

Pogribny, I.P., James, S.J., Jernigan, S., and Pogribna, M. (2004). Genomic hypomethylation is specific for preneoplastic liver in folate/methyl deficient rats and does not occur in non-target tissues. Mutat Res. *548*, 53–59.

Rand, T.A., Petersen, S., Du, F., and Wang, X. (2005). Argonaute2 cleaves the anti-guide strand of siRNA during RISC activation. Cell *123*, 621–629.

Randall, G., Panis, M., Cooper, J.D., Tellinghuisen, T.L., Sukhodolets, K.E., Pfeffer, S., Landthaler, M., Landgraf, P., Kan, S., Lindenbach, B.D., *et al.* (2007). Cellular cofactors affecting hepatitis C virus infection and replication. Proc. Natl. Acad. Sci. U.S.A. *104*, 12884–12889.

Rodriguez, A., Griffiths-Jones, S., Ashurst, J.L., and Bradley, A. (2004). Identification of mammalian microRNA host genes and transcription units. Genome Res. *14*, 1902–1910.

Saetrom, P., Heale, B.S., Snove, O., Jr., Aagaard, L., Alluin, J., and Rossi, J.J. (2007). Distance constraints between microRNA target sites dictate efficacy and cooperativity. Nucleic Acids Res. *35*, 2333–2342.

Saito, T., Hirai, R., Loo, Y.M., Owen, D., Johnson, C.L., Sinha, S.C., Akira, S., Fujita, T., and Gale, M., Jr. (2007). Regulation of innate antiviral defenses through a shared repressor domain in RIG-I and LGP2. Proc. Natl. Acad. Sci. U.S.A. *104*, 582–587.

Sarasin-Filipowicz, M., Krol, J., Markiewicz, I., Heim, M.H., and Filipowicz, W. (2009). Decreased levels of microRNA miR-122 in individuals with hepatitis C responding poorly to interferon therapy. Nat. Med. *15*, 31–33.

Sasaki, T., Shiohama, A., Minoshima, S., and Shimizu, N. (2003). Identification of eight members of the Argonaute family in the human genome small star, filled. Genomics *82*, 323–330.

Shiohama, A., Sasaki, T., Noda, S., Minoshima, S., and Shimizu, N. (2003). Molecular cloning and expression analysis of a novel gene DGCR8 located in the DiGeorge syndrome chromosomal region. Biochem. Biophys. Res. Commun. *304*, 184–190.

Singh, S.K., Kumar, R., and Wengel, J. (1998). Synthesis of novel bicyclo[2.2.1] ribonucleosides: 2′-amino- and 2′-thio-LNA monomeric nucleosides. J. Org. Chem. *63*, 6078–6079.

Stark, K.L., Xu, B., Bagchi, A., Lai, W.S., Liu, H., Hsu, R., Wan, X., Pavlidis, P., Mills, A.A., Karayiorgou, M., *et al.* (2008). Altered brain microRNA biogenesis contributes to phenotypic deficits in a 22q11-deletion mouse model. Nat. Genet. *40*, 751–760.

Tanaka, T., Kato, N., Cho, M.-J., and Shimotohno, K. (1995). A novel sequence found at the 3′ terminus of hepatitis C virus genome. Biochem. Biophys.l Res. Commun. *215*, 744–749.

Tanaka, T., Kato, N., Cho, M.-J., Sugiyama, K., and Shimotohno, K. (1996). Structure of the 3′ terminus of the hepatitis C virus genome. J. Virol. *70*, 3307–3312.

Triboulet, R., Mari, B., Lin, Y.L., Chable-Bessia, C., Bennasser, Y., Lebrigand, K., Cardinaud, B., Maurin, T., Barbry, P., Baillat, V., *et al.* (2007). Suppression of microRNA-silencing pathway by HIV-1 during virus replication. Science *315*, 1579–1582.

Turner, J.J., Arzumanov, A.A., and Gait, M.J. (2005). Synthesis, cellular uptake and HIV-1 Tat-dependent trans-activation inhibition activity of oligonucleotide analogues disulphide-conjugated to cell-penetrating peptides. Nucleic Acids Res. 33, 27–42.

Vasudevan, S., Tong, Y., and Steitz, J.A. (2007). Switching from repression to activation: microRNAs can up-regulate translation. Science *318*, 1931–1934.

Wahlestedt, C., Salmi, P., Good, L., Kela, J., Johnsson, T., Hokfelt, T., Broberger, C., Porreca, F., Lai, J., Ren, K., *et al.* (2000). Potent and nontoxic antisense oligonucleotides containing locked nucleic acids. Proc. Natl. Acad. Sci. U.S.A. *97*, 5633–5638.

Wang, B., Love, T.M., Call, M.E., Doench, J.G., and Novina, C.D. (2006). Recapitulation of short RNA-directed translational gene silencing in vitro. Mol. Cell *22*, 553–560.

Wang, C., Sarnow, P., and Siddiqui, A. (1993). Translation of human hepatitis C virus RNA in cultured cells is mediated by an internal ribosome-binding mechanism. J. Virol. *67*, 3338–3344.

Wightman, B., Ha, I., and Ruvkun, G. (1993). Posttranscriptional regulation of the heterochronic gene lin-14 by lin-4 mediates temporal pattern formation in *C. elegans*. Cell *75*, 855–862.

Witting, S.R., Brown, M., Saxena, R., Nabinger, S., and Morral, N. (2008). Helper-dependent adenovirus-mediated short hairpin RNA expression in the liver activates the interferon response. J. Biol. Chem. *283*, 2120–2128.

Wu, L., and Belasco, J.G. (2008). Let me count the ways: mechanisms of gene regulation by miRNAs and siRNAs. Mol. Cell *29*, 1–7.

Wu, L., Fan, J., and Belasco, J.G. (2006). MicroRNAs direct rapid deadenylation of mRNA. Proc. Natl. Acad. Sci. U.S.A. *103*, 4034–4039.

Yi, M., and Lemon, S.M. (2003a). 3′ Nontranslated RNA signals required for replication of hepatitis C virus RNA. J. Virol. *77*, 3557–3568.

Yi, M., and Lemon, S.M. (2003b). Structure-function analysis of the 3′ stem–loop of hepatitis C virus genomic RNA and its role in viral RNA replication. Rna *9*, 331–345.

Yi, R., Qin, Y., Macara, I.G., and Cullen, B.R. (2003). Exportin-5 mediates the nuclear export of pre-microRNAs and short hairpin RNAs. Genes Dev. *17*, 3011–3016.

Yu, J.Y., DeRuiter, S.L., and Turner, D.L. (2002). RNA interference by expression of short-interfering RNAs and hairpin RNAs in mammalian cells. Proc. Natl. Acad. Sci. U.S.A. *99*, 6047–6052.

Zamecnik, P.C., and Stephenson, M.L. (1978). Inhibition of Rous sarcoma virus replication and cell transformation by a specific oligodeoxynucleotide. Proc. Natl. Acad. Sci. U.S.A. *75*, 280–284.

# Viral Escape from RNAi in Mammalian Cells

# 6

Maria Nevot and Miguel Angel Martínez

## Abstract

Fire, Mello, and coworkers first discovered that introducing long double-stranded RNA (dsRNA) into the nematode *Caenorhabditis elegans* led to the targeted degradation of homologous messenger (m)RNA. This discovery revealed the existence of a fundamental mechanism for gene expression regulation, now known as RNA interference (RNAi) (Fire *et al.*, 1998). Later, Elbashir *et al.* (2001) showed that RNAi also occurs in mammalians cells. Importantly, they performed the extraordinary demonstration by cell transfection that synthetic, short, 21-basepair (bp) duplexes of interfering RNA (siRNA) could mediate RNAi in a sequence-specific manner. This finding enabled the specific regulation of gene expression in a variety of biological systems, including diseased cells. Recent studies regarding the utility of RNAi to specifically inhibit virus replication have opened new possibilities for the development of novel therapies against viral infection. However, viruses appear to be capable of escaping RNAi inhibition in mammalian cells. Viral mechanisms for escaping RNAi may include suppression of RNAi, mutational escape from RNAi, and modulation of the cell's microRNA (miRNA)/RNAi profile. Here, we summarize these viral mechanisms and discuss potential strategies for neutralizing viral escape from RNAi.

## Introduction

RNAi is a highly conserved eukaryotic regulatory mechanism triggered by dsRNAs that can induce sequence-specific gene silencing (Fire *et al.*, 1998). Once in the cytoplasm, long endogenous or exogenous dsRNAs are cleaved into fragments by an RNase III-like enzyme termed Dicer. Dicer cleavage generates 3′ single-stranded overhangs of two nucleotides at each end of small ~20–30 bp RNA molecules. There are two primary categories of these small RNAs: siRNAs and miRNAs (Elbashir *et al.*, 2001; Lagos-Quintana *et al.*, 2001; Lau *et al.*, 2001; Lee and Ambros, 2001). Although they can be functionally equivalent, siRNAs and miRNAs differ in their biogenesis (Cullen, 2009a; Dykxhoorn and Lieberman, 2006; Matranga and Zamore, 2007). The miRNAs are small RNAs that are derived from larger RNA polymerase II transcripts, termed primary miRNAs (pri-miRNAs). These pri-miRNAs contain imperfect RNA hairpins that are hundreds to thousands of bp long. Pri-miRNAs are processed in the nucleus by Drosha, an RNAse III enzyme, and its cofactor DGCR8. These convert the pri-miRNA to a ~60–70 bp stem–loop precursor miRNA (pre-miRNA), which is then exported into the cytoplasm by exportin 5 (Lee and Ambros, 2001; Lund *et al.*, 2004). In the cytoplasm, Dicer cleavage produces the miRNA form. The small RNA duplexes combine with nucleases to form complexes known as RNA-induced silencing complexes (RISC). One RNA strand, the siRNA guide strand (the antisense strand) or miRNA strand, is used to target the complex to homologous RNAs. When the match between the siRNA and target-RNA is perfect, the target RNA is cleaved and degraded; when the match is partial, translation is inhibited (Zeng *et al.*, 2003). Because miRNAs and siRNAs presumably inhibit mRNA expression by identical mechanisms, the exact criteria for recognition is not completely understood. It has been shown that the pairing of seven or eight nucleotides on the 5′ end of the miRNA (termed the seed region) with multiple sites in the 3′ untranslated region (UTR) of a target mRNA is sufficient to inhibit translation (Grimson *et al.*, 2007).

The discovery that gene expression can be controlled by the pairing of small RNAs with mRNAs that contain a complementary sequence has provided a powerful tool for studying gene regulation and function (Aigner, 2007). In addition, RNAi has become a potential new approach for developing therapies against diseases associated with dysfunction of gene expression or regulation, including cancer, infectious diseases, and genetic disorders (Castanotto and Rossi, 2009). Given that exogenous RNA can be targeted without affecting cellular functions, one of the most promising applications of RNAi is in the treatment of infectious diseases. Nevertheless, the exquisite sequence specificity of RNAi gives it little flexibility for surmounting the problem of viral escape, particularly when the target is highly mutable, like an RNA virus. Soon after the demonstration that synthetic siRNAs were able to induce the RNAi mechanism in mammalian cells (Elbashir *et al.*, 2001), several studies reported that viral gene expression and replication could be inhibited *ex vivo* by virus-specific synthetic siRNAs or expressed siRNAs.

Presently, there are several well-known viruses and emerging viruses that are increasingly becoming serious threats to public health. Effective vaccines and antiviral drugs are not available for most of these viruses. The limited success in adapting current antiviral therapies to new challenges justifies the continued effort to develop new strategies. Many viruses have been successfully targeted by RNAi (Haasnoot and Berkhout, 2006; Haasnoot *et al.*, 2007) (Chapter 4). These include important human pathogens such as respiratory syncytial virus (RSV) (Bitko and Barik, 2001), human immunodeficiency virus type 1 (HIV-1) (Coburn and Cullen, 2002; Jacque *et al.*, 2002; Lee *et al.*, 2002; Martinez *et al.*, 2002b; Novina *et al.*, 2002), hepatitis C virus (HCV) (Kronke *et al.*, 2004; Yokota *et al.*, 2003), hepatitis B virus (HBV) (McCaffrey *et al.*, 2003; Shlomai and Shaul, 2003), severe acute respiratory syndrome coronavirus (SARSCoV) (He *et al.*, 2003; Li *et al.*, 2005), poliovirus (Gitlin *et al.*, 2002) and influenza A virus (Ge *et al.*, 2003). These studies have shown that RNAi therapeutics can be highly effective for inhibiting viral replication, thus making them outstanding candidates for future clinical use. In fact, several RNAi-based antiviral drugs are currently being tested in clinical trials (Castanotto and Rossi, 2009; Haasnoot *et al.*, 2007) (Table 6.1). Despite these promising results, it was soon demonstrated that human viral pathogens could quickly and readily escape inhibition by single siRNAs (Table 6.2). Coexpression of multiple siRNAs could reduce the emergence of single-siRNA-resistant viruses. This strategy had an effect comparable to that achieved by highly active antiretroviral therapy (HAART), which combines three- or four-anti-HIV-1 drugs into one treatment (Martinez *et al.*, 2002a). An alternative strategy for countering the generation of siRNA-resistant viruses would be to target transcripts of cellular cofactors essential for HIV-1 infectivity or replication (e.g. specific cell-surface receptors). However, the side-effects

**Table 6.1** Current clinical trials testing RNAi-based therapeutics against viruses

| Virus | Inhibitor (name) | Target gene(s) | Stage | Sponsor | References |
|---|---|---|---|---|---|
| RSV | siRNA (ALN-RSV01) | Nucleocapsid | Phase II | Alnylam Pharmaceuticals | NA |
| HBV | siRNA (Nuc B1000) | Pre-gen./pre-C, Pre-S1, Pre-S2/S, X | Phase I | Nucleonics | NA |
| HIV-1 | shRNA, ribozyme and RNA decoy (Triple-Rvector) | Tat/Rev, CCR5, TAR | Phase I | City of Hope National Medical Centre, and Benitec | Castanotto and Rossi (2009) |

NA, not available.

**Table 6.2** Viral strategies for escaping RNAi

| |
|---|
| Point mutation, deletion, or insertion within the target region |
| Point mutation outside the target site that induces an inaccessible RNA structure |
| Viral RNAi suppressors |
| VA RNAs that compete for Dicer and/or RISC |
| Suppressor proteins that inhibit Dicer or other RNAi components |
| RNAi suppressor proteins that sequester siRNAs |
| Target inaccessibility within subcellular compartments |
| Occlusion of the target site by binding of viral proteins |

of down-regulating cellular targets over the long term are currently unknown.

Importantly, viruses are the most abundant form of life, and they infect practically all organisms. Consequently, these obligate parasites compose an enormous virosphere that includes viral species with very different genome compositions (DNA or RNA) or lengths, life cycles, and other characteristics that have distinct effects on virus–host interactions (http://www.ncbi.nlm.nih.gov/ICTVdb/Images/virosphere.htm). The RNA virus is the paradigm of a mutable and variable organism. Nevertheless, DNA viruses, although they do not have the genetic variability of RNA viruses, currently have longer genomes and very sophisticated means of manipulating the host in order to escape host inhibition mechanisms.

Another important factor to keep in mind when considering RNAi as a new therapeutic antiviral tool is the possible interaction of mammalian viruses with host RNAi machinery. In particular, microRNAs are produced by the various members of the herpesvirus family during both the latent stage of the viral life cycle and the lytic (or productive) stage (see Chapter 3). Thus, viral encoded microRNAs may be important for regulating the transition from latent to lytic replication. In addition, viral miRNAs appear to be important for attenuating antiviral host immune responses (Cullen, 2009a,b). So far, more than 100 miRNA genes have been identified in 15 viruses from three viral families, including Herpesvirus, Polyomavirus, and Retrovirus (Ghosh *et al.*, 2009; Grassmann and Jeang, 2008). In general, it appears that most viral miRNAs down-regulate either cellular or viral genes that increase the sensitivity of virally infected cells to host innate or adaptive immune responses (Gottwein and Cullen, 2008). This suggests that inhibition of these viral-encoded miRNAs in infected cells or individuals may help control viral replication. Thus, strategies based on antisense miRNA constructs may provide therapeutic benefit. Finally, viruses may modulate the cellular miRNA expression profile (Triboulet *et al.*, 2007) in order to suppress harmful miRNAs; or, alternatively, they may learn to adapt miRNA-restriction to increase the expression of cellular miRNAs that benefit viral replication (Jopling *et al.*, 2005).

Increasing the knowledge of interactions between virus and host RNAi machinery should lead towards the development of effective, durable, non-toxic RNAi-based antiviral strategies and, possibly, to anticipate and avoid viral escape mechanisms (Table 6.3). This chapter discusses the known mechanisms used by mammalian viruses to escape RNAi inhibition, and suggests possible strategies for minimizing viral escape (Table 6.4).

## Should mammalian viruses worry about RNAi? RNAi suppressor proteins from mammalian viruses

An exogenously induced RNAi response in mammalian cells, mediated by siRNAs or short-hairpin RNAs (shRNAs), can inhibit viral replication; this raises the question of whether viruses interact with the cellular RNAi machinery. Understanding this potential physiological interaction could significantly contribute to the design of more effective RNAi-based therapies. In both plants

**Table 6.3** Important parameters that influence variability and adaptability of RNA virus populations

| |
|---|
| Average number of mutations per genome within the mutant spectrum of an infected individual viral population. Generally, this averages 1–100 (more in some cases) mutations per genome |
| Mutation rate. Error rate has been estimated at between $10^{-4}$ and $10^{-5}$ mutations per nucleotide and cycle of replication |
| Genome length: 3–32 kb |
| Virus population size and fecundity. Variable; but an acutely infected organism may harbour $10^{9}$–$10^{12}$ viral particles at any given time |
| Mutations needed for a phenotypic change. Many recorded adaptive changes depend on one or a few mutations |

**Table 6.4** Strategies for minimizing viral escape

| |
|---|
| Combination of multiple siRNAs that target different conserved genomic viral regions |
| Expression of multiple different shRNAs from a single lentiviral vector |
| Long-hairpin RNAs (lhRNAs), from which multiple siRNAs can be produced |
| Secondary RNAi. A single region is targeted with multiple siRNAs or shRNAs |
| Combination of shRNAs together with other nucleic acid-based inhibitors (ribozymes, decoys, and aptamers) |
| Targeting cellular co-factors involved in the viral infection |

and invertebrate animals, RNAi constitutes a key component of the innate immune response to viral infection (Cullen, 2006); it has been postulated that RNAi has a similar protective function in mammals (Grassmann and Jeang, 2008; Haasnoot *et al.*, 2007; Voinnet, 2005). In plants and invertebrate animals, long dsRNA serves as the initial trigger for initiating RNAi; the dsRNA is then processed by Dicer into siRNAs. However, in mammalian cells, long dsRNA sequences (more than 30 bp in length) are potent inducers of the interferon (IFN) response (Garcia-Sastre and Biron, 2006). It has been proposed that in mammalian cells, the extremely potent IFN system has displaced RNAi as the key defence against virus infection, and that RNAi now exists only as a mechanism of post-transcriptional regulation programmed by endogenously encoded miRNA. Thus, only artificial conditions, for instance, the introduction of exogenous nucleic acids with precise structural characteristics, can cause reprogramming of mammalian RNAi machinery to resist virus replication (Cullen, 2006). Nevertheless, it is also worth mentioning that through miRNAs, RNAi might regulate components of the antiviral IFN pathway, and thus provide a possible link between the RNAi and IFN pathways (Sonkoly *et al.*, 2008). In fact, certain intermingling occurs between the RNAi machinery and the IFN pathway in HCV infected cells, because IFN induction leads to the expression of antiviral cellular miRNAs (Pedersen *et al.*, 2007).

The human genome encodes several hundred different miRNA molecules that are believed to be key players in the posttranscriptional regulation of many aspects of cellular differentiation. RNAi can be induced in mammalian cells by endogenous nuclear expression of RNAs that are intermediates in the miRNA biogenesis pathway (i.e. siRNAs and shRNAs). Nevertheless, it remains controversial whether the introduction of long dsRNA into mammalian somatic cells is capable of producing siRNAs. Thus, it has been postulated that RNAi exists in mammals only as a mechanism of post-transcriptional regulation programmed by endogenously encoded miRNA, and is not typically involved in intrinsic antiviral immunity (Cullen, 2006, 2009a,b).

All RNA viruses, except retroviruses, produce long dsRNA molecules in infected cells that represent essential intermediates in genomic RNA production. Many DNA viruses also generate long dsRNA, as their small, packed genomes have convergent transcription promoters. Some studies have attempted to identify siRNAs in virus-infected human cells. Pfeffer *et al.* (Pfeffer, 2007; Pfeffer *et al.*, 2004) failed to identify any viral siRNA in cells infected by DNA viruses, including human cytomegalovirus, Kaposi sarcoma-associated herpesvirus, Epstein–Barr virus and mouse herpesvirus. The same result was found for human cells infected with HIV-1 and the RNA viruses for yellow fever and hepatitis C. However, viral siRNA molecules have been described more recently for several endogenous and exogenous viruses, including HIV-1 (Bennasser *et al.*, 2005). The significance of these findings is still under debate (Lin and Cullen, 2007).

There is strong evidence that RNAi as part of the antiviral innate immune response in plants. Almost all plant viruses encode one or more RNAi silencing suppressor (RSS) proteins, which target several key steps in the RNAi response (Ding and Voinnet, 2007; Voinnet, 2005). Similarly, several viruses that target invertebrates (specifically nematodes and insects) also encode RSS proteins. Flock house virus, a member of the nodavirus family that infects both insects and vertebrate cells, encodes a viral RSS that inhibits Dicer function (Li *et al.*, 2002). Moreover, in insects like *Drosophila melanogaster*, RNAi mediates antiviral immunity. Based on observations that endogenously expressed RNA hairpins do not spread from cell to cell, it was previously thought that *Drosophila melanogaster* was unable to mount a systemic RNAi response. However, a recent study demonstrated that, upon virus infection, infected cells secrete a silencing signal that spreads systemically and elicits protective RNAi-dependent immunity throughout the organism (Saleh *et al.*, 2009) (see Chapter 1). Consequently, like protein-based immunity in vertebrates, the antiviral RNAi response in *Drosophila* relies on the systemic spread of a virus-specific immunity signal (Saleh *et al.*, 2009). A recent study reported an interesting finding that Sindbis virus, a virus that infects humans, but is maintained in nature by transmission between vertebrate (bird) hosts and invertebrate (mosquito) vectors, replicated with similar kinetics in Dicer-defective and wild-type murine fibroblasts; yet this virus is controlled with endogenous RNAi in *Drosophila* and other insects (van Rij, 2008). Arthropod-borne viruses, like the arbovirus Semliki Forest virus (SFV), generally counteract innate defences in vertebrate hosts and trigger cell death. In contrast, in mosquito cells, an early phase of efficient virus production is followed by a persistent infection with low levels of virus production, because SFV cannot prevent the establishment of an antiviral RNAi response or the cell to cell spread of protective antiviral dsRNA/siRNA that inhibits replication of incoming virus (Attarzadeh-Yazdi *et al.*, 2009). Thus, the spread of the RNAi signal contributes to limiting virus dissemination.

In contrast, it has been shown that HIV-1 replication can be restricted by RNAi machinery (Bennasser *et al.*, 2005; Triboulet *et al.*, 2007). Triboulet *et al.* (Triboulet *et al.*, 2007) provided evidence that RNAi silencing machinery played a physiological role in controlling HIV-1 replication. They showed that HIV-1 infection and replication occurred efficiently in peripheral blood mononuclear cells from HIV-1-infected donors and latently infected cells knocked-down for Dicer or Drosha. Moreover, HIV-1-mediated restriction of RNA silencing has been attributed to the viral transcriptional transactivator, Tat (Bennasser *et al.*, 2005). Benasser *et al.* determined that Tat RSS activity was genetically separable from Tat transcriptional activity, but segregated with the arginine-rich double-stranded RNA-binding domain (Bennasser *et al.*, 2005). A similar RNA binding domain is conserved in plant virus RSS that confers interaction with miRNA duplexes in a sequence nonspecific manner and blocks programming of RISC for RNA silencing (Ding and Voinnet, 2007). However, in another report, the stable expression of physiological levels of Tat did not globally inhibit miRNA production or expression in HIV-1 infected human cells (Lin and Cullen, 2007).

In support of the notion that the RNAi mechanism is part of the innate immune system in mammals, an increasing number of mammalian viruses have been shown to encode RSS proteins. For example, the HCV core and envelope protein 2 (Ji *et al.*, 2008; Wang *et al.*, 2006), vaccinia virus E3L (Li *et al.*, 2004), Ebola virus VP35

(Haasnoot *et al.*, 2007), primate foamy virus Tas (Lecellier *et al.*, 2005), 2005), and influenza A virus NS1 (Bucher *et al.*, 2004). HIV-1 Tat and HCV core have been proposed to block the activity of Dicer (Bennasser *et al.*, 2005; Chen *et al.*, 2008), while NS1, E3L, and VP35 probably act by sequestering dsRNA (Bucher *et al.*, 2004) (Haasnoot *et al.*, 2007; Li *et al.*, 2004). NS1 and VP35 can also trans-complement the production of a Tat-negative HIV-1 variant (Haasnoot *et al.*, 2007). Moreover, common features have been found between the RNA silencing suppression of plant and animal viruses. Recent studies have demonstrated that HIV-1 Tat and tomato bushy stunt virus P19 function equivalently in plant protoplasts and animal cells to suppress RNA silencing at a step downstream of the dsRNA processing, most likely by sequestering mature siRNAs or miRNAs (Qian *et al.*, 2009). Similarly, the non-structural protein 3 (NS3) of rice hoja blanca virus is a RSS that exclusively binds to small dsRNA molecules and lacks IFN antagonistic activity; yet it is able to substitute for the Tat protein in the RSS function of HIV-1 (Schnettler *et al.*, 2009). Overall, these studies support the significance of the innate antiviral RNAi response in mammalian cells.

Virally encoded miRNAs have been discovered in herpesviruses and human cytomegalovirus that help the viruses evade the host immune system by targeting host genes (Cullen, 2009a; Cullen, 2009b). The γ-herpesvirus Epstein–Barr virus encodes at least 23, the human γ-herpesvirus Kaposi's sarcoma-associated herpesvirus encodes 12, the β-herpesvirus human cytomegalovirus encodes 11, and the human α-herpesvirus herpes simplex virus 1 (HSV-1) encodes at least 6 miRNAs. Simian virus 40 (SV40) (a polyomavirus) encodes miRNAs that regulate viral gene expression, thus it reduces the recognition of infected cells by the immune system (Sullivan *et al.*, 2005). The therapeutic implications of this molecular feature are intriguing. Targeting these viral miRNAs might constitute an antiviral therapy, but mimicking their role could provide a means of immunosuppressive therapy (Cullen, 2006). Other, unrelated, DNA viruses also appear to encode either one or two miRNAs (for example, primate polyoma viruses and human adenoviruses) or none at all (Cullen, 2009b). It remains controversial whether RNA viruses also encode miRNAs (Cullen, 2006).

The retrovirus primate foamy virus type 1 (PFV-1) and vesicular stomatitis virus (VSV) are inhibited by cellular miRNAs known as miR-32 and miR-24 + miR-93, respectively (Lecellier *et al.*, 2005; Otsuka *et al.*, 2007). The expression of miR-24 and miR-93 is reduced in Dicer-deficient mice, thus VSV replication is greatly increased (Otsuka *et al.*, 2007). HIV-1 actively suppressed the expression of the polycistronic miRNA cluster, miR-17/92, and this suppression was found to be required for efficient viral replication (Triboulet *et al.*, 2007). As suggested in that study, these findings may be useful in addressing the current challenge of how to activate latent viral reservoirs in HIV-1 therapy. Latent infection is one of the most important characteristics required for the *in vivo* survival of all HIV-1 strains, and it presents a major obstacle in eradicating virus infection with HAART (Finzi *et al.*, 1997; Wong *et al.*, 1997). A recent study showed that cellular miRNAs potently inhibited HIV-1 production in resting primary CD4+ T cells (Huang *et al.*, 2007). The 3′ ends of HIV-1 mRNAs were found to be targeted by a cluster of cellular miRNAs, including miR-28, miR-125b, miR-150, miR-223, and miR-382; the expression of these miRNAs is elevated in resting CD4+ T cells compared to activated CD4+ T cells. Specific inhibitors of these miRNAs substantially counteracted their effects on the target mRNAs (Huang *et al.*, 2007); thus, a combined miRNA inhibitor panel could be used to activate latent HIV-1 for therapeutic purposes. In addition, the cellular miRNAs, miR-196, miR-296, miR-351, miR-431, and miR-448 have been shown to be capable of inhibiting HCV replication and infection (Pedersen *et al.*, 2007). These miRNAs are up-regulated by IFNα/β, suggesting that cellular miRNAs may be components of the mammalian innate immune response (Pedersen *et al.*, 2007). Human T-lymphotropic virus type 1 (HTLV-1) is the retroviral aetiological agent of adult T cell leukaemia, a severe and fatal lymphoproliferative disease of mainly CD4+ T cell origin that develops after prolonged viral persistence. HTLV-1 can down-regulate several cellular miRNAs, including 21, 24, 93, 130b, 146a, 155, and 223 (Pichler *et al.*, 2008; Yeung *et al.*, 2008). The pathways modulated by these miRNAs are critically involved in apoptosis, proliferation, and innate immune response. Thus, interference with cellular miRNAs may be crucial to maintaining persistence or may facilitate the transformation of host cells.

Another example of the intricate interplay between mammalian viruses and the host cell RNAi is the finding that cellular miR-122 stimulates HCV replication through a yet unknown mechanism (Jopling *et al.*, 2005). New antiviral therapeutics may involve targeting this miR-122 cofactor with antisense oligonucleotides that specifically inhibit its function. Because miRNA function is believed to be highly redundant, silencing a single miRNA might be well tolerated by a cell. Another potential therapeutic approach that relies on manipulation of the cellular miRNA pool might be the overexpression of cellular miRNAs that are harmful to the virus. As discussed above, HIV-1 inhibits the expression of the polycistronic miRNA cluster, miR-17/92; this results in increased expression of histone acetylase PCAF (P300/CBP [cAMP response element binding protein]-associated factor), which is required for optimal HIV-1 transcription (Triboulet *et al.*, 2007). Overexpression of miR20 has been shown to reduce PCAF expression, resulting in reduced virus replication. Furthermore, it has been suggested that overexpression of natural miRNAs may cause fewer adverse side-effects than treatments with man-made siRNAs (Haasnoot *et al.*, 2007). Similarly, if cellular miRNAs can target viral mRNAs, a therapy that silences the viral RNAi suppressor could rescue these miRNAs, resulting in a multiple miRNA attack (Lecellier *et al.*, 2005; Otsuka *et al.*, 2007).

Although it is unclear whether RSS proteins can interfere with the activity of RNAi-based therapeutics, the fact that mammalian viruses are efficiently inhibited by exogenous siRNAs or shRNAs indicates that viral suppression of RNAi activity does not preclude the use of RNAi as a therapeutic tool.

## RNA virus escape from RNA interference

This book highlights the evidence that supports RNAi as a robust method for specifically inhibiting the expression of targeted viral genes and its promise as a novel and broadly applicable approach to antiviral therapy. However, the clinical application of RNAi faces several challenges, particularly the potential for viral escape. One of the main advantages of the RNAi mechanism is that it is highly sequence specific. Pioneer studies confirmed the specificity of RNAi by evaluating the activity of an siRNA with one or more mismatches relative to the target RNA sequence (Jacque *et al.*, 2002). These studies showed that a single nucleotide mismatch may be sufficient to reduce the silencing effect, suggesting that the silencing effect did not involve a non-specific antiviral response induced by siRNA. However, this also showed that an RNA virus like HIV-1 may easily escape siRNA inhibition by small mutations. Indeed, RNA viruses have been observed to promptly escape suppression by effective siRNAs (Boden *et al.*, 2003; Das *et al.*, 2004; Gitlin *et al.*, 2002; Gitlin *et al.*, 2005; Haasnoot *et al.*, 2007; Kusov *et al.*, 2006; Soldan *et al.*, 2005; Westerhout *et al.*, 2005; Wilson and Richardson, 2005; Wu *et al.*, 2005; Zheng *et al.*, 2005). Among the RNA viruses that have exhibited escape from RNAi, there are important human pathogens, including HIV-1, HCV, HBV, HAV, and poliovirus.

Unlike eukaryotic DNA polymerases, RNA viruses lack proofreading activity; thus, the error rate during replication has been estimated at $10^{-4}$ to $10^{-5}$ mutations per nucleotide each cycle (Table 6.3) (Domingo *et al.*, 2006). If one assumes that an acutely infected organism may include $10^{9}$–$10^{12}$ viral particles at any given time, these must be the product of at least $10^{7}$–$10^{8}$ replication cycles. Given the length of the RNA virus genome (approximately 10 000 nucleotides), it is likely that every possible single point mutation ($10^{4}$) (and probably many double mutations) will occur by the time the population attains the size of many natural virus populations. In contrast, the total number of possible single mutations for a mammalian genome is about $10^{10}$, well above the population size of mammalian species. Although specific combinations of multiple mutations may be rare, it is clear that the degree of potential genetic change drives the diversification of RNA viruses in response to selective pressure of host immune responses or antiviral therapies (Table 6.3). An important example of the high adaptability of RNA viruses is the high frequency of virus mutants with one or a few amino acid substitutions that confer reduced sensitivity to antiviral inhibitors. This is a general phenomenon that has been documented for many viruses in the last decades, and it has made it very difficult to treat several viral diseases (Domingo, 2003). The best example of adaptive selection has been the HIV-1 virus mutants that are resistant to antiviral inhibitors. All currently available classes of antiretroviral therapy, i.e., reverse transcriptase, fusion, co-receptor antagonists, and integrase inhibitors exert selective pressure for mutations within the target gene that confers high-level drug resistance (Johnson *et al.*, 2008). Moreover, the capacity of novel compounds to exert selective pressure for a mutation is now used as evidence that the inhibitor has anti-HIV-1 activity. Experimental studies of HIV-1 populations have demonstrated that many resistant mutants may preexist in HIV-1 populations before they have been exposed to the inhibitors (Najera *et al.*, 1995). These resistant mutants may exist at very low frequencies in the naive viral population, but then selectively multiply in the presence of the inhibitor. The relative fitness values of wild-type and resistant mutants in the absence and presence of the inhibitor will determine the kinetics and degree of dominance of resistant mutants (Coffin, 1995).

Most RNAi escape studies have been performed with HIV-1. Boden *et al.* (Boden *et al.*, 2003) characterized the potency and durability of virus-specific RNAi in cell lines that stably expressed shRNA that targeted the HIV-1 transactivator protein gene Tat. They found that the antiviral activity of Tat shRNA was evaded by the emergence of viral species harbouring a point mutation in the shRNA target region. During the first three weeks of treatment, HIV-1 replication was reduced by 95% in cells expressing Tat shRNA compared to control cells. By day 25, however, viral titres increased, indicating a loss of Tat shRNA-mediated antiviral activity. The sequencing of this virus stock revealed that the emergent viral species contained a non-synonymous mutation at nucleotide position 9 of the targeted sequence. Das *et al.* (2004) used retroviral transduction to stably introduce vectors that expressed siRNAs directed against the HIV-1 Nef gene into human T cells. HIV-1 escape variants that were resistant to siRNA-Nef appeared after several weeks of culture, and these RNAi-resistant viruses contained point mutations, double point mutations, or partial or complete deletion of the Nef gene target sequence. The complete inactivation of the accessory Nef gene has a relatively minor impact on the replication capacity of HIV-1 *ex vivo*. Interestingly, Westerhout *et al.* (2005) used human T cells that stably expressed an siRNA-Nef target sequence to show that HIV-1 can also escape siRNA-mediated Nef inhibition by a point mutation outside the target sequence. This mutation changed local RNA folding into a structure that impeded the accessibility of the target sequence to the RNAi machinery. Recently, Sabariegos *et al.* (2006) showed that optimal HIV-1 gene silencing by siRNA requires precise complementarity with most of the target sequence and that only a few substitutions at the 5′ and 3′ ends are partially tolerated. Viral escape was simulated by systematically introducing single-nucleotide substitutions in all 19 HIV-1 residues targeted by an effective siRNA directed against the reverse transcriptase coding region (siRNA reverse transcriptase). All mutant viruses that were tested replicated better in the presence of the siRNA- reverse transcriptase than in the presence of the wild type virus. The antiviral activity of the siRNA- reverse transcriptase was completely neutralized by single substitutions in 10 positions (4 to 11, 14, and 15) out of 16 positions tested (substitutions at 3 of the 19 positions rendered nonviable viruses). With the exception of one position, the viral substitutions at either the 5′ or 3′ end of the target sequence were tolerated by the RNA interference machinery and allowed partial siRNA- reverse transcriptase inhibition. More recently, viral escape routes were analysed when essential and highly conserved sequences were targeted in the Gag, protease, integrase, and Tat-Rev regions of HIV-1 (von Eije *et al.*, 2008). Target sequences were analysed from more than 500 escape viruses that had mutated in T cells expressing individual shRNAs. Viruses acquired single point mutations, occasionally secondary mutations, but, in contrast to the mutations observed in the escape of Nef targeting, no deletions were detected. Mutations occurred predominantly at target positions 6, 8, 9, 14, and 15, but none were observed at positions 1, 2, 5, 18, and 19. The types of mismatch that occurred in the siRNA-target RNA duplexes were also analysed, and G–U mismatched pairs were most frequently observed. These results provided insight into the sequence requirements for optimal RNAi inhibition (von Eije *et al.*, 2008). Similar to HIV-1, other viruses have been able to escape from transient siRNA treatments, including poliovirus (Gitlin *et al.*, 2005), HCV (Wilson and Richardson, 2005), and hepatitis A virus (HAV) (Kusov *et al.*, 2006). In one study, siRNA directed against a 21-nucleotide segment within the poliovirus capsid or the viral polymerase coding sequences resulted in effective inhibition of viral replication. However, at 30 to 40h post-infection, resistant viruses emerged that contained silent mutations within the siRNA-target sequence (Gitlin *et al.*, 2005). Analysis of these viral escape mutants showed that RNAi recognition was sensitive to subtle point mutations within the central region and the 3′ end of the target RNA. Even single mutations resulting in G–U mismatches were sufficient for the virus to overcome inhibition by defined siRNAs. In another study, siRNA molecules designed to target the HCV genome blocked gene expression and RNA synthesis from hepatitis C replicons propagated in human liver cells. However, this block was not complete. After several treatments with a highly effective siRNA, growth of replicon RNAs was observed that were resistant to subsequent treatment with the same siRNA (Wilson and Richardson, 2005). As demonstrated with other viral systems, these replicon RNAs were not resistant to siRNAs that targeted another part of the genome. Sequence analysis of the siRNA-resistant replicons showed the generation of point mutations within the siRNA target sequence. Finally, siRNAs that targeted various sites in the HAV non-structural genes were transfected in human liver cells and resulted in an efficient and sustained silencing effect. Again, in most instances, multiple applications of a single siRNA led to the emergence of viral escape mutants with mutations in the target sites that rendered these genomes resistant to RNAi (Kusov *et al.*, 2006).

siRNAs may also exert selection pressure on pre-existing resistant mutants of HBV (Wu *et al.*, 2005). HBV is a DNA virus, but its DNA replicates through a genomic RNA intermediate and utilizes a virally encoded reverse transcriptase. Consequently, a significant amount of diversity, similar to that seen in RNA viruses, occurs in the sequences of HBV isolates. For example, a mutant was isolated that contained a substitution in the 3′ portion of the targeted region that rendered the virus more resistant to shRNA-mediated silencing than the wild-type HBV. In human liver cells co-transfected with mutant and wild-type HBV in the presence of shRNA, the observed mutant–wild type HBV expression was 9:1 at day 5 post-transfection. Knowledge regarding RNAi escape routes seems essential in designing and selecting siRNAs or shRNAs for the development of an effective RNAi therapy against RNA virus infections.

## DNA virus escape from RNA interference

In contrast to RNA viruses, which have a high rate of mutation during genome replication, the genomes of DNA organisms have lower mutation rates, ranging from $10^{-7}$ to $10^{-11}$ substitutions per incorporated nucleotide (Drake, 1991; Drake *et al.*, 1998). Data primarily from DNA bacteriophages has shown that DNA viruses may have a mutation rate in the neighbourhood of 0.003 mutations per genome replication (Drake and Hwang, 2005), which is much lower than that estimated for RNA viruses (0.1–1 per genome replication). The pathogenic human DNA virus, HSV-1, has an estimated genomic mutation rate compatible with the value of 0.003, which is close to the values observed in multicellular species. Nevertheless, estimations of mutation rates are scarce for mammalian and human DNA viruses. Similarly, low mutation rates have been estimated in some small dsDNA viruses, including the vertebrate-infecting papillomaviruses (Duffy *et al.*, 2008). Interestingly, some DNA viruses evolve rapidly. Mounting evidence shows that single-stranded (ss)DNA viruses (all with genomes smaller than ~13 kb) evolve at rates approaching those observed in their RNA counterparts (Duffy *et al.*, 2008). The first precise estimates for the rate of ssDNA virus evolution came from a study on canine parvovirus (CPV-2); a substitution rate of approximately $10^{-4}$ substitution/site/year was estimated (Shackelton *et al.*, 2005). This value is within the range observed in RNA viruses (Domingo *et al.*, 2006). Resistance to RNAi-mediated viral gene silencing through mutation of the target sequence has not been observed in DNA virus infections, although there have probably been very few attempts to search for RNAi escape mutants in DNA viruses.

Adenovirus infection may also result in the inhibition of RNAi. Adenoviruses are associated with highly structured, approximately 160-nucleotide long, non-coding RNAs that can inhibit RNAi at physiological levels of expression (Andersson *et al.*, 2005; Lu and Cullen, 2004). These viral-associated RNAs (VAs) are similar in structure to pre-miRNAs and are expressed at very high levels in adenovirus-infected cells. VAs have been shown to potently inhibit RNAi induced by shRNAs or human miRNA precursors, but did not affect RNAi induced by artificial siRNA duplexes. Inhibition appears to be due to a block in the nuclear export of shRNA or premiRNA precursors, competition for the Exportin 5 nuclear export factor, and inhibition of Dicer function by direct binding. Together, these data show that adenovirus infection can result in inhibition of RNAi through the action of VA RNAs in human cells. Blocking VA RNA-mediated inhibition has been shown to affect efficient adenovirus production, indicating that VA RNAs are required for virus viability. Thus, VA RNA-mediated silencing could represent a mechanism used by adenoviruses to control cellular or viral gene expression (Aparicio *et al.*, 2006).

RNA silencing of human papillomavirus 16 (HPV16) in cervical cancer cells has provided evidence on how a DNA virus might escape RNAi. A synthetic siRNA based on a sequence motif of 21 nucleotides from HPV16 E6E7 bicistronic RNA was found to potently suppress expression of both the E6 and E7 oncogenes in CaSki and SiHa cervical cancer cell lines; this led to stabilization of p53 and increased expression of p21 (Tang *et al.*, 2006). When stably expressed as a shRNA in these cells, however, efficient silencing of E6 and E7 expression only occurred in the early cell passages; silencing became inefficient with increased cell passages, despite the continued expression of the siRNA at the same level. The cells that were resistant to siRNA function retained normal siRNA processing, duplex unwinding, degradation of the unwound sense strand, and RNA-silencing-induced complex formation. This suggested that loss of the siRNA function occurred at a later step. Finally, the siRNA-resistant cells were found to express a cytoplasmic protein of approximately 50 kDa that specifically and characteristically interacted with the unwound, antisense strand of the E7 siRNA. These data indicate that potent siRNA targeting to an essential or regulatory gene might induce cells to develop siRNA-suppressive functional resistance through the production of a specific protein (Tang *et al.*, 2006; Zheng *et al.*, 2005).

## Other factors that might lead to RNAi resistance in mammalian cells

As discussed above, some viruses can escape from RNAi inhibition by developing active strategies (e.g. RSS proteins). However, the replication cycle of a virus may also provide an intrinsic protection against RNAi attack; this may be classified as a passive countermeasure (Haasnoot *et al.*, 2007). An important question arose soon after the demonstration that synthetic siRNAs were able to induce the RNAi mechanism and specifically inhibit viral replication: is the genomic viral RNA targeted by RNAi? Targeting the incoming viral RNA may have several advantages, because the RNAi machinery would only have to deal with two or a few (in the case of superinfection) genomic viral RNAs. The data has been conflicting with respect to whether RNAi can target the incoming genomic viral RNA of infecting HIV-1 particles. Several studies have reported degradation of the incoming RNA genome in cells transfected with artificial siRNAs or that stably express siRNAs (Coburn and Cullen, 2002; Jacque *et al.*, 2002). However, other studies have reported an absence of RNAi-mediated degradation of the genomic viral RNA in siRNA-transfected or siRNA-producing cells (Hu *et al.*, 2002; Surabhi and Gaynor, 2002). Moreover, a pioneer study with the non-segmented, negative-strand RNA of RSV showed that genomic and antigenomic viral RNAs encapsidated within the virus nucleocapsid proteins were not susceptible to siRNA-mediated silencing; but, viral mRNAs were highly susceptible to silencing (Bitko and Barik, 2001). Similarly, the nucleocapsid particle of Rous sarcoma virus (a lytic retrovirus) shielded the incoming viral RNA from RNAi-mediated degradation (Hu *et al.*, 2002). West Nile virus (WNV) replication and infection was greatly reduced when siRNAs were introduced by cytoplasmic-targeted transfection prior to, but not after, the establishment of viral replication, suggesting that WNV evade, rather than actively block, the RNAi machinery (Geiss *et al.*, 2005).

The mechanisms underlying the efficiency of specific siRNAs in targeting genes are not completely understood. RNAi efficiency is known to be influenced by the local RNA structure of the target sequence (Westerhout and Berkhout, 2007). Randomly selected siRNAs against a target sequence showed a large variation in their efficiency (Holen *et al.*, 2002). It has been argued that some regions of genomic viral RNA might be less accessible to siRNAs when the RNA is in the virus reverse transcription complex; therefore, some siRNAs might be more effective than others at inhibiting this early step in virus replication (Stevenson, 2003). Recently, a more exhaustive study was performed to address this issue (Westerhout *et al.*, 2006). In this study, the transduction efficiency of a lentiviral vector, measured by the number of successful integration events, was determined in a cell line that stably expressed a siRNA targeting the HIV-1 Nef sequence. The results indicated that the vector exhibited similar transduction efficiency regardless of whether it contained the Nef target sequence in the genome. Similar results were observed in the presence of multiple other stably expressed siRNAs that targeted the lentiviral vector genome, or when a synthetic siRNA targeting Nef was transiently transfected prior to lentiviral transduction. These results highlight the difficulties of designing therapeutic RNAi strategies aimed at preventing HIV-1 proviral integration. HIV-1 is capable of being integrated in the host genome and infecting resting CD4+T cells to create a pool of latently infected cells (Finzi *et al.*, 1997). Thus, the capacity to target incoming viral RNA by siRNAs has important therapeutic implications (Stevenson, 2003). If it is not possible to target incoming viral genomic RNA, it will be virtually impossible to prevent the formation of integrated provirus and, as a result, to sterilize cells from infection. Another layer of complexity was revealed in a study of picornavirus (coxsackievirus) silencing that showed the siRNA inhibitory efficiency correlated exclusively with the activity directed against the viral (+)-strand (Schubert *et al.*, 2007). The finding that only (+)-strand specific siRNAs exert significant antiviral potency may hold true for other RNA viruses with (+)-stranded genomes. Poliovirus replication has been observed to occur in complexes that are associated with membranous vesicles within cells (Egger *et al.*, 2000). It is therefore tempting to speculate that the (−)-strand is shielded by the membrane system proteins, or by the nascent RNA (+)-strands in the replication intermediate complex and therefore might not be accessible to RISC.

Another viral mechanism that has evolved for counteracting RNAi is the formation of highly structured RNAs that are inaccessible to Dicer or RISC and thus confer resistance to RNAi. It has been found that only siRNA targeted against hepatitis delta virus (HDV) mRNA sequences could interfere with HDV genome replication, which requires HBV as a helper virus for its replication. The siRNA targeted against genomic and antigenomic HDV RNA sequences had no detectable effect on the accumulation of HDV (Chang and Taylor, 2003). This may due, in part, to the fact that the circular, 1679-nuclotide genome of HDV and its exact complement, the antigenomic RNA, can fold into a rod-like structure with 74% pairing. In an *in vitro* cleavage assay with purified recombinant Dicer, less than 0.2% of unit-length HDV RNAs were cleaved, even though Dicer was able to digest fragments of the genome (Chang *et al.*, 2003). The data from these *in vitro* Dicer digestion experiments are consistent with the finding that small RNAs of 21 nucleotides could not be detected during HDV replication in hepatocytes (Chang *et al.*, 2003). The resistance of highly structured HDV RNAs against Dicer cleavage suggests the some viral RNAs may have evolved to prevent initiation of RNAi by resisting Dicer. Another example of RNAi resistance based on RNA structure has been suggested for rhinovirus, and similar mechanisms may occur in other picornaviruses (Haasnoot *et al.*, 2007). In fact, Gitlin *et al.* had difficulties in targeting siRNAs to the 5′ end of the poliovirus genome (Gitlin *et al.*, 2005).

## Strategies for minimizing viral escape

Not all viruses are equally inhibited by RNAi. Similarly, not all viral targets are equally suitable for effective inhibition by RNAi. In addition, siRNAs require effective delivery to the target cells and tissues for therapeutic applications. The two main strategies are the delivery of chemically synthesized siRNAs (non-viral delivery), and the delivery of engineered viruses that encode shRNA genes that will ultimately generate siRNAs by transcription in the target cells. Intranasal delivery has proved an effective means of delivering siRNA in mice (Bitko *et al.*, 2005) and in non-human primates (Li *et al.*, 2005) as a therapy for blocking RSV infection of the upper respiratory tract. The delivery of siRNAs to mucosal membranes has proven to be an effective approach in general (Castanotto and Rossi, 2009). Alnylam Pharmaceuticals has a leading candidate siRNA, ALN-RSV01, now in a phase II clinical trial (Table 6.1), that is targeted to the RSV nucleocapsid N gene, which is essential to viral replication. ALN-RSV01 was the first antiviral siRNA to enter clinical trials, and trials will soon be expanded to paediatric patients. In mice, intravaginal delivery of lipid-encapsulated siRNAs that target HSV-2 provided protection against lethal viral infections in more than two-thirds of the siRNA-treated group (Palliser *et al.*, 2006). Likewise, selective delivery of siRNAs to the central nervous system of mice by systemic intravenous injection was shown to be effective against Japanese encephalitis virus (Kumar *et al.*, 2007). These siRNA were conjugated with a short peptide derived from the rabies virus glycoprotein, which binds to the neuronal cell acetylcholine receptor. Eighty per cent of the treated mice survived infection with Japanese encephalitis virus (Kumar *et al.*, 2007). In contrast to acute virus infections, chronic infections with viruses like HIV-1 or HCV require long-term RNAi treatment to ensure a constant supply of intracellularly expressed antiviral siRNAs. Therefore, gene therapy approaches should aim to stably transduce cells with siRNA expression constructs. Stable transduction has been achieved with different viral vectors, like lentiviruses and AAVs (adeno-associated viruses), that carry shRNA expression constructs.

Because the emergence of resistant virus variants poses a serious problem for using RNAi in a therapeutic setting, different strategies have been developed to counteract viral escape (Table 6.4). Clearly, the first strategy for simultaneously inhibiting RNA virus diversity and protecting against the emergence of viral escape would be to identify conserved viral sequences for targeting. Targeting highly conserved sequences may compromise viral viability or development of escape mutants. Indeed, Lee *et al.* (Lee *et al.*, 2005) targeted highly conserved HIV-1 Vif sequences and successfully suppressed a variety of primary viral isolates from five different viral clades. This study also showed that tolerance to target sequence mismatches may depend on the sequence of the siRNA tested (Lee *et al.*, 2005). Nevertheless, our knowledge of HIV-1 variability and experience from over one decade of antiretroviral therapy has taught us that it is unrealistic to try to target RNA viruses with a unique siRNA, because the virus will rapidly develop resistance. The high replication and error rates of RNA viruses provide an advantage for viral mutation; thus, several potential escape variants of each of the 19 nucleotides targeted by a siRNA or shRNA are likely to be present before the initiation of therapy (Coffin, 1995). Moreover, RNA viruses may escape RNAi through silent mutations that do not affect viral fitness. To counteract this strategic advantage, therapies that involve co-expression of multiple siRNAs or shRNAs that target conserved RNA sequences could reduce the emergence of single siRNA-resistant viruses. This may have an effect comparable to that achieved by the multiple anti-HIV drug combination approach employed by HAART. Delivery of multiple siRNAs could induce highly active antiretroviral gene silencing (HAAGS) (Martinez *et al.*, 2002a).

Expression of an anti-HBV shRNA in the livers of HBV-transgenic mice resulted in efficient and persistent viral suppression (Grimm *et al.*, 2006). Notably, a combination of four shRNAs (NUC B1000) that target different sequences of the HBV genome has entered a phase 1 clinical trial (Table 6.1) (Castanotto and Rossi, 2009; Haasnoot and Berkhout, 2006). The combination of multiple shRNAs against conserved HIV-1 genomic regions effectively inhibits HIV-1 replication (Table 6.4). Expression of three different shRNAs from a single lentiviral vector resulted in similar levels of inhibition per shRNA compared to single shRNA vectors (ter Brake *et al.*, 2006). In that study, the three shRNAs targeted Gag and Pol HIV-1 coding regions. Moreover, when cells transduced with a double shRNA viral vector were infected, virus escape was delayed (ter Brake *et al.*, 2006). There are several methods currently in use that enable the expression of multiple effective siRNAs. One method is to insert multiple shRNA-expression cassettes into a viral vector. However, repeats of the same regulatory sequences (e.g. the U6 or H1 polymerase III promoter) may cause genetic instability and reduced titre of the vector system (Liu *et al.*, 2007). Moreover, some studies have shown that it is possible to saturate the RNAi pathway with high levels of shRNA expression (An *et al.*, 2006; Grimm *et al.*, 2006). This can result in cellular toxicity, particularly with saturated levels of the U6 promoter (An *et al.*, 2006). Polymerase III promoters other than U6 and H1 have also been used (Scherer *et al.*, 2007). Another method is to use long-hairpin RNAs (lhRNAs), from which multiple siRNAs can be produced. lhRNAs greater than 50 bp in length can be expressed in cells and create multiple siRNAs via Dicer-mediated processing without apparently inducing the IFN pathway (Akashi *et al.*, 2005). Akashi *et al.* used a plasmid that encoded a 51-bp, U6-driven shRNA for the efficient co-targeting of two NS5B genes that differed in nine nucleotides, derived from two distinct HCV strains. This lhRNA was able to suppress both isolates and also provided a more rapid knockdown than that observed with a conventional 20 nucleotide shRNA. Several reports described efficient RNAi induction by lhRNAs against HIV-1(Konstantinova *et al.*, 2006; Nishitsuji *et al.*, 2006). Nishitsuji *et al.* (Nishitsuji *et al.*, 2006) showed that a 50 bp lhRNA against a conserved HIV-1 integrase region could suppress replication of a HIV-1 variant that was resistant to a shorter shRNA against the same target. Similarly, lentiviral vectors expressing 50, 53, or 80 bp lhRNAs that targeted contiguous sequences within the Tat and Rev genes inhibited viral replication against both non-mutant and mutant variants of HIV-1 (Sano *et al.*, 2007).

An alternative to targeting multiple conserved viral genomic regions is to use second generation siRNAs that recognize the mutated target sites (Table 6.4) (Nishitsuji *et al.*, 2006; Sabariegos *et al.*, 2006; ter Brake and Berkhout, 2005; von Eije *et al.*, 2008); this has been accomplished by using siRNAs or shRNAs that target the most likely escape variants. Sabariegos *et al.* tested a second generation siRNA that targeted the reverse transcriptase gene of HIV-1, which counteracts a fully resistant mutation at position 9 of the target sequence (Sabariegos *et al.*, 2006). Nishitsuji *et al.* isolated single point mutations that conferred resistance to viruses that emerged in lentiviral vector transduced cells expressing shRNAs against the HIV-1 U3 region, integrase, and Tat genes (Nishitsuji *et al.*, 2006). They restored viral inhibition by using a second generation shRNA that matched the escape mutants (Nishitsuji *et al.*, 2006). Cell culture infections can be used to determine which target site mutations result in replication-competent escape variants; this will allow the design of siRNAs capable of inhibiting the escape variants (von Eije *et al.*, 2008). To reduce the number of second generation siRNAs that need to be designed, this strategy should target a conserved region of the viral genome sequence and be capable of inducing RNAi. Recently, *Escherichia coli* endoribonuclease III (RNase III) or mammalian Dicer was used to cleave dsRNA into endoribonuclease-prepared siRNA (esiRNA) (Kawasaki *et al.*, 2003; Myers *et al.*, 2003). This generated a variety of siRNAs that efficiently and specifically targeted multiple sites in the cognate RNA. To date, esiRNAs have been described that target poliovirus, HCV, and HIV (Gimenez-Barcons *et al.*, 2007; Gitlin *et al.*, 2005; Kronke *et al.*, 2004). Kronke *et al.* used esiRNAs directed against various regions of the HCV coding sequence and the 5′ untranslated region to efficiently block the replication of subgenomic and genomic HCV replicons (Kronke *et al.*, 2004). Similarly, esiRNAs derived from a poliovirus capsid dsRNA that was 1kb long protected HeLa cells against type 1 poliovirus, but not against type 3 (Gitlin *et al.*, 2005). Interestingly, treatment with esiRNAs did not allow selective expression of escape mutants during six to eight rounds of replication (Gitlin *et al.*, 2005). esiRNAs that targeted the HIV-1 reverse transcriptase coding region also reduced viral replication by 90% in a dose dependent and sequence specific manner (Gimenez-Barcons *et al.*, 2007). Importantly, esiRNAs derived from the prototypic reverse transcriptase sequence of the HXB2 strain and from highly mutated reverse transcriptase sequences showed similar degrees of viral inhibition, suggesting that the heterogeneous population of esiRNAs may overcome individual mismatches in the reverse transcriptase sequence (Gimenez-Barcons *et al.*, 2007). Overall, these studies have demonstrated the ability of esiRNAs to function as potent viral inhibitors. Moreover, the target sequences do not need to be highly conserved to attain a high level of viral replication inhibition. However, although this approach seemed useful as a way to inhibit multiple viral genotypes and to avert escape, its translation to an *in vivo* setting may be difficult.

Another recently explored strategy has been to combine a shRNA with an anti-HIV-1 ribozyme and an RNA decoy (Table 6.4) (Castanotto and Rossi, 2009). This mixture of RNAi and unrelated silencers may minimize the potential risks associated with high-level shRNA expression in the cell (Grimm and Kay, 2007). A triple combination lentiviral construct has been recently described that was comprised of (1) a U6-driven transactivation response region (TAR) RNA decoy with a U16 snoRNA for nucleolar localization, (2) a U6 promoted shRNA targeted to both Tat and Rev, and (3) a VA1 promoted chimeric anti-CCR5 trans-cleaving hammerhead ribozyme. This construct efficiently transduced human progenitor CD34+ cells and improved suppression of HIV-1 over 42 days compared to a single anti-Tat/Rev shRNA, double combinations of shRNA/ribozyme, or decoy (Anderson *et al.*, 2007; Li and Rossi, 2005). This triple combination is currently in a pilot feasibility study of a human clinical trial for AIDS/lymphoma patients (Table 6.3) (Castanotto and Rossi, 2009). In this clinical trial, autologous haematopoietic progenitor cells were used as the targets for vector insertion. A second trial in which the same construct will be inserted in autologous T lymphocytes will most likely be initiated in the near future (Scherer *et al.*, 2007).

In order to augment the genetic barrier for viral escape, cellular co-factors that are involved in the viral infection may be targeted. The combination of siRNAs or shRNAs directed against both viral and cellular transcripts may result in a more potent and robust inhibition of viral replication (Table 6.4). Nevertheless, targeting a cellular cofactor required by the virus should be performed with caution, because this approach has the potential to harm host cells. Several cellular factors involved in the HIV-1 life-cycle have been explored as possible RNAi targets, including the CD4 receptor and coreceptors CCR5 and CXCR4 (Martinez *et al.*, 2002b; Novina *et al.*, 2002; Qin *et al.*, 2003); integration factors like BAF-1, Emerin, and LEDGF/75 (Jacque and Stevenson, 2006; Llano *et al.*, 2006; Maertens *et al.*, 2003); transcriptional factors like NFκB, PAK-1, and cyclin (Chiu *et al.*, 2004; Nguyen *et al.*, 2006; Surabhi and Gaynor, 2002); and Furin, which is involved in Env maturation (Nguyen *et al.*, 2006). Because HIV-1 actively suppressed expression of the polycistronic miRNA cluster miR-17/92 and this suppression was found to be required for efficient viral replication (Triboulet *et al.*, 2007), another alternative for inhibiting HIV-1 replication may be to induce the nuclear expression of miRNAs from this cluster (Scherer *et al.*, 2007). *In vivo* suppression of genes like CD4 may have limited value, due to their roles in normal immune function. Thus, targeting HIV-1 coreceptors is a more attractive approach for host protein suppression. CCR5 may be an effective coreceptor target; a homozygous mutation in CCR5 effectively conferred protection from HIV-1 without any serious deleterious effects in human immune function, and heterozygous individuals with a 50% reduction in CCR5 surface expression had a lower plasma viral load and a substantially prolonged course of disease (Berger *et al.*, 1999). A potent and non-cytotoxic shRNA directed against CCR5 stably down-regulated CCR5 when introduced via CD34+ haematopoietic stem cell transplantation in non-human primates (An *et al.*, 2007). These transduced cells were less susceptible to simian immunodeficiency virus (SIV) infection *ex vivo*. Recently, several CCR5 antagonists examined in clinical trials were able to reduce plasma viral loads by more than one order of magnitude during treatment (Este and Telenti, 2007). Although CCR5 antagonists hold great promise, the long-term toxicity associated with the impairment of CCR5 function remains to be addressed. For HCV replication, inhibition of cellular cofactors proteasome alpha-subunit 7 (PSMA7) or Hu antigen R (HuR) successfully diminished HCV replicon RNA (Korf *et al.*, 2005). Interestingly, combinations of PSMA7- and HuR-directed siRNAs with HCV-directed siRNAs revealed additive HCV RNA inhibitory effects in monocistronic replicon cells. This suggested that a dual approach of direct- and co-factor-mediated inhibition of HCV replication might block the selective replication of mutants and thereby become a powerful therapeutic strategy against HCV (Korf *et al.*, 2005). Expression of shRNAs that simultaneously targeted two regions of the HCV 1b genome (IRES or NS5) and the host cell receptor CD81 were effective in inhibiting HCV replication by 82 and 98%, respectively (Henry *et al.*, 2006). These studies demonstrate that lentiviral delivery of multiple shRNAs that simultaneously target viral and host cell elements could increase the potency of anti-HCV gene therapies. DNA virus replication can be also inhibited by using RNAi to target cellular co-factors essential for virus viability, as shown with HSV (Henry *et al.*, 2006). RNAi-mediated knockdown of DNA ligase IV and its co-factor, XRCC4, caused a hundred-fold reduction of virus yield, a small plaque phenotype, and reduced DNA synthesis. The effect was specific, because RNAi against DNA ligase I or DNA ligase III failed to reduce HSV-1 replication. Furthermore, RNAi against DNA ligase IV and XRCC4 did not affect replication of adenovirus (Henry *et al.*, 2006). However, a fundamental concern with this strategy is that knocking down endogenous genes could create an unacceptable loss-of-function that is pathological for the cell. Therefore, targeting cellular factors should require extensive toxicity studies.

## Concluding remarks

The discovery of RNAi has provided us with powerful new tools for biological research and drug discovery. Moreover, RNAi-based therapies are currently advancing from basic research to clinical trials. RNAi has also expanded the direction of a new field of therapeutic strategies with the potential to treat a wide range of different diseases, including viral diseases. Several decades ago, the discovery of adaptive immunity allowed the potential for preventing the spread of many pathogenic human viruses. Nevertheless, the discovery of an effective preventive or therapeutic vaccine against several viruses remains elusive. RNAi therapeutics is rapidly arriving into clinical studies for many diseases, including several that are currently untreatable or difficult to treat with conventional drugs. However, viruses appear to be capable of responding to RNAi in different ways, including suppression of RNAi, mutational escape from RNAi, and modulation of the cell's miRNA/RNAi profile. Thus, the question remains as to whether RNAi will be suitable for treating or preventing virus infections.

## Acknowledgements

This study was supported by the Spanish Ministerio de Ciencia e Innovación (MICINN) project BFU2006-01066/BMC, Fondo de Investigaciones Sanitarias project PI07/0098, and Fundación para la Investigación y Prevención del SIDA en España (FIPSE) project 36557/06.

## References

Aigner, A. (2007). Applications of RNA interference: current state and prospects for siRNA-based strategies *in vivo*. DOI: 10.1007/s00253-007-0984-y. Appl Microbiol Biotechnol *76*, 9–21.

Akashi, H., Miyagishi, M., Yokota, T., Watanabe, T., Hino, T., Nishina, K., Kohara, M., and Taira, K. (2005). Escape from the interferon response associated with RNA interference using vectors that encode long modified hairpin-RNA. Mol Biosyst *1*, 382–390.

An, D.S., Donahue, R.E., Kamata, M., Poon, B., Metzger, M., Mao, S.H., Bonifacino, A., Krouse, A.E., Darlix, J.L., Baltimore, D., *et al.* (2007). Stable reduction of CCR5 by RNAi through hematopoietic stem cell transplant in non-human primates. Proc. Natl. Acad. Sci. U.S.A. *104*, 13110–13115.

An, D.S., Qin, F.X., Auyeung, V.C., Mao, S.H., Kung, S.K., Baltimore, D., and Chen, I.S. (2006). Optimization and functional effects of stable short hairpin RNA expression in primary human lymphocytes via lentiviral vectors. DOI: 10.1016/j.ymthe.2006.05.015. Mol. Ther. *14*, 494–504.

Anderson, J., Li, M.J., Palmer, B., Remling, L., Li, S., Yam, P., Yee, J.K., Rossi, J., Zaia, J., and Akkina, R. (2007). Safety and efficacy of a lentiviral vector containing three anti-HIV genes--CCR5 ribozyme, tat-rev siRNA, and TAR decoy--in SCID-hu mouse-derived T cells. Mol. Ther. *15*, 1182–1188.

Andersson, M.G., Haasnoot, P.C., Xu, N., Berenjian, S., Berkhout, B., and Akusjarvi, G. (2005). Suppression of RNA interference by adenovirus virus-associated RNA. J. Virol. *79*, 9556–9565.

Aparicio, O., Razquin, N., Zaratiegui, M., Narvaiza, I., and Fortes, P. (2006). Adenovirus virus-associated RNA is processed to functional interfering RNAs involved in virus production. J. Virol. *80*, 1376–1384.

Attarzadeh-Yazdi, G., Fragkoudis, R., Chi, Y., Siu, R.W., Ulper, L., Barry, G., Rodriguez-Andres, J., Nash, A.A., Bouloy, M., Merits, A., *et al.* (2009). Cell to cell spread of the RNAi response suppresses Semliki Forest virus infection of mosquito cell cultures and cannot be antagonised by this virus. J. Virol. *83*, 5735–5748.

Bennasser, Y., Le, S.Y., Benkirane, M., and Jeang, K.T. (2005). Evidence that HIV-1 encodes an siRNA and a suppressor of RNA silencing. DOI: 10.1016/j.immuni.2005.03.010. Immunity *22*, 607–619.

Berger, E.A., Murphy, P.M., and Farber, J.M. (1999). Chemokine receptors as HIV-1 coreceptors: roles in viral entry, tropism, and disease. Annu. Rev. Immunol. *17*, 657–700.

Bitko, V., and Barik, S. (2001). Phenotypic silencing of cytoplasmic genes using sequence-specific double-stranded short interfering RNA and its application in the reverse genetics of wild type negative-strand RNA viruses. BMC Microbiol. *1*, 34.

Bitko, V., Musiyenko, A., Shulyayeva, O., and Barik, S. (2005). Inhibition of respiratory viruses by nasally administered siRNA. Nat. Med. *11*, 50–55.

Boden, D., Pusch, O., Lee, F., Tucker, L., and Ramratnam, B. (2003). Human immunodeficiency virus type 1 escape from RNA interference. J. Virol. *77*, 11531–11535.

Bucher, E., Hemmes, H., de Haan, P., Goldbach, R., and Prins, M. (2004). The influenza A virus NS1 protein binds small interfering RNAs and suppresses RNA silencing in plants. J. Gen. Virol. *85*, 983–991.

Castanotto, D., and Rossi, J.J. (2009). The promises and pitfalls of RNA-interference-based therapeutics. Nature *457*, 426–433.

Chang, J., Provost, P., and Taylor, J.M. (2003). Resistance of human hepatitis delta virus RNAs to dicer activity. J. Virol. *77*, 11910–11917.

Chang, J., and Taylor, J.M. (2003). Susceptibility of human hepatitis delta virus RNAs to small interfering RNA action. J. Virol. *77*, 9728–9731.

Chen, W., Zhang, Z., Chen, J., Zhang, J., Wu, Y., Huang, Y., Cai, X., and Huang, A. (2008). HCV core protein interacts with Dicer to antagonize RNA silencing. Virus Res. *133*, 250–258.

Chiu, Y.L., Cao, H., Jacque, J.M., Stevenson, M., and Rana, T.M. (2004). Inhibition of human immunodeficiency virus type 1 replication by RNA interference directed against human transcription elongation factor P-TEFb (CDK9/CyclinT1). J. Virol. *78*, 2517–2529.

Coburn, G.A., and Cullen, B.R. (2002). Potent and specific inhibition of human immunodeficiency virus type 1 replication by RNA interference. J. Virol. *76*, 9225–9231.

Coffin, J.M. (1995). HIV population dynamics *in vivo*: implications for genetic variation, pathogenesis, and therapy. Science *267*, 483–489.

Cullen, B.R. (2006). Is RNA interference involved in intrinsic antiviral immunity in mammals? DOI:10.1038/ni1352. Nat. Immunol. *7*, 563–567.

Cullen, B.R. (2009a). Viral and cellular messenger RNA targets of viral microRNAs. Nature *457*, 421–425.

Cullen, B.R. (2009b). Viral RNAs: lessons from the enemy. Cell *136*, 592–597.

Das, A.T., Brummelkamp, T.R., Westerhout, E.M., Vink, M., Madiredjo, M., Bernards, R., and Berkhout, B. (2004). Human immunodeficiency virus type 1 escapes from RNA interference-mediated inhibition. J. Virol. *78*, 2601–2605.

Ding, S.W., and Voinnet, O. (2007). Antiviral immunity directed by small RNAs. DOI:10.1016/j.cell.2007.07.039. Cell *130*, 413–426.

Domingo, E. (2003). Quasispecies and the development of new antiviral strategies. Prog Drug Res. *60*, 133–158.

Domingo, E., Martin, V., Perales, C., Grande-Perez, A., Garcia-Arriaza, J., and Arias, A. (2006). Viruses as quasispecies: biological implications. Curr. Top. Microbiol. Immunol. *299*, 51–82.

Drake, J.W. (1991). A constant rate of spontaneous mutation in DNA-based microbes. Proc. Natl. Acad. Sci. U.S.A. *88*, 7160–7164.

Drake, J.W., Charlesworth, B., Charlesworth, D., and Crow, J.F. (1998). Rates of spontaneous mutation. Genetics *148*, 1667–1686.

Drake, J.W., and Hwang, C.B. (2005). On the mutation rate of herpes simplex virus type 1. Genetics *170*, 969–970.

Duffy, S., Shackelton, L.A., and Holmes, E.C. (2008). Rates of evolutionary change in viruses: patterns and determinants. Nat Rev Genet *9*, 267–276.

Dykxhoorn, D.M., and Lieberman, J. (2006). Running interference: prospects and obstacles to using small interfering RNAs as small molecule drugs. DOI:10.1146/annurev.bioeng.8.061505.095848. Annu Rev Biomed Eng *8*, 377–402.

Egger, D., Teterina, N., Ehrenfeld, E., and Bienz, K. (2000). Formation of the poliovirus replication complex requires coupled viral translation, vesicle production, and viral RNA synthesis. J. Virol. *74*, 6570–6580.

Elbashir, S.M., Harborth, J., Lendeckel, W., Yalcin, A., Weber, K., and Tuschl, T. (2001). Duplexes of 21-nucleotide RNAs mediate RNA interference in cultured mammalian cells. DOI:10.1038/35078107. Nature *411*, 494–498.

Este, J.A., and Telenti, A. (2007). HIV entry inhibitors. Lancet *370*, 81–88.

Finzi, D., Hermankova, M., Pierson, T., Carruth, L.M., Buck, C., Chaisson, R.E., Quinn, T.C., Chadwick, K., Margolick, J., Brookmeyer, R., *et al.* (1997). Identification of a reservoir for HIV-1 in patients on highly active antiretroviral therapy. Science *278*, 1295–1300.

Fire, A., Xu, S., Montgomery, M.K., Kostas, S.A., Driver, S.E., and Mello, C.C. (1998). Potent and specific genetic interference by double-stranded RNA in Caenorhabditis elegans. DOI:10.1038/35888. Nature *391*, 806–811.

Garcia-Sastre, A., and Biron, C.A. (2006). Type 1 interferons and the virus–host relationship: a lesson in detente. Science *312*, 879–882.

Ge, Q., McManus, M.T., Nguyen, T., Shen, C.H., Sharp, P.A., Eisen, H.N., and Chen, J. (2003). RNA interference of influenza virus production by directly targeting mRNA for degradation and indirectly inhibiting all viral RNA transcription. DOI: 10.1073/pnas.0437841100. Proc. Natl. Acad. Sci. U.S.A. *100*, 2718–2723.

Geiss, B.J., Pierson, T.C., and Diamond, M.S. (2005). Actively replicating West Nile virus is resistant to cytoplasmic delivery of siRNA. Virol J *2*, 53.

Ghosh, Z., Mallick, B., and Chakrabarti, J. (2009). Cellular versus viral microRNAs in host–virus interaction. Nucleic Acids Res. *37*, 1035–1048.

Gimenez-Barcons, M., Clotet, B., and Martinez, M.A. (2007). Endoribonuclease-prepared short interfering RNAs induce effective and specific inhibition of human immunodeficiency virus type 1 replication. J. Virol. *81*, 10680–10686.

Gitlin, L., Karelsky, S., and Andino, R. (2002). Short interfering RNA confers intracellular antiviral immunity in human cells. DOI: 10.1038/nature00873. Nature *418*, 430–434.

Gitlin, L., Stone, J.K., and Andino, R. (2005). Poliovirus escape from RNA interference: short interfering RNA-target recognition and implications for therapeutic approaches DOI: 10.1128/JVI.79.2.1027-1035.2005. J. Virol. *79*, 1027–1035.

Gottwein, E., and Cullen, B.R. (2008). Viral and cellular microRNAs as determinants of viral pathogenesis and immunityDOI: 10.1016/j.chom.2008.05.002. Cell Host Microbe *3*, 375–387.

Grassmann, R., and Jeang, K.T. (2008). The roles of microRNAs in mammalian virus infection DOI: 10.1016/j.bbagrm.2008.05.005. Biochim. Biophys. Acta *1779*, 706–711.

Grimm, D., and Kay, M.A. (2007). Combinatorial RNAi: a winning strategy for the race against evolving targets? Mol. Ther. *15*, 878–888.

Grimm, D., Streetz, K.L., Jopling, C.L., Storm, T.A., Pandey, K., Davis, C.R., Marion, P., Salazar, F., and Kay, M.A. (2006). Fatality in mice due to oversaturation of cellular microRNA/short hairpin RNA pathways. Nature *441*, 537–541.

Grimson, A., Farh, K.K., Johnston, W.K., Garrett-Engele, P., Lim, L.P., and Bartel, D.P. (2007). MicroRNA targeting specificity in mammals: determinants beyond seed pairing DOI: 10.1016/j.molcel.2007.06.017. Mol. Cell *27*, 91–105.

Haasnoot, J., and Berkhout, B. (2006). RNA interference: its use as antiviral therapy. Handb Exp Pharmacol, 117–150.

Haasnoot, J., Westerhout, E.M., and Berkhout, B. (2007). RNA interference against viruses: strike and counterstrike. Nat. Biotechnol. *25*, 1435–1443.

He, M.L., Zheng, B., Peng, Y., Peiris, J.S., Poon, L.L., Yuen, K.Y., Lin, M.C., Kung, H.F., and Guan, Y. (2003). Inhibition of SARS-associated coronavirus infection and replication by RNA interference. Jama *290*, 2665–2666.

Henry, S.D., van der Wegen, P., Metselaar, H.J., Tilanus, H.W., Scholte, B.J., and van der Laan, L.J. (2006). Simultaneous targeting of HCV replication and viral binding with a single lentiviral vector containing multiple RNA interference expression cassettes. Mol. Ther. *14*, 485–493.

Holen, T., Amarzguioui, M., Wiiger, M.T., Babaie, E., and Prydz, H. (2002). Positional effects of short interfering RNAs targeting the human coagulation trigger Tissue Factor. Nucleic Acids Res. *30*, 1757–1766.

Hu, W.Y., Myers, C.P., Kilzer, J.M., Pfaff, S.L., and Bushman, F.D. (2002). Inhibition of retroviral pathogenesis by RNA interference. Curr Biol *12*, 1301–1311.

Huang, J., Wang, F., Argyris, E., Chen, K., Liang, Z., Tian, H., Huang, W., Squires, K., Verlinghieri, G., and Zhang, H. (2007). Cellular microRNAs contribute to HIV-1 latency in resting primary CD4+ T lymphocytes DOI: 10.1038/nm1639. Nat. Med. *13*, 1241–1247.

Jacque, J.M., and Stevenson, M. (2006). The inner-nuclear-envelope protein emerin regulates HIV-1 infectivity. Nature *441*, 641–645.

Jacque, J.M., Triques, K., and Stevenson, M. (2002). Modulation of HIV-1 replication by RNA interference. Nature *418*, 435–438.

Ji, J., Glaser, A., Wernli, M., Berke, J.M., Moradpour, D., and Erb, P. (2008). Suppression of short interfering RNA-mediated gene silencing by the structural proteins of hepatitis C virus. J. Gen. Virol. *89*, 2761–2766.

Johnson, V.A., Brun-Vezinet, F., Clotet, B., Gunthard, H.F., Kuritzkes, D.R., Pillay, D., Schapiro, J.M., and Richman, D.D. (2008). Update of the Drug Resistance Mutations in HIV-1. Top HIV Med *16*, 138–145.

Jopling, C.L., Yi, M., Lancaster, A.M., Lemon, S.M., and Sarnow, P. (2005). Modulation of hepatitis C virus RNA abundance by a liver-specific MicroRNA. DOI: 10.1126/science.1113329. Science *309*, 1577–1581.

Kawasaki, H., Suyama, E., Iyo, M., and Taira, K. (2003). siRNAs generated by recombinant human Dicer induce specific and significant but target site-independent gene silencing in human cells. Nucleic Acids Res. *31*, 981–987.

Konstantinova, P., de Vries, W., Haasnoot, J., ter Brake, O., de Haan, P., and Berkhout, B. (2006). Inhibition of human immunodeficiency virus type 1 by RNA interference using long-hairpin RNA. Gene Ther. *13*, 1403–1413.

Korf, M., Jarczak, D., Beger, C., Manns, M.P., and Kruger, M. (2005). Inhibition of hepatitis C virus translation and subgenomic replication by siRNAs directed against highly conserved HCV sequence and cellular HCV cofactors. J Hepatol 43, 225–234.

Kronke, J., Kittler, R., Buchholz, F., Windisch, M.P., Pietschmann, T., Bartenschlager, R., and Frese, M. (2004). Alternative approaches for efficient inhibition of hepatitis C virus RNA replication by small interfering RNAs. J. Virol. *78*, 3436–3446.

Kumar, P., Wu, H., McBride, J.L., Jung, K.E., Kim, M.H., Davidson, B.L., Lee, S.K., Shankar, P., and Manjunath, N. (2007). Transvascular delivery of small interfering RNA to the central nervous system. Nature *448*, 39–43.

Kusov, Y., Kanda, T., Palmenberg, A., Sgro, J.Y., and Gauss-Muller, V. (2006). Silencing of hepatitis A virus infection by small interfering RNAs. J. Virol. *80*, 5599–5610.

Lagos-Quintana, M., Rauhut, R., Lendeckel, W., and Tuschl, T. (2001). Identification of novel genes coding for small expressed RNAs. Science *294*, 853–858.

Lau, N.C., Lim, L.P., Weinstein, E.G., and Bartel, D.P. (2001). An abundant class of tiny RNAs with probable regulatory roles in Caenorhabditis elegans. Science *294*, 858–862.

Lecellier, C.H., Dunoyer, P., Arar, K., Lehmann-Che, J., Eyquem, S., Himber, C., Saib, A., and Voinnet, O. (2005). A cellular microRNA mediates antiviral defense in human cells. DOI: 10.1126/science.1108784. Science *308*, 557–560.

Lee, N.S., Dohjima, T., Bauer, G., Li, H., Li, M.J., Ehsani, A., Salvaterra, P., and Rossi, J. (2002). Expression of small interfering RNAs targeted against HIV-1 rev transcripts in human cells. Nat. Biotechnol. *20*, 500–505.

Lee, R.C., and Ambros, V. (2001). An extensive class of small RNAs in Caenorhabditis elegans. Science *294*, 862–864.

Lee, S.K., Dykxhoorn, D.M., Kumar, P., Ranjbar, S., Song, E., Maliszewski, L.E., Francois-Bongarcon, V., Goldfeld, A., Swamy, N.M., Lieberman, J., and Shankar, P. (2005). Lentiviral delivery of short hairpin RNAs protects CD4 T cells from multiple clades and primary isolates of HIV. Blood *106*, 818–826.

Li, B.J., Tang, Q., Cheng, D., Qin, C., Xie, F.Y., Wei, Q., Xu, J., Liu, Y., Zheng, B.J., Woodle, M.C., *et al.* (2005). Using siRNA in prophylactic and therapeutic regimens against SARS coronavirus in Rhesus macaque. Nat. Med. *11*, 944–951.

Li, H., Li, W.X., and Ding, S.W. (2002). Induction and suppression of RNA silencing by an animal virus. Science *296*, 1319–1321.

Li, M., and Rossi, J.J. (2005). Lentiviral vector delivery of siRNA and shRNA encoding genes into cultured and primary hematopoietic cells. Methods Mol Biol *309*, 261–272.

Li, W.X., Li, H., Lu, R., Li, F., Dus, M., Atkinson, P., Brydon, E.W., Johnson, K.L., Garcia-Sastre, A., Ball, L.A., *et al.* (2004). Interferon antagonist proteins of influenza and vaccinia viruses are suppressors of RNA silencing. Proc. Natl. Acad. Sci. U.S.A. *101*, 1350–1355.

Lin, J., and Cullen, B.R. (2007). Analysis of the Interaction of Primate Retroviruses with the Human RNA Interference Machinery. J. Virol. *81*, 12218–12226.

Liu, Y.P., Haasnoot, J., and Berkhout, B. (2007). Design of extended short hairpin RNAs for HIV-1 inhibition. Nucleic Acids Res. *35*, 5683–5693.

Llano, M., Saenz, D.T., Meehan, A., Wongthida, P., Peretz, M., Walker, W.H., Teo, W., and Poeschla, E.M. (2006). An essential role for LEDGF/p75 in HIV integration. Science *314*, 461–464.

Lu, S., and Cullen, B.R. (2004). Adenovirus VA1 noncoding RNA can inhibit small interfering RNA and MicroRNA biogenesis. J. Virol. *78*, 12868–12876.

Lund, E., Guttinger, S., Calado, A., Dahlberg, J.E., and Kutay, U. (2004). Nuclear export of microRNA precursors. DOI: 10.1126/science.1090599. Science *303*, 95–98.

Maertens, G., Cherepanov, P., Pluymers, W., Busschots, K., De Clercq, E., Debyser, Z., and Engelborghs, Y. (2003). LEDGF/p75 is essential for nuclear and chromosomal targeting of HIV-1 integrase in human cells. J. Biol. Chem. *278*, 33528–33539.

Martinez, M.A., Clotet, B., and Este, J.A. (2002a). RNA interference of HIV replication. Trends Immunol 23, 559–561.

Martinez, M.A., Gutierrez, A., Armand-Ugon, M., Blanco, J., Parera, M., Gomez, J., Clotet, B., and Este, J.A. (2002b). Suppression of chemokine receptor expression by RNA interference allows for inhibition of HIV-1 replication. Aids *16*, 2385–2390.

Matranga, C., and Zamore, P.D. (2007). Small silencing RNAs. DOI:10.1016/j.cub.2007.07.014. Curr. Biol. *17*, R789–793.

McCaffrey, A.P., Nakai, H., Pandey, K., Huang, Z., Salazar, F.H., Xu, H., Wieland, S.F., Marion, P.L., and Kay, M.A. (2003). Inhibition of hepatitis B virus in mice by RNA interference. Nat. Biotechnol. *21*, 639–644.

Myers, J.W., Jones, J.T., Meyer, T., and Ferrell, J.E., Jr. (2003). Recombinant Dicer efficiently converts large dsRNAs into siRNAs suitable for gene silencing. Nat. Biotechnol. *21*, 324–328.

Najera, I., Holguin, A., Quinones-Mateu, M.E., Munoz-Fernandez, M.A., Najera, R., Lopez-Galindez, C., and Domingo, E. (1995). Pol gene quasispecies of human immunodeficiency virus: mutations associated with drug resistance in virus from patients undergoing no drug therapy. J. Virol. *69*, 23–31.

Nguyen, D.G., Wolff, K.C., Yin, H., Caldwell, J.S., and Kuhen, K.L. (2006). 'UnPAKing' human immunodeficiency virus (HIV) replication: using small interfering RNA screening to identify novel cofactors and elucidate the role of group I PAKs in HIV infection. J. Virol. *80*, 130–137.

Nishitsuji, H., Kohara, M., Kannagi, M., and Masuda, T. (2006). Effective suppression of human immunodeficiency virus type 1 through a combination of short- or long-hairpin RNAs targeting essential sequences for retroviral integration. J. Virol. *80*, 7658–7666.

Novina, C.D., Murray, M.F., Dykxhoorn, D.M., Beresford, P.J., Riess, J., Lee, S.K., Collman, R.G., Lieberman, J., Shankar, P., and Sharp, P.A. (2002). siRNA-directed inhibition of HIV-1 infection. Nat. Med. *8*, 681–686.

Otsuka, M., Jing, Q., Georgel, P., New, L., Chen, J., Mols, J., Kang, Y.J., Jiang, Z., Du, X., Cook, R., *et al.* (2007). Hypersusceptibility to vesicular stomatitis virus infection in Dicer1-deficient mice is due to impaired miR24 and miR93 expression. Immunity 27, 123–134.

Palliser, D., Chowdhury, D., Wang, Q.Y., Lee, S.J., Bronson, R.T., Knipe, D.M., and Lieberman, J. (2006). An siRNA-based microbicide protects mice from lethal herpes simplex virus 2 infection. Nature *439*, 89–94.

Pedersen, I.M., Cheng, G., Wieland, S., Volinia, S., Croce, C.M., Chisari, F.V., and David, M. (2007). Interferon modulation of cellular microRNAs as an antiviral mechanism. DOI: 10.1038/nature06205. Nature *449*, 919–922.

Pfeffer, S. (2007). Identification of virally encoded microRNAs. DOI: 10.1016/S0076-6879(07)27003-X. Methods Enzymol 427, 51–63.

Pfeffer, S., Zavolan, M., Grasser, F.A., Chien, M., Russo, J.J., Ju, J., John, B., Enright, A.J., Marks, D., Sander, C., and Tuschl, T. (2004). Identification of virus-encoded microRNAs. DOI: 10.1126/science.1096781. Science *304*, 734–736.

Pichler, K., Schneider, G., and Grassmann, R. (2008). MicroRNA miR-146a and further oncogenesis-related cellular microRNAs are dysregulated in HTLV-1-transformed T lymphocytes. DOI: 10.1186/1742-4690-5-100. Retrovirology *5*, 100.

Qian, S., Zhong, X., Yu, L., Ding, B., de Haan, P., and Boris-Lawrie, K. (2009). HIV-1 Tat RNA silencing suppressor activity is conserved across kingdoms and counteracts translational repression of HIV-1. DOI: 10.1073/pnas.0806822106. Proc. Natl. Acad. Sci. U.S.A. *106*, 605–610.

Qin, X.F., An, D.S., Chen, I.S., and Baltimore, D. (2003). Inhibiting HIV-1 infection in human T cells by lentiviral-mediated delivery of small interfering RNA against CCR5. Proc. Natl. Acad. Sci. U.S.A. *100*, 183–188.

Sabariegos, R., Gimenez-Barcons, M., Tapia, N., Clotet, B., and Martinez, M.A. (2006). Sequence homology required by human immunodeficiency virus type 1 to escape from short interfering RNAs. DOI: 10.1128/JVI.80.2.571-577.2006. J. Virol. *80*, 571–577.

Saleh, M.C., Tassetto, M., van Rij, R.P., Goic, B., Gausson, V., Berry, B., Jacquier, C., Antoniewski, C., and Andino, R. (2009). Antiviral immunity in *Drosophila* requires systemic RNA interference spread. Nature *458*, 346–350.

Sano, M., Li, H., Nakanishi, M., and Rossi, J.J. (2007). Expression of Long Anti-HIV-1 Hairpin RNAs for the Generation of Multiple siRNAs: Advantages and Limitations. Mol. Ther. *16*, 170–177.

Scherer, L.J., Frank, R., and Rossi, J.J. (2007). Optimization and characterization of tRNA-shRNA expression constructs. Nucleic Acids Res. 35, 2620–2628.

Schnettler, E., de Vries, W., Hemmes, H., Haasnoot, J., Kormelink, R., Goldbach, R., and Berkhout, B. (2009). The NS3 protein of rice hoja blanca virus complements the RNAi suppressor function of HIV-1 Tat. DOI: 10.1038/embor.2009.6. EMBO Rep *10*, 258–263.

Schubert, S., Rothe, D., Werk, D., Grunert, H.P., Zeichhardt, H., Erdmann, V.A., and Kurreck, J. (2007). Strand-specific silencing of a picornavirus by RNA interference: evidence for the superiority of plus-strand specific siRNAs. Antiviral Res. 73, 197–205.

Shackelton, L.A., Parrish, C.R., Truyen, U., and Holmes, E.C. (2005). High rate of viral evolution associated with the emergence of carnivore parvovirus. Proc. Natl. Acad. Sci. U.S.A. *102*, 379–384.

Shlomai, A., and Shaul, Y. (2003). Inhibition of hepatitis B virus expression and replication by RNA interference. Hepatology *37*, 764–770.

Soldan, S.S., Plassmeyer, M.L., Matukonis, M.K., and Gonzalez-Scarano, F. (2005). La Crosse virus nonstructural protein NSs counteracts the effects of short interfering RNA. J. Virol. *79*, 234–244.

Sonkoly, E., Stahle, M., and Pivarcsi, A. (2008). MicroRNAs and immunity: novel players in the regulation of normal immune function and inflammation. Semin Cancer Biol *18*, 131–140.

Stevenson, M. (2003). Dissecting HIV-1 through RNA interference. Nat. Rev. Immunol. *3*, 851–858.

Sullivan, C.S., Grundhoff, A.T., Tevethia, S., Pipas, J.M., and Ganem, D. (2005). SV40-encoded microRNAs regulate viral gene expression and reduce susceptibility to cytotoxic T cells. Nature *435*, 682–686.

Surabhi, R.M., and Gaynor, R.B. (2002). RNA interference directed against viral and cellular targets inhibits human immunodeficiency Virus Type 1 replication. J. Virol. *76*, 12963–12973.

Tang, S., Tao, M., McCoy, J.P., Jr., and Zheng, Z.M. (2006). Short-term induction and long-term suppression of HPV16 oncogene silencing by RNA interference in cervical cancer cells. Oncogene *25*, 2094–2104.

ter Brake, O., and Berkhout, B. (2005). A novel approach for inhibition of HIV-1 by RNA interference: counteracting viral escape with a second generation of siRNAs. Journal of RNAi and Gene Silencing *1*, 56–65.

ter Brake, O., Konstantinova, P., Ceylan, M., and Berkhout, B. (2006). Silencing of HIV-1 with RNA interference: a multiple shRNA approach. DOI: 10.1016/j.ymthe.2006.07.007. Mol. Ther. *14*, 883–892.

Triboulet, R., Mari, B., Lin, Y.L., Chable-Bessia, C., Bennasser, Y., Lebrigand, K., Cardinaud, B., Maurin, T., Barbry, P., Baillat, V., *et al.* (2007). Suppression of microRNA-silencing pathway by HIV-1 during virus replication. DOI: 10.1126/science.1136319. Science *315*, 1579–1582.

van Rij, R.P. (2008). Virus meets RNAi. Symposium on antiviral applications of RNA interference. EMBO Rep *9*, 725–729.

Voinnet, O. (2005). Induction and suppression of RNA silencing: insights from viral infections. Nat Rev Genet *6*, 206–220.

von Eije, K.J., ter Brake, O., and Berkhout, B. (2008). Human immunodeficiency virus type 1 escape is restricted when conserved genome sequences are targeted by RNA interference. DOI: 10.1128/JVI.02035-07. J. Virol. *82*, 2895–2903.

Wang, Y., Kato, N., Jazag, A., Dharel, N., Otsuka, M., Taniguchi, H., Kawabe, T., and Omata, M. (2006). Hepatitis C virus core protein is a potent inhibitor of RNA silencing-based antiviral response. Gastroenterology *130*, 883–892.

Westerhout, E.M., and Berkhout, B. (2007). A systematic analysis of the effect of target RNA structure on RNA interference. Nucleic Acids Res. *35*, 4322–4330.

Westerhout, E.M., Ooms, M., Vink, M., Das, A.T., and Berkhout, B. (2005). HIV-1 can escape from RNA interference by evolving an alternative structure in its RNA genome. DOI:10.1093/nar/gki220. Nucleic Acids Res. *33*, 796–804.

Westerhout, E.M., ter Brake, O., and Berkhout, B. (2006). The virion-associated incoming HIV-1 RNA genome is not targeted by RNA interference. Retrovirology *3*, 57.

Wilson, J.A., and Richardson, C.D. (2005). Hepatitis C virus replicons escape RNA interference induced by a short interfering RNA directed against the NS5b coding region. DOI: 10.1128/JVI.79.11.7050-7058.2005. J. Virol. *79*, 7050–7058.

Wong, J.K., Hezareh, M., Gunthard, H.F., Havlir, D.V., Ignacio, C.C., Spina, C.A., and Richman, D.D. (1997). Recovery of replication-competent HIV despite prolonged suppression of plasma viremia. Science *278*, 1291–1295.

Wu, H.L., Huang, L.R., Huang, C.C., Lai, H.L., Liu, C.J., Huang, Y.T., Hsu, Y.W., Lu, C.Y., Chen, D.S., and Chen, P.J. (2005). RNA interference-mediated control of hepatitis B virus and emergence of resistant mutant. Gastroenterology *128*, 708–716.

Yeung, M.L., Yasunaga, J., Bennasser, Y., Dusetti, N., Harris, D., Ahmad, N., Matsuoka, M., and Jeang, K.T. (2008). Roles for microRNAs, miR-93 and miR-130b, and tumor protein 53-induced nuclear protein 1 tumor suppressor in cell growth dysregulation by human T-cell lymphotrophic virus 1 68/21/8976 [pii] 10.1158/0008-5472.CAN-08-0769. Cancer Res. *68*, 8976–8985.

Yokota, T., Sakamoto, N., Enomoto, N., Tanabe, Y., Miyagishi, M., Maekawa, S., Yi, L., Kurosaki, M., Taira, K., Watanabe, M., and Mizusawa, H. (2003). Inhibition of intracellular hepatitis C virus replication by synthetic and vector-derived small interfering RNAs. EMBO Rep *4*, 602–608.

Zeng, Y., Yi, R., and Cullen, B.R. (2003). MicroRNAs and small interfering RNAs can inhibit mRNA expression by similar mechanisms DOI: 10.1073/pnas.1630797100. Proc. Natl. Acad. Sci. U.S.A. *100*, 9779–9784.

Zheng, Z.M., Tang, S., and Tao, M. (2005). Development of resistance to RNAi in mammalian cells. Ann. N.Y. Acad. Sci. *1058*, 105–118.

# RNAi Gene Therapy to Control HIV-1 Infection

7

Ben Berkhout and Olivier ter Brake

Abstract

RNA interference (RNAi) was discovered as cellular gene regulation mechanism in 1998, but several RNAi-based applications for gene silencing have already made it into clinical trials. We will discuss RNAi approaches to target pathogenic human viruses causing acute or chronic infections, with a focus on persistent HIV-1 infection that would most likely require an RNAi-based gene therapy. Viruses like HIV-1 are particularly difficult targets for RNAi-attack because they are escape-prone, which requires combinatorial RNAi strategies to prevent viral escape. The future of antiviral RNAi therapeutics is very promising, but it remains of critical importance to include many controls in pre-clinical test models to unequivocally demonstrate sequence-specific action of the RNAi inducers.

## Introduction: the RNA interference mechanism

RNAi is an ancient gene regulation mechanism that operates in plants, insects and mammals. Its role in mammals and humans is the processing of small non-coding microRNAs (miRNAs) that regulate cellular gene expression to control cell differentiation and development (He and Hannon, 2004). The RNAi pathway can also be induced by artificial substrates, either transiently by transfection of small interfering RNA (siRNA) (Elbashir *et al.*, 2001) or stably by intracellularly expressed short hairpin RNAs (shRNAs) that are processed into effective siRNAs (Brummelkamp *et al.*, 2002). Perfect complementarity of the siRNA with a target sequence in an mRNA results in cleavage of that transcript (Hammond *et al.*, 2000). Since its discovery, RNAi has been widely used in gene knock-down studies to analyse gene function. Many excellent reviews have recently been written on the mechanistic details of RNAi action (Kim and Rossi, 2008; Dykxhoorn *et al.*, 2008; Jaskiewicz and Filipowicz, 2008; Paddison, 2008).

### RNA interference as antiviral strategy

It seems an attractive idea to use the RNAi technology for the specific inhibition of human pathogenic viruses in a therapeutic setting. For acute virus infections like influenza virus A and respiratory syncytial virus (RSV), siRNA therapy is particularly attractive since local delivery of the RNA drug to the lungs is feasible (Bitko *et al.*, 2005; Haasnoot *et al.*, 2003). Another major advantage of targeting acute virus infections is that a single siRNA administration should be sufficient to reduce the viral load and thereby dampen or avoid the induction of disease symptoms. A transient effect is all that is needed because the immune system will quickly kick in and clear the pathogenic virus. For chronic infections with human immunodeficiency virus type 1 (HIV-1), hepatitis B (HBV) and hepatitis C virus (HCV), a lifelong supply of siRNAs will probably be required for a sustained antiviral effect. This scenario seems much more difficult than the acute infections, also because the persistent viruses will tend to escape from the imposed RNAi-inhibition. As a durable alternative, one can propose a gene therapy strategy in which cells are made resistant to HIV-1 due to constitutive shRNA expression. Several reviews have recently been written on antiviral RNAi approaches (Haasnoot *et al.*, 2007; Grimm and Kay, 2007a; Li and Ding, 2005; Leonard and Schaffer, 2006; Liu and Berkhout, 2008; Scherer *et al.*, 2007; Stram and Kuzntzova, 2006; Haasnoot and Berkhout, 2006). We will briefly discuss the therapeutic approaches for acute virus infections and then focus in more detail on gene therapeutic strategies against chronic infections.

### Acute virus infections

Transient treatment with synthetic siRNAs seems an appropriate means to control acute virus infections. Respiratory viruses are ideal targets for this therapy approach because the upper airways and lungs are relatively easy to target. The RNAi therapeutic compounds could for instance be administered via aerosol delivery devices similar to the ones used for delivery of asthma therapeutics. Even a partial reduction of the peak viral load will significantly reduce or avoid disease symptoms, as the virus is subsequently controlled and eventually cleared by the immune system. Mice treated with siRNAs against influenza virus, RSV and SARS coronavirus showed reduced virus titres and reduced virus-induced mortality, both with prophylactic siRNAs and in treatment of established infections (Ge *et al.*, 2004; Bitko *et al.*, 2005; Li *et al.*, 2005a). Interestingly, these studies indicated that siRNA treatment is effective when applied with or without transfection reagent (Bitko *et al.*, 2005). The effect of siRNA prophylaxis on the primary and memory immune response to RSV infection was analysed in mice (Zhang and Tripp, 2008). The results support the hypothesis that siRNAs can be developed as effective antiviral drugs that can be used to reduce the viral load and parameters of pathogenesis without limiting the induction of the memory immune response. Intranasal administration of DNA plasmids with a shRNA expression cassette against RSV also significantly decreased the viral titres (Zhang *et al.*, 2005). Potent siRNA inhibitors against the spike and polymerase genes of the SARS coronavirus reduced viral RNA levels, but also the SARS-like symptoms and lung histopathology in rhesus macaques (Li *et al.*, 2005a). The efficiency, safety and long-term duration of siRNA delivery remains the major challenge to its clinical application. Recently, Alnylam pharmaceuticals initiated a phase II clinical trial with an siRNA for the treatment of RSV. No adverse effects were observed in the initial phase I trial, and the safety, tolerability and antiviral activity of the siRNA will subsequently be tested in adults experimentally infected with RSV.

## The need to address the specificity of siRNA therapeutics

As with many drugs, delivery of siRNA therapeutics remains a major hurdle. The first RNAi therapy that could reach patients aims at a debilitating eye disease called age-related macular degeneration that afflicts 30–50 million people globally. The siRNAs can be injected directly into the eye. This local delivery to the diseased tissue minimizes systemic exposure and makes it less likely that the siRNA drug will have unanticipated, harmful effects elsewhere in the body. Macular degeneration is triggered by a protein called vascular endothelial growth factor (VEGF) that promotes blood vessel growth. The new siRNA drug should either knock down VEGF expression or that of its receptor VEGFR1. However, many alarm bells went off when new results indicated that the preclinical efficacy as observed in mouse models of macular degeneration probably results from non-specific side-effects rather than RNAi-mediated sequence-specific gene knockdown (Kleinman *et al.*, 2008).

Mammalian cells do not spontaneously take up 'naked' siRNA molecules without cell-permeating entities. Nevertheless, siRNAs targeting the VEGF system that were injected directly into an affected eye did reduce angiogenesis in preclinical studies with mice (Reich *et al.*, 2003; Shen *et al.*, 2005). Based on these promising results, the clinical trials were approved that exploit RNAi to shut down the VEGF signaling pathway. The naked, unmodified siRNAs were injected directly into the confined space of ocular tissue, but it remained obscure how the siRNAs entered the target cells. Recently, Kleinman *et al* (Kleinman *et al.*, 2008) described that the anti-angiogenesis effect in the mouse can also be induced with unrelated siRNAs that differ significantly in sequence. In their search for possible cell surface molecules that could mediate such a general siRNA effect, the Toll-like receptor 3 (TLR3) surfaced as possible candidate. TLR3 is an innate immune system regulator that is present on the surface of many cell types. Any double-stranded RNA (dsRNA) of at least 21 nucleotides can bind TLR3, inducing receptor dimerization and an intracellular signaling pathway that eventually leads to induction of cytokines that are the likely inducers of the anti-angiogenesis program.

A recent study by Robbins *et al.* confirms that non-specific siRNA effects on the innate immune responses can be mistaken for sequence-specific therapeutic effects (Robbins *et al.*, 2008). In fact, this study calls into question the results presented in several milestone papers involving *in vivo* siRNA studies. In the Robbins study, the siRNAs were packaged into lipid or other formulations. It was shown that interferon (IFN)-α induction accounted for the therapeutic effect of antiviral siRNAs in an influenza mouse model. Thus, sequence specificity that is characteristic for RNAi action was not observed. Intriguingly, this problem was simply not noticed in earlier studies because the particular GFP-based siRNA control exhibits unusually low immunostimulatory properties. Unfortunately, other *in vivo* studies did use the identical GFP control siRNA. Active siRNAs directed against viral, oncogenic and angiogenic mRNA targets were shown to vigorously stimulate IFN-α. These sobering results indicate that the claimed therapeutic RNAi effects were probably caused by activation of the innate immune system. Further research should address these serious problems, for which solutions may be presented. For instance, the use of modified siRNA molecules may neutralize off-target effects without affecting the RNAi activity. Alternatively, the siRNAs could be packaged in liposomes that shield them from recognition by the innate immune system. Anyhow, the recent siRNA results do indicate that the development of safe siRNA-based therapies might be more challenging than initially expected.

It remains important to realize that these siRNA problems do not necessarily relate to other therapeutic RNAi approaches. Specifically, the recent siRNA findings are not relevant for gene therapy approaches in which shRNAs are intracellularly expressed as the precursor for siRNAs, e.g. with antiviral activity. The recent findings do emphasize the importance of appropriate controls in pre-clinical RNAi experiments. The validation of candidate siRNA/shRNA inhibitors should be accompanied by multiple siRNA/shRNA controls to test for sequence-specificity, off-target effects against other mRNAs, saturation of the RNAi pathway and non-specific activation of the innate immune system.

## Chronic virus infection: do we really need a gene therapy?

It can be reasoned that a sustained antiviral RNAi effect against chronic viral infections is best achieved by means of a gene therapy approach, that is by constitutive shRNA expression in the cells that are specifically infected by the virus (Berkhout, 2004; Lee and Rossi, 2004). In fact, a phase I clinical trial against HIV-AIDS has recently been launched with a viral vector that delivers three inhibitor genes including an anti-tat/rev shRNA (Anderson *et al.*, 2007). The use of viral gene therapy vectors obviously carries risks. This includes the induction of unwanted immune responses and – for retroviral and lentiviral vectors – the unknown consequences of random integration of the viral vector into the human genome. In addition, concerns remain about the toxicity related to shRNA expression and IFN induction by these dsRNA molecules (Grimm and Kay, 2007b).

A recent study did however advocate the repeated delivery of exogenous siRNAs as anti-HIV therapy (Kumar *et al.*, 2008). Using a mouse model with a humanized immune system, these researchers were able to control virus replication and to prevent the loss of CD4-positive T cells. An obvious challenge remains the *in vivo* delivery of these siRNAs to the relevant cell types. Systemic delivery to T cells, the major cell lineage that supports HIV-1 replication, is particularly difficult because these cells resist siRNA uptake, even by conventional lipid-based transfection in vitro (Goffinet and Keppler, 2006). Kumar et al coupled the siRNA to a cell-targeting nona-arginine peptide and a single-chain antibody against the CD7 protein that is present on the surface of most human T cells (Kumar *et al.*, 2008). This preparation did not induce TLR activation. Nevertheless, it remains doubtful whether one can achieve and maintain sufficiently high siRNA levels in infected humans to durably control virus replication. In case the virus is not inhibited in certain body compartments, this will provide the means for subsequent viral escape and therapy failure. Kumar et al argued that siRNA therapy provides the flexibility to change the siRNAs cocktail in order to keep pace with the mutating virus. Indeed, frequent adaptation of the siRNA drug regimen seems much more simple than changing the shRNA-encoding viral vector, although the latter adjustment is not impossible either (Ter Brake and Berkhout, 2005). However, this therapy adjustment concept is more complex as a simple change in siRNA regimen can block the virus escape variant, but not the original virus. Thus, one should ideally add new siRNAs to the existing siRNA cocktail once virus escape is noticed. This strategy will soon fail because of the increasing siRNA load and the concomitant increase in toxicity, e.g. due to saturation of the RNAi machinery. We would argue that it is of key importance to prevent viral escape in the first place, which can be achieved by stable expression of four potent shRNAs in a combinatorial RNAi approach (Ter Brake *et al.*, 2008a).

## Optimization of RNAi attack on HIV-1

Several viruses were shown to be susceptible to RNAi-mediated inhibition. However, it is important to realize that not all viral RNAs will be good targets for RNAi attack. Proteins bound to the viral RNA may block RNAi action, the viral transcript may adopt an obstructive RNA structure, the viral RNA may be protected inside a virion particle, or the viral RNA may reside in a subcellular compartment that is not surveyed by the RNAi machinery (Haasnoot *et al.*, 2007). RNAi is very effective in inhibiting HIV-1 replication in stably transduced T cell lines, either using an antiviral siRNA/shRNA/miRNA against a viral transcript or a cellular mRNA that encodes an important co-factor for virus replication (Boden *et al.*, 2003a; Boden *et al.*, 2004a; Boden *et al.*, 2004b; Das *et al.*, 2004; Westerhout *et al.*, 2005, 2006b; Liu *et al.*, 2008; Ter Brake and Berkhout, 2005; Ter Brake *et al.*, 2006, 2008a; von Eije *et al.*, 2008; Martinez *et al.*, 2002a,b; Sabariegos *et al.*, 2006; Haasnoot *et al.*, 2007; Anderson *et al.*, 2003; Barichievy *et al.*, 2007; Boden *et al.*, 2003b; Capodici *et al.*, 2002; Chiu *et al.*, 2004; Huff, 2004; Konstantinova *et al.*, 2007; Kumar *et al.*, 2008; Lawrence, 2002; Lee and Rossi, 2004; Li *et al.*, 2005b; Liu *et al.*, 2007, 2008; Martinez *et al.*, 2002b; Nishitsuji *et al.*, 2004; Omoto *et al.*, 2004; Ping *et al.*, 2004; Rossi, 2006; Saayman *et al.*, 2008; Son *et al.*, 2008; Ter Brake and Berkhout, 2005; Yamamoto *et al.*, 2002; Yamamoto *et al.*, 2006; Zhou *et al.*, 2004; Ter Brake *et al.*, 2008b; Konstantinova *et al.*, 2007; Naito *et al.*, 2007). Theoretically, two forms of the HIV-1 RNA can be targeted by RNAi, which reflect different phases of the HIV-1 replication cycle. The 'incoming' RNA genome that is present in infectious virion particles is a putative target when it is delivered to the cytoplasm of an infected cell. Upon establishment of the integrated provirus, the *de novo* synthesized viral transcripts are obvious mRNA targets for RNAi attack. Several studies suggested that the incoming RNA genome is a target for RNAi (Coburn and Cullen, 2002; Jacque *et al.*, 2002; Capodici *et al.*, 2002; Joshi *et al.*, 2005; Nishitsuji *et al.*, 2006), but other studies have presented good evidence against this possibility (Surabhi and Gaynor, 2002; Hu *et al.*, 2002; Nishitsuji *et al.*, 2004; Westerhout *et al.*, 2006a; Gao *et al.*, 2008). We measured complete protection of the incoming RNA genome and reasoned that the RNAi machinery – that is the RISC complex – is unable to access this RNA genome because it remains encapsulated in a core particle (Westerhout *et al.*, 2006a). Alternatively, coating by the viral nucleocapsid protein may shield this genomic RNA from RNAi attack. The inability to inactivate the incoming HIV-1 RNA genome is unfortunate since it would prevent the establishment of a stably integrated proviral DNA. Thus, only *de novo* synthesized HIV-1 transcripts seem vulnerable to RNAi attack.

Several criteria can be formulated to find optimal target sites within the 9-kb HIV-1 RNA genome. It may be beneficial to select target sequences in the early spliced mRNAs encoding the early proteins Tat, Rev and Nef. Alternatively, it may be a good idea to select targets that are present in all subgenomic HIV-1 RNAs. For instance, sequences in the untranslated leader and the 3′-terminal nef gene qualify for this because they are present in all spliced RNA forms and obviously also in the full-length genomic

RNA. Special domains of the HIV-1 genome may be targeted to induce transcriptional gene silencing (TGS) of the viral long terminal repeat (LTR) promoter (Morris, 2008; Kawasaki and Taira, 2004; Suzuki *et al.*, 2005). The molecular mechanism of TGS requires further description, but transcriptional silencing of HIV-1 remains an attractive possibility. Epigenetic modifications through the use of promoter-specific siRNAs could be an interesting method of achieving a more robust and particularly a prolonged gene silencing effect. Another important consideration for target selection is the sequence variation in natural HIV-1 isolates. This can simply be analysed using the available HIV-1 sequence databases. This screen can be performed for isolates that belong to a certain HIV-1 subtype, but one can also identify conserved genome segments across the subtypes. The idea to attack well-conserved viral sequences is also relevant for preventing viral escape, as natural sequence variation provides an obvious indication of the allowed sequence variation. Despite all these considerations, it remains of utmost importance to experimentally test a set of siRNA/shRNA candidates and to select the most potent inhibitors.

## HIV-1 as moving target

Antiretroviral drug treatment of HIV-infected individuals usually fails when a single drug is used. Likewise, RNAi-resistant viruses will probably emerge when a single RNAi inhibitor is used. Several studies have described the selection of such RNAi-escape mutants (Boden *et al.*, 2003a; Das *et al.*, 2004; Lee *et al.*, 2005; Nishitsuji *et al.*, 2006; Unwalla *et al.*, 2006). Interestingly, the majority of HIV-1 escape mutants acquired a single point mutation within the target sequence. These results underscore the strict sequence-specificity of the RNAi mechanism. On the other hand, there is also some evidence that prolonged RNAi inhibition can select for multiple point mutations, indicating that a single mutation may not provide complete resistance (Das *et al.*, 2004; Westerhout *et al.*, 2005). Intriguingly, mutations outside the target sequence that result in an RNA structural change have also been observed as an exotic escape route (Westerhout *et al.*, 2005). Deletion of non-essential viral sequences forms another escape route for some targets. Deletions were described for a shRNA directed against the non-essential Nef gene, which allows the virus more freedom in the selection of escape variants. A more indirect viral escape route was recently reported in an RNAi study that targeted the TAR hairpin motif that is involved in HIV-1 transcriptional activation by the viral Tat protein (Leonard *et al.*, 2008). Viral escape coincided with duplication of the number of binding sites for the Sp1 transcription factor in the viral LTR promoter. However, it could be argued that this adaptation reflects a general improvement of virus replication and not a specific escape from RNAi attack. Consistent with this idea is the selection of this identical Sp1-adaptation in unrelated HIV-1 evolution studies (Berkhout *et al.*, 1999).

For the development of an effective RNAi gene therapy, it is therefore preferred to target highly conserved HIV-1 sequences that allow only a few escape routes. The many other theoretically possible RNAi resistance mutations will not be selected because they impose a high cost in terms of viral replication capacity. Indeed, no deletion-based escape mutations were observed in such a restricted setting (von Eije *et al.*, 2008; von Eije *et al.*, 2009). Furthermore, silent codon changes were preferentially selected for some protein-encoding target sequences, arguing that HIV-1 can only move in a restricted domain of total sequence space. A surprising finding is that the RNAi-induced sequence changes mimic the sequence variation present in natural HIV-1 isolates (Ter Brake *et al.*, 2008c). We argued that this result indicates that the sequence space available for natural virus isolates is also restricted by replication/fitness parameters. This novel insight does suggest that inspection of natural sequence variation can provide a first indication of the type of escape mutations one can expect, but experimental validation remains important.

## Combinatorial RNAi approaches

Based on the clinical success of highly active antiretroviral therapy (HAART), in which multiple antiretroviral drugs are combined, a similar strategy to counteract viral escape is the use of multiple shRNA inhibitors that target different regions of the HIV-1 RNA genome (Liu and Berkhout, 2008; Grimm and Kay, 2007a;

Berkhout, 2004; Hannon and Rossi, 2004; Shankar *et al.*, 2005; Stevenson, 2003; Ter Brake and Berkhout, 2005). In a large screen, we identified 21 potent shRNAs against highly conserved HIV-1 targets, which enabled us to design such a combinatorial gene therapy (Ter Brake *et al.*, 2006). First, the combined expression of multiple shRNAs results in additive inhibition as compared to a single shRNA inhibitor (Ter Brake *et al.*, 2006). Second and most importantly, viral escape will be more difficult because the virus has to acquire resistance mutations in the multiple targets simultaneously. The number and type of mutations do similarly dictate the evolution of drug-resistant HIV-1 variants (Keulen *et al.*, 1997; Keulen *et al.*, 1996). Compared to single shRNA treatment, we observed delayed viral escape with two shRNAs and no viral escape with four shRNAs (Ter Brake *et al.*, 2008a). These findings seem very important for the development of a durable multi-shRNA-based gene therapy.

We previously tried to estimate the chance of viral escape with a single versus multiple shRNA inhibitors (Berkhout, 2004; Ter Brake and Berkhout, 2005). Upfront, we made the assumption that escape by means of a deletion is not an option for the virus. This seems appropriate when essential viral genes are targeted. Thus, only point mutations are allowed to occur, and a single point mutation can in theory make the virus insensitive to RNAi. The error rate of the reverse transcriptase of HIV-1 is $3 \times 10^{-5}$ (Mansky, 1996) and consequently the chance of viral escape for a 19-nucleotide target in a single infection is $19 \times (3 \times 10^{-5}) = 5.7 \times 10^{-4}$. Studies in the field of drug resistance indicate that an untreated HIV infected individual has a virus population size of $10^4$ to $10^5$. This means that for each shRNA, several potential escape variants will be present even before the start of therapy, and the emergence of drug-resistant variants seems inevitable when a single shRNA is used. When multiple shRNAs ($N$) are used simultaneously, the likelihood of obtaining a drug resistant variant drops exponentially with the number of inhibitors $(5.7 \times 10^{-4})^N$. If we assume that there is already resistance to at least one of the shRNAs, the chance of a resistant variant emerging is $(5.7 \times 10^{-4})^{N-1}$. For instance, if four shRNAs are used simultaneously, the chance of escape is $1.9 \times 10^{-10}$. This chance seems remote given the average viral load in a patient. However, it cannot formally be excluded that multi-shRNA resistant mutants can evolve. For instance, virus replication may continue in cells that were not modified by the gene therapy, although these cells are expected to be quickly removed by the immune system upon virus-infection and the presentation of viral epitopes on the cell surface. Once virus replication reaches levels to generate several resistance mutations, it seems fairly easy for the virus to combine these genetic markers by means of recombination.

## Targeting cellular co-factors as anti-escape strategy

Cellular factors that are closely involved in the virus replication cycle can be targeted to prevent virus spread. A major reason to consider this alternative strategy has to do with the concept that the genetic barrier for viral escape may be much higher when cellular factors are targeted. For instance, one expects that HIV-1 adaptation to alternative cellular co-factors is more complicated or even impossible when no alternative cellular functions are available. We do stipulate that this concept has not yet been proven experimentally, and the properties may vary widely for different cellular targets.

The attack on cellular functions is obviously not without danger. Host gene targets need to be carefully selected because the knock-down of cellular factors that are essential for HIV-1 may not only impair virus replication but may also be detrimental to the cell and the host. Since an RNAi-based gene therapy will preferably use a low multiplicity of infection of the lentiviral vector, such that the cellular factors will only be knocked down partially, problems may be less severe compared with a complete knock-out. Alternatively, an exceptionally high lentiviral vector copy number ('intensified RNAi') may be required to achieve nearly complete knock-down and a therapeutic effect, as demonstrated for LEDGF/p75 that facilitates HIV-1 integration (Llano *et al.*, 2006).

Several cellular factors involved in the HIV-1 life cycle have been considered as possible targets for an antiviral RNAi attack: the CD4 receptor and co-receptors CCR5 and CXCR4 (Anderson and Akkina, 2005; Anderson *et al.*,

2003), integration factors like BAF1, Emerin and LEDGF/p75 (Llano *et al.*, 2006; Maertens *et al.*, 2003; Jacque and Stevenson, 2006), transcriptional factors like NF-κB, PAK-1, and cyclin T1 (Surabhi and Gaynor, 2002; Nguyen *et al.*, 2006; Chiu *et al.*, 2004), or Furin, which is involved in envelope protein maturation (Nguyen *et al.*, 2006). Not all cellular co-factors are ideal targets. For instance, knocking down CD4 is not desirable in a therapeutic setting because this cell surface receptor plays essential roles in the immune system. The CCR5 receptor is an obvious and attractive target. HIV-1 infected people that carry a defective CCR5 gene, CCR5-Δ32, show delayed disease progression and people homozygous for CCR5-Δ32 are healthy and largely protected from HIV-1 infection (Liu *et al.*, 1996; de Roda Husman *et al.*, 1999). Targeting of CCR5, resulting in perhaps only a partial knock-down, is expected to provide a therapeutic benefit for HIV-infected patients. This potential was recently highlighted in a proof of principle study in non-human primates (An *et al.*, 2007). Primates received blood stem cells treated with an SIV lentiviral vector expressing a shRNA against CCR5. CCR5 expression was knocked down and T cells from these primates were found to be less susceptible to SIV infection as compared to the appropriate control cells. The primates exhibited normal haematopoietic reconstitution, an important indication that the treatment is safe.

The potential of such an anti-CCR5 gene therapy is further supported by the functional cure of an HIV-infected patient, who had leukaemia in addition to AIDS, with a special bone marrow transplant (CROI 2008, abstract 719: Hütter *et al.*, Treatment of HIV-1 infection by allogeneic CCR5-Δ32/Δ32 stem cell transplantation: a promising approach). This condition warranted the high risk of a blood stem cell transplant, although 10% to 30% of people who receive bone marrow transplants die. Although the odds are extremely small, a donor was identified who was both a good tissue match and a carrier of the CCR5-inactivating mutation. Prior to the transplant, a standard regimen of drugs and radiation were administered to kill the patient's bone marrow cells and many cells of the immune system. Antiretroviral treatment was stopped when the donor cells were transfused because of concerns about their survival ability. Although the plan was to resume the antiretroviral regimen once HIV-1 re-emerged in the patient's blood, more than 600 days later, standard tests have not detected HIV-1 in his blood. As said, this treatment is unthinkable for the millions of people living with HIV/AIDS. Nevertheless, the results form a virtual 'proof of principle' for CCR5-targeting gene therapy approaches.

These examples highlight that only targets can be selected that are dispensable for the host cell and for which a partial knock-down does sufficiently impair HIV-1 replication. The recent development of RNAi knock-down libraries that cover nearly the complete human genome, either using siRNAs that need to be transfected or shRNA-expressing viral vectors, will assist in the identification of novel cellular factors that are involved in HIV-1 replication (Root *et al.*, 2006; Huesken *et al.*, 2005; Miyagishi *et al.*, 2004; Paddison *et al.*, 2004; Berns *et al.*, 2004; Brass *et al.*, 2008; König *et al.*, 2008). Proteomics studies on HIV-infected T cells should facilitate the identification of additional co-factors (Ringrose *et al.*, 2008; Robbins *et al.*, 2008). All cellular factors that show up as potential therapeutic targets will require extensive validation and in particular screening for unwanted side-effects.

## Targeting miRNAs that are involved in virus replication

miRNAs are small non-coding RNAs that inhibit the expression of target genes by promoting mRNA degradation or inhibiting mRNA translation. In plants and invertebrates, miRNAs are involved in the innate response to invading viruses. Recent reports indicate similar antiviral effects of human miRNAs against viral pathogens. Other studies highlight the complex interplay between human viruses and the cellular RNAi machinery. Several viruses encode their own miRNAs and viruses like HCV do actually employ cellular miRNAs as co-factor in their replication cycle (Jopling *et al.*, 2005; Jopling *et al.*, 2008; Gottwein *et al.*, 2007; Gottwein and Cullen, 2008; Umbach *et al.*, 2008; Berkhout and Jeang, 2007; Berkhout, 2008; Ter Brake *et al.*, 2008b; Henke *et al.*, 2008). These new molecular insights in the virus–cell interactions at the level of the RNAi mechanism present new therapeutic options (Haasnoot *et al.*, 2007; Houzet *et al.*, 2008). The miRNAs that specifically suppress pathogenic viruses could be used to design miRNA-mimics as a new class of antivirals. Alternatively, the cellular miRNAs that act as viral co-factors or the virally encoded miRNAs could be blocked and inactivated by antisense inhibitors.

HIV-1 has been proposed by Huang *et al.* (2007) to use miRNA-mediated down-regulation of gene expression to its own advantage for the induction of the state of proviral latency. HIV-1 latency means that the provirus is stably integrated into the host genome, but not producing viral transcripts and proteins to support new rounds of infection. This reservoir of infected, provirus-containing cells poses a major problem in attempts to clear the virus from infected individuals. Huang and colleagues showed that a set of miRNAs is enriched in resting T cells that harbour these latent proviruses. Five of these up-regulated miRNAs have predicted binding sites in the HIV-1 RNA genome. When these miRNAs were blocked with inhibitors in resting T cells from patients, viral gene expression was boosted. These results show that blocking of miRNA-mediated HIV-1 latency has the potential to activate the virus in the reservoir of latently infected T cells, followed by immune recognition and cell destruction. Of course, more detailed mechanistic insight is needed to produce safe therapeutics, in particular because all miRNAs can be expected to execute important cellular functions.

## Gene therapy for chronic HIV-1 infection

A gene therapy should ideally reach the many different cell types that are infected by HIV-1. Since CD4-positive T cells constitute the major cell population implicated in HIV infection and progression to AIDS, making these cells resistant to HIV-1 should be a key aspect of any anti-HIV gene therapy. Currently, two optional strategies have been proposed to make T cells resistant to HIV-1. The simple scenario is to isolate peripheral blood mononuclear cells from a patient's blood and to purify the CD4-positive T cells. These cells are subsequently treated *ex vivo* with a viral vector with the anti-HIV gene that provides resistance against HIV-1. The transduced cells are subsequently engrafted back into the patient, where they will survive and improve the immunity of the patient. However, these T cells will have a limited life span, such that repeated infusions may be necessary. The more complicated, but hopefully more durable therapeutic scenario focuses at the hematopoietic blood stem cells. Blood stem cells continuously populate the myeloid and lymphoid cell lineages in the periphery. Engraftment of genetically modified, but autologous blood stem cells will result in a steady production of new T cells, but also monocytes, macrophages and dendritic cells. The targeting of the CD34-positive stem cells with a single gene therapy treatment may suffice for a sustained therapeutic effect, although there remain many unknowns that should be addressed in pilot clinical studies. In HIV-infected individuals, the resistant cells will preferentially survive over unprotected cells, which are either killed directly by the virus or removed by the immune system upon recognition of viral epitopes on the cell surface. This stem cell gene therapy may lead to partial reconstitution of the immune system, although it is currently difficult to specify this repair in quantitative and temporal terms.

## Advantages and problems of RNAi delivery by a lentiviral vector

The appropriate choice of the viral vector system for delivery of the RNAi-inducing transgene is of crucial importance for the clinical development of a gene therapy. The lentiviral vector is ideally suited for a durable gene transfer because it stably integrates in the target cell, resulting in a constant supply of antiviral RNAi molecules. The lentiviral vector is also very efficient in transduction of the target cells for an HIV-AIDS therapy, the CD34-positive blood stem cells or the CD4-positive T cells. However, this vector system is largely based on HIV-1 sequences, and expression of antiviral shRNAs by the lentiviral vector may cause unwanted complications during vector production in the packaging cell. Indeed, targeting of the Gag-Pol and Rev mRNAs of the packaging system can reduce the transduction titre (Banerjea *et al.*, 2003; Chang *et al.*, 2005). We systematically addressed this issue with a large set of anti-HIV shRNAs by investigating all possible routes by which the shRNAs can interfere with lentiviral vector production (Ter Brake and Berkhout, 2007). Both lentivirus production (capsid titre) and the transduction titre were reduced with shRNAs against the Gag-Pol mRNA. However, this problem could simply be avoided with a human codon-optimized Gag-Pol version, which is not recognized by the shRNAs. Targeting of the Rev mRNA in the packaging cell did not cause problems in lentivirus production, probably because only little Rev protein is needed to support its nuclear RNA export function. RNAi attack on remaining HIV-1 sequences in the vector genome is a serious possibility, which can be avoided by selecting shRNAs for which the target is absent in the lentiviral vector. Alternatively, one could actually modify the target sequence in the lentiviral vector, but obviously without affecting vector sequences that play a role in lentiviral transduction. A priori, it would seem impossible to produce a shRNA-encoding lentiviral vector because the shRNA-encoding sequence in the vector genome should be attacked by the shRNA that is also expressed in the packaging cell (self-targeting). However, the vector RNA genome is protected from RNAi-attack due to folding of a stable hairpin structure that masks the target sequence (Westerhout *et al.*, 2005; Westerhout and Berkhout, 2007). This absence of self-targeting has allowed the construction of numerous lentiviral vector systems for RNAi approaches that are available from a variety of academic and commercial sources (Li and Rossi, 2008; Szulc and Aebischer, 2008).

Other specific problems may be encountered in combinatorial approaches that use the lentiviral vector system. HIV-1 is recombination prone and the presence of multiple shRNA expression cassettes with repeated promoter sequences within a single lentiviral vector posed serious problems. The lentiviral vector was found to recombine at the repeat sequences during the transduction process, resulting in deletion of one or multiple expression cassettes (An and Telesnitsky, 2001; Ter Brake *et al.*, 2006). This is rather dangerous because cells that express only a single inhibitor may form the breeding ground for virus variants that are resistant to that particular inhibitor, but several of these viruses may recombine later to generate multi-shRNA-resistance. In order to avoid recombination-mediated deletion within the lentiviral vector, multiple shRNAs should be expressed from different promoter elements. Indeed, vector genome stability was improved and no deletions were observed anymore when four shRNAs were expressed from four different promoters (Ter Brake *et al.*, 2006). Alternative shRNA designs may avoid the use of multiple promoter elements. For instance, one could stack multiple shRNA modules on top of each other in a single transcript driven by a single promoter element. We designed such anti-HIV molecules, the so-called extended shRNAs or e-shRNAs, but this approach requires a careful design of the extended hairpin molecule (Liu *et al.*, 2007; Weinberg *et al.*, 2007). One could also generate a single miRNA-like transcript that encodes multiple antiviral miRNAs/siRNAs (Liu *et al.*, 2008). The incorporation of miRNA sequences, e.g. Drosha cleavage sites, within the context of the lentiviral RNA vector genome may also have a negative impact on lentiviral vector production or transduction. Further studies are needed to optimize vector systems, in particular for the more complicated combinatorial RNAi approaches.

## Preclinical studies

Although promising anti-HIV effects have been described for combinatorial RNAi strategies in cell culture infections, important issues in relation to safety and efficacy need to be addressed in appropriate animal models before clinical trials can be considered. Several potential side-effects have been reported for shRNAs. For instance, the *in vivo* expression of shRNAs resulted in the death of mice due to saturation of the RNAi machinery (Grimm *et al.*, 2006). The observed fatal side-effects were primarily ascribed to saturation of Exportin-5, leading to interference with nuclear export of miRNA precursors and miRNA function. The mice that died were actually treated with very high doses of the viral vector, resulting in extremely high copy numbers per cell. This result demonstrates in a dramatic manner that shRNA overdosing can be dangerous. We should however not forget that the mice that received a more moderate RNAi dose were fine. There are also indications that shRNA expression can induce the IFN pathway. The original paradigm is that only dsRNA molecules that are larger than 30 basepairs can induce this pathway, but small dsRNA were also able to evoke this reaction (Bridge *et al.*, 2003; Hornung *et al.*, 2005; Pebernard and Iggo, 2004; Robbins and Rossi, 2005; Sioud, 2005). This IFN induction is dose-dependent, and some sequence motifs have been implicated (Pebernard and Iggo, 2004; Sioud, 2005). Finally, RNAi could also induce off-target effects in which siRNAs silence partially complementary transcripts through an miRNA-like mechanism. Such an effect requires complementarity between the siRNA seed region and the 3′ untranslated region of a target gene (Birmingham *et al.*, 2006; Jackson *et al.*, 2006; Lin *et al.*, 2005). As such, any effective siRNA will in theory have numerous potential off-target genes. When multiple shRNAs are combined, the number of potential off-target genes and the chance of undesired off-target effects will increase significantly. The use of inducible or tissue-specific shRNA expression cassettes may be helpful in this respect.

In our experiments, we always used the lentiviral vector at a low multiplicity of infection, thus obtaining cells with a single vector copy to avoid high expression levels that may induce unwanted side-effects. The potent HIV-1 inhibition was observed in this single copy context, and the RNAi effect remained stable for at least hundred days of cell passaging. Furthermore, the growth rates of transduced cell lines and primary T cells were not affected, and we did neither score induction of IFN-β and PKR phosphorylation in transient transfection experiments with the multiple shRNA lentiviral vectors. These potential side-effects remain a genuine concern for the development of a multiple-shRNA approach against HIV-1 and the potential risks should be properly assessed in pre-clinical evaluations.

A thorough pre-clinical evaluation is required to properly assess the safety and efficacy of a multiple shRNA gene therapy. The first *in vivo* studies on therapy efficacy and safety were performed in a mouse model by Akkina *et al.* (Anderson *et al.*, 2007; Banerjea *et al.*, 2003). Recently, a new humanized mouse model was developed that is ideally suited for such preclinical assessment. This humanized Rag2$^{-/-}\gamma c^{-/-}$ mouse sustains long-term multi-lineage human haematopoiesis and is capable of mounting immune responses (Legrand *et al.*, 2006). We have engrafted human CD34-positive cells after *ex vivo* transduction with a lentiviral vector that expresses a shRNA against the HIV-1 Nef gene (Ter Brake *et al.*, 2009). The shRNA expression did not affect the development of the CD34-positive stem cells into various mature leucocyte subsets, including CD4-positive T cells. HIV-1 replication was tested *ex vivo*, demonstrating sequence-specific inhibition, which forms an important proof of concept in the clinical development of an anti-HIV RNAi-based gene therapy. These results indicate that the procedure should in principle be safe, at least with single shRNA expression vectors. The next step in the preclinical evaluation is to challenge these mice with HIV-1, which normally results in viraemia and depletion of the CD4-positive cells, some of the main features of human HIV-1 infection (Baenziger *et al.*, 2006; Berges *et al.*, 2006; Zhang *et al.*, 2007).

## Concluding remarks

The gene suppression effects of RNAi were first noticed about 20 years ago in plant transgenesis studies, but only described in mechanistic terms in 1998 by Fire and Mello in studies with *Caenorhabditis elegans* (Fire *et al.*, 1998). More recent studies underscored the importance of the RNAi mechanism for regulated gene expression in mammalian cell biology. Antiviral RNAi strategies may be more efficient than other nucleic acid-based antiviral approaches because they tap in on an existing cell mechanism. Several antiviral RNAi applications are currently being tested in clinical trials, ranging from a transient siRNA therapy for RSV infection to a gene therapy for HIV-infected individuals. Since we are still in the early days of development, it seems obvious that there is ample room for further improvement of the activity and the specificity of the RNAi inducers and the means of delivery. Perhaps more potent antivirals can be designed by mimicking the miRNA transcripts. On the target side, it is crucial to realize that not all viral targets are equally suitable for RNAi-mediated inhibition, and it will be important to identify the viral Achilles heel for an RNAi attack. For instance, because the structure of the viral RNA genome can significantly affect the efficiency of RNAi-inhibition (Westerhout *et al.*, 2005), novel high throughput RNA structure mapping methods like SHAPE (Wilkinson *et al.*,

2006) could speed up the selection of candidate shRNA inhibitors.

For antiviral therapy of chronic infections, the evolution of drug-resistant virus variants is a serious problem and RNAi therapy forms no exception. Special measures can be designed to counter viral escape. For example, when preferred viral escape routes are frequently observed with a single shRNA inhibitor, one could consider a multiple shRNA approach in which these escape mutations are selectively counteracted in advance by 'second-generation' shRNAs (Ter Brake and Berkhout, 2005). The inclusion of cellular co-factors as RNAi targets is also expected to restrict viral escape. However, we should realize that viruses and their hosts have co-evolved for millions of years and viruses have developed mechanisms to neutralize RNAi action. Viral proteins may frustrate an RNAi attack by acting as RNAi suppressor or by binding to the target sequence. Although the presence of a viral RNAi suppressor protein may seem bad news, promising results have been reported against suppressor-encoding viruses like HCV and HIV-1 in gene therapy settings (Henry *et al.*, 2006; Ter Brake *et al.*, 2008a). One may in fact consider to specifically target the viral mRNAs that encode such suppressor functions. Thus, a better understanding of the natural virus–RNAi interactions will be important for future fine-tuning of RNAi-based therapies. Given the immense interest in RNAi as a therapeutic, the coming years are likely to see an increasing range of potential clinical applications, of which the development of treatments for human pathogenic viruses will be in the forefront and perhaps even among the first successes.

## Acknowledgements

We thank all members of the RNAi group for stimulating discussions. RNAi research in the Berkhout lab is sponsored by NWO-CW (TOP program), ZonMw (Translational gene therapy program) and the Dutch AIDS Fund.

## References

An, D.S., Donahue, R.E., Kamata, M., Poon, B., Metzger, M., Mao, S.H., Bonifacino, A., Krouse, A.E., Darlix, J.L., Baltimore, D., Qin, F.X., and Chen, I.S. (2007). Stable reduction of CCR5 by RNAi through hematopoietic stem cell transplant in non-human primates. Proc. Natl. Acad. Sci. U.S.A. *104*, 13110–13115.

An, W., and Telesnitsky, A. (2001). Frequency of direct repeat deletion in a human immunodeficiency virus type 1 vector during reverse transcription in human cells. Virol. *286*, 475–482.

Anderson, J., and Akkina, R. (2005). HIV-1 resistance conferred by siRNA cosuppression of CXCR4 and CCR5 coreceptors by a bispecific lentiviral vector. AIDS Res. Ther. *2*, 1.

Anderson, J., Banerjea, A., and Akkina, R. (2003). Bispecific short hairpin siRNA constructs targeted to CD4, CXCR4, and CCR5 confer HIV-1 resistance. Oligonucleotides *13*, 303–312.

Anderson, J., Li, M.J., Palmer, B., Remling, L., Li, S., Yam, P., Yee, J.K., Rossi, J., Zaia, J., and Akkina, R. (2007). Safety and efficacy of a lentiviral vector containing three anti-HIV genes-CCR5 ribozyme, tat-rev siRNA, and TAR decoy-in SCID-hu mouse-derived T cells. Mol .Ther. *14*, 1182–1188.

Baenziger, S., Tussiwand, R., Schlaepfer, E., Mazzucchelli, L., Heikenwalder, M., Kurrer, M.O., Behnke, S., Frey, J., Oxenius, A., Joller, H., Aguzzi, A., Manz, M.G., and Speck, R.F. (2006). Disseminated and sustained HIV infection in CD34+ cord blood cell-transplanted Rag2$^{-/-}$gamma c$^{-/-}$ mice. Proc. Natl. Acad. Sci. U.S.A. *103*, 15951–15956.

Banerjea, A., Li, M.J., Bauer, G., Remling, L., Lee, N.S., Rossi, J., and Akkina, R. (2003). Inhibition of HIV-1 by lentiviral vector-transduced siRNAs in T lymphocytes differentiated in SCID-hu mice and CD34+ progenitor cell-derived macrophages. Mol. Ther. *8*, 62–71.

Barichievy, S., Saayman, S., von Eije, K.J., Morris, K.V., Arbuthnot, P., and Weinberg, M.S. (2007). The inhibitory efficacy of RNA POL III-expressed long hairpin RNAs targeted to untranslated regions of the HIV-1 5′ long terminal repeat. Oligonucleotides *17*, 419–431.

Berges, B.K., Wheat, W.H., Palmer, B.E., Connick, E., and Akkina, R. (2006). HIV-1 infection and CD4 T cell depletion in the humanized Rag2$^{-/-}$gamma c$^{-/-}$ (RAG-hu) mouse model. Retrovirology 3, 76.

Berkhout, B. (2004). RNA interference as an antiviral approach: targeting HIV-1. Curr. Opin. Mol. Ther. *6*, 141–145.

Berkhout, B. (2008). A balancing act: viruses and miRNAs. J. Formosan Med. Ass. *107*, 1–3.

Berkhout, B., and Jeang, K.T. (2007). RISCy business: microRNAs, pathogenesis, and viruses. J. Biol. Chem. *282*, 26641–26645.

Berkhout, B., Verhoef, K., van Wamel, J.L.B., and Back, B. (1999). Genetic instability of live-attenuated HIV-1 vaccine strains. J. Virol. 73, 1138–1145.

Berns, K., Hijmans, E.M., Mullenders, J., Brummelkamp, T.R., Velds, A., Heimerikx, M., Kerkhoven, R.M., Madiredjo, M., Nijkamp, W., Weigelt, B., Agami, R., Ge, W., Cavet, G., Linsley, P.S., Beijersbergen, R.L., and Bernards, R. (2004). A large-scale RNAi screen in human cells identifies new components of the p53 pathway. Nature *428*, 431–437.

Birmingham, A., Anderson, E.M., Reynolds, A., Ilsley-Tyree, D., Leake, D., Fedorov, Y., Baskerville, S., Maksimova, E., Robinson, K., Karpilow, J., Marshall, W.S., and Khvorova, A. (2006). 3′-UTR seed matches, but not overall identity, are associated with RNAi off-targets. Nat. Methods 3, 199–204.

Bitko, V., Musiyenko, A., Shulyayeva, O., and Barik, S. (2005). Inhibition of respiratory viruses by nasally administered siRNA. Nat. Med *11*, 50–55.

Boden, D., Pusch, O., Lee, F., Tucker, L., and Ramratnam, B. (2004a). Efficient gene transfer of HIV-1-specific short hairpin RNA into human lymphocytic cells using recombinant adeno-associated virus vectors. Mol. Ther. *9*, 396–402.

Boden, D., Pusch, O., Lee, F., Tucker, L., and Ramratnam, B. (2003a). Human immunodeficiency virus type 1 escape from RNA interference. J. Virol. *77*, 11531–11535.

Boden, D., Pusch, O., Lee, F., Tucker, L., Shank, P.R., and Ramratnam, B. (2003b). Promoter choice affects the potency of HIV-1 specific RNA interference. Nucleic Acids Res. *31*, 5033–5038.

Boden, D., Pusch, O., Silbermann, R., Lee, F., Tucker, L., and Ramratnam, B. (2004b). Enhanced gene silencing of HIV-1 specific siRNA using microRNA designed hairpins. Nucleic Acids Res. 32, 1154–1158.

Brass, A.L., Dykxhoorn, D.M., Benita, Y., Yan, N., Engelman, A., Xavier, R.J., Lieberman, J., and Elledge, S.J. (2008). Identification of host proteins required for HIV infection through a functional genomic screen. Science *319*, 921–926.

Bridge, A.J., Pebernard, S., Ducraux, A., Nicoulaz, A.L., and Iggo, R. (2003). Induction of an interferon response by RNAi vectors in mammalian cells. Nat. Genet. *34*, 263–264.

Brummelkamp, T.R., Bernards, R., and Agami, R. (2002). A system for stable expression of short interfering RNAs in mammalian cells. Science *296*, 550–553.

Capodici, J., Kariko, K., and Weissman, D. (2002). Inhibition of HIV-1 infection by small interfering RNA-mediated RNA interference. J. Immunol. *169*, 5196–5201.

Chang, L.J., Liu, X., and He, J. (2005). Lentiviral siRNAs targeting multiple highly conserved RNA sequences of human immunodeficiency virus type 1. Gene Ther. *12*, 1133–1144.

Chiu, Y.L., Cao, H., Jacque, J.M., Stevenson, M., and Rana, T.M. (2004). Inhibition of human immunodeficiency virus type 1 replication by RNA interference directed against human transcription elongation factor P-TEFb (CDK9/CyclinT1). J. Virol. *78*, 2517–2529.

Coburn, G.A., and Cullen, B.R. (2002). Potent and specific inhibition of human immunodeficiency virus type 1 replication by RNA interference. J. Virol. *76*, 9225–9231.

Das, A.T., Brummelkamp, T.R., Westerhout, E.M., Vink, M., Madiredjo, M., Bernards, R., and Berkhout, B. (2004). Human immunodeficiency virus type 1 escapes from RNA interference-mediated inhibition. J. Virol. *78*, 2601–2605.

de Roda Husman, A.M., Blaak, H., Brouwer, M., and Schuitemaker, H. (1999). CC chemokine receptor 5 cell-surface expression in relation to CC chemokine receptor 5 genotype and the clinical course of HIV-1 infection. J. Immunol. *163*, 4597–4603.

Dykxhoorn, D.M., Chowdhury, D., and Lieberman, J. (2008). RNA interference and cancer: endogenous pathways and therapeutic approaches. Adv. Exp. Med. Biol. *615*, 299–329.

Elbashir, S.M., Harborth, J., Lendeckel, W., Yalcin, A., Weber, K., and Tuschl, T. (2001). Duplexes of 21-nucleotide RNAs mediate RNA interference in cultured mammalian cells. Nature *411*, 494–498.

Fire, A., Xu, S., Montgomery, M.K., Kostas, S.A., Driver, S.E., and Mello, C.C. (1998). Potent and specific genetic interference by double-stranded RNA in Caenorhabditis elegans. Nature *391*, 806–811.

Gao, Y., Lobritz, M.A., Roth, J., Abreha, M., Nelson, K.N., Nankya, I., Moore-Dudley, D.M., Abraha, A., Gerson, S.L., and Arts, E.J. (2008). Targets of small interfering RNA restriction during human immunodeficiency virus type 1 replication. J. Virol. *82*, 2938–2951.

Ge, Q., Filip, L., Bai, A., Nguyen, T., Eisen, H.N., and Chen, J. (2004). Inhibition of influenza virus production in virus-infected mice by RNA interference. Proc. Natl. Acad. Sci. U.S.A. *101*, 8676–8681.

Goffinet, C., and Keppler, O.T. (2006). Efficient nonviral gene delivery into primary lymphocytes from rats and mice. FASEB J. *20*, 500–502.

Gottwein, E., and Cullen, B.R. (2008). Viral and cellular microRNAs as determinants of viral pathogenesis and immunity. Cell Host. Microbe 3, 375–387.

Gottwein, E., Mukherjee, N., Sachse, C., Frenzel, C., Majoros, W.H., Chi, J.T., Braich, R., Manoharan, M., Soutschek, J., Ohler, U., and Cullen, B.R. (2007). A viral microRNA functions as an orthologue of cellular miR-155. Nature *450*, 1096–1099.

Grimm, D., and Kay, M.A. (2007a). Combinatorial RNAi: a winning strategy for the race against evolving targets? Mol. Ther. *15*, 878–888.

Grimm, D., and Kay, M.A. (2007b). Therapeutic application of RNAi: is mRNA targeting finally ready for prime time? J. Clin. Invest. *117*, 3633–3641.

Grimm, D., Streetz, K.L., Jopling, C.L., Storm, T.A., Pandey, K., Davis, C.R., Marion, P., Salazar, F., and Kay, M.A. (2006). Fatality in mice due to oversaturation of cellular microRNA/short hairpin RNA pathways. Nature *441*, 537–541.

Haasnoot, J., Westerhout, E.M., and Berkhout, B. (2007). RNA interference against viruses: strike and counterstrike. Nat. Biotechnol. 25, 1435–1443.

Haasnoot, P.C.J., and Berkhout, B. (2006). RNA interference: Its use as antiviral therapy. In Handbook of Experimental Pharmacology, (Heidelberg: Springer-Verlag Berlin Heidelberg), pp. 117–150.

Haasnoot, P.C.J., Cupac, D., and Berkhout, B. (2003). Inhibition of virus replication by RNA interference. J. Biomed. Sc. *10*, 607–616.

Hammond, S.M., Bernstein, E., Beach, D., and Hannon, G.J. (2000). An RNA-directed nuclease mediates post-transcriptional gene silencing in *Drosophila* cells. Nature *404*, 293–296.

Hannon, G.J., and Rossi, J.J. (2004). Unlocking the potential of the human genome with RNA interference. Nature *431*, 371–378.

He, L., and Hannon, G.J. (2004). MicroRNAs: small RNAs with a big role in gene regulation. Nat. Rev. Genet. 5, 522–531.

Henke, J.I., Goergen, D., Zheng, J., Song, Y., Schuttler, C.G., Fehr, C., Junemann, C., and Niepmann, M. (2008). microRNA-122 stimulates translation of hepatitis C virus RNA. EMBO J. *27*, 3300–3310.

Henry, S.D., van der Wegen, P., Metselaar, H.J., Tilanus, H.W., Scholte, B.J., and van der Laan, L.J. (2006). Simultaneous targeting of HCV replication and viral binding with a single lentiviral vector containing multiple RNA interference expression cassettes. Mol. Ther. *14*, 485–493.

Hornung, V., Guenthner-Biller, M., Bourquin, C., Ablasser, A., Schlee, M., Uematsu, S., Noronha, A., Manoharan, M., Akira, S., de, F.A., Endres, S., and Hartmann, G. (2005). Sequence-specific potent induction of IFN-alpha by short interfering RNA in plasmacytoid dendritic cells through TLR7. Nat. Med. *11*, 263–270.

Houzet, L., Yeung, M.L., de Lame, V., Desai, D., Smith, S.M., and Jeang, K.T. (2008). MicroRNA profile changes in human immunodeficiency virus type 1 (HIV-1) seropositive individuals. Retrovirology *5*, 118.

Hu, W.Y., Myers, C.P., Kilzer, J.M., Pfaff, S.L., and Bushman, F.D. (2002). Inhibition of retroviral pathogenesis by RNA interference. Curr. Biol. *12*, 1301–1311.

Huang, J., Wang, F., Argyris, E., Chen, K., Liang, Z., Tian, H., Huang, W., Squires, K., Verlinghieri, G., and Zhang, H. (2007). Cellular microRNAs contribute to HIV-1 latency in resting primary CD4(+) T lymphocytes. Nat. Med. *13*, 1241–1247.

Huesken, D., Lange, J., Mickanin, C., Weiler, J., Asselbergs, F., Warner, J., Meloon, B., Engel, S., Rosenberg, A., Cohen, D., Labow, M., Reinhardt, M., Natt, F., and Hall, J. (2005). Design of a genome-wide siRNA library using an artificial neural network. Nat. Biotechnol. *23*, 995–1001.

Huff, B. (2004). News from the bench. Assay for less-fit phenotype? GMHC. Treat. Issues *18*, 14.

Jackson, A.L., Burchard, J., Schelter, J., Chau, B.N., Cleary, M., Lim, L., and Linsley, P.S. (2006). Widespread siRNA 'off-target' transcript silencing mediated by seed region sequence complementarity. RNA. *12*, 1179–1187.

Jacque, J.M., and Stevenson, M. (2006). The inner-nuclear-envelope protein emerin regulates HIV-1 infectivity. Nature *441*, 641–645.

Jacque, J.M., Triques, K., and Stevenson, M. (2002). Modulation of HIV-1 replication by RNA interference. Nature *418*, 435–438.

Jaskiewicz, L., and Filipowicz, W. (2008). Role of Dicer in posttranscriptional RNA silencing. Curr. Top. Microbiol. Immunol. *320*, 77–97.

Jopling, C.L., Schütz, S., and Sarnow, P. (2008). Position-dependent function for a tandem microRNA miR-122-binding site located in the hepatitis C virus RNA genome. Cell Host. Microbe *4*, 77–85.

Jopling, C.L., Yi, M., Lancaster, A.M., Lemon, S.M., and Sarnow, P. (2005). Modulation of hepatitis C virus RNA abundance by a liver-specific microRNA. Science *309*, 1577–1581.

Joshi, P.J., North, T.W., and Prasad, V.R. (2005). Aptamers directed to HIV-1 reverse transcriptase display greater efficacy over small hairpin RNAs targeted to viral RNA in blocking HIV-1 replication. Mol. Ther. *11*, 677–686.

Kawasaki, H., and Taira, K. (2004). Induction of DNA methylation and gene silencing by short interfering RNAs in human cells. Nature *431*, 211–217.

Keulen, W., Back, N.K.T., van Wijk, A., Boucher, C.A.B., and Berkhout, B. (1997). Initial appearance of the 184Ile variant in lamivudine-treated patients is caused by the mutational bias of the human immunodeficiency virus type 1 Reverse Transcriptase. J. Virol. *71*, 3346–3350.

Keulen, W., Boucher, C., and Berkhout, B. (1996). Nucleotide substitution patterns can predict the requirements for drug-resistance of HIV-1 proteins. Antiviral Res. *31*, 45–57.

Kim, D., and Rossi, J. (2008). RNAi mechanisms and applications. Biotechnology *44*, 613–616.

Kleinman, M.E., Yamada, K., Takeda, A., Chandrasekaran, V., Nozaki, M., Baffi, J.Z., Albuquerque, R.J., Yamasaki, S., Itaya, M., Pan, Y., Appukuttan, B., Gibbs, D., Yang, Z., Kariko, K., Ambati, B.K., Wilgus, T.A., DiPietro, L.A., Sakurai, E., Zhang, K., Smith, J.R., Taylor, E.W., and Ambati, J. (2008). Sequence- and target-independent angiogenesis suppression by siRNA via TLR3. Nature *452*, 591–597.

König, R., Zhou, Y., Elleder, D., Diamond, T.L., Bonamy, G.M.C., Irelan, J.T., Chiang, C.Y., Tu, B.P., De Jesus, P.D., Lilley, C.E., Seidel, S., Opaluch, A.M., Caldwell, J.S., Weitzman, M.D., Kuhen, K.L., Bandyopadhyay, S., Ideker, T., Orth, A.P., Miraglia, L.J., Bushman, F.D., Young, J.A., and Chanda, S.K. (2008). Global analysis of host–pathogen interactions that regulate early-stage HIV-1 replication. Cell *135*, 49–60.

Konstantinova, P., Ter Brake, O., Haasnoot, J., de Haan, P., and Berkhout, B. (2007). Trans-inhibition of HIV-1 by a long hairpin RNA expressed within the viral genome. Retrovirology *4*, 15.

Kumar, P., Ban, H.S., Kim, S.S., Wu, H., Pearson, T., Greiner, D.L., Laouar, A., Yao, J., Haridas, V., Habiro, K., Yang, Y.G., Jeong, J.H., Lee, K.Y., Kim, Y.H., Kim, S.W., Peipp, M., Fey, G.H., Manjunath, N., Shultz, L.D., Lee, S.K., and Shankar, P. (2008). T cell-specific siRNA delivery suppresses HIV-1 infection in humanized mice. Cell *134*, 577–586.

Lawrence, D. (2002). RNAi could hold promise in the treatment of HIV. Lancet *359*, 2007.

Lee, N.S., and Rossi, J.J. (2004). Control of HIV-1 replication by RNA interference. Virus Res. *102*, 53–58.

Lee, S.K., Dykxhoorn, D.M., Kumar, P., Ranjbar, S., Song, E., Maliszewski, L.E., Francois-Bongarcon, V., Goldfeld, A., Swamy, N.M., Lieberman, J., and Shankar, P. (2005). Lentiviral delivery of short hairpin RNAs protects CD4 T cells from multiple clades and primary isolates of HIV. Blood *106*, 818–826.

Legrand, N., Weijer, K., and Spits, H. (2006). Experimental models to study development and function of the human immune system *in vivo*. J. Immunol. *176*, 2053–2058.

Leonard, J.N., and Schaffer, D.V. (2006). Antiviral RNAi therapy: emerging approaches for hitting a moving target. Gene Ther. *13*, 532–540.

Leonard, J.N., Shah, P.S., Burnett, J.C., and Schaffer, D.V. (2008). HIV evades RNA interference directed at TAR by an indirect compensatory mechanism. Cell Host. Microbe *4*, 484–494.

Li, B.J., Tang, Q., Cheng, D., Qin, C., Xie, F.Y., Wei, Q., Xu, J., Liu, Y., Zheng, B.J., Woodle, M.C., Zhong, N., and Lu, P.Y. (2005a). Using siRNA in prophylactic and therapeutic regimens against SARS coronavirus in Rhesus macaque. Nat. Med *11*, 944–951.

Li, H.W., and Ding, S.W. (2005). Antiviral silencing in animals. FEBS Lett. *579*, 5965–5973.

Li, M., and Rossi, J.J. (2008). Lentiviral vector delivery of siRNA and shRNA encoding genes into cultured and primary hematopoietic cells. Methods Mol. Biol. *433*, 287–299.

Li, Z., Xiong, Y., Peng, Y., Pan, J., Chen, Y., Wu, X., Hussain, S., Tien, P., and Guo, D. (2005b). Specific inhibition of HIV-1 replication by short hairpin RNAs targeting human cyclin T1 without inducing apoptosis. FEBS Lett. *579*, 3100–3106.

Lin, X., Ruan, X., Anderson, M.G., McDowell, J.A., Kroeger, P.E., Fesik, S.W., and Shen, Y. (2005). siRNA-mediated off-target gene silencing triggered by a 7 nt complementation. Nucleic Acids Res. *33*, 4527–4535.

Liu, R., Paxton, W.A., Choe, S., Ceradini, D., Martin, S.R., Horuk, R., MacDonald, M.E., Stuhlmann, H., Koup, R.A., and Landau, N.R. (1996). Homozygous defect in HIV-1 coreceptor accounts for resistance of some multiply-exposed individuals to HIV-1 infection. Cell *86*, 367–377.

Liu, Y.P., and Berkhout, B. (2008). Combinatorial RNAi strategies against HIV-1 and other escape-prone viruses. Int. J. BioSci. Technol. *1*, 1–10.

Liu, Y.P., Haasnoot, J., and Berkhout, B. (2007). Design of extended short hairpin RNAs for HIV-1 inhibition. Nucleic Acids Res. *35*, 5683–5693.

Liu, Y.P., Haasnoot, J., Ter Brake, O., Berkhout, B., and Konstantinova, P. (2008). Inhibition of HIV-1 by multiple siRNAs expressed from a single microRNA polycistron. Nucleic Acids Res. *36*, 2811–2824.

Llano, M., Saenz, D.T., Meehan, A., Wongthida, P., Peretz, M., Walker, W.H., Teo, W., and Poeschla, E.M. (2006). An essential role for LEDGF/p75 in HIV integration. Science *314*, 461–464.

Maertens, G., Cherepanov, P., Pluymers, W., Busschots, K., De, C.E., Debyser, Z., and Engelborghs, Y. (2003). LEDGF/p75 is essential for nuclear and chromosomal targeting of HIV-1 integrase in human cells. J. Biol. Chem. *278*, 33528–33539.

Mansky, L.M. (1996). The mutation rate of human immunodeficiency virus type 1 is influenced by the vpr gene. Virology *222*, 391–400.

Martinez, M.A., Clotet, B., and Este, J.A. (2002a). RNA interference of HIV replication. Trends Immunol. *23*, 559–561.

Martinez, M.A., Gutierrez, A., Armand-Ugon, M., Blanco, J., Parera, M., Gomez, J., Clotet, B., and Este, J.A. (2002b). Suppression of chemokine receptor expression by RNA interference allows for inhibition of HIV-1 replication. AIDS *16*, 2385–2390.

Miyagishi, M., Matsumoto, S., and Taira, K. (2004). Generation of an shRNAi expression library against the whole human transcripts. Virus Res. *102*, 117–124.

Morris, K.V. (2008). RNA-mediated transcriptional gene silencing in human cells. Curr. Top. Microbiol. Immunol. *320*, 211–224.

Naito, Y., Nohtomi, K., Onogi, T., Uenishi, R., Ui-Tei, K., Saigo, K., and Takebe, Y. (2007). Optimal design and validation of antiviral siRNA for targeting HIV-1. Retrovirology *4*, 80.

Nguyen, D.G., Wolff, K.C., Yin, H., Caldwell, J.S., and Kuhen, K.L. (2006). 'UnPAKing' human immunodeficiency virus (HIV) replication: using small interfering RNA screening to identify novel cofactors and elucidate the role of group I PAKs in HIV Infection. J. Virol. *80*, 130–137.

Nishitsuji, H., Ikeda, T., Miyoshi, H., Ohashi, T., Kannagi, M., and Masuda, T. (2004). Expression of small hairpin RNA by lentivirus-based vector confers efficient and stable gene-suppression of HIV-1 on human cells including primary non-dividing cells. Microbes Infect. *6*, 76–85.

Nishitsuji, H., Kohara, M., Kannagi, M., and Masuda, T. (2006). Effective suppression of human immunodeficiency virus type 1 through a combination of short- or long-hairpin RNAs targeting essential sequences for retroviral integration. J. Virol. *80*, 7658–7666.

Omoto, S., Ito, M., Tsutsumi, Y., Ichikawa, Y., Okuyama, H., Andi, B.E., Saksena, N.K., and Fuji, Y. (2004). HIV-1 nef suppression by virally encoded microRNA. Retrovirology *1*, 44–55.

Paddison, P.J. (2008). RNA interference in mammalian cell systems. Curr. Top. Microbiol. Immunol. *320*, 1–19.

Paddison, P.J., Silva, J.M., Conklin, D.S., Schlabach, M., Li, M., Aruleba, S., Balija, V., O'Shaughnessy, A., Gnoj, L., Scobie, K., Chang, K., Westbrook, T., Cleary, M., Sachidanandam, R., McCombie, W.R., Elledge, S.J., and Hannon, G.J. (2004). A resource for large-scale RNA-interference-based screens in mammals. Nature *428*, 427–431.

Pebernard, S., and Iggo, R.D. (2004). Determinants of interferon-stimulated gene induction by RNAi vectors. Differentiation *72*, 103–111.

Ping, Y.H., Chu, C.Y., Cao, H., Jacque, J.M., Stevenson, M., and Rana, T.M. (2004). Modulating HIV-1 replication by RNA interference directed against human transcription elongation factor SPT5. Retrovirology *1*, 46.

Reich, S.J., Fosnot, J., Kuroki, A., Tang, W., Yang, X., Maguire, A.M., Bennett, J., and Tolentino, M.J. (2003). Small interfering RNA (siRNA) targeting VEGF effectively inhibits ocular neovascularization in a mouse model. Molecular Vision *9*, 210–216.

Ringrose, J., Jeeninga, R.E., Berkhout, B., and Speijer, D. (2008). Proteomic studies reveal coordinated changes in T cell expression patterns upon HIV-1 infection. J. Virol. *82*, 4320–4330.

Robbins, M., Judge, A., Ambegia, E., Choi, C., Yaworski, E., Palmer, L., McClintock, K., and Maclachlan, I. (2008). Misinterpreting the therapeutic effects of siRNA caused by immune stimulation. Hum. Gene Ther. *19*, 991–999.

Robbins, M.A., and Rossi, J.J. (2005). Sensing the danger in RNA. Nat. Med. *11*, 250–251.

Root, D.E., Hacohen, N., Hahn, W.C., Lander, E.S., and Sabatini, D.M. (2006). Genome-scale loss-of-function screening with a lentiviral RNAi library. Nat. Methods 3, 715–719.

Rossi, J.J. (2006). RNAi as a treatment for HIV-1 infection. Biotechn. (*Suppl.*), 25–29.

Saayman, S., Barichievy, S., Capovilla, A., Morris, K.V., Arbuthnot, P., and Weinberg, M.S. (2008). The efficacy of generating three independent anti-HIV-1 siRNAs from a single U6 RNA Pol III-expressed long hairpin RNA. PLoS. ONE 3, e2602.

Sabariegos, R., Gimenez-Barcons, M., Tapia, N., Clotet, B., and Martinez, M.A. (2006). Sequence homology required by human immunodeficiency virus type 1 to escape from short interfering RNAs. J. Virol. *80*, 571–577.

Scherer, L., Rossi, J.J., and Weinberg, M.S. (2007). Progress and prospects: RNA-based therapies for treatment of HIV infection. Gene Ther. *14*, 1057–1064.

Shankar, P., Manjunath, N., and Lieberman, J. (2005). The prospect of silencing disease using RNA interference. JAMA *293*, 1367–1373.

Shen, J., Samul, R., Silva, R.L., Akiyama, H., Liu, H., Saishin, Y., Hackett, S.F., Zinnen, S., Kossen, K., Fosnaugh, K., Vargeese, C., Gomez, A., Bouhana, K., Aitchison, R., Pavco, P., and Campochiaro, P.A. (2005). Suppression of ocular neovascularization with siRNA targeting VEGF receptor 1. Gene Ther. *13*, 225–234.

Sioud, M. (2005). Induction of inflammatory cytokines and interferon responses by double-stranded and single-stranded siRNAs is sequence-dependent and requires endosomal localization. J. Mol. Biol. *348*, 1079–1090.

Son, J., Uchil, P.D., Kim, Y.B., Shankar, P., Kumar, P., and Lee, S.K. (2008). Effective suppression of HIV-1 by artificial bispecific miRNA targeting conserved sequences with tolerance for wobble basepairing. Biochem. Biophys. Res. Commun. *374*, 214–218.

Stevenson, M. (2003). Dissecting HIV-1 through RNA interference. Nat. Rev. Immunol. 3, 851–858.

Stram, Y., and Kuzntzova, L. (2006). Inhibition of viruses by RNA interference. Virus Gen. *32*, 299–306.

Surabhi, R.M., and Gaynor, R.B. (2002). RNA interference directed against viral and cellular targets inhibits human immunodeficiency virus type 1 replication. J. Virol. *76*, 12963–12973.

Suzuki, K., Shijuuku, T., Fukamachi, T., Zaunders, J.J., Guillemin, G., Cooper, D., and Kelleher, A.D. (2005). Prolonged transcriptional silencing and CpG methylation induced by dsRNAs targeted to the HIV-1 promoter region. J. RNAi and Gene Silencing *1*, 66–78.

Szulc, J., and Aebischer, P. (2008). Conditional gene expression and knockdown using lentivirus vectors encoding shRNA. Methods Mol. Biol. *434*, 291–309.

Ter Brake, O., 't Hooft, K., Liu, Y.P., Centlivre, M., von Eije, K.J., and Berkhout, B. (2008a). Lentiviral vector design for multiple shRNA expression and durable HIV-1 inhibition. Mol Ther. *16*, 557–564.

Ter Brake, O., and Berkhout, B. (2007). Lentiviral vectors that carry anti-HIV shRNAs: problems and solutions. J. Gene Med. *9*, 743–750.

Ter Brake, O., and Berkhout, B. (2005). A novel approach for inhibition of HIV-1 by RNA interference: counteracting viral escape with a second generation of siRNAs. J. RNAi Gene Silencing *1*, 56–65.

Ter Brake, O., Haasnoot, J., Kurreck, J., and Berkhout, B. (2008b). ESF-EMBO symposium: antiviral applications of RNA interference. Retrovirology *5*, 81.

Ter Brake, O., Konstantinova, P., Ceylan, M., and Berkhout, B. (2006). Silencing of HIV-1 with RNA interference: a multiple shRNA approach. Mol. Ther. *14*, 883–892.

Ter Brake, O., von Eije, K.J., and Berkhout, B. (2008c). Probing the sequence space available for HIV-1 evolution. AIDS *22*, 1875–1877.

Ter Brake, O., Legrand, N., von Eije, K.J., Centlivre, M., Spits, H., Weijer, K., Blom, B., and Berkhout, B. (2009). Evaluation of safety and efficacy of RNAi against HIV-1 in the human immune system (Rag-2$(^{-/-})$(c)$(^{-/-})$) mouse model. Gene Ther. *16*, 148–153.

Umbach, J.L., Kramer, M.F., Jurak, I., Karnowski, H.W., Coen, D.M., and Cullen, B.R. (2008). MicroRNAs expressed by herpes simplex virus 1 during latent infection regulate viral mRNAs. Nature *454*, 780–783.

Unwalla, H.J., Li, H.T., Bahner, I., Li, M.J., Kohn, D., and Rossi, J.J. (2006). Novel Pol II fusion promoter directs human immunodeficiency virus type 1-inducible coexpression of a short hairpin RNA and protein. J. Virol. *80*, 1863–1873.

von Eije, K.J., Ter Brake, O., and Berkhout, B. (2008). Human immunodeficiency virus type 1 escape is restricted when conserved genome sequences are targeted by RNA interference. J. Virol. *82*, 2895–2903.

von Eije, K.J., Ter Brake, O., and Berkhout, B. (2009). Stringent testing identifies highly potent and escape-proof anti-HIV shRNAs. J. Gene Med. *11*, 459–467.

Weinberg, M.S., Ely, A., Barichievy, S., Crowther, C., Mufamadi, S., Carmona, S., and Arbuthnot, P. (2007). Specific inhibition of HBV replication in vitro and *in vivo* with expressed long hairpin RNA. Mol Ther. *15*, 534–541.

Westerhout, E.M., and Berkhout, B. (2007). A systematic analysis of the effect of target RNA structure on RNA interference. Nucleic Acids Res. *35*, 4322–4330.

Westerhout, E.M., Ooms, M., Vink, M., Das, A.T., and Berkhout, B. (2005). HIV-1 can escape from RNA interference by evolving an alternative structure in its RNA genome. Nucleic Acids Res. *33*, 796–804.

Westerhout, E.M., Ter Brake, O., and Berkhout, B. (2006a). The virion-associated incoming HIV-1 RNA genome is not targeted by RNA interference. Retrovirology *3*, 57–65.

Westerhout, E.M., Vink, M., Haasnoot, P.C., Das, A.T., and Berkhout, B. (2006b). A conditionally replicating HIV-based vector that stably expresses an antiviral shRNA against HIV-1 replication. Mol. Ther. *14*, 268–275.

Wilkinson, K.A., Merino, E.J., and Weeks, K.M. (2006). Selective 2′-hydroxyl acylation analyzed by primer extension (SHAPE): quantitative RNA structure analysis at single nucleotide resolution. Nat. Protoc. *1*, 1610–1616.

Yamamoto, T., Miyoshi, H., Yamamoto, N., Yamamoto, N., Inoue, J., and Tsunetsugu-Yokota, Y. (2006). Lentivirus vectors expressing short hairpin RNAs against the U3-overlapping region of HIV nef inhibit HIV replication and infectivity in primary macrophages. Blood *108*, 3305–3312.

Yamamoto, T., Omoto, S., Mizuguchi, M., Mizukami, H., Okuyama, H., Okada, N., Saksena, N.K., Brisibe, E.A., Otake, K., and Fuji, Y.R. (2002). Double-stranded nef RNA interferes with human immunodeficiency virus type 1 replication. Microbiol. Immunol *46*, 809–817.

Zhang, L., Kovalev, G.I., and Su, L. (2007). HIV-1 infection and pathogenesis in a novel humanized mouse model. Blood *109*, 2978–2981.

Zhang, W., and Tripp, R.A. (2008). RNA interference inhibits respiratory syncytial virus replication and disease pathogenesis without inhibiting priming of the memory immune response. J. Virol. *82*, 12221–12231.

Zhang, W., Yang, H., Kong, X., Mohapatra, S., Juan-Vergara, H.S., Hellermann, G., Behera, S., Singam, R., Lockey, R.F., and Mohapatra, S.S. (2005). Inhibition of respiratory syncytial virus infection with intranasal siRNA nanoparticles targeting the viral NS1 gene. Nat. Med *11*, 56–62.

Zhou, N., Fang, J., Mukhtar, M., Acheampong, E., and Pomerantz, R.J. (2004). Inhibition of HIV-1 fusion with small interfering RNAs targeting the chemokine coreceptor CXCR4. Gene Ther. *11*, 1703–1712.

# Advances in the Use of RNAi to Treat Chronic Hepatitis B Virus Infection

8

Abdullah Ely and Patrick Arbuthnot

Abstract

Chronic infection with the hepatitis B virus (HBV) occurs in approximately 6% of the world's population and carriers of the virus are at risk for complicating hepatocellular carcinoma (HCC) and cirrhosis. Although effective vaccination is available, it is prophylactic and of little use to individuals who are already infected with the virus. Furthermore, current treatment options have limited efficacy and chronic HBV infection is likely to be a significant global medical problem for many years to come. Silencing HBV gene expression by harnessing RNA interference (RNAi) presents an attractive option for development of novel and effective anti HBV agents. Numerous studies have reported highly successful suppression of viral replication, which bodes well for employing this approach to counter HBV infection. However, despite significant and rapid progress, further refinement of existing technologies is necessary before clinical application of RNAi-based HBV therapies is realized. Improvement of delivery efficiency, dose regulation, limiting of off target effects and preventing reactivation of viral replication are some of the hurdles that need to be overcome. Nevertheless, the vast potential of RNAi-based therapeutics will continue to drive innovative research, and this promises to surmount the obstacles that face this exciting field.

## Introduction

There are approximately 387 million chronically infected carriers of HBV in the world and these individuals are at high risk for the serious complications of cirrhosis and HCC (Chisari and Ferrari, 1995; Arbuthnot and Kew, 2001). Several mechanisms of HBV-induced HCC have been proposed and these include roles for integration of HBV DNA into the genome of hepatocytes, the HBx protein and the necroinflammatory hepatic disease that often accompanies infection (reviewed in Arbuthnot and Kew, 2001). Most HBV carriers are from sub-Saharan Africa, east and south east Asia as well as the Western Pacific islands. Factors that influence development of HBV-related cirrhosis and HCC are complex and as yet incompletely understood. For example, HBV genotype, age at which viral infection occurs, gender of carriers and their geographical location are some of the contributors to the likelihood of HCC development. A landmark cohort study from Taiwan estimated that HBV carriers face a lifetime relative risk for developing HCC of more than 100 and approximately 40% of these carriers die from HCC, cirrhosis, or both diseases (Beasley *et al.*, 1981). The virus is transmitted parenterally, which is often through sexual contact in adults, perinatally in neonates and through poorly defined mechanisms of horizontal spread amongst children. Infection at a young age is associated with persistence and associated high risk of complications while HBV infection among adults is typically acute and self-limiting with rare long-term sequelae (Kew *et al.*, 1987). Babies who acquire HBV perinatally have a high risk (80–90%) of becoming chronic carriers, and virus persistence is lower (20–30%) when HBV is transmitted during childhood. Based on sequence divergence, eight HBV genotypes (A–H) have been described, and each of them has distinct geographical distribution and clinical course. HBV subgenotype A1 is hyperendemic to South Africa and is associated with a particularly high risk for liver cancer (Kew *et al.*, 2005; Kramvis and Kew, 2007).

## Molecular biology of HBV replication

The DNA genome of HBV, as with other hepadnaviruses, has a relaxed circular (rcDNA) partly double-stranded arrangement (Mason *et al.*, 1982; Weiser, 1983) (Fig. 8.1). The long (minus) strand encompasses the entire genome and is approximately 3 200 bases in length. The plus strand is of variable length and maintains the circular structure of the genome by formation of a cohesive duplex that spans the 5′ and 3′ ends of the minus strand. After infecting hepatocytes, the virion DNA is converted to covalently closed circular DNA (cccDNA), which is the template for transcription by cellular RNA polymerase II (Mason *et al.*, 1982; Mason *et al.*, 1983; Weiser, 1983). Elimination of episomal cccDNA has proved to be difficult to achieve and is an important reason for the poor efficacy of many of the currently available anti HBV treatment regimens. Transcripts are initiated from four different promoters on the cccDNA and terminate at a single polyadenylation site (Fig. 8.1). The pregenomic RNA, which is longer than the cccDNA, comprises approximately 3.5 kb with duplicated terminal sequences that are essential for HBV replication and HBV DNA circularization during

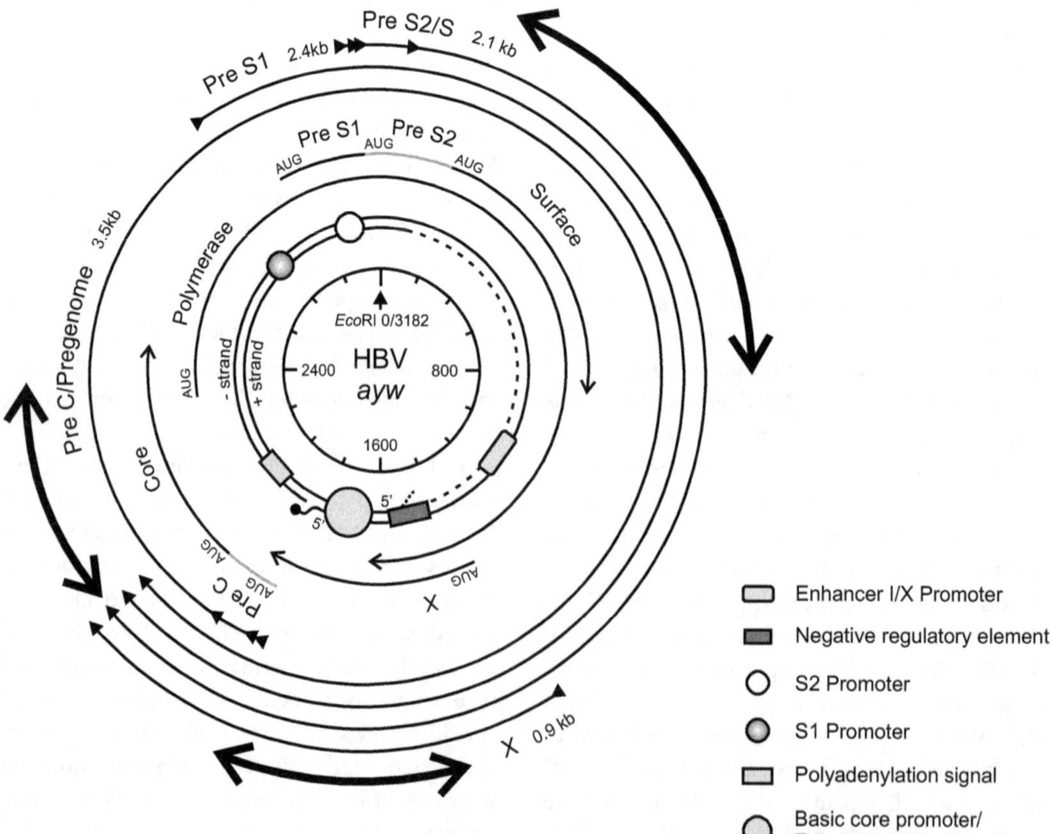


**Figure 8.1** Schematic representation of HBV genome organization. The partially double-stranded genome is indicated with transcription regulatory elements and nucleotide co-ordinates (relative to *Eco*RI restriction site). Upon infection the partially dsDNA genome is converted to cccDNA, which functions as the template for transcription from four viral promoters. Four major viral RNA species are transcribed from the cccDNA template and terminate at a single polyadenylation signal. Viral open reading frames are shown as arrows immediately surrounding the genome and code for seven viral proteins namely HBV e antigen (HBeAg), core, polymerase, three surface antigens and HBx. Regions of the viral genome that have been successfully targeted using RNAi are indicated by the bold double arrow line.

reverse transcription (Seeger *et al.*, 1986). The HBV genome, which is remarkably compact, encodes *precore/core (C), polymerase (pol), envelope (surface, S)* and *X (HBx)* open reading frames (ORFs) (Tiollais *et al.*, 1985). Each ORF overlaps at least one other, and together they cover the entire genome. The HBV *cis*-elements, which are required for regulating viral replication, are therefore included within the protein-coding regions. The C ORF contains two in-phase start codons that encode two overlapping proteins. Translation initiated from the precore AUG codon generates a precursor protein that is modified by proteolysis and secreted as a marker of HBV replication, which is the HBV e antigen (HBeAg). Translation initiated at the second C ORF initiation codon gives rise to core protein that forms the nucleocapsid. The *Pol* ORF is the largest of the virus and encodes the enzyme that is responsible for priming and reverse transcription of the pregenomic HBV RNA. Pre S1, pre S2 and S in-phase translational start codons of the *S* ORF initiate translation of the large, middle and major surface proteins, respectively. All three HBV surface antigens (HBsAgs) are required for the formation of the virion envelope. *HBx* encodes a protein of 154 amino acids, HBx, and has a molecular weight of 17 kDa. The protein is required for natural infection in the woodchuck and presumably for the replication of other mammalian hepadnaviruses *in vivo* (Chen *et al.*, 1993; Zoulim *et al.*, 1994). Importantly, HBx is conserved in mammalian hepadnaviruses associated with HCC and several lines of evidence implicate this protein in the hepatocarcinogenic process (reviewed in Arbuthnot *et al.*, 2000).

## HBV treatment currently available and under development

The primary objective of HBV therapy is to achieve sustained suppression of viral replication to improve long term patient outcome. Typically, HBeAg, HBsAg and circulating viral DNA are used as markers of viral replication. For therapy to be effective, clearance of antigens with development of antibodies (antiHBs and antiHBe) and elimination of both circulating and intrahepatic cccDNA should be achieved. Currently licensed therapy for HBV infection falls into two broad categories: nucleoside/nucleotide analogues and immunomodulators such as interferon-α (IFN-α).

### Nucleoside and nucleotide analogues

Nucleoside and nucleotide analogues have been widely used to treat chronic HBV infection (reviewed in (Arora and Keeffe, 2007; Loomba and Liang, 2007; Ferir *et al.*, 2008; Keeffe *et al.*, 2008)). Members of this class of antiviral have similar mechanisms of action. After intracellular phosphorylation, they function as mimics of naturally occurring nucleotides and are competitive inhibitors of HBV Pol during viral reverse transcription. When incorporated into HBV DNA, these drugs cause premature termination of DNA synthesis and in some cases inhibition of the priming step of HBV DNA synthesis. Although nucleoside and nucleotide analogues are well tolerated and easily administered by the oral route, emergence of viral resistance is a problem of their use (Keeffe *et al.*, 2008). Currently, lamivudine (3TC), adefovir, entecovir and telbivudine are approved for use against HBV. Lamivudine has been most widely used and treatment typically causes an initial profound decrease (3–4 $\log_{10}$ fold) in circulating HBV DNA (Lai *et al.*, 1998; Dienstag *et al.*, 1999). This drop is also associated with seroconversion from HBeAg positive to anti HBe positive status and normalization of serum alanine transaminase (ALT) activity. However, prolonged lamivudine monotherapy commonly leads to emergence of viral resistance to the drug (Allen *et al.*, 1998). Resistance mutations occur within the YMDD (tyrosine-methionine-aspartate-aspartate) motif of HBV Pol. Commonly, valine or isoleucine is substituted for methionine (M204V/I) and this mutation is also often combined with an additional L180M mutation (Fu and Cheng, 1998; Delaney *et al.*, 2003). These substitutions result in steric hindrance of HBV polymerase interaction with lamivudine. Although cross-resistance to other nucleoside/nucleotide analogues such as entecavir has been demonstrated, combination therapy limits emergence of resistance and is the preferred first line of therapy.

### Immunomodulators

An adequate immune response to HBV infection is crucial for sustained control of the infection. This is supported by observations that show reactivation of HBV infections in patients who receive immunosuppressive therapy, and that the T-cell response in HBV chronic carriers is usually highly specific and transient (Rehermann *et al.*, 1996; Roggendorf *et al.*, 2007). To augment the anti HBV immune response, IFN-α has been in widespread use for many years (Greenberg *et al.*, 1976; Weimar *et al.*, 1977). Currently, unmodified and polyethylene glycol (PEG)-conjugated IFN-α are licensed for therapy of HBV (Loomba and Liang, 2007). Use of IFN-α is, however ,often complicated by side-effects, flares of hepatitis and is contraindicated in patients with cirrhosis. Other disadvantages include its expense and lack of availability in resource-poor settings.

### HBV therapies under development

The inadequacies of existing HBV therapy have prompted investigation of novel approaches to countering this important viral infection. In addition to novel immunomodulatory agents and nucleoside/nucleotide analogues, other distinct and interesting classes of antivirals show promise. Destabilizing formation of nucleocapsids by heteroaryldihydropyrimidines (Weber *et al.*, 2002; Stray *et al.*, 2005) and phenylpropenamides (King *et al.*, 1998) is effective against the virus *in vivo* and in cell culture-based assays. Inhibitors of glycosidation of HBV impede viral particle assembly and are effective in the woodchuck hepatitis virus (WHV) model (Block *et al.*, 1998). Following discovery of RNAi, there has been considerable enthusiasm for the potential of harnessing this pathway for therapeutic suppression of HBV replication. Many studies have recently been carried out on this topic (Giladi *et al.*, 2003; Klein *et al.*, 2003; McCaffrey *et al.*, 2003; Shlomai and Shaul, 2003; Carmona *et al.*, 2006). Proof of principle has been demonstrated *in vivo* and an early clinical trial is in progress (http://www.nucleonicsinc.com/news/pdfs/FDAClearance-HepBTrial_may2007.pdf).

## RNAi against HBV

### Activating the RNAi pathway

Endogenous RNAi functions to regulate gene expression involved in diverse biological processes. Interaction between small RNAs, called microRNAs (miRs), underlie the regulatory mechanism (reviewed in (Kim, 2005)). miR biogenesis (Fig. 8.2) involves a two-step process during which an RNA polymerase II (Pol II)-derived transcript, the hairpin motif-containing primary precursor-miR (pri-miR), is processed to form the final mature miR. Drosha together with DGCR8, which is a double-stranded RNA (dsRNA) binding protein, form the microprocessor complex. This microprocessor cleaves the pri-miR into an approximately 70 nt stem–loop structure known as a precursor miR (pre-miR). The exportin-5 karyopherin mediates export of the pre-miR from the nucleus to the cytoplasm. Here it undergoes further processing by Dicer and TAR RNA-binding protein (TRBP) to form an approximately 22-nt miR duplex with characteristic 2-nt 3′ overhangs. The miR duplex is incorporated into the RNA induced silencing complex (RISC) following which removal of the passenger strand and retention of the guide sequence activates the complex. The guide functions to direct RISC to target sequences by Watson-Crick basepairing. Gene silencing is typically effected by transcript cleavage or translational suppression. Hybridization of the guide seed sequence, nucleotides 2–8 from the 5′ end, is all that is required to cause translational suppression. This low stringency sequence requirement is important for therapeutic application of exogenous RNAi effecters, where minimizing non specific gene silencing is crucial. When the guide has near perfect complementarity to its cognate, cleavage of the target occurs to effect efficient gene silencing. The enzymatic nature of the RISC mechanism enables highly efficient silencing through action on multiple substrates.

Exploiting RNAi typically involves introducing RNA sequences that are exogenous activators of the pathway at pri-miR, pre-miR or miR duplex stages (Cullen, 2005) (Figs. 8.2 and 8.3). Demonstration that RNAi activation in *Drosophila melanogaster* embryo lysates requires characteristic 21- to 23-nt RNA duplexes with 2-nt 3′ overhangs, was fundamental to development of exogenous RNAi activators. This led to the landmark study by Elbashir *et al.* (Elbashir *et al.*, 2001a), which revealed that chemically synthesized short interfering RNAs (siRNAs) could be used to induce RNAi in mammalian cells. Numerous reports followed this study and

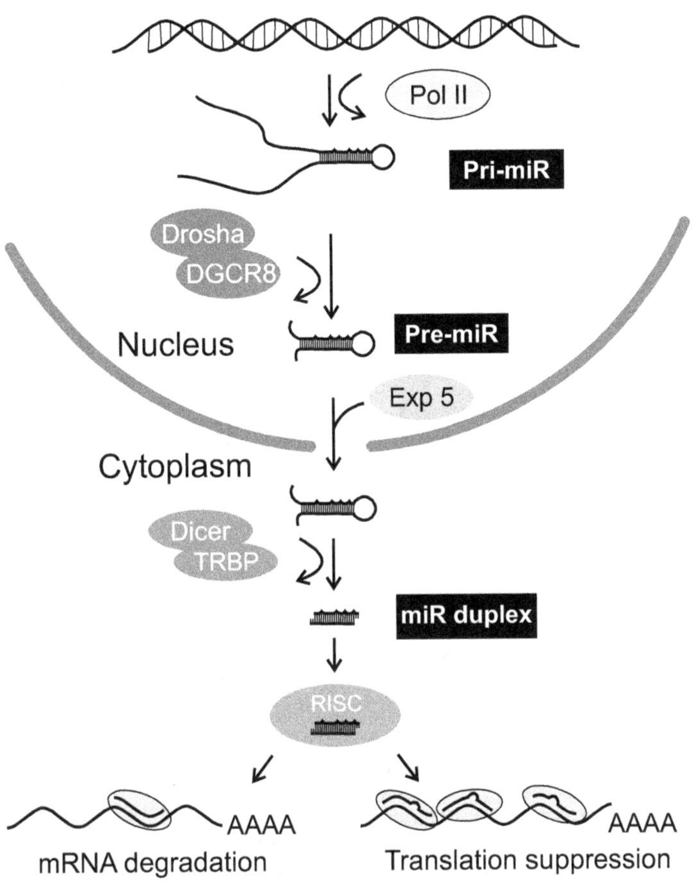


**Figure 8.2** MicroRNA biogenesis pathway. Pri-miR transcripts that exist as polycistronic and monocistronic units are processed in the nucleus by Drosha and its double-stranded RNA binding partner, DGCR8, into approximately 70 nt pre-miR. Exportin-5 mediates transport of the pre-miR to the cytoplasm where it undergoes a second round of processing by Dicer and TRBP to yield the mature miR duplex of approximately 22 bp. Activation of RISC follows incorporation of the duplex into the complex. The passenger strand is removed and the retained strand guides the complex to target complementary mRNA. Translational repression occurs when complementarity between the target mRNA and guide strand is limited to the seed region. Complete complementarity between the target and guide sequences results in mRNA degradation. Mimics of pri-miR, pre-miR and miR duplex intermediates of the miR pathway, shown in boxes, have been used to inhibit HBV replication.

have shown that exogenous synthetic or expressed sequences can be used to effect specific and powerful gene silencing. Advancement of potentially therapeutic knockdown of human pathology-causing viral genes has since been a highly active field of research. The feasibility of this approach for treating viral infection is supported by knowledge that RNAi functions as an antiviral defence in plants. HBV is ideally suited to therapy based on nucleic acid hybridization. The highly compact arrangement of the viral genome, together with the requirement for a pregenomic RNA intermediate, make this virus susceptible to inhibition of replication that is based on RNAi-mediated gene silencing. Both synthetic (Giladi *et al.*, 2003; Hamasaki *et al.*, 2003; Klein *et al.*, 2003; Konishi *et al.*, 2003; Qian *et al.*, 2005) and expressed (Chen *et al.*, 2003; McCaffrey *et al.*, 2003; Shlomai and Shaul, 2003; Zhang *et al.*, 2004; Carmona *et al.*, 2006; Ely *et al.*, 2008; Giering *et al.*, 2008; Keck *et al.*, 2009) RNAi activators targeted to all four major HBV transcripts have been shown to limit viral replication in cell culture and *in vivo*.

A difficulty of assessing usefulness of RNAi against HBV has been the lack of convenient

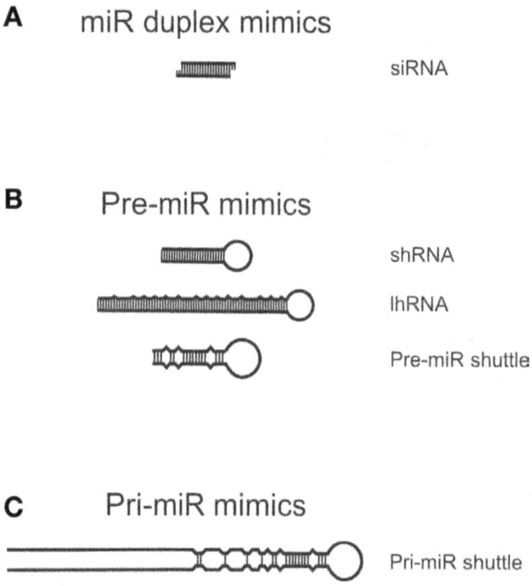


**Figure 8.3** RNAi effecters. RNAi can be activated with synthetic or expressed RN sequences that mimic intermediates the miR biogenesis pathway. (A) Synthetic or expressed siRNA effecters activate RNAi by entering RISC directly. (B) Expressed shRNA, lhRNA and pre-miR shuttles mimic endogenous pre-miR and enter the RNAi pathway as Dicer substrates. (C) Expressed pri-miR shuttles enter the RNAi pathway at the Drosha-DGCR8 stage as they structurally mimic endogenous pri-miR.

small animal models of viral infection and also the limited infectiveness of cells in culture. To overcome this problem, alternative approaches have been used to simulate viral replication in cultured cells and *in vivo*. These include transfection of cells with HBV replication-competent DNA, use of the murine hydrodynamic injection procedure (Liu *et al.*, 1999) and studies on HBV transgenic mice. Infection of cells in culture that are permissive for HBV (Gripon *et al.*, 2002), and HBV infection of mice that are xenografted (Turrini *et al.*, 2006) with human hepatocytes are potentially useful models for developing RNAi-based HBV therapy. The murine hydrodynamic tail vein injection procedure is a convenient means of nucleic acid delivery to mouse hepatocytes (Liu *et al.*, 1999) and has been widely used to simulate HBV replication *in vivo* (Yang *et al.*, 2002b). The technique involves tail vein injection of a large volume of solution containing HBV replication-competent DNA into mice over a period of 5–10 seconds. Efficient transfection of mouse hepatocytes activates HBV replication *in vivo* (Yang *et al.*, 2002b). Viral proliferation is however transient and the procedure itself causes damage to hepatocytes. HBV transgenic mice simulate the human condition of HBV persistence more closely and represent a more stringent model to assess efficacy of RNAi effecters. To date, delivering exogenous RNAi sequences to hepatocytes has relied mainly on use of recombinant viruses or non-viral formulations. The utility of each type of approach for delivering synthetic or expressed anti HBV RNAi effecters is discussed in detail below.

## Synthetic and expressed anti HBV RNAi activators

Both synthetic and expressed RNAi-inducing sequences have been used successfully to silence HBV gene expression. DNA expression cassettes are compatible with recombinant viral vectors, they naturally achieve more sustained knockdown and obviate problems of degradation by RNases. However, synthetic RNA has the important advantages of easier cellular delivery as well as small molecule drug-like properties that make regulation of dosage of silencing sequences easier to achieve. The general properties of synthetic and expressed RNAi effecters are discussed below. Although not all observations have been derived from studies on HBV silencing, general principles are relevant to advancement of RNAi-based therapy for this important viral infection.

### *General properties of synthetic siRNAs*

Synthetic siRNAs typically comprise RNA duplexes of 21–23 bp with 2 nt overhangs at their 3′ ends (Fig. 8.3). The large size (approximately 13 kDa), negatively charged phosphodiester backbone and susceptibility to degradation by ubiquitous RNases necessitate the use of vectors and chemical modifications to enhance siRNA efficacy (reviewed in Behlke, 2006, 2008). As a result of their potential effects on siRNA structure and RISC interaction, chemical modifications are usually restricted to only a few nucleotides within the siRNA duplexes. Indeed, substitution of the either or both siRNA strands entirely with 2′-deoxy or 2′-*O*-methyl (2′-O-Me) residues completely abolishes silencing function (Elbashir *et al.*, 2001b). A variety of modified siRNAs enhance stability, potency and specificity. Changes to ribose at the 2′ position, which include addition of 2′-O-Me RNA and 2′-fluoro (2′-F) moieties, are commonly used to achieve these improvements (Morrissey *et al.*, 2005b; Muhonen *et al.*, 2007; Robbins *et al.*, 2007). Other modifications have also been employed and include the use of locked nucleic acid (LNA) siRNA derivatives that provide nuclease protection and increase siRNA duplex thermodynamic stability (Mook *et al.*, 2007).

Sequence-independent non specific stimulation of the innate immune response by siRNAs has also been shown to be attenuated by chemical modification (Morrissey *et al.*, 2005b; Robbins *et al.*, 2007). Sequence-dependent immunostimulation may also occur during cellular uptake of siRNAs and interaction with Toll-like receptor 3 (TLR3), TLR7 and TLR8 (Hornung *et al.*, 2005; Judge *et al.*, 2005). These 'danger' motifs (e.g. GU-rich sequences, 5′ GUCCUUCAA 3′ and 5′ UGUGU 3′) should be excluded from potentially therapeutic siRNAs. Cholesterol conjugation of siRNA, which has been shown to improve liver uptake of siRNA (Soutschek *et al.*, 2004), may also diminish TLR3 recognition by facilitating cytoplasmic uptake and incorporation into RISC. The importance of ensuring that immunostimulation is avoided for development of RNAi-based therapy was emphasized in a mechanistic analysis of siRNAs that were in an advanced stage of development for treatment of the wet form of age-related macular degeneration (Kleinman *et al.*, 2008). This study demonstrated sequence-independent activation of the interferon response through cell-surface TLR3 activation and suppression of angiogenesis by intraocular injection of naked siRNA. Thus, although a therapeutic effect was apparently observed, it was independent of RNAi-mediated silencing, but rather a consequence of non specific innate immunostimulation.

The seed region of any siRNA guide sequence is likely to be complementary to non targeted cellular sequences and result in non specific translational suppression. Interestingly, a 2′-*O*-Me modification at nucleotide 2 from the 5′ end has been found to attenuate the off target silencing by siRNAs (Jackson *et al.*, 2006). Ensuring that antisense strands are preferentially selected as the guide is important to diminish off target silencing that may be caused by a sense strand bias. Design of siRNAs such that H bonding at the 5′ end of the intended guide is weakened may facilitate this (Khvorova *et al.*, 2003). Recently, it was shown that a 5′-O-methyl group on the terminal ribose of a sense strand helps direct the unmethylated strand into RISC and may be used to influence guide strand bias (Chen *et al.*, 2008b).

*Delivery of synthetic siRNAs*

Cationic lipid-containing nucleic acid complexes (lipoplexes), lipid conjugates, polymer nucleic acid complexes (polyplexes) and polymer-modified lipoplexes (lipopolyplexes) are the main types of non viral vector (NVV) that have been used for hepatic delivery of synthetic RNAi activators. Capacity for large scale synthesis, chemical modification to alter biological properties and low immunostimulation are useful properties that make NVVs well-suited to therapeutic application. Importantly, limitations of low NVV delivery efficiency are at least partially offset by the cytoplasmic, and not nuclear, site of synthetic siRNA action. NVVs should be capable of condensing siRNAs to form small particles that traverse hepatic fenestrations and do not aggregate (Spagnou *et al.*, 2004). They should be stable, avoid interaction with blood components and prevent degradation of their nucleic acid payloads. Additionally, NVVs should facilitate endosomal release without TLR activation. Various approaches have been used to achieve this, which include incorporation of polycations with proton sponge properties (e.g. polyethylene imine (PEI)), pH-sensitive fusogenic lipids and fusogenic peptides. To target NVVs to the liver, incorporation of galactose moieties on the surface of these vectors has been used. This modification facilitates hepatotropism through direct interaction of the vectors with the liver specific asialoglycoprotein receptor. Since first demonstrating this 20 years ago (Wu and Wu, 1988), several different formulations using carbohydrate moieties attached to lipids and polymer components have been used. In addition to incorporating targeting moieties, NVVs have also been designed with 'stealth' molecules such as PEG to improve vector stability and passive hepatotropism. siRNAs may also be directly conjugated to carrier molecules for delivery. For example, liver-targeting of siRNA has been achieved with cholesterol, α-tocopherol (vitamin E) (Nishina *et al.*, 2008), apolipoprotein A1 (Kim *et al.*, 2007) and vitamin A-coupled liposome complexes (Sato *et al.*, 2008). 'Stabilized nucleic acid–lipid particles' (SNALPs) have also been used to deliver functional siRNA *in vivo* (Morrissey *et al.*, 2005b; Zimmermann *et al.*, 2006). Utility of some of these particles in the development of anti HBV RNAi-based therapy is discussed below in more detail.

*Use of synthetic siRNAs against HBV*

Initial reports demonstrated the feasibility of employing chemically synthesized siRNA to limit HBV proliferation in cell culture and *in vivo* (Giladi *et al.*, 2003; Hamasaki *et al.*, 2003; Klein *et al.*, 2003; Konishi *et al.*, 2003; Ying *et al.*, 2003; Carmona *et al.*, 2009). Hydrodynamic injection (Giladi *et al.*, 2003; Klein *et al.*, 2003) and more stringent HBV transgenic mouse models (Carmona *et al.*, 2009) were used to demonstrate efficacy *in vivo* of synthetic siRNAs. Numerous target sites within all four viral transcripts were found to be susceptible to siRNA-mediated inhibition (Fig. 8.1). Furthermore knockdown was observed at protein, RNA and DNA levels, which indicated that HBV gene expression and viral replication was efficiently disrupted. Unlike the case with nucleotide and nucleoside analogues, which require viral replication to exert their inhibitory effects, siRNA suppressed viral gene expression in the absence of active viral replication. Since HBV transcription is terminated at a single polyadenylation site, there is significant sequence overlap of all of the viral mRNAs. An important consequence of this is that multiple viral transcripts may be suppressed by a single siRNA targeted to a region shared by all the transcripts.

A complex variety of chemical modifications to anti HBV siRNA was recently reported to improve inhibition of viral replication *in vivo* (Morrissey *et al.*, 2005a). The changes to normal RNA structure included inverted abasic residues at the 3′ and 5′ ends of the sense strand, phosphothiorate linkages, 2′-fluoro, 2′-O-Me and deoxy nucleotides. Although intravenous administration of naked siRNA caused inhibition of viral replication, the high dose required to achieve a significant effect made the approach impractical. Lipid encapsulation of the siRNAs was therefore employed to improve anti HBV efficacy *in vivo* (Morrissey *et al.*, 2005b). These SNALP formulations were capable of reducing markers of HBV replication over a 6 week period following murine hydrodynamic injection. Importantly this study also demonstrated that chemically modified siRNAs within SNALPs did not cause non-specific inhibition of HBV replication, which was observed with similar formulations containing unmodified siRNAs. The authors ascribed the non-specific antiviral activity to induction of inflammatory cytokines. A subsequent recent study showed anti HBV efficacy of modular lipoplex formulations containing unmodified siRNAs (Carmona *et al.*, 2009). NVVs were generated in a two step procedure that initially involved condensing siRNAs with cationic lipids to form core particles. These complexes included an aminoxy cholesteryl lipid to enable convenient postcoupling of polyethylene glycol to act as a stealth/biocompatibility component. After intravenous administration, the vectors showed passive hepatotropism and were capable of silencing HBV replication in cell culture and *in vivo*. This effect, which was observed over 4 weeks, compared favourably to knockdown of HBV achieved by lamivudine treatment of transgenic mice, and was detectable without induction of an innate immunostimulatory effect. Collectively, the observations from a number of studies show that synthetic siRNA-containing anti HBV formulations will be valuable in the development of RNAi-based HBV therapy.

Endoribonuclease-prepared siRNAs (esiRNAs) have been developed as an interesting alternative to chemically synthesized siRNA (Yang *et al.*, 2002a; Myers *et al.*, 2003). These gene-regulating sequences are distinct from the recently described new class of endogenous siRNAs (also termed esiRNAs) that have been found in *Drosophila melanogaster* (Ghildiyal *et al.*, 2008) and oocytes of mice (Tam *et al.*, 2008). Generation of exogenous esiRNAs involves the hydrolysis of large dsRNA by recombinant Dicer or *E. coli* RNase III to produce short RNA duplexes that are structurally similar to siRNA. The procedure is cost effective and silencing efficacy is improved by the large pool of different siRNAs targeted to a specific viral sequence.

Such esiRNAs, which targeted HBV *C*, *HBx* and *pol*, inhibited HBV replication in culture and *in vivo* (Xuan *et al.*, 2006). An intriguing observation was that esiRNA produced from C dsRNA (esiHBC) diminishes the concentration of viral cccDNA templates. Other researchers (Zhang *et al.*, 2004; Wu *et al.*, 2005c) report similar findings and ascribe them to a feedback mechanism where knockdown of reverse transcription of pregenomic RNA is effected by inhibition of HBsAg production. However, non specific effects may also contribute to causing this outcome (Hamasaki *et al.*, 2003; McCaffrey *et al.*, 2003). Although improved knockdown and prevention of viral escape may be achieved with a combination of siRNAs (Chen *et al.*, 2005; Wu *et al.*, 2005b), the pool of esiRNA may have the undesirable consequence of increasing off-target silencing. Although esiRNAs exhibit potent antiviral activity, chemical modification to aid stability and efficiency has not yet been tested. Also, large scale synthesis for therapeutic application is not yet feasible.

## General properties of expressed RNAi activators

Inserting short hairpin RNA (shRNA)-encoding DNA sequences downstream of Pol III promoters has been a widely used method of generating exogenous intracellular interfering RNA molecules (Brummelkamp *et al.*, 2002; Lee *et al.*, 2002; Miyagishi and Taira, 2002; Paul *et al.*, 2002). shRNA sequences enter the RNAi pathway as Dicer substrates, and mimic the pre-miR products of microprocessor cleavage (Figs. 8.2 and 8.3). Small nuclear RNA U6 or human RNase H1 Pol III promoters have been used commonly to generate shRNAs and these transcription regulators have the advantage of containing all of their *cis* elements, with the exception of the first transcribed nucleotide, upstream of the transcription initiation site (Brummelkamp *et al.*, 2002; Lee *et al.*, 2002; Miyagishi and Taira, 2002; Paul *et al.*, 2002). Transcription termination can be conveniently achieved by inserting a poly dT sequence in the sense sequence of the transcription template. Expression systems that transcribe the individual strands of a siRNA from two separate Pol III promoters have also been described (Lee *et al.*, 2002; Miyagishi and Taira, 2002). Since demonstration that over expression of shRNAs may result in serious toxicity *in vivo* (Grimm *et al.*, 2006) a focus of research has been the improvement of control of expression of RNAi effecters. Refining transcription control with Pol II promoters is therefore gaining favour for the development of HBV silencing expression cassettes.

Expressed RNAi-activating sequences do not have the same immunostimulatory properties of synthetic dsRNA. Importantly, RNA that is transcribed within cells does not typically traverse the endosomal compartment and interact with TLR3, TLR7 or TLR8 to stimulate the innate immune response. Moreover, products of Dicer processing of expressed sequences have characteristic structural features, such as 2-nt 3′ overhangs, which suppress immunostimulation by RIG-I (retinoic acid-inducible protein I) (Marques *et al.*, 2006). These factors, and possibly others such as presence of G:U wobbles in the hairpin duplex (Akashi *et al.*, 2005; Watanabe *et al.*, 2006), are likely to limit induction of the IFN response by expressed silencing sequences.

### *Anti-HBV Pol III expression cassettes*

Numerous studies describe potent and specific inhibition of markers of HBV replication *in vitro* and *in vivo* when using H1- or U6-driven anti HBV shRNA expression cassettes (McCaffrey *et al.*, 2003; Shlomai and Shaul, 2003; Zhang *et al.*, 2004; Cheng *et al.*, 2005; Ren *et al.*, 2005; Shin *et al.*, 2006; Ying *et al.*, 2007). These studies demonstrated efficient knockdown of transient viral replication when anti HBV shRNA cassettes were co-transfected with an HBV expressing plasmid. Two groups described potent inhibition in HBV transgenic mice after infection with recombinant adenoviruses (Ads) carrying U6 promoter-driven anti HBV shRNA cassettes (Uprichard *et al.*, 2005; Carmona *et al.*, 2006). Significant and sustained reduction in viral DNA, RNA and antigen levels were observed in mice that had received anti HBV shRNA Ad. This important observation demonstrated efficacy of RNAi against HBV in a context that is similar to established chronic infection.

Although results from these studies are impressive, targeting single sites within the HBV genome may create a selective pressure for the emergence of viral escape mutants. Indeed several reports describe the emergence of HIV-1, HCV and poliovirus mutants that are resistant to RNAi (Das *et al.*, 2004; Gitlin *et al.*, 2005; Wilson and Richardson, 2005). Wu *et al.* (2005a) identified an HBV mutant isolated from a chronically infected individual that was resistant to site-specific RNAi-mediated silencing. A single point mutation in an H1 shRNA target was identified which conferred resistance to RNAi while maintaining the amino acid sequence of the polymerase and surface proteins. The authors artificially reproduced natural infection *in vitro* with a heterogeneous HBV population and demonstrated that survival of the mutant isolate was selected after the viral population was challenged with the anti HBV H1 shRNA. Importantly, adjusting the shRNA to match the mutant sequence perfectly restored antiviral activity. Although the reports summarized above demonstrated efficient gene silencing by Pol III promoter-driven cassettes, the severe toxicity associated with use of U6 anti HBV shRNAs *in vivo* has raised concerns (Grimm *et al.*, 2006). Lethality was observed in HBV transgenic mice that had received recombinant adeno-associated virus expressing U6-driven anti HBV shRNA. The cause of death was shown to result from saturation of exportin-5 through the overexpression of shRNA. Dysregulation of the miR biogenesis pathway, specifically export of pre-miR to the cytoplasm, caused the loss of liver miR functionality and eventually led to the death of mice. Importantly this step is bypassed with synthetic siRNAs and suggests that disruption of the endogenous miR pathway is not as serious a concern when using synthetic RNAi activators (John *et al.*, 2007). Development of expression cassettes with tissue-specific and inducible transcriptional control of RNAi effecters would therefore be desirable to limit toxicity of this promising therapeutic.

In an attempt to generate multitargeting expression cassettes that improve efficacy and overcome problems of viral escape mutation, long hairpin RNA (lhRNA)-expressing sequences have been assessed against HBV (Weinberg *et al.*, 2007). The 62-bp stem regions of the lhRNAs, which included G:U wobbles, were intended to generate Dicer substrates that are processed to form three independent HBV targeting siRNAs. The lhRNAs exhibited potent *in vitro* and *in vivo* efficacy without induction of interferon response genes. Although duplex RNA that is longer than 30 bp has been shown to activate the IFN response through stimulation of dsRNA-dependent protein kinase R (PKR) and interaction with endosomal TLRs (Caplen *et al.*, 2001), this effect may be of lesser significance when using lhRNA expression cassettes to effect gene silencing. As indicated above, RNAi effecters that are transcribed within cells do not typically traverse the endosomal compartment and also have characteristic structural features that suppress immunostimulation by RIG-I (Marques *et al.*, 2006). Further analysis of Dicer processing of anti HBV lhRNAs revealed that siRNAs generated from the stem base were produced more efficiently than siRNA from the loop side (Weinberg *et al.*, 2007). Incomplete intracellular processing of lhRNAs may therefore limit the applicability of this approach to simultaneous multitarget silencing. Recent successful use of Pol II polycistronic pri-miR shuttles suggests that this approach may be preferable to the use of lhRNAs for multitarget silencing and this is discussed in more detail below.

### *Pol II expression cassettes*

Using Pol II promoters to enable better transcriptional control of expressed RNAi effecters would assist to overcome many of the problems of toxicity related to saturation of the endogenous miR pathway. Natural pri-miR expression occurs under transcriptional control of Pol II regulatory elements (Lee *et al.*, 2004) and use of engineered mimics of endogenous pri-miRs to silence HBV replication provides a means of enabling Pol II promoter compatibility. A recent study demonstrated that pri-miR-31- and pri-miR-122-based anti-HBV shuttles were capable of potent antiviral activity (Ely *et al.*, 2008). Guides produced from the pri-miR shuttles were present at considerably lower concentrations than the equivalent sequence derived from a U6 shRNA cassette. However markers of viral replication were silenced with equivalent efficiency in cultured cells and *in vivo*, which indicates that the mature antiviral RNAi effecters used in this study acted potently. Importantly, the pri-miR shuttles did not interfere with independent RNAi-mediated gene silencing *in vitro* or *in vivo* and supports the notion that Pol II-expressed potent silencing sequences do not saturate the endogenous miRNA biogenesis pathway.

In addition to enabling improved transcription control with Pol II promoters, pri-miR shuttle sequences are amenable to polycistronic expression (Aagaard *et al.*, 2008; Liu *et al.*, 2008). Snyder *et al.* demonstrated these properties in multimeric anti HBV miR-30 mimics that were produced specifically in liver-derived cells (Snyder *et al.*, 2008). Expression cassettes were generated with single or multiple anti HBV miR-30 shuttles that included naturally occurring non structured flanking RNA. Liver-specific transcriptional control was achieved with a regulatory element comprising the alpha-fetoprotein enhancer and a minimal albumin promoter. Silencing function was not significantly altered by insertion of miR-30 shuttles at intronic or exonic regions of the cassettes. Polycistronic cassettes combining four discrete miR shuttles were processed according to intended design and were capable of silencing individual target sequences with similar efficacy to the monocistronic miR shuttles. Although this is a promising approach to preventing HBV escape, effects on disruption of the endogenous miR pathway by this simultaneous production of multiple miR sequences needs to be characterized.

*Delivery of RNAi expression cassettes*

Expression cassettes that generate anti HBV silencing sequences, unlike synthetic siRNAs, have the important distinction of being compatible with incorporation into recombinant viruses. Amongst these vectors, Ads and certain serotypes of adeno-associated viruses (AAVs) have been most widely used and show promise for therapeutic application. Efficient HBV knockdown *in vitro* was also observed in studies using recombinant human foamy virus (Sun *et al.*, 2007) and recombinant retroviral vectors (Jia *et al.*, 2006, 2007). Properties of recombinant Ads that make them suitable for therapeutic delivery of anti HBV sequences are that they transport genes to the liver with high efficiency *in vivo*, do not integrate into host genome, are stable and can be prepared in large scale using well-established methods (reviewed in (Bangari and Mittal, 2006)). Highly efficient natural hepatotropism of serotype 5 recombinant Ads was explained in a recent study, which showed that liver specificity is dependent on interaction of the capsid hexon protein with blood clotting factor X (Waddington *et al.*, 2008). Although Ads have features that are suited to delivery of expressed RNAi effecters to the liver, a limitation to their widespread use has been attenuation of delivery efficiency caused by preexisting immunity and powerful capsid protein-mediated induction of innate and adaptive immune responses. Initial studies using first generation Ads demonstrated that these vectors are capable of efficiently transducing the liver with Pol III-driven anti HBV shRNA cassettes and suppression of viral replication in HBV transgenic mice (Uprichard *et al.*, 2005; Carmona *et al.*, 2006). To attenuate immune responses of vectors delivering expressed anti HBV sequences, polymer modification (Crowther *et al.*, 2008) and generation of helper-dependent (HD) Ads (Rauschhuber *et al.*, 2008) have been employed. Given the importance of Ad capsid proteins in mediating immune responses, their modification with polymers is particularly useful for attenuating immunostimulation (Kreppel and Kochanek, 2008). A recent report showed improved HBV silencing efficacy *in vivo* and attenuated immunostimulation after polymer modification (Crowther *et al.*, 2008). These anti HBV Ads were conjugated with monomethoxy PEG-succinimidyl propionate (mPEG). The polymer binds covalently to lysine residues of the hexon protein (Kreppel and Kochanek, 2008), which is also the capsid component that is responsible for Ad vectors' hepatotropism (Waddington *et al.*, 2008). Retention of anti HBV bioactivity suggests that conditions of mPEG modification ensure attenuation of Ad immunostimulation but do not to interfere with factor X-dependent hepatotropism. HD Ads also provide a means of diminishing immunostimulation. These delivery agents have all viral vector protein-encoding sequences removed. HD Ads do not express viral proteins after target cell infection and are capable of sustained transgene expression. There has been one study to date that reported on the efficacy of HD anti HBV Ads (Rauschhuber *et al.*, 2008). Although this class of vector has promising application for long-term silencing of HBV replication, it is clear that the expressed cassette within the context of helper-dependent Ads needs to be carefully selected for optimal anti HBV efficacy. Importantly, the genomes of HD vectors are packaged in standard capsid structures comprising hexon, penton and fibre proteins, which are capable of innate immune response stimulation (Schnell *et al.*, 2001). PEGylation of HD vectors to attenuate the initial innate immune response may be well suited to treatment of persistent HBV infection, which requires production of RNAi effectors over a long period of time.

The safety and low immunogenicity of recombinant AAVs has generated considerable interest in this class of vector for delivery of anti HBV RNAi effecters. Hepatotropic AAVs have been generated by pseudotyping the widely used AAV-2 genome with the capsid of the liver-specific AAV-8 (Chen *et al.*, 2007). Pseudotyped AAV-2/8 vectors efficiently delivered Pol III-driven anti HBV shRNA to the livers of transgenic mice and resulted in potent and persistent suppression of viral replication. Suppression declined after the first 21 days but persisted for up to 120 days and was independent of cytokine stimulation. The diminishing inhibition was attributed to an immune response with associated gradual loss of transgenes from hepatocytes. Chen and colleagues subsequently attempted to extend use of anti HBV AAVs by employing an approach that entailed sequential administration of different pseudotyped vectors (Chen *et al.*, 2009). In this study, shRNA-expressing recombinant AAV-8 caused silencing of viral replication for up to 22 weeks in HBV transgenic mice. Subsequent administration of AAV9 further silenced HBV replication, which was not achieved after AAV8 readministration. Using AAVs from different serotypes is a useful means of avoiding the host immune response and achieving more sustained silencing efficacy. Interestingly, these studies did not report adverse effects caused by saturating the endogenous miR pathway. This was despite administration of high doses of AAV8, but the expression cassettes contained the weaker H1 Pol III promoter to express anti HBV shRNAs (Chen *et al.*, 2009). Diminished promoter activity is likely to attenuate toxicity caused by saturation of the endogenous pathway.

As with synthetic siRNA, DNA expression cassettes are amenable to delivery with NVVs and studies have been carried out to investigate efficacy of anti HBV expression cassettes in these formulations. Apo I liposomal nanoparticles described above have also been used for liver-specific delivery of anti HBV cassettes expressed from plasmid DNA (Kim *et al.*, 2007). However, a limitation of these vectors may be the low gene transfer efficiency that is achieved. To compensate for this partially, a useful approach would be to utilize long-term expression vector systems such as the recently described pEPI-1 plasmid (Jenke *et al.*, 2004). This expression vector contains a scaffold/matrix attachment region (S/MAR) downstream of a transcription unit to allow episomal replication and therefore enhance long-term maintenance of the expression cassette. This vector was used to deliver anti HBV U6 shRNA sequences to HepG2.2.15 cells (Jenke *et al.*, 2008). Remarkably, suppression of markers of HBV replication was maintained for up to 8 months. The therapeutic usefulness of these sequences will however be dependent on improvement of efficiency of available transgene-delivering NVVs and proof of efficacy *in vivo*.

## Concluding remarks

The progress that has been made in advancing RNAi-based therapeutics for the management of chronic HBV infection has been rapid and impressive. However, before this approach contributes significantly to improving existing HBV treatment, there are several hurdles that need to be overcome. Refining dose, limiting off-target effects and improving delivery of RNAi effecters are particularly important. The highly charged nature of nucleic acids and the requirement for vectors complicate regulation of the dose of RNAi activators. Off target effects, which may result from immune stimulation, saturation of the RNAi pathway and seed region complementarity, are well documented complications of introducing silencing sequences into cells. Immunostimulation by synthetic sequences is now better understood, and ingenious approaches that utilize chemical modification of siRNAs have been devised to overcome the problem. Complications resulting from immune stimulation by expressed RNAi activators appear to be less significant, and this is thought to result from differences in their mechanisms of action. However, therapeutic application of expressed RNAi activators has other attendant difficulties. This class of RNAi effector is more likely to cause toxic disruption of miR biogenesis, which has emphasized the need to improve transcription control of RNAi expression cassettes. Unintended knockdown of cellular genes that results from seed region interaction will also need to be addressed. Recent demonstration that chemical modification of synthetic siRNAs improves guide strand selectivity and unintended translational suppression is an important development (Jackson *et al.*, 2006). Although promising viral and non-viral vectors have been developed for delivery of nucleic acids to cells, improvements in safety and efficacy will be necessary before their eventual clinical application. A recent study demonstrating that shRNA-mediated suppression did not alter established cccDNA adds another difficulty to advancing RNAi therapy for chronic HBV infection (Starkey *et al.*, 2009). The persistence of HBV cccDNA in infected hepatocytes has been a major obstacle to eliminating this virus infection. Treatment regimens that use RNAi effecters in combination with established licensed drugs will be interesting to assess and may augment RNAi-based anti HBV efficacy. Despite the difficulties summarized above, advances in the development of efficient RNAi-based HBV silencing continue to be rapid. The momentum in the field is likely to surmount existing impediments and prospects for advancing RNAi-based therapy for chronic HBV infection are exciting.

## Acknowledgements

Work in the authors' laboratory is supported by funding from the Sixth Research Framework Programme of the European Union, Project RIGHT (LSHB-CT-2004–005276), from CANSA, the South African National Research Foundation (NRF GUN 68339 and 65495), ESASTAP and the South African Poliomyelitis Research Foundation.

## References

Aagaard, L.A., Zhang, J., von Eije, K.J., Li, H., Saetrom, P., Amarzguioui, M., and Rossi, J.J. (2008). Engineering and optimization of the miR-106b cluster for ectopic expression of multiplexed anti-HIV RNAs. Gene Ther. *15*, 1536–1549.

Akashi, H., Miyagishi, M., Yokota, T., Watanabe, T., Hino, T., Nishina, K., Kohara, M., and K., T. (2005). Escape from the interferon response associated with RNA interference using vectors that encode long modified hairpin-RNA. Mol. BioSystems *1*, 382–390.

Allen, M.I., Deslauriers, M., Andrews, C.W., Tipples, G.A., Walters, K.A., Tyrrell, D.L., Brown, N., and Condreay, L.D. (1998). Identification and characterization of mutations in hepatitis B virus resistant to lamivudine. Lamivudine Clinical Investigation Group. Hepatology 27, 1670–1677.

Arbuthnot, P., Capovilla, A., and Kew, M. (2000). Putative role of hepatitis B virus X protein in hepatocarcinogenesis: effects on apoptosis, DNA repair, mitogen-activated protein kinase and JAK/STAT pathways. J. Gastroenterol. Hepatol. *15*, 357–368.

Arbuthnot, P., and Kew, M. (2001). Hepatitis B virus and hepatocellular carcinoma. Int. J.Exp. Pathol.*82*, 77–100.

Arora, G., and Keeffe, E.B. (2007). Chronic hepatitis B with advanced fibrosis or cirrhosis: impact of antiviral therapy. Rev. Gastroenterol Disord. 7, 63–73.

Bangari, D.S., and Mittal, S.K. (2006). Current strategies and future directions for eluding adenoviral vector immunity. Curr. Gene Ther. *6*, 215–226.

Beasley, R.P., Hwang, L.Y., Lin, C.C., and Chien, C.S. (1981). Hepatocellular carcinoma and hepatitis B virus. A prospective study of 22 707 men in Taiwan. Lancet 2, 1129–1133.

Behlke, M.A. (2006). Progress towards *in vivo* use of siRNAs. Mol. Ther. *13*, 644–670.

Behlke, M.A. (2008). Chemical modification of siRNAs for *in vivo* use. Oligonucleotides *18*, 305–319.

Block, T.M., Lu, X., Mehta, A.S., Blumberg, B.S., Tennant, B., Ebling, M., Korba, B., Lansky, D.M., Jacob, G.S., and Dwek, R.A. (1998). Treatment of chronic hepadnavirus infection in a woodchuck animal model with an inhibitor of protein folding and trafficking. Nat. Med. 4, 610–614.

Brummelkamp, T.R., Bernards, R., and Agami, R. (2002). A system for stable expression of short interfering RNAs in mammalian cells. Science *296*, 550–553.

Caplen, N.J., Parrish, S., Imani, F., Fire, A., and Morgan, R.A. (2001). Specific inhibition of gene expression by small double-stranded RNAs in invertebrate and vertebrate systems. Proc. Natl. Acad. Sci. U.S.A. *98*, 9742–9747.

Carmona, S., Ely, A., Crowther, C., Moolla, N., Salazar, F.H., Marion, P.L., Ferry, N., Weinberg, M.S., and Arbuthnot, P. (2006). Effective inhibition of HBV replication *in vivo* by anti-HBx short hairpin RNAs. Mol. Ther. *13*, 411–421.

Carmona, S., Jorgensen, M.R., Kolli, S., Crowther, C., Salazar, F.H., Marion, P.L., Fujino, M., Natori, Y., Thanou, M., Arbuthnot, P., *et al.* (2009). Controlling HBV replication in vivo by intravenous administration of triggered PEGylated siRNA-nanoparticles. Mol. Pharm. 3, 706–717.

Chen, C.C., Ko, T.M., Ma, H.I., Wu, H.L., Xiao, X., Li, J., Chang, C.M., Wu, P.Y., Chen, C.H., Han, J.M., *et al.* (2007). Long-term inhibition of hepatitis B virus in transgenic mice by double-stranded adeno-associated virus 8-delivered short hairpin RNA. Gene Ther. *14*, 11–19.

Chen, C.C., Sun, C.P., Ma, H.I., Fang, C.C., Wu, P.Y., Xiao, X., and Tao, M.H. (2009). Comparative study of anti-hepatitis B virus RNA Interference by double-stranded adeno-associated virus serotypes 7, 8, and 9. Mol. Ther. *17*, 352–359.

Chen, H.S., Kaneko, S., Girones, R., Anderson, R.W., Hornbuckle, W.E., Tennant, B.C., Cote, P.J., Gerin, J.L., Purcell, R.H., and Miller, R.H. (1993). The woodchuck hepatitis virus X gene is important for establishment of virus infection in woodchucks. J. Virol. *67*, 1218–1226.

Chen, P.Y., Weinmann, L., Gaidatzis, D., Pei, Y., Zavolan, M., Tuschl, T., and Meister, G. (2008b). Strand-specific 5′-O-methylation of siRNA duplexes controls guide strand selection and targeting specificity. Rna *14*, 263–274.

Chen, Y., Du, D., Wu, J., Chan, C.P., Tan, Y., Kung, H.F., and He, M.L. (2003). Inhibition of hepatitis B virus replication by stably expressed shRNA. Biochem. Biophys. Res. Commun. *311*, 398–404.

Chen, Z., Xu, Z.F., Ye, J.J., Yao, H.P., Zheng, S., and Ding, J.Y. (2005). Combination of small interfering RNAs mediates greater inhibition of human hepatitis B virus replication and antigen expression. J. Zhejiang Univ. Sci. B *6*, 236–241.

Cheng, T.L., Chang, W.W., Su, I.J., Lai, M.D., Huang, W., Lei, H.Y., and Chang, W.T. (2005). Therapeutic inhibition of hepatitis B virus surface antigen expression by RNA interference. Biochem. Biophys. Res. Commun. *336*, 820–830.

Chisari, F.V., and Ferrari, C. (1995). Hepatitis B virus immunopathogenesis. Annu. Rev. Immunol. *13*, 29–60.

Crowther, C., Ely, A., Hornby, J., Mufamadi, M.S., Salazar, F., Marion, P., and Arbuthnot, P. (2008). efficient inhibition of hepatitis B virus replication *in vivo* using Peg-modified adenovirus vectors. Hum. Gene Ther. *19*, 1325–1331.

Cullen, B.R. (2005). RNAi the natural way. Nat. Genet. *37*, 1163–1165.

Das, A., Brummelkamp, T., Westerhout, E., Vink, M., Madiredjo, M., Bernards, R., and Berkhout, B. (2004). Human Immunodeficiency Virus Type 1 Escapes from RNA Interference-Mediated Inhibition. J. Virol. *78*, 2601–2605.

Delaney, W.E.t., Yang, H., Westland, C.E., Das, K., Arnold, E., Gibbs, C.S., Miller, M.D., and Xiong, S. (2003). The hepatitis B virus polymerase mutation rtV173L is selected during lamivudine therapy and enhances viral replication in vitro. J. Virol. *77*, 11833–11841.

Dienstag, J.L., Schiff, E.R., Wright, T.L., Perrillo, R.P., Hann, H.W., Goodman, Z., Crowther, L., Condreay, L.D., Woessner, M., Rubin, M., *et al.* (1999). Lamivudine as initial treatment for chronic hepatitis B in the United States. N. Engl. J. Med. *341*, 1256–1263.

Elbashir, S.M., Harborth, J., Lendeckel, W., Yalcin, A., Weber, K., and Tuschl, T. (2001a). Duplexes of 21-nucleotide RNAs mediate RNA interference in cultured mammalian cells. Nature *411*, 494–498.

Elbashir, S.M., Martinez, J., Patkaniowska, A., Lendeckel, W., and Tuschl, T. (2001b). Functional anatomy of siRNAs for mediating efficient RNAi in *Drosophila melanogaster* embryo lysate. EMBO J. *20*, 6877–6888.

Ely, A., Naidoo, T., Mufamadi, S., Crowther, C., and Arbuthnot, P. (2008). Expressed anti-HBV primary microRNA shuttles inhibit viral replication efficiently in vitro and *in vivo*. Mol. Ther. *16*, 1105–1112.

Ferir, G., Kaptein, S., Neyts, J., and De Clercq, E. (2008). Antiviral treatment of chronic hepatitis B virus infections: the past, the present and the future. Rev. Med. Virol. *18*, 19–34.

Fu, L., and Cheng, Y.C. (1998). Role of additional mutations outside the YMDD motif of hepatitis B virus polymerase in L(−)SddC (3TC) resistance. Biochem. Pharmacol. *55*, 1567–1572.

Ghildiyal, M., Seitz, H., Horwich, M.D., Li, C., Du, T., Lee, S., Xu, J., Kittler, E.L., Zapp, M.L., Weng, Z., *et al.* (2008). Endogenous siRNAs derived from transposons and mRNAs in *Drosophila* somatic cells. Science *320*, 1077–1081.

Giering, J.C., Grimm, D., Storm, T.A., and Kay, M.A. (2008). Expression of shRNA from a tissue-specific pol II promoter is an effective and safe RNAi therapeutic. Mol. Ther. *16*, 1630–1636.

Giladi, H., Ketzinel-Gilad, M., Rivkin, L., Felig, Y., Nussbaum, O., and Galun, E. (2003). Small interfering RNA inhibits hepatitis B virus replication in mice. Mol. Ther. *8*, 769–776.

Gitlin, L., Stone, J., and Andino, R. (2005). Poliovirus Escape from RNA Interference: Short Interfering RNA-Target Recognition and Implications for Therapeutic Approaches. J. Virol. *79*, 1027–1035.

Greenberg, H.B., Pollard, R.B., Lutwick, L.I., Gregory, P.B., Robinson, W.S., and Merigan, T.C. (1976). Effect of human leukocyte interferon on hepatitis B virus infection in patients with chronic active hepatitis. N. Engl. J. Med. *295*, 517–522.

Grimm, D., Streetz, K.L., Jopling, C.L., Storm, T.A., Pandey, K., Davis, C.R., Marion, P., Salazar, F., and Kay, M.A. (2006). Fatality in mice due to oversaturation of cellular microRNA/short hairpin RNA pathways. Nature *441*, 537–541.

Gripon, P., Rumin, S., Urban, S., Le Seyec, J., Glaise, D., Cannie, I., Guyomard, C., Lucas, J., Trepo, C., and Guguen-Guillouzo, C. (2002). Infection of a human hepatoma cell line by hepatitis B virus. Proc. Natl. Acad. Sci. U.S.A. *99*, 15655–15660.

Hamasaki, K., Nakao, K., Matsumoto, K., Ichikawa, T., Ishikawa, H., and Eguchi, K. (2003). Short interfering RNA-directed inhibition of hepatitis B virus replication. FEBS Lett. *543*, 51–54.

Hornung, V., Guenthner-Biller, M., Bourquin, C., Ablasser, A., Schlee, M., Uematsu, S., Noronha, A., Manoharan, M., Akira, S., de Fougerolles, A., *et al.* (2005). Sequence-specific potent induction of IFN-alpha by short interfering RNA in plasmacytoid dendritic cells through TLR7. Nat. Med. *11*, 263–270.

Jackson, A.L., Burchard, J., Leake, D., Reynolds, A., Schelter, J., Guo, J., Johnson, J.M., Lim, L., Karpilow, J., Nichols, K., *et al.* (2006). Position-specific chemical modification of siRNAs reduces 'off-target' transcript silencing. RNA *12*, 1197–1205.

Jenke, A., Stehle, I., Herrmann, F., Eisenberger, T., Baiker, A., Bode, J., Fackelmayer, F., and Lipps, H. (2004). Nuclear scaffold/matrix attached region modules linked to a transcription unit are sufficient for replication and maintenance of a mammalian episome. Proceedings of the National Academy of Sciences *101*, 11322–11327.

Jenke, A.C., Wilhelm, A.D., Orth, V., Lipps, H.J., Protzer, U., and Wirth, S. (2008). Long-term suppression of hepatitis B virus replication by short hairpin RNA expression using the scaffold/matrix attachment region-based replicating vector system pEPI-1. Antimicrob Agents Chemother *52*, 2355–2359.

Jia, F., Zhang, Y.Z., and Liu, C.M. (2006). A retrovirus-based system to stably silence hepatitis B virus genes by RNA interference. Biotechnol Lett *28*, 1679–1685.

Jia, F., Zhang, Y.Z., and Liu, C.M. (2007). Stable inhibition of hepatitis B virus expression and replication in HepG2.2.15 cells by RNA interference based on retrovirus delivery. J Biotechnol *128*, 32–40.

John, M., Constien, R., Akinc, A., Goldberg, M., Moon, Y.A., Spranger, M., Hadwiger, P., Soutschek, J., Vornlocher, H.P., Manoharan, M., *et al.* (2007). Effective RNAi-mediated gene silencing without interruption of the endogenous microRNA pathway. Nature *449*, 745–747.

Judge, A.D., Sood, V., Shaw, J.R., Fang, D., McClintock, K., and MacLachlan, I. (2005). Sequence-dependent stimulation of the mammalian innate immune response by synthetic siRNA. Nat. Biotechnol. *23*, 457–462.

Keck, K., Volper, E.M., Spengler, R.M., Long, D.D., Chan, C.Y., Ding, Y., and McCaffrey, A.P. (2009). Rational design leads to more potent RNA interference against hepatitis B virus: factors effecting silencing efficiency. Mol. Ther. *17*, 538–547.

Keeffe, E.B., Dieterich, D.T., Pawlotsky, J.M., and Benhamou, Y. (2008). Chronic hepatitis B: preventing, detecting, and managing viral resistance. Clin. Gastroenterol. Hepatol. *6*, 268–274.

Kew, M.C., Kassianides, C., Berger, E.L., Song, E., and Dusheiko, G.M. (1987). Prevalence of chronic hepatitis B virus infection in pregnant black women living in Soweto. J. Med. Virol. *22*, 263–268.

Kew, M.C., Kramvis, A., Yu, M.C., Arakawa, K., and Hodkinson, J. (2005). Increased hepatocarcinogenic potential of hepatitis B virus genotype A in Bantu-speaking sub-saharan Africans. J. Med. Virol. *75*, 513–521.

Khvorova, A., Reynolds, A., and Jayasena, S.D. (2003). Functional siRNAs and miRNAs exhibit strand bias. Cell *115*, 209–216.

Kim, S.I., Shin, D., Choi, T.H., Lee, J.C., Cheon, G.J., Kim, K.Y., Park, M., and Kim, M. (2007). Systemic and specific delivery of small interfering RNAs to the liver mediated by apolipoprotein A-I. Mol. Ther. *15*, 1145–1152.

Kim, V.N. (2005). MicroRNA biogenesis: coordinated cropping and dicing. Nat. Rev. Mol. Cell Biol. *6*, 376–385.

King, R.W., Ladner, S.K., Miller, T.J., Zaifert, K., Perni, R.B., Conway, S.C., and Otto, M.J. (1998). Inhibition of human hepatitis B virus replication by AT-61, a phenylpropenamide derivative, alone and in combination with (−)beta-L-2′, 3′-dideoxy-3′-thiacytidine. Antimicrob. Agents Chemother. *42*, 3179–3186.

Klein, C., Bock, C.T., Wedemeyer, H., Wustefeld, T., Locarnini, S., Dienes, H.P., Kubicka, S., Manns, M.P., and Trautwein, C. (2003). Inhibition of hepatitis B virus replication *in vivo* by nucleoside analogues and siRNA. Gastroenterology *125*, 9–18.

Kleinman, M.E., Yamada, K., Takeda, A., Chandrasekaran, V., Nozaki, M., Baffi, J.Z., Albuquerque, R.J., Yamasaki, S., Itaya, M., Pan, Y., *et al.* (2008). Sequence- and target-independent angiogenesis suppression by siRNA via TLR3. Nature *452*, 591–597.

Konishi, M., Wu, C.H., and Wu, G.Y. (2003). Inhibition of HBV replication by siRNA in a stable HBV-producing cell line. Hepatology *38*, 842–850.

Kramvis, A., and Kew, M.C. (2007). Epidemiology of hepatitis B virus in Africa, its genotypes and clinical associations of genotypes. Hepatol. Res. *37*, S9–S19.

Kreppel, F., and Kochanek, S. (2008). Modification of adenovirus gene transfer vectors with synthetic polymers: a scientific review and technical guide. Mol. Ther. *16*, 16–29.

Lai, C.L., Chien, R.N., Leung, N.W., Chang, T.T., Guan, R., Tai, D.I., Ng, K.Y., Wu, P.C., Dent, J.C., Barber, J., *et al.* (1998). A one-year trial of lamivudine for chronic hepatitis B. Asia Hepatitis Lamivudine Study Group. N. Engl. J. Med. *339*, 61–68.

Lee, N.S., Dohjima, T., Bauer, G., Li, H., Li, M.J., Ehsani, A., Salvaterra, P., and Rossi, J. (2002). Expression of small interfering RNAs targeted against HIV-1 rev transcripts in human cells. Nat. Biotechnol. *20*, 500–505.

Lee, Y., Kim, M., Han, J., Yeom, K.H., Lee, S., Baek, S.H., and Kim, V.N. (2004). MicroRNA genes are transcribed by RNA polymerase II. EMBO J. *23*, 4051–4060.

Liu, F., Song, Y., and Liu, D. (1999). Hydrodynamics-based transfection in animals by systemic administration of plasmid DNA. Gene Ther. *6*, 1258–1266.

Liu, Y.P., Haasnoot, J., ter Brake, O., Berkhout, B., and Konstantinova, P. (2008). Inhibition of HIV-1 by multiple siRNAs expressed from a single microRNA polycistron. Nucleic Acids Res. *36*, 2811–2824.

Loomba, R., and Liang, T.J. (2007). Treatment of chronic hepatitis B. Antivir Ther *12* Suppl. 3, H33–41.

Marques, J.T., Devosse, T., Wang, D., Zamanian-Daryoush, M., Serbinowski, P., Hartmann, R., Fujita, T., Behlke, M.A., and Williams, B.R. (2006). A structural basis for discriminating between self and nonself double-stranded RNAs in mammalian cells. Nat. Biotechnol. *24*, 559–565.

Mason, W.S., Aldrich, C., Summers, J., and Taylor, J.M. (1982). Asymmetric replication of duck hepatitis B virus DNA in liver cells: Free minus-strand DNA. Proc. Natl. Acad. Sci. U.S.A. *79*, 3997–4001.

Mason, W.S., Halpern, M.S., England, J.M., Seal, G., Egan, J., Coates, L., Aldrich, C., and Summers, J. (1983). Experimental transmission of duck hepatitis B virus. Virology *131*, 375–384.

McCaffrey, A.P., Nakai, H., Pandey, K., Huang, Z., Salazar, F.H., Xu, H., Wieland, S.F., Marion, P.L., and Kay,

M.A. (2003). Inhibition of hepatitis B virus in mice by RNA interference. Nat. Biotechnol. *21*, 639–644.

Miyagishi, M., and Taira, K. (2002). U6 promoter-driven siRNAs with four uridine 3′ overhangs efficiently suppress targeted gene expression in mammalian cells. Nat. Biotechnol. *20*, 497–500.

Mook, O.R., Baas, F., de Wissel, M.B., and Fluiter, K. (2007). Evaluation of locked nucleic acid-modified small interfering RNA in vitro and *in vivo*. Mol. Cancer Ther. *6*, 833–843.

Morrissey, D.V., Blanchard, K., Shaw, L., Jensen, K., Lockridge, J.A., Dickinson, B., McSwiggen, J.A., Vargeese, C., Bowman, K., Shaffer, C.S., *et al.* (2005a). Activity of stabilized short interfering RNA in a mouse model of hepatitis B virus replication. Hepatology *41*, 1349–1356.

Morrissey, D.V., Lockridge, J.A., Shaw, L., Blanchard, K., Jensen, K., Breen, W., Hartsough, K., Machemer, L., Radka, S., Jadhav, V., *et al.* (2005b). Potent and persistent *in vivo* anti-HBV activity of chemically modified siRNAs. Nat. Biotechnol. *23*, 1002–1007.

Muhonen, P., Tennila, T., Azhayeva, E., Parthasarathy, R.N., Janckila, A.J., Vaananen, H.K., Azhayev, A., and Laitala-Leinonen, T. (2007). RNA interference tolerates 2′-fluoro modifications at the Argonaute2 cleavage site. Chem Biodivers *4*, 858–873.

Myers, J.W., Jones, J.T., Meyer, T., and Ferrell, J.E., Jr. (2003). Recombinant Dicer efficiently converts large dsRNAs into siRNAs suitable for gene silencing. Nat. Biotechnol. *21*, 324–328.

Nishina, K., Unno, T., Uno, Y., Kubodera, T., Kanouchi, T., Mizusawa, H., and Yokota, T. (2008). Efficient *in vivo* delivery of siRNA to the liver by conjugation of alpha-tocopherol. Mol. Ther. *16*, 734–740.

Paul, C.P., Good, P.D., Winer, I., and Engelke, D.R. (2002). Effective expression of small interfering RNA in human cells. Nat. Biotechnol. *20*, 505–508.

Qian, Z.K., Xuan, B.Q., Min, T.S., Xu, J.F., Li, L., and Huang, W.D. (2005). Cost-effective method of siRNA preparation and its application to inhibit hepatitis B virus replication in HepG2 cells. World J. Gastroenterol. *11*, 1297–1302.

Rauschhuber, C., Xu, H., Salazar, F.H., Marion, P.L., and Ehrhardt, A. (2008). Exploring gene-deleted adenoviral vectors for delivery of short hairpin RNAs and reduction of hepatitis B virus infection in mice. J. Gene Med. *10*, 878–889.

Rehermann, B., Ferrari, C., Pasquinelli, C., and Chisari, F.V. (1996). The hepatitis B virus persists for decades after patients' recovery from acute viral hepatitis despite active maintenance of a cytotoxic T-lymphocyte response. Nat. Med. *2*, 1104–1108.

Ren, X.R., Zhou, L.J., Luo, G.B., Lin, B., and Xu, A. (2005). Inhibition of hepatitis B virus replication in 2.2.15 cells by expressed shRNA. J. Viral. Hepat. *12*, 236–242.

Robbins, M., Judge, A., Liang, L., McClintock, K., Yaworski, E., and MacLachlan, I. (2007). 2′-O-methyl-modified RNAs act as TLR7 antagonists. Mol. Ther. *15*, 1663–1669.

Roggendorf, M., Schulte, I., Xu, Y., and Lu, M. (2007). Therapeutic vaccination in chronic hepatitis B: preclinical studies in the woodchuck model. J. Viral. Hepat. *14* Suppl. *1*, 51–57.

Sato, Y., Murase, K., Kato, J., Kobune, M., Sato, T., Kawano, Y., Takimoto, R., Takada, K., Miyanishi, K., Matsunaga, T., *et al.* (2008). Resolution of liver cirrhosis using vitamin A-coupled liposomes to deliver siRNA against a collagen-specific chaperone. Nat. Biotechnol. *26*, 431–442.

Schnell, M.A., Zhang, Y., Tazelaar, J., Gao, G.P., Yu, Q.C., Qian, R., Chen, S.J., Varnavski, A.N., LeClair, C., Raper, S.E., *et al.* (2001). Activation of innate immunity in nonhuman primates following intraportal administration of adenoviral vectors. Mol. Ther. *3*, 708–722.

Seeger, C., Ganem, D., and Varmus, H.E. (1986). Biochemical and genetic evidence for the hepatitis B virus replication strategy. Science *232*, 477–484.

Shin, D., Kim, S.I., Kim, M., and Park, M. (2006). Efficient inhibition of hepatitis B virus replication by small interfering RNAs targeted to the viral X gene in mice. Virus Res. *119*, 146–153.

Shlomai, A., and Shaul, Y. (2003). Inhibition of hepatitis B virus expression and replication by RNA interference. Hepatology *37*, 764–770.

Snyder, L.L., Esser, J.M., Pachuk, C.J., and Steel, L.F. (2008). Vector design for liver-specific expression of multiple interfering RNAs that target hepatitis B virus transcripts. Antiviral Res. *80*, 36–44.

Soutschek, J., Akinc, A., Bramlage, B., Charisse, K., Constien, R., Donoghue, M., Elbashir, S., Geick, A., Hadwiger, P., Harborth, J., *et al.* (2004). Therapeutic silencing of an endogenous gene by systemic administration of modified siRNAs. Nature *432*, 173–178.

Spagnou, S., Miller, A.D., and Keller, M. (2004). Lipidic carriers of siRNA: differences in the formulation, cellular uptake, and delivery with plasmid DNA. Biochemistry *43*, 13348–13356.

Starkey, J.L., Chiari, E.F., and Isom, H.C. (2009). Hepatitis B virus (HBV)-specific short hairpin RNA is capable of reducing the formation of HBV covalently closed circular (CCC) DNA but has no effect on established CCC DNA in vitro. J. Gen. Virol. *90*, 115–126.

Stray, S.J., Bourne, C.R., Punna, S., Lewis, W.G., Finn, M.G., and Zlotnick, A. (2005). A heteroaryldihydropyrimidine activates and can misdirect hepatitis B virus capsid assembly. Proc. Natl. Acad. Sci. U.S.A. *102*, 8138–8143.

Sun, Y., Li, Z., Li, L., Li, J., Liu, X., and Li, W. (2007). Effective inhibition of hepatitis B virus replication by small interfering RNAs expressed from human foamy virus vectors. Int J Mol Med *19*, 705–711.

Tam, O.H., Aravin, A.A., Stein, P., Girard, A., Murchison, E.P., Cheloufi, S., Hodges, E., Anger, M., Sachidanandam, R., Schultz, R.M., *et al.* (2008). Pseudogene-derived small interfering RNAs regulate gene expression in mouse oocytes. Nature *453*, 534–538.

Tiollais, P., Pourcel, C., and Dejean, A. (1985). The hepatitis B virus. Nature *317*, 489–495.

Turrini, P., Sasso, R., Germoni, S., Marcucci, I., Celluci, A., Di Marco, A., Marra, E., Paonessa, G., Eutropi, A., Laufer, R., *et al.* (2006). Development of humanized mice for the study of hepatitis C virus infection. Transplant Proc *38*, 1181–1184.

Uprichard, S.L., Boyd, B., Althage, A., and Chisari, F.V. (2005). Clearance of hepatitis B virus from the liver of transgenic mice by short hairpin RNAs. Proc. Natl. Acad. Sci. U.S.A. *102*, 773–778.

Waddington, S.N., McVey, J.H., Bhella, D., Parker, A.L., Barker, K., Atoda, H., Pink, R., Buckley, S.M., Greig, J.A., Denby, L., *et al.* (2008). Adenovirus serotype 5 hexon mediates liver gene transfer. Cell *132*, 397–409.

Watanabe, T., Sudoh, M., Miyagishi, M., Akashi, H., Arai, M., Inoue, K., Taira, K., Yoshiba, M., and Kohara, M. (2006). Intracellular-diced dsRNA has enhanced efficacy for silencing HCV RNA and overcomes variation in the viral genotype. Gene Ther. *13*, 883–892.

Weber, O., Schlemmer, K.H., Hartmann, E., Hagelschuer, I., Paessens, A., Graef, E., Deres, K., Goldmann, S., Niewoehner, U., Stoltefuss, J., *et al.* (2002). Inhibition of human hepatitis B virus (HBV) by a novel non-nucleosidic compound in a transgenic mouse model. Antiviral Res. *54*, 69–78.

Weimar, W., Heijtink, R.A., Schalm, S.W., van Blankenstein, M., Schellekens, H., Masurel, N., Edy, V.G., Billiau, A., and De Somer, P. (1977). Fibroblast interferon in HBsAg-positive chronic active hepatitis. Lancet 2, 1282.

Weinberg, M.S., Ely, A., Barichievy, S., Crowther, C., Mufamadi, S., Carmona, S., and Arbuthnot, P. (2007). Specific inhibition of HBV replication in vitro and *in vivo* with expressed long hairpin RNA. Mol. Ther. *15*, 534–541.

Weiser, B. (1983). As they lay dying, Part 1. The doctor's agony: deciding to end life-saving therapy. Washington Post, A1, A16.

Wilson, J., and Richardson, C. (2005). Hepatitis C virus replicons escape RNA interference induced by a short interfering RNA DIRECTçED against the NS5b coding region. J. Virol. 79, 7050–7058.

Wu, G.Y., and Wu, C.H. (1988). Receptor-mediated gene delivery and expression *in vivo*. J. Biol. Chem. *263*, 14621–14624.

Wu, H.L., Huang, L.R., Huang, C.C., Lai, H.L., Liu, C.J., Huang, Y.T., Hsu, Y.W., Lu, C.Y., Chen, D.S., and Chen, P.J. (2005a). RNA interference-mediated control of hepatitis B virus and emergence of resistant mutant. Gastroenterology *128*, 708–716.

Wu, K.L., Zhang, X., Zhang, J., Yang, Y., Mu, Y.X., Liu, M., Lu, L., Li, Y., Zhu, Y., and Wu, J. (2005b). Inhibition of Hepatitis B virus gene expression by single and dual small interfering RNA treatment. Virus Res. *112*, 100–107.

Wu, Y., Huang, A.L., Tang, N., Zhang, B.Q., and Lu, N.F. (2005c). Specific anti-viral effects of RNA interference on replication and expression of hepatitis B virus in mice. Chin. Med. J. (Engl.) *118*, 1351–1356.

Xuan, B., Qian, Z., Hong, J., and Huang, W. (2006). EsiRNAs inhibit hepatitis B virus replication in mice model more efficiently than synthesized siRNAs. Virus Res. *118*, 150–155.

Yang, D., Buchholz, F., Huang, Z., Goga, A., Chen, C.Y., Brodsky, F.M., and Bishop, J.M. (2002a). Short RNA duplexes produced by hydrolysis with *Escherichia coli* RNase III mediate effective RNA interference in mammalian cells. Proc. Natl. Acad. Sci. U.S.A. *99*, 9942–9947.

Yang, P.L., Althage, A., Chung, J., and Chisari, F.V. (2002b). Hydrodynamic injection of viral DNA: a mouse model of acute hepatitis B virus infection. Proc. Natl. Acad. Sci. U.S.A. *99*, 13825–13830.

Ying, C., De Clercq, E., and Neyts, J. (2003). Selective inhibition of hepatitis B virus replication by RNA interference. Biochem. Biophys. Res. Commun. *309*, 482–484.

Ying, R.S., Zhu, C., Fan, X.G., Li, N., Tian, X.F., Liu, H.B., and Zhang, B.X. (2007). Hepatitis B virus is inhibited by RNA interference in cell culture and in mice. Antiviral Res. 73, 24–30.

Zhang, X.N., Xiong, W., Wang, J.D., Hu, Y.W., Xiang, L., and Yuan, Z.H. (2004). siRNA-mediated inhibition of HBV replication and expression. World J. Gastroenterol. *10*, 2967–2971.

Zimmermann, T.S., Lee, A.C., Akinc, A., Bramlage, B., Bumcrot, D., Fedoruk, M.N., Harborth, J., Heyes, J.A., Jeffs, L.B., John, M., *et al.* (2006). RNAi-mediated gene silencing in non-human primates. Nature *441*, 111–114.

Zoulim, F., Saputelli, J., and Seeger, C. (1994). Woodchuck hepatitis virus X protein is required for viral infection *in vivo*. J. Virol. *68*, 2026–2030.

# Hepatitis C: New Insights and Therapeutics by RNAi

9

Qiuwei Pan and Luc J.W. van der Laan

Abstract

Hepatitis C virus (HCV) is a leading cause of chronic hepatitis, the sequelae of which, end-stage liver disease and hepatocellular carcinoma, are the most common indications for liver transplantation. The current standard treatment, pegylated-interferon-alpha in combination with ribavirin, represents a milestone in therapy, but eradicates the virus in only half of the patients, leaving tens of millions without hope for a cure. Therefore, it is urgent to explore novel and more effective therapeutic options. Such advancements could be derived from a better understanding of HCV biology. Progress in this area has accelerated with the application of RNA interference (RNAi) as a discovery tool to investigate viral and host cell factors involved in viral infection. Furthermore, RNAi has high potential as a therapeutic avenue for treatment of HCV. In this chapter, we aim to provide a comprehensive overview of the current developments and the potential application of RNAi in research and therapy of HCV infection.

## Introduction

HCV is a single-stranded positive-sense RNA virus that was first identified in 1989 (Choo *et al.*, 1989). HCV belongs to the Flaviviridae family and has six major genotypes (Simmonds *et al.*, 1994). According to World Health Organization estimates, approximately 170 million people, 3% of the world population, are HCV positive with 3–4 million *de novo* infections each year. Of the newly infected individuals only approximately 15–40% will effectively clear the virus and those who fail to do so will develop a chronic and progressive infection. Chronic hepatitis C is a slowly progressive disease causing no or few symptoms in the initial phase, but 10–20% of the patients develop liver cirrhosis over a period of 10–30 years. Patients with liver cirrhosis, have an annual risk of 1–5% of developing liver cancer, particularly hepatocellular carcinoma (HCC) (Sarbah and Younossi, 2000). Currently the only effective treatment for end-stage liver disease and HCC caused by chronic HCV infection is liver transplantation and HCV has now become the major indication for liver transplantation worldwide. However, HCV universally reinfects the liver graft after transplantation causing an accelerated recurrent liver disease, resulting in cirrhosis of the donor liver in 10–41% of the patients after 5–10 years (Brown, 2005; Yilmaz *et al.*, 2007). Despite extensive research efforts, there is currently no effective vaccine or immunoglobulin cocktail available for prevention of the infection.

Therapy against HCV has achieved substantial success since the introduction of treatment with interferon-α (IFN-α). IFN-α-based therapy generally leads to a rapid decline in viral genomic RNA levels in serum, and long-term responses are marked by sustained loss of HCV RNA from the serum and liver and resolution of the chronic infection (Lau *et al.*, 1998). Further therapeutic improvements have been achieved by using a pegylated (peg) form of IFN-α causing an extended serum half-life and by combinational treatment with ribavirin (Manns *et al.*, 2001). Peg-IFN-α with ribavirin is now the standard of care, but the success of the therapy depends on the genotype and the viral load at the start of therapy. Still, half of the patients fail to develop a sustained virological response, i.e. to clear the virus (Davis *et al.*, 2003; Zeuzem *et al.*, 2006). Furthermore, both IFN-α and ribavirin cause severe side-effects contributing to discontinuation and failure of treatment (Fried, 2002; Fried *et al.*, 2002; Manns *et al.*, 2001). Therapy in the setting of liver transplantation to prevent HCV recurrence is less efficient, with only 20–30% of patients completely clearing the virus (Garcia-Retortillo and Forns, 2004; Samuel *et al.*, 2003).

Owing to the accelerated accumulation of basic knowledge of viral and host cellular factors involved in the HCV life cycle as well as the recent technical advances in viral cell culture systems, opportunities for the development of new therapies have dramatically increased. Several small-molecule inhibitors, immune modulators, monoclonal and polyclonal antibodies, antisense RNA and other therapeutic agents are currently being tested in clinical trials (McHutchison *et al.*, 2006; Pawlotsky, 2006). Among the new technologies, RNAi is one of the most promising avenues currently being explored as novel antiviral therapy. Furthermore, the use of RNAi as research tool has recently led to several important discoveries on HCV biology, providing new therapeutic targets for treatment and prevention of infection.

As outlined in previous chapters, RNAi is a sequence-specific inhibition of gene expression at post-transcriptional level. It is triggered by small interfering RNAs (siRNAs) of 21 nucleotides, which can be introduced into cells directly as synthetic siRNAs or indirectly as precursor long double-stranded RNAs (dsRNAs) or short-hairpin RNAs (shRNAs). These siRNAs are assembled into a multicomponent complex, known as the RNA-induced silencing complex (RISC), which unwinds the siRNA and uses its single-stranded form as a guide to the homologous target mRNA sequences, resulting in degradation of the target mRNA (Hannon, 2002; Tomari and Zamore, 2005). Indeed, such a simple, potent and specific gene silencing technology has greatly fuelled the study of HCV biology and the development of new therapeutic entities. In this chapter, we will provide a comprehensive overview of current developments and the potential applications of RNAi in HCV biology and therapy.

## Molecular biology of HCV

### HCV culture models

Since the discovery of HCV in 1989, the major progress of understanding HCV biology was made by the development of cell culture systems. In the early years it was proven difficult to get reproducible HCV viral replication *in vitro* using primary hepatocytes or hepatoma cell lines. A first breakthrough was the establishment of subgenomic and selectable replicons (Lohmann *et al.*, 1999), capable of autonomous replication due to tissue culture adaptive mutations in the non-structural region of the genome. Due to deletion of the structural sequence of the genome these replicon models do not result in the production of infectious virions. Nowadays, replicons derived from more than eight different clinical isolates, covering genotypes 1a, 1b and 2a, are available (Bartenschlager and Sparacio, 2007).

In addition to the replicons, other cell-culture systems have been developed. The HCV pseudotyped particle (HCVpp) system has been developed to study viral binding and entry. HCVpp are retroviral nucleocapsids surrounded by a lipid envelope that contains authentic HCV glycoprotein complexes E1/E2 (Bartosch *et al.*, 2003b). It has been used intensively to characterize neutralizing antibodies. HCV-like particles (HCV-LPs), such as trans-complemented HCV particles containing subgenomic RNA, are particular useful to study the viral entry. Due to the deletion of structural sequences, the system is not able to produce virus particles (Steinmann *et al.*, 2008). Based on structural similarity, pestiviruses, like bovine viral diarrhoea virus and GB virus B, have also been considered as alternative models to study HCV (Beames *et al.*, 2001; Buckwold *et al.*, 2003).

Currently the most advanced HCV model was developed in 2005 (Wakita *et al.*, 2005) and is based on a genotype 2a HCV isolate (termed JFH-1) from a Japanese patient with fulminant hepatitis (Kato *et al.*, 2001, 2003). It was shown that, in contrast to genotype 1 viruses, JFH-1 is capable of high-level replication *in vitro* without requiring cell culture-adaptive mutations. Furthermore, transfection of JFH-1 genome in the human hepatoma cell line Huh-7, results in the production of infectious virus particles which can establish a productive infection, both in cell culture and animal models such as the chimpanzee or mice with a humanized liver (Lindenbach *et al.*, 2006; Wakita *et al.*, 2005). HCV particles produced from this system are termed cell culture-derived HCV particles (HCVcc).

## The HCV genome

HCV particles contain a single positive-stranded RNA genome of approximately 9600 nucleotides in length. To date, six major genotypes and over 100 subtypes of HCV have been identified that differ in their nucleotide sequence up to 34% between genotypes and up to 25% between subtypes (Simmonds *et al.*, 2005). The whole genome composes of a single open reading frame, encoding a polyprotein precursor of approximately 3000 amino acids, flanked by 5′ and 3′ non-coding region (NCR) (Fig. 9.1). The genomic RNA acts both as a template for the production of the negative-stranded RNA replication intermediate and as an mRNA template for the synthesis of the viral polyprotein.

Translation of the viral polyprotein is mediated by an internal ribosome entry site (IRES) located within the highly conserved 5′ NCR. The IRES initiates the first step of translation by directly binding the 40S ribosomal subunit to form a stable pre–initiation complex (Penin *et al.*, 2004b; Spahn *et al.*, 2001). The 40S subunit assembles with eIF3, followed by eIF2, guanosine triphosphate, and the initiator transfer RNA to form a 48S-ribosomal RNA complex. Finally, the 48S complex is converted into functional 80S ribosome that initiates viral protein synthesis (Ji *et al.*, 2004; Otto and Puglisi, 2004; Pawlotsky *et al.*, 2007).

The polyprotein is subsequently cleaved into four structural (core, E1, E2 and p7) and six non-structural (NS2, NS3, NS4A, NS4B, NS5A, and NS5B) proteins (Giannini and Brechot, 2003;

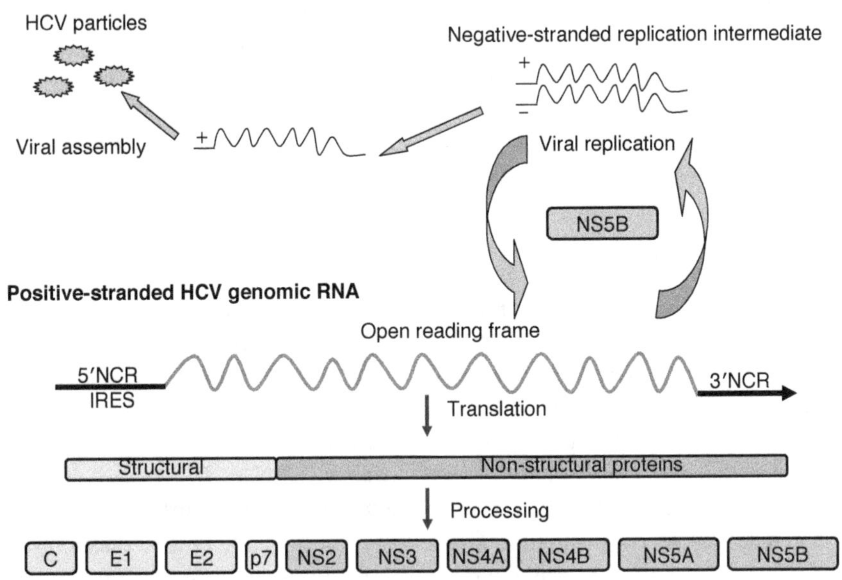


**Figure 9.1** The HCV genome and life-cycle. HCV is a positive single-stranded RNA virus of about 9.6kb nucleotides, consisting of two non-coding regions (NCRs) and a single open reading frame encoding a polyprotein precursor of approximately 3000 amino acids. Translation of the polyprotein is mediated by an internal ribosome entry site (IRES) located in the 5′ NCR. The structural region contains the core protein, two envelope proteins (E1 and E2) and the p7 protein, which are cleaved into individual proteins by host signal peptidases. Processing of the six non-structural proteins (NS2, NS3, NS4A, NS4B, NS5A, and NS5B) are carried out by two viral proteases, NS2 and NS3/4A. The HCV viral genome consists of a positive-stranded RNA molecule, which provides a template for the synthesis of a negative-stranded RNA replication intermediate by the viral RNA-depended RNA polymerase, NS5B. In turn, the negative strand acts as a template for the *de novo* synthesis of positive-stranded RNA which can be packaged in newly formed HCV virions during viral assembly.

Penin *et al.*, 2004a). Processing of structural proteins, including the p7/NS2 junction, is carried out by at least two host signal peptidases and non-structural proteins are matured by two viral proteases, NS2 and NS3/4A (Bartenschlager *et al.*, 1994; Lemberg and Martoglio, 2002; Okamoto *et al.*, 2004; Pallaoro *et al.*, 2001; Penin *et al.*, 2004b).

### Viral entry

Viral entry involves HCV envelope protein interaction with membrane proteins expressed on the surface of the host cell. The exact nature of the host factors contributing to viral binding and entry is still not completely known and is currently the focus of much research in the field. The viral factors involved in binding and entry, namely the envelope glycoproteins E1 and E2, are well established in several model systems. These include HCV-LPs expressing E1–E2 heterodimers instead of glycosylated individual E1 and E2 (Barth *et al.*, 2005; Steinmann *et al.*, 2004; Triyatni, 2002; Wellnitz *et al.*, 2002), HCVpp consisting of unmodified HCV envelope glycoproteins E1 and E2 assembled onto retroviral or lentiviral core particles (Bartosch *et al.*, 2003a; Lavillette *et al.*, 2005; Logvinoff *et al.*, 2004; Meunier *et al.*, 2005; Schofield *et al.*, 2005), vesicular stomatitis virus (VSV)/HCV pseudotypes expressing HCV E1 or E2 chimeric proteins containing transmembrane and cytoplasmic domains of the VSV G glycoprotein, or HCVcc neutralization assays with E1 or E2 antibody (Blight *et al.*, 2000; Buonocore *et al.*, 2002; Dreux *et al.*, 2006; Lagging *et al.*, 1998). The lack of either E1 or E2 significantly decreases HCV infection activity whereas deletion of the whole envelope protein coding sequence abolishes the particle infectivity.

Over the past decade, several host cell surface molecules have been identified as potential binding partners for the HCV envelope proteins E1/E2 and considered important for viral binding and entry. The 25-kDa transmembrane protein, CD81, belonging to the tetraspanin family has been best characterized. The extracellular loop of CD81 is able to bind to HCV E2 protein with high affinity (Pileri *et al.*, 1998). However, CD81 is not uniquely expressed by hepatocytes and ectopic or intrinsic expression of CD81 in many other cell types cannot render them to be susceptible to HCV infection (Bartosch *et al.*, 2003c; Levy *et al.*, 1998). Therefore, CD81 alone is not sufficient and additional host factors are possibly required for HCV infection, such as glycosaminoglycan (Barth *et al.*, 2003), scavenger receptor class B type-1 (SR-BI) (Scarselli *et al.*, 2002), low-density lipoprotein receptor (Agnello *et al.*, 1999), DC-SIGN (Pohlmann *et al.*, 2003), and L-SIGN (Gardner *et al.*, 2003). Particularly, Claudin-1, a tight junction protein belonging to the claudin family, has been recently shown to be essential for HCV infection of human hepatoma cell lines and is the first factor to confer susceptibility to HCV when ectopically expressed in non-hepatic cells (Evans *et al.*, 2007). However, several cell lines remained resistant to HCVpp infection, even upon the ectopic expression of Claudin-1. It is therefore reasonable to assume that Claudin-1 functions as a coreceptor rather than a primary receptor. The role of claudin-1 in HCV entry has been confirmed by other studies and additional Claudin members such as claudin-6 and -9 also appear to be coreceptors for HCV (Meertens *et al.*, 2008; Zheng *et al.*, 2007). More recently, human occludin (OCLN) was identified as a new HCV entry factor that is able to render murine cells infectable with HCVpp, but efficient infection requires both OCLN and CD81 of human origin (Ploss *et al.*, 2009).

### Viral replication

Since HCV is a single positive-strand RNA virus, the genomic replication is achieved by synthesis of a full-length negative-strand RNA intermediate. Negative stranded RNA in turn provides a template for the *de novo* production of positive-stranded RNA. In general, positive-stranded template is much more abundant than the negative one, because of its utilization of translation, replication, and package of progeny viruses.

Similar to viral entry, HCV replication requires both viral and cellular factors. Although our current knowledge of the HCV life cycle still remains limited, several critical viral components and host cell factors have been identified. The 5′ NCR, in particular the IRES sequence, plays an important function in ribosomal assembly. The non-structural proteins, NS3, NS4A, NS4B, NS5A and NS5B, appear to be essential for replication (Lohmann *et al.*, 1999). For instance,

NS4B is a membrane protein modelling the endoplasmic reticulum membrane in order to facilitate viral replication (Egger *et al.*, 2002). Adaptive mutations in NS4B or NS5A have been found to greatly enhance replication of HCV replicons (Blight *et al.*, 2000; Lohmann *et al.*, 1999). NS5B is the active subunit of the replication complex known as an RNA-dependent RNA polymerase for the *de novo* generation of positive and negative stranded RNA (Behrens *et al.*, 1996).

Interestingly, much of the current knowledge on host proteins that participate in HCV replication came from studies with immunosuppressive drugs. Cyclosporine A (CsA) possesses potent immunosuppressive capacity and is a widely used drug in organ transplantation for the prevention of rejection. CsA interacts with cyclophilins, which are molecular chaperones with peptidyl-proline isomerase activity (Takahashi *et al.*, 1989). The binding of CsA to cyclophilins inhibits the activation of calcineurin, one of the critical signalling molecules involved in T lymphocyte activation (Liu *et al.*, 1991). The anti-viral activity of CsA was first shown in human immunodeficiency virus (HIV), via interaction of host factor cyclophilin A (Towers *et al.*, 2003). Subsequently, CsA and its derivatives were also shown to be capable of inhibiting HCV replication in culture models (Inoue *et al.*, 2003; Watashi *et al.*, 2003). The anti-HCV activity of CsA was suggested as disrupting the functional association of cyclophilin B with NS5B, which participated in the initiation of replication by facilitating the interaction of NS5B and the 3′ NCR of the viral RNA (Watashi *et al.*, 2005). Further investigation revealed that probably other cyclophilins are also involved, since RNAi knockdown of cyclophilin A or C suppressed HCV replication (Nakagawa *et al.*, 2005). Cyclophilins are part of the larger family of immunophilins, which include the FK506-binding proteins (FKBP) members. FKBP8 was identified as a regulator of the HCV protein NS5A. FKBP8 and NS5A were shown to form a complex with heat-shock protein, Hsp90, and that this complex is critical for viral replication (Okamoto *et al.*, 2006). Another immunosuppressive drug which is commonly used in transplant recipients is mycophenolic acid (MPA). MPA inhibits the enzyme inosine monophosphate dehydrogenase (IMPDH), which is responsible for the *de novo* synthesis of GTP nucleotides (Allison and Eugui, 2000). Inhibition of IMPDH by MPA decreases levels of intracellular guanosine nucleotide pools resulting in inadequate quantities for nominal DNA duplication (Carter *et al.*, 1969; Mitsui and Suzuki, 1969). The anti-viral action of MPA or the less potent IMPDH inhibitor ribavirin has been shown for many viruses, including HCV (Henry *et al.*, 2006a). Surprisingly, the anti-HCV activity of MPA seems largely independent of GTP depletion since it cannot be overcome by exogenous guanosine supplementation. This is distinct from the effects of IMPDH inhibitors on, for instance, West Nile virus and yellow fever virus, where inhibition can be overcome by exogenous guanosine (Leyssen *et al.*, 2005; Morrey *et al.*, 2002).

More recently, using HCVcc models, new light has been shed on the interplay between cellular membranes, lipid metabolism and HCV replication. Lipid components are required for the assembly, budding and replication of several viruses (Suzuki and Suzuki, 2006; Teissier and Pecheur, 2007). Increases in saturated and monounsaturated fatty acids enhance HCV RNA replication, in contrast to suppression by polyunsaturated fatty acids, suggesting that enzymes associated with lipid biosynthesis are also involved in HCV replication (Kapadia and Chisari, 2005). In addition, inhibition of protein geranylgeranylation leads to disassembly of the HCV replication complexes and inhibition of viral replication (Ye *et al.*, 2003). FBL-2, the geranylgeranylated host protein, was shown to interact with NS5A and to be required for HCV RNA replication (Wang *et al.*, 2005). The FBL-2 contains an F-box motif, which is generally essential for the formation of the ubiquitin ligase complex. This suggests that FBL2 regulates the ubiquitination of host or viral proteins through the interaction with NS5A (Ilyin *et al.*, 1999). Moreover, a more recent study showed that the primary location for the assembly of HCV virions in hepatocytes are the small lipid droplets located in close proximity of the endoplasmic reticulum (Miyanari *et al.*, 2007). They demonstrated that HCV core protein initiate the recruitment of non-structural proteins and replication complexes to lipid droplet-associated membranes.

In addition, other host factors, such as vesicle-associated membrane protein-associated proteins VAP-A and VAP-B, were also identified as interaction partners of NS5A and NS5B (Evans *et al.*, 2004; Hamamoto *et al.*, 2005). Some of the host factors involved in HCV infection have been identified using different RNAi-based techniques, as will be further discussed in the following paragraph.

## RNAi as a research tool for HCV

RNAi-mediated gene silencing creates a viable alternative strategy for gene functional analysis through the specific knockdown of target gene expression. A major step forward is the application of RNAi libraries, which allows performing functional genome-wide screening. Indeed, the process of understanding HCV biology has been greatly speeded up by successfully applying these techniques.

The following paragraphs provide some recent examples of the use of RNAi to identify or confirm host factors in HCV life cycle. Early studies have shown that HCV NS5B and NS5A proteins bind to two different domains of a vesicle membrane protein hVAP-33 (Tu *et al.*, 1999). Further confirmation was obtained by siRNA-mediated knockdown of hVAP-33, which could inhibit these interactions and correspondingly inhibit HCV RNA replication (Gao *et al.*, 2004). NS5B has been shown to interact with a cellular RNA Helicase p68 and siRNA-mediated knockdown of p68 reduces the transcription of negative-strand HCV RNA from its positive-strand template, indicating the potential role of p68 in HCV replication (Goh *et al.*, 2004). Ras-GTPase-activating protein binding protein 1 (G3BP1), was found to be associated with NS5B and knockdown of G3BP1 by siRNA in HCV replicon cells significantly reduced viral replication, suggesting G3BP1 as a potential component of HCV replication complexes (Yi *et al.*, 2006). A cellular protein polypyrimidine-tract-binding protein (PTB) has shown to bind to HCV RNA at several different sites and modulates HCV translation in several ways. sRNA-mediated knockdown of PTB revealed its positive regulation in HCV RNA synthesis (Aizaki *et al.*, 2006). NF/NFAR proteins, a group of predominantly nuclear proteins, were reported to be actively involved in the HCV replication process (Isken *et al.*, 2007). FUSE (far upstream element of the c-myc proto-oncogene) binding protein (FBP) was reported to interact with HCV 3′ NCR and siRNA knockdown of FBP severely reduced viral replication, indicating that it's a cellular factor required for efficient replication of HCV (Chen *et al.*, 2008). Finally, a similar approach was used in a recent study to confirm the new HCV entry factor, OCLN (Ploss *et al.*, 2009).

In contrast to individual siRNA silencing, RNAi libraries provide a high-throughput approach to identify novel host factors. By using a mini RNAi library, Randall *et al.* (Randall *et al.*, 2007) re-evaluated the significance of 62 host genes, which physically interact with HCV RNA or proteins or belong to cellular pathways thought to modulate HCV infection. siRNA targeting 26 cellular host proteins modulated the production of infectious HCV by at least 3-fold; siRNA targeting 25 of these proteins decreased HCV infectious titres by as much as 42-fold. Interestingly, dicer was found to be one the factors required for viral replication (Randall *et al.*, 2007), however, this was not consistent with an earlier study (Wang *et al.*, 2006b). By using an RNAi library that targets approximately 4000 human genes in an HCV subgenomic replicon cell culture model, screening revealed nine cellular genes that potentially regulate viral replication. Silencing of these genes resulted in inhibition of viral replication by more than 60%. These genes included those that encoded a G-protein-coupled receptor (TBXA2R), a membrane protein (LTbeta), an adapter protein (TRAF2), two transcription factors (RelA and NFkappaB2), two protein kinases (MKK7 and SNARK), and two closely related transporter proteins (SLC12A4 and SLC12A5) (Ng *et al.*, 2007). By using a mini library directed against ~500 human genes (including 380 protein kinases), three kinases, Csk, Jak1, and Vrk1, were identified that reproducibly reduce viral RNA and viral protein levels in HCV replicon cells (Supekova *et al.*, 2008).

In summary, RNAi, and in particular the use of RNAi libraries, has led to a revolution in the understanding of HCV host factors. As a result of this, the function of previously identified putative factors has been confirmed by RNAi and many novel cellular factors are currently emerging.

## Interplay of HCV with RNAi pathway

### miRNA and HCV infection

Mature microRNAs (miRNAs) are approximately 22-nucleotide non-coding RNAs that can down-regulate various gene products by inducing either cleavage or a reduction in the translational efficiency of the target mRNA (Ali *et al.*, 2004; Bartel, 2004). In the last few years, over 3000 miRNAs have been identified in vertebrates, flies, worms, plants and even viruses. Most miRNAs have been shown to participate in essential biological processes, such as cell proliferation, apoptosis, differentiation and metabolism (Harfe, 2005). Most of the miRNAs are transcribed by RNA polymerase II as long primary polyadenylated transcripts (pri-miRNAs) (Lee *et al.*, 2004). These can be recognized and cleaved at a specific processing site by the RNase III enzyme, drosha, to produce a miRNA precursor (pre-miRNA) of approximately 70–90 nt (Lee *et al.*, 2003). The pre-miRNA is recognized by the Exportin-5–Ran-GTP heterodimer and exported to the cytoplasm (Yi *et al.*, 2003). Subsequently, the miRNA precursor is recognized and processed into a mature miRNA by the enzyme dicer, an RNase type III. Dicer recognizes the 2-nt 3′ overhang at one end as a guide for site-specific cleavage at the second processing site (Siolas *et al.*, 2005).

As detailed in other chapters in this book, there is increasing evidence suggesting a general involvement of miRNAs in viral infection. Several DNA viruses can regulate both viral and cellular transcripts by expressing miRNAs, and some host miRNAs can directly target and repress viruses, such as primate foamy virus 1 and vesicular-stomatitis virus (Gottwein and Cullen, 2008). The first miRNA linked to HCV replication is miR-122 (Jopling *et al.*, 2005). The 22-nt mature miR-122, derived from a non-coding polyadenylated RNA transcript of the *hcr* gene, is a highly abundant liver-specific miRNA. It is already expressed before birth and slowly increased up to 70% of the total miRNA population in adult liver with approx. 66 000 copies per cell (Chang *et al.*, 2004; Lagos-Quintana *et al.*, 2002).

The first evidence of miR-122 in HCV replication was discovered by Jopling *et al* (Jopling *et al.*, 2005), linking differential expression of this miRNA between hepatoma cell lines to their ability to support HCV replication. In permissive hepatoma cells, down regulation of miR-122 using antisense oligonucleotides (ASO) resulted in an 80% reduction of HCV replication. They went on to demonstrate that this is direct effect and located to putative binding sites in the 5′ NCR as targets for miR-122. Both miR-122 binding sites were found to be necessary for viral replication. The two sites are highly conserved between genotypes and are closely located separated by a short spacer (Jopling *et al.*, 2008). The binding site requirements for miR-122 to positively regulate HCV replication are notably very different from the common observation that miRNA targets the 3′ NCR of mRNA, leading to suppression of mRNA translation or degradation. Interestingly, insertion of the 5′ NCR miR-122 binding site into the 3′ NCR of a reporter gene leads to down-regulation of mRNA expression indicating that the location rather than the sequence of the miR-122 binding site is crucial in gene regulation (Jopling *et al.*, 2008). In addition to the direct interaction, miR-122 can also indirectly facilitate HCV replication by regulation of haem oxygenase-1 expression, a key cytoprotective enzyme capable of suppressing HCV replication (Shan *et al.*, 2007; Zhu *et al.*, 2008).

Despite the crucial role of miR-122 in HCV replication, its absolute requirement appears to be dependent on the viral genotype. Subgenomic replicons derived from both genotypes 1b and 2a are capable of replicating in a variety of non-hepatic cell lines, such as HeLa, HEK293 and mouse embryonic fibroblasts, which do not express detectable levels of miR-122 (Ali *et al.*, 2004; Chang *et al.*, 2006b; Kato *et al.*, 2005; Zhu *et al.*, 2003). Consistently, overexpression of miR-122 can dramatically enhance HCV replication in HEK293 cells (Chang *et al.*, 2008). More recently, miR-122 has been shown to stimulate HCV translation by enhancing the association of ribosomes with the viral RNA at an early initiation stage and independent of viral RNA synthesis (Henke *et al.*, 2008). Although the mechanism of miR-122 in HCV infection has not been well established, it indeed has provided a novel target for the development of anti-HCV therapies. Moreover, IFN-β has been shown to rapidly modulate the expression of numerous cellular miRNAs including miR-122 (Pedersen *et al.*, 2007). Eight of these miRNAs have sequence-predicted targets within the HCV genome leading to the hypothesis that mammalian species can use cellular miRNAs to combat viral infections through the interferon system. A recent clinical study showed that miR-122 expression was markedly decreased in HCV patients who did not develop a sustained virological response during IFN-α therapy, confirming the link between the IFN response and the miRNA pathway (Sarasin-Filipowicz *et al.*, 2009).

### Inhibition of RNAi by HCV

It is generally accepted that RNAi is involved in the inhibition of viral infections and silencing of transposable elements in plants, insects, fungi and nematodes. Primarily this involves the generation of siRNA from the dsRNA viral replication intermediates, resulting in degradation or repression of viruses via complementary paring with target viral RNAs (Waterhouse *et al.*, 2001) (Voinnet, 2001) (Wilkins *et al.*, 2005) (Segers *et al.*, 2007; Wang *et al.*, 2006a). Still, it remains controversial whether RNAi acts as an antiviral defence mechanism in mammalian species like rodents, primates and humans (Cullen, 2006). However, the interplay of viral infection with RNAi pathway is less controversial (de Vries and Berkhout, 2008). Given the fact that the replication of the HCV genome involves a dsRNA intermediate step during genome replication and the IRES region contains multiple hairpin structures, it is feasible that RNAi could function as an innate defence against HCV. It has been shown that either incubation of HCV IRES with dicer enzyme or transfection of HCV sequence into cells can result in generation of IRES-related siRNA, indicating both exogenous and endogenous dicer can target HCV IRES. Furthermore, NS5B-related siRNA can be detected in a HCV subgenomic replicon, suggesting that a dsRNA intermediate may also be targeted by the RNAi pathway (Wang *et al.*, 2006b).

Like many other viruses, HCV has a potential arsenal of mechanisms to avoid the antiviral action of RNAi. Using a cell line that constitutively expresses the reporter protein green fluorescent protein, it was found that siRNA-mediated silencing of the reporter protein was inhibited when the entire HCV polyprotein was expressed. The structural protein E2 was found to be essential for this inhibition by its ability to bind Argonaute-2, a component of RISC (Ji *et al.*, 2008). HCV core protein was also found to inhibit shRNA but not siRNA-mediated gene silencing, suggesting functional inhibition of dicer enzyme without affecting the activity of RISC (Chen *et al.*, 2008; Wang *et al.*, 2006b). Furthermore, the N-terminal 62 amino acids of the core were required for RNAi suppression (Chen *et al.*, 2008).

## Liver-targeted delivery of RNAi

In terms of *in vivo* application of RNAi, particularly for the treatment of hepatitis C, the success will greatly depend on the effective delivery systems to target the liver. RNAi effects can be achieved by delivery of chemically synthesized siRNA or small hairpin RNA (shRNA) each requiring different delivery systems to effectively target the liver (Fig. 9.2). The many aspects of RNAi delivery will be discussed in the next sections. The issues of delivery of modified antisense oligonucleotides, which can be applied to down-regulate miRNA like miR-122, will not be discussed here but in the section of 'target miRNA'.

### Delivery of synthetic siRNA

Chemically synthesized siRNAs are quick and easy to obtain and have been widely used at the early stage for functional genomic analysis (Elbashir *et al.*, 2001; Fire *et al.*, 1998). Although several studies have emphasized the possible applications of synthetic siRNA for therapy (Behlke, 2006; Uprichard, 2005), the pharmacological usage faces a number of limitations that need to be circumvented. First are the potential off-target effects caused by high concentration of siRNA. Induction of an IFN response, leading to a global perturbation of gene expression, has been reported in certain applications of raw siRNA (Persengiev *et al.*, 2004; Robbins and Rossi, 2005; Sledz *et al.*, 2003). A recent study found that suppression of neovascularization is a generic property of siRNAs independent of sequence, target and internalization. The anti-angiogenesis effect is induced through the Toll-like receptor (TLR) 3, an immune-system regulator present on the surface of many cells (Kleinman *et al.*, 2008). These provocative findings raised a host of questions for both biologists and clinicians. Because clinical trials of siRNA-mediated

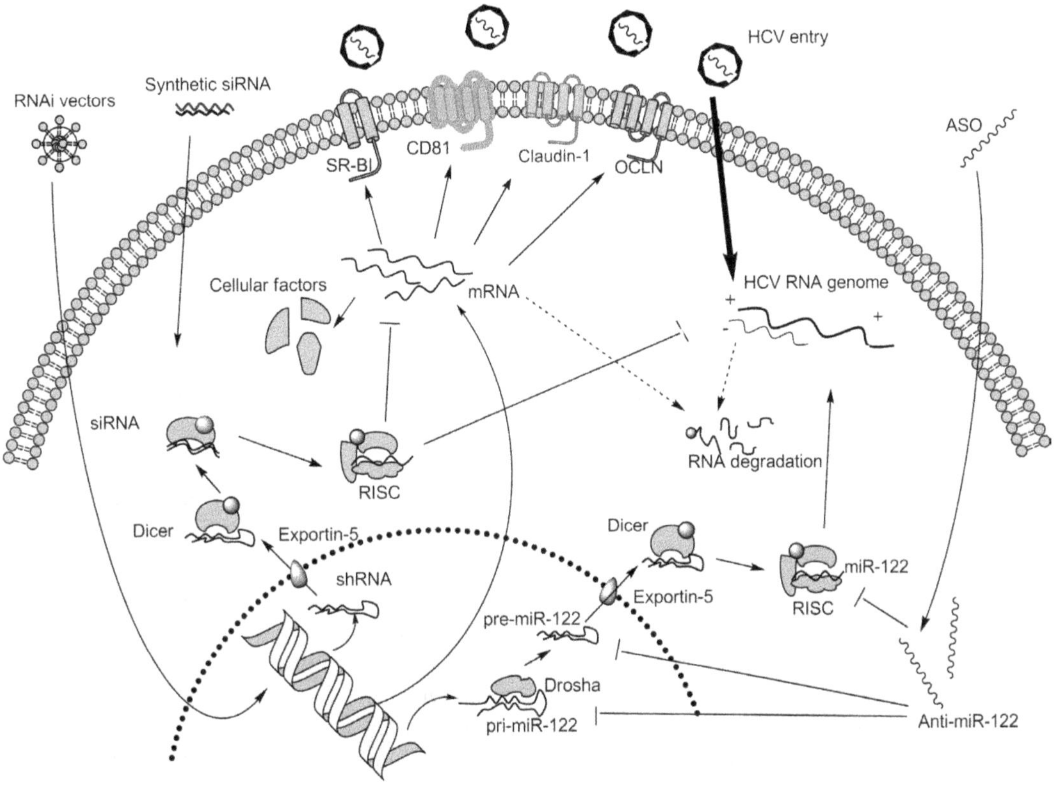


**Figure 9.2** RNAi-based novel antiviral strategies. RNAi effects can be achieved by delivery of chemically synthetic siRNA or by vector-expression of shRNA. shRNA can be expressed in the host cell using non-integrating, adenoviral or adeno-associated viral vectors, or integrating, lentiviral vectors. shRNA is cleaved into active siRNA by the cellular enzyme, dicer. RNAi-mediated gene silencing can be applied to treat HCV infection by targeting either viral RNA or mRNA of host cell factors, such as surface receptor CD81. MiR-122, a liver specific miRNA, plays a positive role in HCV replication and therefore silencing of miR-122 by modified antisense oligonucleotides (ASO), such as the locked nucleic acid probes, could be considered as a potential therapeutic modality.

anti-angiogenesis therapy in patients with blinding choroidal neovascularization (CNV) from age-related macular degeneration, are based on the principle of intracellular targeting of vascular endothelial growth factor-A or its receptor VEGFR1 by RNAi (source: http://www.wiley.co.uk/genmed/clinical/). Additional limitations of synthetic siRNA are low bioavailability and limited delivery due to attack by plasmatic nucleases, short blood persistence and restricted cell penetration. Therefore, proper modification and conjugation of 'naked' raw siRNA are required to achieve targeted delivery to the liver.

Chemical modifications have been developed to improve the *in vivo* properties of nucleic acids, which offer potentials for tailoring and improving the properties of siRNAs to render them more suitable for *in vivo* and clinical applications (Bumcrot *et al.*, 2006). The forms of modification are variable and include phosphorothioate linkages, boranophosphate linkages, locked nucleic acid, 2′-modified RNA, 4′-thio modified RNA, ribo-difluorotoluyl nucleotide and uncharged nucleic acid mimics (Corey, 2007). Chemically modified siRNAs target hepatitis B virus (HBV) have been shown significantly more stable in human serum (Morrissey *et al.*, 2005a). These modified siRNAs were shown to be efficiently delivered to the mouse liver leading to a reduction of HBV DNA levels in both serum and the liver. Furthermore, modified siRNA had reduced toxicity and immunostimulatory effects and had an increased serum half-life (Morrissey *et al.*, 2005b).

Additional to chemical modification, conjugation of siRNA to small molecules or peptides is another strategy for improving the function of siRNA *in vivo*. There is extensive literature on the use of various forms of conjugation for delivery of nucleic acids and therefore here we only include a limited number of recent studies specifically referring to targeting of the liver. Cationic liposome-conjugation with siRNAs has been used with some success in animals. Intravenously injected siRNAs with cationic liposomes containing lactose residues accumulated primarily in the mice liver and were widespread throughout the hepatic parenchymal cells where they efficiently suppressed intrahepatic HCV expression in transgenic mice (Watanabe *et al.*, 2007). Liver-targeted delivery of siRNAs in mice was also achieved by conjugation with galactosylated cationic liposomes (Sato *et al.*, 2007). Antibody conjugation of siRNA was first applied to target HIV infected cells. A fusion protein was constructed with the protamine coding sequence linked to the C-terminus of the heavy chain Fab fragment of an antibody directed against the HIV-1 envelope. siRNAs bound to the fusion protein specifically targeted HIV-1 envelope positive cells both *in vitro* and in mice. Similarly, an ErbB2 single-chain antibody fused with protamine delivered siRNAs specifically into ErbB2-expressing cancer cells (Song *et al.*, 2005). To target the liver and inhibit HBV gene expression and replication, siRNAs were conjugated with a single chain of the human Fab fragment directed against HBV surface antigen fused with protamine (Wen *et al.*, 2007). Based on the key feature of hepatic stellate cells, namely uptake and storage of vitamin A, vitamin A-coupled liposomes to deliver siRNA against gp46 exerted the efficacy of suppressing collagen secretion and treating fibrosis (Sato *et al.*, 2008).

## Delivery of vector-based shRNA

If RNAi is to be utilized as an effective treatment for chronic infectious disease like HCV infection, long-term and stable silencing needs to be achieved. Raw siRNA transfections elicit only transient silencing, whereas vector-based encoding of shRNA can provide relative long-term and continuous silencing. This can be achieved safely using transcriptional regulation of shRNA with tissue-specific promoter elements. The first-generation shRNA was based on the structure of pre-miRNA and expressed under a polymerase-III type promoter (such as H1 or U6) and cleaved into biologically active siRNA by dicer. The second-generation shRNA mimics the structure of pri-miRNA, which therefore can be expressed using polymerase-II type promoters (Chang *et al.*, 2006a). Polymerase-III promoters, in particular U6, often overly express shRNA inducing acute cytotoxicity (Grimm *et al.*, 2006). Using liver-specific polymerase II promoters to drive shRNA expression were shown to improve tolerability (Giering *et al.*, 2008).

Hydrodynamic delivery of plasmid encoded shRNA has shown some efficacy for delivery to the liver in rodents. Gene silencing in the livers of adult mice lasted for at least 20 weeks after a single administration of a plasmid containing shRNA cassette with peak knockdown levels between 7 and 14 days (Wooddell *et al.*, 2005). Although there is even a phase I clinical trial using plasmid expressed siRNA for the treatment of chronic hepatitis B (source: http://www.wiley.co.uk/genmed/clinical/), plasmid-based delivery systems may not be sufficient to achieve final success against a liver-hosted and long-term persistent HCV in patients. Virus-derived vectors possess various advantages as transgenic vehicles and have been widely used in experimental and clinical applications. Indeed, extensive numbers of studies have demonstrated promising results with viral vectors to deliver siRNAs. We will focus on three of the most widely used systems in the application of liver-targeted delivery of RNAi: the adenoviral, adeno-associated viral and lentiviral vectors.

Adenoviral vectors. Adenoviral vectors are derived from a non-enveloped double-stranded linear DNA virus, adenovirus. The 36 kb adenoviral genome encodes a set of 'early' genes (E) and 'late' genes (L). In a first generation of adenoviral vectors the immediate early gene E1 was replaced by cargo sequences generating a replication-defective virus which was effective in gene transfer to the liver in various animal models (Douglas, 2007; Kay *et al.*, 1995). Recombinant adenoviral vector expressed siRNA has shown to suppress HBV gene expression and replication in HBV transgenic mice (Uprichard *et al.*, 2005).

However, immune responses to adenoviral vectors are the major obstacles for the application in patients, largely owing to the prevalence of pre-existing immunity against adenoviruses (Schulick *et al.*, 1997; Yang *et al.*, 1994). In the setting of delivering shRNA, another issue is that adenovirus highly express non-coding RNAs, VA RNAI and VA RNAII, which can inhibit RNAi pathway by acting as competitive substrates which saturate the Dicer and RISC machinery (Andersson *et al.*, 2005; Lu and Cullen, 2004). A third generation adenoviral vector, termed as gutless or helper-dependent adenoviral (HDAd) vector, can potentially circumvent these limitations, since all viral encoding sequences are replaced by non-coding DNA and cargo sequences (Kochanek *et al.*, 1996). The proof of concept has been provided by recent studies that HDAd-mediated shRNA expression allows inhibition of target gene expression in liver, although high level shRNA expression seems to result in induction of IFN production (Ruiz *et al.*, 2009; Witting *et al.*, 2008).

Adeno-associated viral vectors. Recombinant adeno-associated virus (AAV) vectors are developed from the single-stranded DNA genome of wild type AAVs. The ability to infect both non-dividing and dividing cells with persistent expression and lack of apparent pathogenicity, have made these attractive vectors for shRNA delivery. AAV serotype 2 is the first AAV that was vectored for gene delivery and was tested in several *in vitro* and *in vivo* models as well as a series of phase-I/II gene therapy trials. However, pre-existing immunity for AAV2 is prevalent in humans and probably the cause for the long-term failure of a haemophilia B gene therapy trial (Manno *et al.*, 2006). AAV serotype 8, a new member of the AAV family isolated from rhesus monkeys, is less prone to recognition by prevailing antibodies in human (Gao *et al.*, 2004). It is an attractive candidate for hepatic-directed shRNA transfer also because of 10- to 100-fold increased transduction efficiency in mouse liver models, compared with the previous AAV2-based vectors (Nakai *et al.*, 2005). Another important impediment to improve efficient gene transfer with AAV vectors is the conversion step of the single-stranded AAV genome to a transcriptionally active double-stranded form. A solution to this problem has been provided by constructing self-complementary AAV vectors (scAAVs) that complementary proviral single-stranded molecules are packaged within a single virion, and subsequently can anneal to form a stable transcriptionally active double-stranded template after uncoating (Wang *et al.*, 2003).

Lentiviral vectors. Recombinant lentiviral vectors are based on HIV-1 genome and, when coexpressed with the VSV-G envelope protein, have been shown to effectively transduce liver, muscle, and haematopoietic cells. These vectors integrate their payloads into the host genome ensuring transmission to progeny cells (Naldini *et al.*, 1996). Although lentivirus-mediated short hairpin RNA (shRNA, precursor of siRNA) delivery has been widely developed for therapeutic application, there are only a few reports referring to HCV therapy (Henry *et al.*, 2006b; Takigawa *et al.*, 2004). Some current limitations for the use of lentiviral vectors in clinical applications are: (1) production efficiency limits large scale for clinical use, (2) possibility of insertional mutagenesis and (3) generation of wild-type virus leading to safety considerations. To circumvent these drawbacks, certain strategies have been used to achieve further improvement. Firstly, new generation vectors have been developed, such as the third generation VSV-G pseudotyped HIV-1 vectors, HIV-2 vectors, simian immunodeficiency virus (SIV) vectors, or even immunodeficiency viruses derived from felines and equines (Johnston *et al.*, 1999; Poeschla *et al.*, 1998; Rizvi and Panganiban, 1993). Secondly, targeted approaches such as modification of envelope protein with a HBV peptide showed preferential gene transfer to liver-derived cells (Markusic *et al.*, 2007).

## Delivery during isolated liver perfusion

With end-stage liver disease due to chronic HCV infection being the leading indication for liver transplantation worldwide, management and treatment of recurrent disease remains one of the major clinical challenges today. RNAi, in particular delivered by viral vectors, can be ideal to modify the new graft and to protect it from HCV recurrence. During the process of liver transplantation, the graft undergoes multiple perfusions in order to clear out blood and to preserve the tissue from ischaemia during shipping and reperfusion injury after transplantation. Therefore, graft preservation provides a unique opportunity for *ex vivo* liver-targeted delivery of gene- and RNAi-based therapies. This *ex vivo* approach can target specific cells or whole organs while avoiding off-target integrations commonly occurring with systemic administration.

Early studies have demonstrated that *ex vivo* perfusion of the cold-preserved rat liver graft with replication-defective adenoviral vectors in the organ preservation solution, resulted in uptake and expression of transgenes within 8 hours after transplantation (Shaked *et al.*, 1994). In this setting, the transduction efficiency was directly correlated with the vector-to-cell ratio (multiplicity of infection) and the time of cold ischaemia (Drazan *et al.*, 1995b). Transgenic expression lasted up to 2 weeks without evidence of liver injury or detectable immune responses against vector transduced cells (Drazan *et al.*, 1995a). Using a similar approach, adenovirus expressing CTLA4, an immunosuppressive gene, was delivered into a cold-preserved rat liver graft. This resulted in rapid transient local expression shortly after revascularization, leading to the development of donor-specific unresponsiveness and long term graft survival (Olthoff *et al.*, 1998). Similar approaches were shown to be effective in protecting renal or cardiac grafts against acute rejection (Gong *et al.*, 2006) (Dudler *et al.*, 2007).

To effectively use RNAi for protection of the liver graft against re-infection by HCV, integrating self-inactivating lentiviral vectors would be more suitable, because of their capability of stable and long-term expression of shRNA (Henry *et al.*, 2006b; Li *et al.*, 2003). Interestingly, a commonly used graft preservation solution, the University of Wisconsin solution (UW), seems to create a superior transduction environment, improving the lentiviral transduction of hepatocytes. UW provided a much better environment that facilitated a greater number of transduced cells under both normothermic and hypothermic conditions. A major component of UW, hydroxyethyl starch (Viastarch), was responsible for the increase in transduction (Henry *et al.*, 2008). In the context of liver transplantation, this opens a distinct window of opportunity to increase the *ex vivo* transfer of vectors, even under cold conditions.

## RNAi-based therapeutic strategies

Armed with the rapid progress of understanding the host factors involved in viral life cycle, numerous novel therapeutic targets have emerged which can be explored to combat HCV (Fig. 9.2). Host cell factors are less likely targets to develop therapeutic resistance than the highly variable viral sequences. In case viral factors are targeted by RNAi, combinatory strategies will have to be considered to prevent mutational escape and therapeutic resistance, as will be discussed in the next sections.

### Targeting host cellular factors

Distinct from viral elements, host cellular factors are much less prone to mutation. As discussed previously, CD81 is considered an essential receptor for HCV entry. Both by ectopically expressing CD81 in Huh7-Lunet cells, which have low CD81 expression, or by reduction of CD81 expression in Huh-7.5 cells, which highly expression CD81, it was revealed that the density of cell surface-exposed CD81 is a key determinant for HCV entry (Koutsoudakis *et al.*, 2007). Knockdown of CD81 by RNAi significantly prevented the binding of Huh7 cells to E2 protein (Henry *et al.*, 2006a). A definitive proof of the therapeutic value of CD81 *in vivo* was provided by a recent study showing that treatment with anti-CD81 antibodies completely protected uPA-SCID mice with humanized livers from a subsequent challenge with HCV (Meuleman *et al.*, 2008). Another potential HCV receptor, SR-BI, is primarily expressed in the liver and steroidogenic tissues. A 90% down-regulation of SR-BI expression in Huh7 cells by RNAi caused a 30–90% inhibition of HCVpp infection, depending on the HCV genotype (Lavillette *et al.*, 2005).

To date, many host factors participating in viral entry, translation, replication or assembly have been identified as discussed. In theory, most of these factors can be considered therapeutic targets for RNAi. However, their physiological function is a crucial factor when considering them as therapeutic targets in the light of potential consequences of their down regulation. CD81 knockout mice appear to display normal liver function, but some dysfunction in the immune system have been reported (Maecker and Levy, 1997; Silvie *et al.*, 2003). Host factors offer numerous opportunities for RNAi-based therapeutic strategies, but extensive investigation of their exact physiological role and possible consequences of down regulation need to be further investigated in appropriate pre-clinical models before application in hepatitis C patients can be considered.

## Targeting miRNA

Distinct from conventional host factors described above, miRNAs represent a novel group of cellular molecules involved in HCV life cycle. The discovery of the positive regulatory function of miR-122 in HCV replication and the existing technology of manipulating miRNA function have provided the basis of developing miRNA-targeted therapeutic strategies for HCV infection. Inhibition of miR-122 function can be achieved through the use of traditional antisense technology like single-stranded reverse complement oligonucleotides (ASO). This approach can theoretically act at multiple levels to affect miR-122, such as by binding to the mature, pre- or pri-miR-122 and block their function. However, the low efficacy of the conventional ASO often requires additional modification to improve their biological activity.

The first report on the successful use of 2′-O-methyl (OMe) ASO was to knockdown the miR let-7 in *Drosophila* (Hutvagner *et al.*, 2004). Subsequently, ASOs termed 'antagomirs', were applied to inhibit miR-122 in mice. Systemic delivery via intravenous injection efficiently reduced miRNA levels and resulted in up-regulation of genes involved in cholesterol biosynthesis leading to a reduction in serum cholesterol levels of 44% (Krutzfeldt *et al.*, 2005). Transfection of miR-122 antagomirs reduced HCV RNA up to 84% in two differently subgenomic HCV models. In contrast, treatment with 2′-O-methyl-mimic miR-122 increased HCV levels up to 2.5-fold (Shan *et al.*, 2007). Another approach, 2′-O-methoxyethyl phosphorothioate modification, has also shown to effectively inhibit miR-122 activity in the liver (Esau *et al.*, 2006).

Locked nucleic acid (LNA) modification represents an advanced approach. It is a class of ASO with the ribose ring 'locked' by a methylene bridge connecting the 2′-O atom with the 4′-C atom. LNA nucleosides containing the six common nucleobases (T, C, G, A, U and mC) that appear in DNA and RNA are able to form basepairs with their complementary nucleosides according to the standard Watson-Crick basepairing rules. However, by 'locking' the molecule with the methylene bridge LNA, ASO is constrained in the ideal conformation for Watson-Crick binding, making the pairing with a complementary nucleotide strand more rapid and increases the stability of the resulting duplex. LNA technology was invented and patented by the Danish company, Exiqon (Valoczi *et al.*, 2004). The first clinical trial using LNA modified ASO to treat patients with severe chronic lymphocytic leukaemia has been conducted by Santaris Pharma. The company is engaged in the clinical development of a series of LNA based-drug candidates against cancer, metabolic diseases and hepatitis C (Frieden and Orum, 2006). A simple systemic delivery of an unconjugated, PBS-formulated LNA-anti-miR has shown to effectively antagonize miR-122 in non-human primates. This treatment led to a long-lasting and reversible decrease in total plasma cholesterol without any evidence for LNA-associated toxicities or histopathological changes in the animals (Elmen *et al.*, 2008a). Similar results were also reported in mice by the same group (Elmen *et al.*, 2008b).

Additionally, miR-122 silencing can be achieved by the modification of ASO by LNA/2′-*O*-methyl mixmer, peptide nucleic acids (PNA), PNA–peptide conjugates, or a chimeric 2′-fluoro/2′-O-Me. An antisense PNA conjugated to a cell-penetrating peptide was shown to inhibit miR-122 without the need for transfection or electroporation of human or rat cells (Fabani and Gait, 2008). The chimeric 2′Fluoro/2′OMe anti-miR-122 ASO displays improved efficacy and 5- to 10 fold improvement in potency compared to LNA/DNA, indicating its potential for further development as anti-HCV therapeutic agent (Bhat *et al.*, 2008).

## Viral targets

Instead of a 5′ cap, the IRES, located at the 5′ NCR of HCV genome, plays an essential role to bind eukaryotic ribosomal subunits and initiates the assembly of the translationally active 80S complex. Consequently, this sequence is more conserved among the six known HCV genotypes than any other part of the viral genome (Kronke *et al.*, 2004; Prabhu *et al.*, 2006). In particular, the region around the AUG translation initiation codon is highly conserved, being invariant at position +8 to −65 (with the exception of a single nucleotide variation at position −2) in over 81 isolates from various geographical locations (Bukh *et al.*, 1992). Thus, IRES seems an ideal candidate target for RNAi-mediated anti-HCV therapy and several groups have demonstrated efficient inhibition of HCV replication by designing siRNAs directed against this region (Hamazaki *et al.*, 2004; Ilves *et al.*, 2006; Kanda *et al.*, 2007). Other regions, such as core, NS3, NS4B, NS5A and NS5B, have also been demonstrated as targets for RNAi (Kapadia *et al.*, 2003; Randall *et al.*, 2003; Wilson *et al.*, 2003; Yokota *et al.*, 2003).

Next to suppression of viral replication, viral genome targeted siRNAs can efficiently prevent HCV infection in hepatocyte-derived cells (Dimitrova *et al.*, 2008). Reduced production of infectious virus particles was observed with siRNA target IRES in an infectious HCV model (Chevalier *et al.*, 2007).

Due to lack of proper animal models for HCV infection, alternative mice models were applied to demonstrate the efficacy of RNAi *in vivo*. The first proof-of-concept was provided by targeting NS5B in mice. Co-expression of an NS5B–luciferase fusion gene with an anti-NS5B siRNA expression plasmid led to a significant reduction of luciferase expression in the mouse liver indicating selective degradation induced by the NS5B siRNA (McCaffrey, 2002). Intravenous delivery of the adenovirus expressing HCV targeted shRNA into transgenic mice expressing HCV structural proteins resulted in specific suppression of virus protein synthesis in the liver (Sakamoto *et al.*, 2008).

Similar to other RNA viruses, potential concerns for RNAi-based therapy for HCV are the accumulation of mutations and the development of resistance. HCV replication is mediated by NS5B, an RNA-dependent RNA polymerase that lacks proofreading abilities. As a result of extremely high mutation rate ($10^3$ per nucleotide per generation) and replication rate ($10^{12}$ virions per day) in patients, HCV quasispecies are generated (Neumann *et al.*, 1998). It is conceivable that the virus can escape sequence-restricted silencing effects of RNAi. Indeed, under high pressure of siRNA, resistant replicons can emerge via the generation of point mutations within the siRNA target sequence (Konishi *et al.*, 2006; Wilson and Richardson, 2005). A straightforward solution for this problem is to design multitarget-directed RNAi, as outlined in the next section.

## Multiple targets and resistance development

The high sequence specificity of RNAi has a potential disadvantage when applying it as antiviral treatment since a single nucleotide mutation in the viral target sequence abrogates recognition by RISC and provides resistance against silencing. Similar to the highly active anti-retroviral therapy HAART treatment for HIV/AIDS, therapeutic efficacy of RNAi can probably be improved by using a combinational approach, since after developing resistance to a given siRNA, HCV can remain susceptible to silencing by a secondary siRNA (Wilson and Richardson, 2005). A library of endoribonuclease-prepared siRNAs (esiRNAs) to simultaneously target multiple sites of the viral sequence were shown to inhibit the replication of subgenomic and genomic HCV replicons, which seemed useful as a way to inhibit multiple viral genotypes and to avert escape (Kronke *et al.*, 2004). Alternative multi-siRNAs can be generated from raw long dsRNA or vector-derived long shRNA (Watanabe *et al.*, 2006). The first clinical candidate to treat HCV infection, BLT-HCV (Benitec), is based on the multi-targeting principle to prevent resistance development using RNAi directed at three viral targets (De Francesco and Migliaccio, 2005).

A second approach could be the simultaneously targeting of both viral and host factors. Cellular proteasome α-subunit 7 (PSMA7) and Hu antigen R (HuR) have been found to interact with HCV (Kruger *et al.*, 2001; Spangberg *et al.*, 2000). Combination of PSMA7- and HuR-directed siRNAs with HCV-directed siRNAs revealed additive anti-viral effects in HCV replicons (Korf *et al.*, 2005). Similarly, both entry and replication can be simultaneously targeted using shRNAs directed against two regions of the HCV RNA and one region of the host cell receptor, CD81. A triple shRNA was effective in concurrently reducing HCV replication, CD81 expression, and E2 binding, comparable to the single shRNA vector (Henry *et al.*, 2006b).

Invention of the second-generation shRNA that mimics the structure of pri-miRNA not only allows the expression under pol II promoter, but also provides the possibility of expressing multiple shRNA in a single cassette. Multimerization of a single miR-155-based cassette up to at least eight copies of anti-luciferase shRNAs can be used to increase the inhibition of a single target mRNA (Chung *et al.*, 2006). HIV-1 replication can be efficiently inhibited by simultaneous expression of four antiviral siRNAs in the single polycistron mir-17–92 context (Liu *et al.*, 2008). Certainly, such an approach could well fit in the design of multi-shRNAs strategies for HCV treatment.

### RNAi in combination with other antivirals

It is becoming clear that HCV can antagonize RNAi induced gene silencing at least via mutational escape and suppression of RNAi pathway. Similar to existing antiviral monotherapies, RNAi monotherapy may fail to completely control the infection, in particular in the hard-to-treat patients. Therefore, a combined strategy seems a necessity for eliminating HCV infection.

IFN-α possesses indirect antiviral activity by stimulating genes that can lead to a non-virus-specific antiviral response, whereas RNAi can directly interfere with viral entry and replication through targeting viral RNA genome or mRNA of cellular factors. Based on their complementary antiviral mechanisms, combining RNAi with IFN-α may prevent therapeutic resistance and exhibit additive antiviral activity. Promisingly, we reported that lentivirus-mediated RNAi and IFN-α act independently on HCV replication showing combined antiviral activity with no evidence of cross-interference. RNAi-mediated gene silencing of GFP or CD81 was not affected by IFN-α treatment. Treatment with IFN-α had no effect on lentiviral vector transduction (Pan *et al.*, 2009). In addition, up-regulation of certain host genes, such as the ubiquitin specific protease 18 (USP18), has been determined to be associated with patients who do not respond to IFN therapy. It has been demonstrated that silencing of USP18 enhanced the anti-HCV activity of IFN, which provided another approach of combining RNAi with IFN-α (Randall *et al.*, 2006).

Additionally, combining RNAi with ribavirin, ribozymes, ASO or small molecule inhibitors seems feasible, but needs to be further evaluated.

## Clinical application and disease categories

### Acute HCV infection

Acute HCV infection is usually asymptomatic and therefore is a rarely studied event. A minority of infected patients have self-limited disease, whereas the majority go on to persistent infection (Orland *et al.*, 2001). The goal of treatment for acute patients is to eradicate the virus in the early phase of infection, thus preventing progression to chronicity. Although there is no standard treatment, early IFN-α monotherapy has been shown to significantly reduce this risk (Barrera *et al.*, 1995; Nomura *et al.*, 2004), and therefore it has been recommended by international guidelines.

Given the fact that 55–85% of individuals with acute infection become chronically infected and the transition from acute to chronic infection is only partly understood, it remains important to develop novel therapies. RNAi is a potential candidate. Since the acute phase is relatively short, synthetic siRNAs, non-integrating viral vector expressed shRNAs or antisense miR-122 inhibitors may be suitable options.

### Chronic HCV infection

The goal of treatment for chronic HCV infection is the achievement of a sustained virological response (Marcellin *et al.*, 1997). With the current standard treatment, peg-IFN-α in combination with ribavirin, a sustained virological response can only be achieved in about 55% of genotype 1 chronically infected patients, while in patients infected with genotype 2 and 3 the response rate is over 80% (Fried *et al.*, 2002; Hadziyannis *et al.*, 2004; Manns *et al.*, 2001).

Not only is the therapeutic efficacy far from satisfactory, current treatment also causes severe side-effects. The most common side-effects of treatment with peg-IFN are fever-like symptoms, fatigue, muscle aches, psychological disorders, such as depression, irritability, anxiety and sleep disturbance, and induction of pancytopenia through its bone marrow-depressing activity. The most common adverse effect of ribavirin is haemolysis and anaemia (Fried, 2002; Fried *et al.*, 2002; Manns *et al.*, 2001).

Considering the number of non-responders, i.e. patients with relapse, and discontinuations due to severe side-effects of the current therapy, RNAi provides several alternatives to combat chronic infection of HCV. As liver regeneration and constant infection of new hepatocytes occur frequently under the chronic condition, it is conceivable that long-term expression of shRNA by integrating vectors, directed at multiple targets, or even in combination with conventional treatments, may be necessary to completely control the chronic infection.

### HCV recurrence after liver transplantation

End-stage liver disease or hepatocellular carcinoma due to chronic HCV infection is currently the most common indication for liver transplantation (Brown, 2005). However, reinfection of HCV is universal and often accompanied by an accelerated recurrence of liver cirrhosis, which reduces patient and graft survival (Feray *et al.*, 1999; Neumann *et al.*, 2004).

After transplantation, immunosuppressants are necessary to avoid rejection of the organ, but undesirably also significantly influences the course of the infection and the outcome of anti-viral therapy (Everson, 2002). Although a combination of peg-IFN-α with ribavirin remains the standard treatment, the sustained virological response rates after transplantation of approximately 20% are significantly lower than in non-transplant patients (Chalasani *et al.*, 2005; Garcia-Retortillo and Forns, 2004; Samuel *et al.*, 2003). Moreover, in many of those patients, IFN and ribavirin are poorly tolerated after liver transplantation and therapy has to be considered cautiously (Gane, 2002). Therefore, new therapies against HCV recurrence after transplantation are urgently required.

RNAi offers a unique prospect for the protection of the new graft from re-infection, as well as for treatment after recurrence. In particular, isolated liver perfusion provides a window of opportunity for targeted delivery of RNAi as described above. However, the influence of immunosuppressive medication on RNAi or vector-based delivery remains to be further evaluated.

## Concluding remarks

An overwhelming number of studies so far have extensively demonstrated the power of RNAi in understanding HCV biology and therapeutic development. Identification of host factors involved at various stages of the viral life cycle, either by single target or by genome-wide RNAi screening, has provided important new knowledge regarding the virus–host interaction, identifying novel therapeutic targets. In particular, the discovery of cellular miR-122 as a positive modulator of HCV replication has attracted broad interest to develop ASO-based therapeutic strategies. However, targeting miRNAs and other host factors has to be considered with caution, before their exact physiological function is known.

Although HCV RNA genome is a prime target for RNAi, it is becoming clear that the therapeutic application is far from straightforward. Single target-based strategies appear to be a challenge, mainly because of the high mutation rate, genetic diversity, complicated genomic structure and inhibition of the RNAi pathway by viral elements. Therefore, multiple viral targets, combined targeting of viral and host factors, or combination with conventional treatments, maybe necessary to achieve the best therapeutic effects. Together with the rapid improvement of gene delivery technology, RNAi-based therapy against HCV has a very promising future. In the scientific realm, the observed active interplay of HCV with the RNAi pathway may re-ignite a smouldering debate on whether RNAi in humans is contributing to the innate immune defence against viral infections.

Before clinical applications are in reach, it is essential to evaluate the long-term *in vivo* efficacy. Since no simple animal models are currently available for HCV, development of small animal models will be critical for bridging laboratory research to clinical application.

### Acknowledgements

We would like to thank Profs Harry L.A. Janssen, Hugo W. Tilanus and Herold J. Metselaar, Dr Jaap

Kwekkeboom and Dr Bob Scholte of the Erasmus MC, Rotterdam, for their general support. This work is financially supported by an 'Erasmus MC Grant' and by the Rotterdam foundation for Hepatology, SLO. We kindly thank Dr Danielle Wolvers for critically reading and editing the manuscript.

## References

Agnello, V., Abel, G., Elfahal, M., Knight, G.B., and Zhang, Q.X. (1999). Hepatitis C virus and other flaviviridae viruses enter cells via low density lipoprotein receptor. Proc. Natl. Acad. Sci. U.S.A. *96*, 12766–12771.

Aizaki, H., Choi, K.S., Liu, M., Li, Y.J., and Lai, M.M. (2006). Polypyrimidine-tract-binding protein is a component of the HCV RNA replication complex and necessary for RNA synthesis. Journal of biomedical science *13*, 469–480.

Ali, S., Pellerin, C., Lamarre, D., and Kukolj, G. (2004). Hepatitis C virus subgenomic replicons in the human embryonic kidney 293 cell line. J. Virol. *78*, 491–501.

Allison, A.C., and Eugui, E.M. (2000). Mycophenolate mofetil and its mechanisms of action. Immunopharmacology *47*, 85–118.

Andersson, M.G., Haasnoot, P.C., Xu, N., Berenjian, S., Berkhout, B., and Akusjarvi, G. (2005). Suppression of RNA interference by adenovirus virus-associated RNA. J. Virol. *79*, 9556–9565.

Barrera, J.M., Bruguera, M., Ercilla, M.G., Gil, C., Celis, R., Gil, M.P., del Valle Onorato, M., Rodes, J., and Ordinas, A. (1995). Persistent hepatitis C viremia after acute self-limiting posttransfusion hepatitis C. Hepatology *21*, 639–644.

Bartel, D.P. (2004). MicroRNAs: genomics, biogenesis, mechanism, and function. Cell *116*, 281–297.

Bartenschlager, R., Ahlborn-Laake, L., Mous, J., and Jacobsen, H. (1994). Kinetic and structural analyses of hepatitis C virus polyprotein processing. J. Virol. *68*, 5045–5055.

Bartenschlager, R., and Sparacio, S. (2007). Hepatitis C virus molecular clones and their replication capacity *in vivo* and in cell culture. Virus research *127*, 195–207.

Barth, H., Schafer, C., Adah, M.I., Zhang, F., Linhardt, R.J., Toyoda, H., Kinoshita-Toyoda, A., Toida, T., Van Kuppevelt, T.H., Depla, E., *et al.* (2003). Cellular binding of hepatitis C virus envelope glycoprotein E2 requires cell surface heparan sulfate. J. Biol. Chem. *278*, 41003–41012.

Barth, H., Ulsenheimer, A., Pape, G.R., Diepolder, H.M., Hoffmann, M., Neumann-Haefelin, C., Thimme, R., Henneke, P., Klein, R., Paranhos-Baccala, G., *et al.* (2005). Uptake and presentation of hepatitis C virus-like particles by human dendritic cells. Blood *105*, 3605–3614.

Bartosch, B., Bukh, J., Meunier, J.C., Granier, C., Engle, R.E., Blackwelder, W.C., Emerson, S.U., Cosset, F.L., and Purcell, R.H. (2003a). In vitro assay for neutralizing antibody to hepatitis C virus: evidence for broadly conserved neutralization epitopes. Proc. Natl. Acad. Sci. U.S.A. *100*, 14199–14204.

Bartosch, B., Dubuisson, J., and Cosset, F.L. (2003b). Infectious hepatitis C virus pseudo-particles containing functional E1–E2 envelope protein complexes. The Journal of experimental medicine *197*, 633–642.

Bartosch, B., Vitelli, A., Granier, C., Goujon, C., Dubuisson, J., Pascale, S., Scarselli, E., Cortese, R., Nicosia, A., and Cosset, F.L. (2003c). Cell entry of hepatitis C virus requires a set of co-receptors that include the CD81 tetraspanin and the SR-B1 scavenger receptor. J. Biol. Chem. *278*, 41624–41630.

Beames, B., Chavez, D., and Lanford, R.E. (2001). GB virus B as a model for hepatitis C virus. ILAR journal / National Research Council, Institute of Laboratory Animal Resources *42*, 152–160.

Behlke, M.A. (2006). Progress towards *in vivo* use of siRNAs. Mol. Ther. *13*, 644–670.

Behrens, S.E., Tomei, L., and De Francesco, R. (1996). Identification and properties of the RNA-dependent RNA polymerase of hepatitis C virus. EMBO J. *15*, 12–22.

Bhat, B., Esau, C., Davis, S., Propp, S., Kinberger, G., Gaus, H., Freier, S., Swayze, E.E., and Bennett, C.F. (2008). 2′-O-Methoxyethyl/2′-Fluoro Modified Oligonucleotides Result in More Potent Inhibition of micro RNA-122 in Vivo: A Target implicatedin HCV Replication. Nucleic acids symposium series (2004), 69.

Blight, K.J., Kolykhalov, A.A., and Rice, C.M. (2000). Efficient initiation of HCV RNA replication in cell culture. Science *290*, 1972–1974.

Brown, R.S. (2005). Hepatitis C and liver transplantation. Nature *436*, 973–978.

Buckwold, V.E., Beer, B.E., and Donis, R.O. (2003). Bovine viral diarrhea virus as a surrogate model of hepatitis C virus for the evaluation of antiviral agents. Antiviral research *60*, 1–15.

Bukh, J., Purcell, R.H., and Miller, R.H. (1992). Sequence analysis of the 5′ noncoding region of hepatitis C virus. Proc. Natl. Acad. Sci. U.S.A. *89*, 4942–4946.

Bumcrot, D., Manoharan, M., Koteliansky, V., and Sah, D.W. (2006). RNAi therapeutics: a potential new class of pharmaceutical drugs. Nature chemical biology *2*, 711–719.

Buonocore, L., Blight, K.J., Rice, C.M., and Rose, J.K. (2002). Characterization of vesicular stomatitis virus recombinants that express and incorporate high levels of hepatitis C virus glycoproteins. J. Virol. *76*, 6865–6872.

Carter, S.B., Franklin, T.J., Jones, D.F., Leonard, B.J., Mills, S.D., Turner, R.W., and Turner, W.B. (1969). Mycophenolic acid: an anti-cancer compound with unusual properties. Nature *223*, 848–850.

Chalasani, N., Manzarbeitia, C., Ferenci, P., Vogel, W., Fontana, R.J., Voigt, M., Riely, C., Martin, P., Teperman, L., Jiao, J., *et al.* (2005). Peginterferon alfa-2a for hepatitis C after liver transplantation: two randomized, controlled trials. Hepatology *41*, 289–298.

Chang, J., Guo, J.T., Jiang, D., Guo, H., Taylor, J.M., and Block, T.M. (2008). Liver-specific microRNA miR-122 enhances the replication of hepatitis C virus in nonhepatic cells. J. Virol. *82*, 8215–8223.

Chang, J., Nicolas, E., Marks, D., Sander, C., Lerro, A., Buendia, M.A., Xu, C., Mason, W.S., Moloshok, T.,

Bort, R., *et al.* (2004). miR-122, a mammalian liver-specific microRNA, is processed from hcr mRNA and may down-regulate the high affinity cationic amino acid transporter CAT-1. RNA Biol. *1*, 106–113.

Chang, K., Elledge, S.J., and Hannon, G.J. (2006a). Lessons from Nature: microRNA-based shRNA libraries. Nat. Methods 3, 707–714.

Chang, K.S., Cai, Z., Zhang, C., Sen, G.C., Williams, B.R., and Luo, G. (2006b). Replication of hepatitis C virus (HCV) RNA in mouse embryonic fibroblasts: protein kinase R (PKR)-dependent and PKR-independent mechanisms for controlling HCV RNA replication and mediating interferon activities. J. Virol. *80*, 7364–7374.

Chen, W., Zhang, Z., Chen, J., Zhang, J., Zhang, J., Wu, Y., Huang, Y., Cai, X., and Huang, A. (2008). HCV core protein interacts with Dicer to antagonize RNA silencing. Virus Res. *133*, 250–258.

Chevalier, C., Saulnier, A., Benureau, Y., Flechet, D., Delgrange, D., Colbere-Garapin, F., Wychowski, C., and Martin, A. (2007). Inhibition of hepatitis C virus infection in cell culture by small interfering RNAs. Mol. Ther. *15*, 1452–1462.

Choo, Q.L., Kuo, G., Weiner, A.J., Overby, L.R., Bradley, D.W., and Houghton, M. (1989). Isolation of a cDNA clone derived from a blood-borne non-A, non-B viral hepatitis genome. Science *244*, 359–362.

Chung, K.H., Hart, C.C., Al-Bassam, S., Avery, A., Taylor, J., Patel, P.D., Vojtek, A.B., and Turner, D.L. (2006). Polycistronic RNA polymerase II expression vectors for RNA interference based on BIC/miR-155. Nucleic acids research *34*, e53.

Corey, D.R. (2007). Chemical modification: the key to clinical application of RNA interference? J. Clin. Invest. *117*, 3615–3622.

Cullen, B.R. (2006). Is RNA interference involved in intrinsic antiviral immunity in mammals? Nature Immunol. 7, 563–567.

Davis, G.L., Wong, J.B., McHutchison, J.G., Manns, M.P., Harvey, J., and Albrecht, J. (2003). Early virologic response to treatment with peginterferon alfa-2b plus ribavirin in patients with chronic hepatitis C. Hepatology 38, 645–652.

De Francesco, R., and Migliaccio, G. (2005). Challenges and successes in developing new therapies for hepatitis C. Nature *436*, 953–960.

de Vries, W., and Berkhout, B. (2008). RNAi suppressors encoded by pathogenic human viruses. Int. J. Biochem. Cell Biol. *40*, 2007–2012.

Dimitrova, M., Affolter, C., Meyer, F., Nguyen, I., Richard, D.G., Schuster, C., Bartenschlager, R., Voegel, J.C., Ogier, J., and Baumert, T.F. (2008). Sustained delivery of siRNAs targeting viral infection by cell-degradable multilayered polyelectrolyte films. Proc. Natl. Acad. Sci. U.S.A. *105*, 16320–16325.

Douglas, J.T. (2007). Adenoviral vectors for gene therapy. Mol. Biotechnol. 36, 71–80.

Drazan, K.E., Csete, M.E., Da Shen, X., Bullington, D., Cottle, G., Busuttil, R.W., and Shaked, A. (1995a). Hepatic function is preserved following liver-directed, adenovirus-mediated gene transfer. J. Surg. Res. 59, 299–304.

Drazan, K.E., Wu, L., Shen, X.D., Bullington, D., Jurim, O., Busuttil, R.W., and Shaked, A. (1995b). Adenovirus-mediated gene transfer in the transplant setting. Part III. Variables affecting gene transfer in liver grafts. Transplantation 59, 670–673.

Dreux, M., Pietschmann, T., Granier, C., Voisset, C., Ricard-Blum, S., Mangeot, P.E., Keck, Z., Foung, S., Vu-Dac, N., Dubuisson, J., *et al.* (2006). High density lipoprotein inhibits hepatitis C virus-neutralizing antibodies by stimulating cell entry via activation of the scavenger receptor BI. J. Biol. Chem. *281*, 18285–18295.

Dudler, J., Simeoni, E., Fleury, S., Li, J., Pagnotta, M., Pascual, M., von Segesser, L.K., and Vassalli, G. (2007). Gene transfer of interleukin-18-binding protein attenuates cardiac allograft rejection. Transpl. Int. *20*, 460–466.

Egger, D., Wolk, B., Gosert, R., Bianchi, L., Blum, H.E., Moradpour, D., and Bienz, K. (2002). Expression of hepatitis C virus proteins induces distinct membrane alterations including a candidate viral replication complex. J. Virol. *76*, 5974–5984.

Elbashir, S.M., Harborth, J., Lendeckel, W., Yalcin, A., Weber, K., and Tuschl, T. (2001). Duplexes of 21-nucleotide RNAs mediate RNA interference in cultured mammalian cells. Nature *411*, 494–498.

Elmen, J., Lindow, M., Schutz, S., Lawrence, M., Petri, A., Obad, S., Lindholm, M., Hedtjarn, M., Hansen, H.F., Berger, U., *et al.* (2008a). LNA-mediated microRNA silencing in non-human primates. Nature *452*, 896–899.

Elmen, J., Lindow, M., Silahtaroglu, A., Bak, M., Christensen, M., Lind-Thomsen, A., Hedtjarn, M., Hansen, J.B., Hansen, H.F., Straarup, E.M., *et al.* (2008b). Antagonism of microRNA-122 in mice by systemically administered LNA-antimiR leads to up-regulation of a large set of predicted target mRNAs in the liver. Nucleic Acids Res. *36*, 1153–1162.

Esau, C., Davis, S., Murray, S.F., Yu, X.X., Pandey, S.K., Pear, M., Watts, L., Booten, S.L., Graham, M., McKay, R., *et al.* (2006). miR-122 regulation of lipid metabolism revealed by *in vivo* antisense targeting. Cell Metabolism 3, 87–98.

Evans, M.J., Rice, C.M., and Goff, S.P. (2004). Genetic interactions between hepatitis C virus replicons. J. Virol. *78*, 12085–12089.

Evans, M.J., von Hahn, T., Tscherne, D.M., Syder, A.J., Panis, M., Wolk, B., Hatziioannou, T., McKeating, J.A., Bieniasz, P.D., and Rice, C.M. (2007). Claudin-1 is a hepatitis C virus co-receptor required for a late step in entry. Nature *446*, 801–805.

Everson, G.T. (2002). Impact of immunosuppressive therapy on recurrence of hepatitis C. Liver Transpl. *8*, S19–27.

Fabani, M.M., and Gait, M.J. (2008). miR-122 targeting with LNA/2′-O-methyl oligonucleotide mixmers, peptide nucleic acids (PNA), and PNA–peptide conjugates. RNA *14*, 336–346.

Feray, C., Caccamo, L., Alexander, G.J., Ducot, B., Gugenheim, J., Casanovas, T., Loinaz, C., Gigou, M., Burra, P., Barkholt, L., *et al.* (1999). European collaborative study on factors influencing outcome after liver transplantation for hepatitis C. European Concerted Action on Viral Hepatitis (EUROHEP) Group. Gastroenterology *117*, 619–625.

Fire, A., Xu, S., Montgomery, M.K., Kostas, S.A., Driver, S.E., and Mello, C.C. (1998). Potent and specific genetic interference by double-stranded RNA in Caenorhabditis elegans. Nature *391*, 806–811.

Fried, M.W. (2002). Side-effects of therapy of hepatitis C and their management. Hepatology *36*, S237–244.

Fried, M.W., Shiffman, M.L., Reddy, K.R., Smith, C., Marinos, G., Goncales, F.L., Jr., Haussinger, D., Diago, M., Carosi, G., Dhumeaux, D., *et al.* (2002). Peginterferon alfa-2a plus ribavirin for chronic hepatitis C virus infection. N. Engl. J. Med. *347*, 975–982.

Frieden, M., and Orum, H. (2006). The application of locked nucleic acids in the treatment of cancer. IDrugs *9*, 706–711.

Gane, E. (2002). Treatment of recurrent hepatitis C. Liver Transpl. *8*, S28–37.

Gao, L., Aizaki, H., He, J.W., and Lai, M.M. (2004). Interactions between viral nonstructural proteins and host protein hVAP-33 mediate the formation of hepatitis C virus RNA replication complex on lipid raft. J. Virol. *78*, 3480–3488.

Garcia-Retortillo, M., and Forns, X. (2004). Prevention and treatment of hepatitis C virus recurrence after liver transplantation. J. Hepatol. *41*, 2–10.

Gardner, J.P., Durso, R.J., Arrigale, R.R., Donovan, G.P., Maddon, P.J., Dragic, T., and Olson, W.C. (2003). L-SIGN (CD 209L) is a liver-specific capture receptor for hepatitis C virus. Proc. Natl. Acad. Sci. U.S.A. *100*, 4498–4503.

Giannini, C., and Brechot, C. (2003). Hepatitis C virus biology. Cell Death Different.*10* Suppl. *1*, S27–38.

Giering, J.C., Grimm, D., Storm, T.A., and Kay, M.A. (2008). Expression of shRNA from a tissue-specific pol II promoter is an effective and safe RNAi therapeutic. Mol. Ther. *16*, 1630–1636.

Goh, P.Y., Tan, Y.J., Lim, S.P., Tan, Y.H., Lim, S.G., Fuller-Pace, F., and Hong, W. (2004). Cellular RNA helicase p68 relocalization and interaction with the hepatitis C virus (HCV) NS5B protein and the potential role of p68 in HCV RNA replication. J. Virol. *78*, 5288–5298.

Gong, N., Dong, C., Chen, Z., Chen, X., Guo, H., Zeng, Z., Ming, C., and Klaus Chen, Z. (2006). Adenovirus-mediated antisense-ERK2 gene therapy attenuates chronic allograft nephropathy. Transpl. Proc. *38*, 3228–3230.

Gottwein, E., and Cullen, B.R. (2008). Viral and cellular microRNAs as determinants of viral pathogenesis and immunity. Cell Host Microbe 3, 375–387.

Grimm, D., Streetz, K.L., Jopling, C.L., Storm, T.A., Pandey, K., Davis, C.R., Marion, P., Salazar, F., and Kay, M.A. (2006). Fatality in mice due to oversaturation of cellular microRNA/short hairpin RNA pathways. Nature *441*, 537–541.

Hadziyannis, S.J., Sette, H., Jr., Morgan, T.R., Balan, V., Diago, M., Marcellin, P., Ramadori, G., Bodenheimer, H., Jr., Bernstein, D., Rizzetto, M., *et al.* (2004). Peginterferon-alpha2a and ribavirin combination therapy in chronic hepatitis C: a randomized study of treatment duration and ribavirin dose. Ann. Intern. Med. *140*, 346–355.

Hamamoto, I., Nishimura, Y., Okamoto, T., Aizaki, H., Liu, M., Mori, Y., Abe, T., Suzuki, T., Lai, M.M., Miyamura, T., *et al.* (2005). Human VAP-B Is involved in hepatitis C virus replication through interaction with NS5A and NS5B. J. Virol. *79*, 13473–13482.

Hamazaki, H., Ujino, S., Abe, E., Miyano-Kurosaki, N., Shimotohno, K., and Takaku, H. (2004). RNAi expression mediated inhibition of HCV replication. Nucleic acids symposium series (2004), 307–308.

Hannon, G.J. (2002). RNA interference. Nature *418*, 244–251.

Harfe, B.D. (2005). MicroRNAs in vertebrate development. Curr. Opin. Genet. Dev. *15*, 410–415.

Henke, J.I., Goergen, D., Zheng, J., Song, Y., Schuttler, C.G., Fehr, C., Junemann, C., and Niepmann, M. (2008). microRNA-122 stimulates translation of hepatitis C virus RNA. EMBO J. *17*, 3300–3310.

Henry, S.D., Metselaar, H.J., Lonsdale, R.C., Kok, A., Haagmans, B.L., Tilanus, H.W., and van der Laan, L.J. (2006a). Mycophenolic acid inhibits hepatitis c virus replication and acts in synergy with cyclosporin A and interferon-alpha. Gastroenterology *131*, 1452–1462.

Henry, S.D., van der Wegen, P., Metselaar, H.J., Scholte, B.J., Tilanus, H.W., and van der Laan, L.J. (2008). Hydroxyethyl starch-based preservation solutions enhance gene therapy vector delivery under hypothermic conditions. Liver Transpl. *14*, 1708–1717.

Henry, S.D., van der Wegen, P., Metselaar, H.J., Tilanus, H.W., Scholte, B.J., and van der Laan, L.J. (2006b). Simultaneous targeting of HCV replication and viral binding with a single lentiviral vector containing multiple RNA interference expression cassettes. Mol. Ther. *14*, 485–493.

Hutvagner, G., Simard, M.J., Mello, C.C., and Zamore, P.D. (2004). Sequence-specific inhibition of small RNA function. PLoS Biol. 2, E98.

Ilves, H., Kaspar, R.L., Wang, Q., Seyhan, A.A., Vlassov, A.V., Contag, C.H., Leake, D., and Johnston, B.H. (2006). Inhibition of hepatitis C IRES-mediated gene expression by small hairpin RNAs in human hepatocytes and mice. Ann. N.Y. Acad. Sci *1082*, 52–55.

Ilyin, G.P., Rialland, M., Glaise, D., and Guguen-Guillouzo, C. (1999). Identification of a novel Skp2-like mammalian protein containing F-box and leucine-rich repeats. FEBS Lett. *459*, 75–79.

Inoue, K., Sekiyama, K., Yamada, M., Watanabe, T., Yasuda, H., and Yoshiba, M. (2003). Combined interferon alpha2b and cyclosporin A in the treatment of chronic hepatitis C: controlled trial. J. Gastroenterol. *38*, 567–572.

Isken, O., Baroth, M., Grassmann, C.W., Weinlich, S., Ostareck, D.H., Ostareck-Lederer, A., and Behrens, S.E. (2007). Nuclear factors are involved in hepatitis C virus RNA replication. RNA *13*, 1675–1692.

Ji, H., Fraser, C.S., Yu, Y., Leary, J., and Doudna, J.A. (2004). Coordinated assembly of human translation initiation complexes by the hepatitis C virus internal ribosome entry site RNA. Proc. Natl. Acad. Sci. U.S.A. *101*, 16990–16995.

Ji, J., Glaser, A., Wernli, M., Berke, J.M., Moradpour, D., and Erb, P. (2008). Suppression of short interfering RNA-mediated gene silencing by the structural proteins of hepatitis C virus. J. Gen. Virol. *89*, 2761–2766.

Johnston, J.C., Gasmi, M., Lim, L.E., Elder, J.H., Yee, J.K., Jolly, D.J., Campbell, K.P., Davidson, B.L., and Sauter, S.L. (1999). Minimum requirements for efficient transduction of dividing and nondividing cells by feline immunodeficiency virus vectors. J. Virol. *73*, 4991–5000.

Jopling, C.L., Schutz, S., and Sarnow, P. (2008). Position-dependent function for a tandem microRNA miR-122-binding site located in the hepatitis C virus RNA genome. Cell Host Microbe *4*, 77–85.

Jopling, C.L., Yi, M., Lancaster, A.M., Lemon, S.M., and Sarnow, P. (2005). Modulation of hepatitis C virus RNA abundance by a liver-specific MicroRNA. Science *309*, 1577–1581.

Kanda, T., Steele, R., Ray, R., and Ray, R.B. (2007). Small interfering RNA targeted to hepatitis C virus 5′ non-translated region exerts potent antiviral effect. J. Virol. *81*, 669–676.

Kapadia, S.B., Brideau-Andersen, A., and Chisari, F.V. (2003). Interference of hepatitis C virus RNA replication by short interfering RNAs. Proc. Natl. Acad. Sci. U.S.A. *100*, 2014–2018.

Kapadia, S.B., and Chisari, F.V. (2005). Hepatitis C virus RNA replication is regulated by host geranylgeranylation and fatty acids. Proc. Natl. Acad. Sci. U.S.A. *102*, 2561–2566.

Kato, T., Date, T., Miyamoto, M., Furusaka, A., Tokushige, K., Mizokami, M., and Wakita, T. (2003). Efficient replication of the genotype 2a hepatitis C virus subgenomic replicon. Gastroenterology *125*, 1808–1817.

Kato, T., Date, T., Miyamoto, M., Zhao, Z., Mizokami, M., and Wakita, T. (2005). Nonhepatic cell lines HeLa and 293 support efficient replication of the hepatitis C virus genotype 2a subgenomic replicon. J. Virol. *79*, 592–596.

Kato, T., Furusaka, A., Miyamoto, M., Date, T., Yasui, K., Hiramoto, J., Nagayama, K., Tanaka, T., and Wakita, T. (2001). Sequence analysis of hepatitis C virus isolated from a fulminant hepatitis patient. J. Med. Virol. *64*, 334–339.

Kay, M.A., Holterman, A.X., Meuse, L., Gown, A., Ochs, H.D., Linsley, P.S., and Wilson, C.B. (1995). Long-term hepatic adenovirus-mediated gene expression in mice following CTLA4Ig administration. Nat. Genet *11*, 191–197.

Kleinman, M.E., Yamada, K., Takeda, A., Chandrasekaran, V., Nozaki, M., Baffi, J.Z., Albuquerque, R.J., Yamasaki, S., Itaya, M., Pan, Y., *et al.* (2008). Sequence- and target-independent angiogenesis suppression by siRNA via TLR3. Nature *452*, 591–597.

Kochanek, S., Clemens, P.R., Mitani, K., Chen, H.H., Chan, S., and Caskey, C.T. (1996). A new adenoviral vector: Replacement of all viral coding sequences with 28 kb of DNA independently expressing both full-length dystrophin and beta-galactosidase. Proc. Natl. Acad. Sci. U.S.A. *93*, 5731–5736.

Konishi, M., Wu, C.H., Kaito, M., Hayashi, K., Watanabe, S., Adachi, Y., and Wu, G.Y. (2006). siRNA-resistance in treated HCV replicon cells is correlated with the development of specific HCV mutations. J. Viral Hepat. *13*, 756–761.

Korf, M., Jarczak, D., Beger, C., Manns, M.P., and Kruger, M. (2005). Inhibition of hepatitis C virus translation and subgenomic replication by siRNAs directed against highly conserved HCV sequence and cellular HCV cofactors. J. Hepatol. *43*, 225–234.

Koutsoudakis, G., Herrmann, E., Kallis, S., Bartenschlager, R., and Pietschmann, T. (2007). The level of CD81 cell surface expression is a key determinant for productive entry of hepatitis C virus into host cells. J. Virol. *81*, 588–598.

Kronke, J., Kittler, R., Buchholz, F., Windisch, M.P., Pietschmann, T., Bartenschlager, R., and Frese, M. (2004). Alternative approaches for efficient inhibition of hepatitis C virus RNA replication by small interfering RNAs. J. Virol. *78*, 3436–3446.

Kruger, M., Beger, C., Welch, P.J., Barber, J.R., Manns, M.P., and Wong-Staal, F. (2001). Involvement of proteasome alpha-subunit PSMA7 in hepatitis C virus internal ribosome entry site-mediated translation. Molecular and cellular biology *21*, 8357–8364.

Krutzfeldt, J., Rajewsky, N., Braich, R., Rajeev, K.G., Tuschl, T., Manoharan, M., and Stoffel, M. (2005). Silencing of microRNAs *in vivo* with 'antagomirs'. Nature *438*, 685–689.

Lagging, L.M., Meyer, K., Owens, R.J., and Ray, R. (1998). Functional role of hepatitis C virus chimeric glycoproteins in the infectivity of pseudotyped virus. J. Virol. *72*, 3539–3546.

Lagos-Quintana, M., Rauhut, R., Yalcin, A., Meyer, J., Lendeckel, W., and Tuschl, T. (2002). Identification of tissue-specific microRNAs from mouse. Curr Biol *12*, 735–739.

Lau, D.T., Kleiner, D.E., Ghany, M.G., Park, Y., Schmid, P., and Hoofnagle, J.H. (1998). 10-Year follow-up after interferon-alpha therapy for chronic hepatitis C. Hepatology *28*, 1121–1127.

Lavillette, D., Morice, Y., Germanidis, G., Donot, P., Soulier, A., Pagkalos, E., Sakellariou, G., Intrator, L., Bartosch, B., Pawlotsky, J.M., *et al.* (2005). Human serum facilitates hepatitis C virus infection, and neutralizing responses inversely correlate with viral replication kinetics at the acute phase of hepatitis C virus infection. J. Virol. *79*, 6023–6034.

Lee, Y., Ahn, C., Han, J., Choi, H., Kim, J., Yim, J., Lee, J., Provost, P., Radmark, O., Kim, S., *et al.* (2003). The nuclear RNase III Drosha initiates microRNA processing. Nature *425*, 415–419.

Lee, Y., Kim, M., Han, J., Yeom, K.H., Lee, S., Baek, S.H., and Kim, V.N. (2004). MicroRNA genes are transcribed by RNA polymerase II. EMBO J. *23*, 4051–4060.

Lemberg, M.K., and Martoglio, B. (2002). Requirements for signal peptide peptidase-catalyzed intramembrane proteolysis. Mol. Cell *10*, 735–744.

Levy, S., Todd, S.C., and Maecker, H.T. (1998). CD81 (TAPA-1): a molecule involved in signal transduction and cell adhesion in the immune system. Annu. Rev. Immunol. *16*, 89–109.

Leyssen, P., Balzarini, J., De Clercq, E., and Neyts, J. (2005). The predominant mechanism by which ribavirin exerts its antiviral activity in vitro against flaviviruses and paramyxoviruses is mediated by inhibition of IMP dehydrogenase. J. Virol. *79*, 1943–1947.

Li, M.J., Bauer, G., Michienzi, A., Yee, J.K., Lee, N.S., Kim, J., Li, S., Castanotto, D., Zaia, J., and Rossi, J.J. (2003).

Inhibition of HIV-1 infection by lentiviral vectors expressing Pol III-promoted anti-HIV RNAs. Mol. Ther. *8*, 196–206.

Lindenbach, B.D., Meuleman, P., Ploss, A., Vanwolleghem, T., Syder, A.J., McKeating, J.A., Lanford, R.E., Feinstone, S.M., Major, M.E., Leroux-Roels, G., *et al.* (2006). Cell culture-grown hepatitis C virus is infectious *in vivo* and can be recultured in vitro. Proc. Natl. Acad. Sci. U.S.A. *103*, 3805–3809.

Liu, J., Farmer, J.D., Jr., Lane, W.S., Friedman, J., Weissman, I., and Schreiber, S.L. (1991). Calcineurin is a common target of cyclophilin-cyclosporin A and FKBP–FK506 complexes. Cell *66*, 807–815.

Liu, Y.P., Haasnoot, J., ter Brake, O., Berkhout, B., and Konstantinova, P. (2008). Inhibition of HIV-1 by multiple siRNAs expressed from a single microRNA polycistron. Nucleic Acids Res. *36*, 2811–2824.

Logvinoff, C., Major, M.E., Oldach, D., Heyward, S., Talal, A., Balfe, P., Feinstone, S.M., Alter, H., Rice, C.M., and McKeating, J.A. (2004). Neutralizing antibody response during acute and chronic hepatitis C virus infection. Proc. Natl. Acad. Sci. U.S.A. *101*, 10149–10154.

Lohmann, V., Korner, F., Koch, J., Herian, U., Theilmann, L., and Bartenschlager, R. (1999). Replication of subgenomic hepatitis C virus RNAs in a hepatoma cell line. Science *285*, 110–113.

Lu, S., and Cullen, B.R. (2004). Adenovirus VA1 noncoding RNA can inhibit small interfering RNA and MicroRNA biogenesis. J. Virol. *78*, 12868–12876.

Maecker, H.T., and Levy, S. (1997). Normal lymphocyte development but delayed humoral immune response in CD81-null mice. J. Exp. Med. *185*, 1505–1510.

Manno, C.S., Pierce, G.F., Arruda, V.R., Glader, B., Ragni, M., Rasko, J.J., Ozelo, M.C., Hoots, K., Blatt, P., Konkle, B., *et al.* (2006). Successful transduction of liver in hemophilia by AAV-Factor IX and limitations imposed by the host immune response. Nat. Med. *12*, 342–347.

Manns, M.P., McHutchison, J.G., Gordon, S.C., Rustgi, V.K., Shiffman, M., Reindollar, R., Goodman, Z.D., Koury, K., Ling, M., and Albrecht, J.K. (2001). Peginterferon alfa-2b plus ribavirin compared with interferon alfa-2b plus ribavirin for initial treatment of chronic hepatitis C: a randomised trial. Lancet *358*, 958–965.

Marcellin, P., Boyer, N., Gervais, A., Martinot, M., Pouteau, M., Castelnau, C., Kilani, A., Areias, J., Auperin, A., Benhamou, J.P., *et al.* (1997). Long-term histologic improvement and loss of detectable intrahepatic HCV RNA in patients with chronic hepatitis C and sustained response to interferon-alpha therapy. Ann. Intern. med. *127*, 875–881.

Markusic, D.M., Kanitz, A., Oude-Elferink, R.P., and Seppen, J. (2007). Preferential gene transfer of lentiviral vectors to liver-derived cells, using a hepatitis B peptide displayed on GP64. Hum. Gene Ther. *18*, 673–679.

McCaffrey, A.P., Meuse, L., Pham, T.T.T., Conklin, D.S., Hannon, G.J., and Kay, M.A. (2002). RNA interference in adult mice. Nature *418*, 38–39.

McHutchison, J.G., Bartenschlager, R., Patel, K., and Pawlotsky, J.M. (2006). The face of future hepatitis C antiviral drug development: recent biological and virologic advances and their translation to drug development and clinical practice. J. Hepatol. *44*, 411–421.

Meertens, L., Bertaux, C., Cukierman, L., Cormier, E., Lavillette, D., Cosset, F.L., and Dragic, T. (2008). The tight junction proteins claudin-1, -6, and -9 are entry cofactors for hepatitis C virus. J. Virol. *82*, 3555–3560.

Meuleman, P., Hesselgesser, J., Paulson, M., Vanwolleghem, T., Desombere, I., Reiser, H., and Leroux-Roels, G. (2008). Anti-CD81 antibodies can prevent a hepatitis C virus infection *in vivo*. Hepatology *48*, 1761–1768.

Meunier, J.C., Engle, R.E., Faulk, K., Zhao, M., Bartosch, B., Alter, H., Emerson, S.U., Cosset, F.L., Purcell, R.H., and Bukh, J. (2005). Evidence for cross-genotype neutralization of hepatitis C virus pseudo-particles and enhancement of infectivity by apolipoprotein C1. Proc. Natl. Acad. Sci. U.S.A. *102*, 4560–4565.

Mitsui, A., and Suzuki, S. (1969). Immunosuppressive effect of mycophenolic acid. J. Antibiot. *22*, 358–363.

Miyanari, Y., Atsuzawa, K., Usuda, N., Watashi, K., Hishiki, T., Zayas, M., Bartenschlager, R., Wakita, T., Hijikata, M., and Shimotohno, K. (2007). The lipid droplet is an important organelle for hepatitis C virus production. Nature cell biology *9*, 1089–1097.

Morrey, J.D., Smee, D.F., Sidwell, R.W., and Tseng, C. (2002). Identification of active antiviral compounds against a New York isolate of West Nile virus. Antiviral Res. *55*, 107–116.

Morrissey, D.V., Blanchard, K., Shaw, L., Jensen, K., Lockridge, J.A., Dickinson, B., McSwiggen, J.A., Vargeese, C., Bowman, K., Shaffer, C.S., *et al.* (2005a). Activity of stabilized short interfering RNA in a mouse model of hepatitis B virus replication. Hepatology *41*, 1349–1356.

Morrissey, D.V., Lockridge, J.A., Shaw, L., Blanchard, K., Jensen, K., Breen, W., Hartsough, K., Machemer, L., Radka, S., Jadhav, V., *et al.* (2005b). Potent and persistent *in vivo* anti-HBV activity of chemically modified siRNAs. Nat. Biotechnol. *23*, 1002–1007.

Nakagawa, M., Sakamoto, N., Tanabe, Y., Koyama, T., Itsui, Y., Takeda, Y., Chen, C.H., Kakinuma, S., Oooka, S., Maekawa, S., *et al.* (2005). Suppression of hepatitis C virus replication by cyclosporin a is mediated by blockade of cyclophilins. Gastroenterology *129*, 1031–1041.

Nakai, H., Fuess, S., Storm, T.A., Muramatsu, S., Nara, Y., and Kay, M.A. (2005). Unrestricted hepatocyte transduction with adeno-associated virus serotype 8 vectors in mice. J. Virol. *79*, 214–224.

Naldini, L., Blomer, U., Gallay, P., Ory, D., Mulligan, R., Gage, F.H., Verma, I.M., and Trono, D. (1996). In vivo gene delivery and stable transduction of nondividing cells by a lentiviral vector. Science *272*, 263–267.

Neumann, A.U., Lam, N.P., Dahari, H., Gretch, D.R., Wiley, T.E., Layden, T.J., and Perelson, A.S. (1998). Hepatitis C viral dynamics *in vivo* and the antiviral efficacy of interferon-alpha therapy. Science *282*, 103–107.

Neumann, U.P., Berg, T., Bahra, M., Seehofer, D., Langrehr, J.M., Neuhaus, R., Radke, C., and Neuhaus, P. (2004). Fibrosis progression after liver transplantation in patients with recurrent hepatitis C. J. Hepatol. *41*, 830–836.

Ng, T.I., Mo, H., Pilot-Matias, T., He, Y., Koev, G., Krishnan, P., Mondal, R., Pithawalla, R., He, W., Dekhtyar, T., *et al.* (2007). Identification of host genes involved in hepatitis C virus replication by small interfering RNA technology. Hepatology *45*, 1413–1421.

Nomura, H., Sou, S., Tanimoto, H., Nagahama, T., Kimura, Y., Hayashi, J., Ishibashi, H., and Kashiwagi, S. (2004). Short-term interferon-alfa therapy for acute hepatitis C: a randomized controlled trial. Hepatology *39*, 1213–1219.

Okamoto, K., Moriishi, K., Miyamura, T., and Matsuura, Y. (2004). Intramembrane proteolysis and endoplasmic reticulum retention of hepatitis C virus core protein. J. Virol. *78*, 6370–6380.

Okamoto, T., Nishimura, Y., Ichimura, T., Suzuki, K., Miyamura, T., Suzuki, T., Moriishi, K., and Matsuura, Y. (2006). Hepatitis C virus RNA replication is regulated by FKBP8 and Hsp90. EMBO J. *25*, 5015–5025.

Olthoff, K.M., Judge, T.A., Gelman, A.E., da Shen, X., Hancock, W.W., Turka, L.A., and Shaked, A. (1998). Adenovirus-mediated gene transfer into cold-preserved liver allografts: survival pattern and unresponsiveness following transduction with CTLA4Ig. Nat. Med. *4*, 194–200.

Orland, J.R., Wright, T.L., and Cooper, S. (2001). Acute hepatitis C. Hepatology 33, 321–327.

Otto, G.A., and Puglisi, J.D. (2004). The pathway of HCV IRES-mediated translation initiation. Cell *119*, 369–380.

Pallaoro, M., Lahm, A., Biasiol, G., Brunetti, M., Nardella, C., Orsatti, L., Bonelli, F., Orru, S., Narjes, F., and Steinkuhler, C. (2001). Characterization of the hepatitis C virus NS2/3 processing reaction by using a purified precursor protein. J. Virol. *75*, 9939–9946.

Pan, Q., Henry, S.D., Metselaar, H.J., Scholte, B., Kwekkeboom, J., Tilanus, H.W., Janssen, H.L., and van der Laan, L.J. (2009). Combined antiviral activity of interferon-alpha and RNA interference directed against hepatitis C without affecting vector delivery and gene silencing. J. Mol. Med. *87*, 713–722.

Pawlotsky, J.M. (2006). Therapy of hepatitis C: from empiricism to eradication. Hepatology *43*, S207–220.

Pawlotsky, J.M., Chevaliez, S., and McHutchison, J.G. (2007). The hepatitis C virus life cycle as a target for new antiviral therapies. Gastroenterology *132*, 1979–1998.

Pedersen, I.M., Cheng, G., Wieland, S., Volinia, S., Croce, C.M., Chisari, F.V., and David, M. (2007). Interferon modulation of cellular microRNAs as an antiviral mechanism. Nature *449*, 919–922.

Penin, F., Brass, V., Appel, N., Ramboarina, S., Montserret, R., Ficheux, D., Blum, H.E., Bartenschlager, R., and Moradpour, D. (2004a). Structure and function of the membrane anchor domain of hepatitis C virus nonstructural protein 5A. J. Biol. Chem. *279*, 40835–40843.

Penin, F., Dubuisson, J., Rey, F.A., Moradpour, D., and Pawlotsky, J.M. (2004b). Structural biology of hepatitis C virus. Hepatology *39*, 5–19.

Persengiev, S.P., Zhu, X., and Green, M.R. (2004). Nonspecific, concentration-dependent stimulation and repression of mammalian gene expression by small interfering RNAs (siRNAs). RNA *10*, 12–18.

Pileri, P., Uematsu, Y., Campagnoli, S., Galli, G., Falugi, F., Petracca, R., Weiner, A.J., Houghton, M., Rosa, D., Grandi, G., *et al.* (1998). Binding of hepatitis C virus to CD81. Science *282*, 938–941.

Ploss, A., Evans, M.J., Gaysinskaya, V.A., Panis, M., You, H., de Jong, Y.P., and Rice, C.M. (2009). Human occludin is a hepatitis C virus entry factor required for infection of mouse cells. Nature *457*, 882–886.

Poeschla, E.M., Wong-Staal, F., and Looney, D.J. (1998). Efficient transduction of nondividing human cells by feline immunodeficiency virus lentiviral vectors. Nat. Med. *4*, 354–357.

Pohlmann, S., Zhang, J., Baribaud, F., Chen, Z., Leslie, G.J., Lin, G., Granelli-Piperno, A., Doms, R.W., Rice, C.M., and McKeating, J.A. (2003). Hepatitis C virus glycoproteins interact with DC-SIGN and DC-SIGNR. J. Virol. *77*, 4070–4080.

Prabhu, R., Garry, R.F., and Dash, S. (2006). Small interfering RNA targeted to stem–loop II of the 5′ untranslated region effectively inhibits expression of six HCV genotypes. Virology journal *3*, 100.

Randall, G., Chen, L., Panis, M., Fischer, A.K., Lindenbach, B.D., Sun, J., Heathcote, J., Rice, C.M., Edwards, A.M., and McGilvray, I.D. (2006). Silencing of USP18 potentiates the antiviral activity of interferon against hepatitis C virus infection. Gastroenterology *131*, 1584–1591.

Randall, G., Grakoui, A., and Rice, C.M. (2003). Clearance of replicating hepatitis C virus replicon RNAs in cell culture by small interfering RNAs. Proc. Natl. Acad. Sci. U.S.A. *100*, 235–240.

Randall, G., Panis, M., Cooper, J.D., Tellinghuisen, T.L., Sukhodolets, K.E., Pfeffer, S., Landthaler, M., Landgraf, P., Kan, S., Lindenbach, B.D., *et al.* (2007). Cellular cofactors affecting hepatitis C virus infection and replication. Proc. Natl. Acad. Sci. U.S.A. *104*, 12884–12889.

Rizvi, T.A., and Panganiban, A.T. (1993). Simian immunodeficiency virus RNA is efficiently encapsidated by human immunodeficiency virus type 1 particles. J. Virol. *67*, 2681–2688.

Robbins, M.A., and Rossi, J.J. (2005). Sensing the danger in RNA. Nat. Med. *11*, 250–251.

Ruiz, R., Witting, S.R., Saxena, R., and Morral, N. (2009). Robust hepatic gene silencing for functional studies using helper-dependent adenovirus vectors. Hum. Gene Ther. *20*, 87–94.

Sakamoto, N., Tanabe, Y., Yokota, T., Satoh, K., Sekine-Osajima, Y., Nakagawa, M., Itsui, Y., Tasaka, M., Sakurai, Y., Cheng-Hsin, C., *et al.* (2008). Inhibition of hepatitis C virus infection and expression in vitro and *in vivo* by recombinant adenovirus expressing short hairpin RNA.J. Gastroenterol. Hepatol. *23*, 1437–1447.

Samuel, D., Bizollon, T., Feray, C., Roche, B., Ahmed, S.N., Lemonnier, C., Cohard, M., Reynes, M., Chevallier, M., Ducerf, C., *et al.* (2003). Interferon-alpha 2b plus ribavirin in patients with chronic hepatitis C after liver transplantation: a randomized study. Gastroenterology *124*, 642–650.

Sarasin-Filipowicz, M., Krol, J., Markiewicz, I., Heim, M.H., and Filipowicz, W. (2009). Decreased levels of microRNA miR-122 in individuals with hepatitis C responding poorly to interferon therapy. Nat. Med. *15*, 31–33.

Sarbah, S.A., and Younossi, Z.M. (2000). Hepatitis C: an update on the silent epidemic. J. Clin. Gastroenterol. *30*, 125–143.

Sato, A., Takagi, M., Shimamoto, A., Kawakami, S., and Hashida, M. (2007). Small interfering RNA delivery to the liver by intravenous administration of galactosylated cationic liposomes in mice. Biomaterials *28*, 1434–1442.

Sato, Y., Murase, K., Kato, J., Kobune, M., Sato, T., Kawano, Y., Takimoto, R., Takada, K., Miyanishi, K., Matsunaga, T., *et al.* (2008). Resolution of liver cirrhosis using vitamin A-coupled liposomes to deliver siRNA against a collagen-specific chaperone. Nat. Biotechnol. *26*, 431–442.

Scarselli, E., Ansuini, H., Cerino, R., Roccasecca, R.M., Acali, S., Filocamo, G., Traboni, C., Nicosia, A., Cortese, R., and Vitelli, A. (2002). The human scavenger receptor class B type I is a novel candidate receptor for the hepatitis C virus. EMBO J. *21*, 5017–5025.

Schofield, D.J., Bartosch, B., Shimizu, Y.K., Allander, T., Alter, H.J., Emerson, S.U., Cosset, F.L., and Purcell, R.H. (2005). Human monoclonal antibodies that react with the E2 glycoprotein of hepatitis C virus and possess neutralizing activity. Hepatology *42*, 1055–1062.

Schulick, A.H., Vassalli, G., Dunn, P.F., Dong, G., Rade, J.J., Zamarron, C., and Dichek, D.A. (1997). Established immunity precludes adenovirus-mediated gene transfer in rat carotid arteries. Potential for immunosuppression and vector engineering to overcome barriers of immunity. J. Clin. Invest. *99*, 209–219.

Segers, G.C., Zhang, X., Deng, F., Sun, Q., and Nuss, D.L. (2007). Evidence that RNA silencing functions as an antiviral defense mechanism in fungi. Proc. Natl. Acad. Sci. U.S.A. *104*, 12902–12906.

Shaked, A., Csete, M.E., Drazan, K.E., Bullington, D., Wu, L., Busuttil, R.W., and Berk, A.J. (1994). Adenovirus-mediated gene transfer in the transplant setting. II. Successful expression of transferred cDNA in syngeneic liver grafts. Transplantation *57*, 1508–1511.

Shan, Y., Zheng, J., Lambrecht, R.W., and Bonkovsky, H.L. (2007). Reciprocal effects of micro-RNA-122 on expression of heme oxygenase-1 and hepatitis C virus genes in human hepatocytes. Gastroenterology *133*, 1166–1174.

Silvie, O., Rubinstein, E., Franetich, J.F., Prenant, M., Belnoue, E., Renia, L., Hannoun, L., Eling, W., Levy, S., Boucheix, C., *et al.* (2003). Hepatocyte CD81 is required for *Plasmodium falciparum* and *Plasmodium yoelii* sporozoite infectivity. Nat. Med. *9*, 93–96.

Simmonds, P., Alberti, A., Alter, H.J., Bonino, F., Bradley, D.W., Brechot, C., Brouwer, J.T., Chan, S.W., Chayama, K., Chen, D.S., *et al.* (1994). A proposed system for the nomenclature of hepatitis C viral genotypes. Hepatology *19*, 1321–1324.

Simmonds, P., Bukh, J., Combet, C., Deleage, G., Enomoto, N., Feinstone, S., Halfon, P., Inchauspe, G., Kuiken, C., Maertens, G., *et al.* (2005). Consensus proposals for a unified system of nomenclature of hepatitis C virus genotypes. Hepatology *42*, 962–973.

Siolas, D., Lerner, C., Burchard, J., Ge, W., Linsley, P.S., Paddison, P.J., Hannon, G.J., and Cleary, M.A. (2005). Synthetic shRNAs as potent RNAi triggers. Nat. Biotechnol. *23*, 227–231.

Sledz, C.A., Holko, M., de Veer, M.J., Silverman, R.H., and Williams, B.R. (2003). Activation of the interferon system by short-interfering RNAs. Nat. Cell Biol. *5*, 834–839.

Song, E., Zhu, P., Lee, S.K., Chowdhury, D., Kussman, S., Dykxhoorn, D.M., Feng, Y., Palliser, D., Weiner, D.B., Shankar, P., *et al.* (2005). Antibody mediated *in vivo* delivery of small interfering RNAs via cell-surface receptors. Nat. Biotechnol. *23*, 709–717.

Spahn, C.M., Kieft, J.S., Grassucci, R.A., Penczek, P.A., Zhou, K., Doudna, J.A., and Frank, J. (2001). Hepatitis C virus IRES RNA-induced changes in the conformation of the 40s ribosomal subunit. Science *291*, 1959–1962.

Spangberg, K., Wiklund, L., and Schwartz, S. (2000). HuR, a protein implicated in oncogene and growth factor mRNA decay, binds to the 3′ ends of hepatitis C virus RNA of both polarities. Virology *274*, 378–390.

Steinmann, D., Barth, H., Gissler, B., Schurmann, P., Adah, M.I., Gerlach, J.T., Pape, G.R., Depla, E., Jacobs, D., Maertens, G., *et al.* (2004). Inhibition of hepatitis C virus-like particle binding to target cells by antiviral antibodies in acute and chronic hepatitis C. J. Virol. *78*, 9030–9040.

Steinmann, E., Brohm, C., Kallis, S., Bartenschlager, R., and Pietschmann, T. (2008). Efficient trans-encapsidation of hepatitis C virus RNAs into infectious virus-like particles. J. Virol. *82*, 7034–7046.

Supekova, L., Supek, F., Lee, J., Chen, S., Gray, N., Pezacki, J.P., Schlapbach, A., and Schultz, P.G. (2008). Identification of human kinases involved in hepatitis C virus replication by small interference RNA library screening. J. Biol. Chem. *283*, 29–36.

Suzuki, T., and Suzuki, Y. (2006). Virus infection and lipid rafts. Biol. Pharmaceut. Bull. *29*, 1538–1541.

Takahashi, N., Hayano, T., and Suzuki, M. (1989). Peptidyl-prolyl cis-trans isomerase is the cyclosporin A-binding protein cyclophilin. Nature *337*, 473–475.

Takigawa, Y., Nagano-Fujii, M., Deng, L., Hidajat, R., Tanaka, M., Mizuta, H., and Hotta, H. (2004). Suppression of hepatitis C virus replicon by RNA interference directed against the NS3 and NS5B regions of the viral genome. Microbiol. Immunol. *48*, 591–598.

Teissier, E., and Pecheur, E.I. (2007). Lipids as modulators of membrane fusion mediated by viral fusion proteins. Eur. Biophys. J. *36*, 887–899.

Tomari, Y., and Zamore, P.D. (2005). Perspective: machines for RNAi. Genes Dev. *19*, 517–529.

Towers, G.J., Hatziioannou, T., Cowan, S., Goff, S.P., Luban, J., and Bieniasz, P.D. (2003). Cyclophilin A modulates the sensitivity of HIV-1 to host restriction factors. Nat. Med. *9*, 1138–1143.

Triyatni, M., Saunier, B., Maruvada, P., Davis, A.R., Ulianich, L., Heller, T., Patel, A., Kohn, L.D., and Liang, T.J. (2002). Interaction of hepatitis C virus-like particles and cells: a model system for studying viral binding and entry. J. Virol. *76*, 9335–9344.

Tu, H., Gao, L., Shi, S.T., Taylor, D.R., Yang, T., Mircheff, A.K., Wen, Y., Gorbalenya, A.E., Hwang, S.B., and Lai, M.M. (1999). Hepatitis C virus RNA polymerase and NS5A complex with a SNARE-like protein. Virology *263*, 30–41.

Uprichard, S.L. (2005). The therapeutic potential of RNA interference. FEBS Lett. *579*, 5996–6007.

Uprichard, S.L., Boyd, B., Althage, A., and Chisari, F.V. (2005). Clearance of hepatitis B virus from the liver of transgenic mice by short hairpin RNAs. Proc. Natl. Acad. Sci. U.S.A. *102*, 773–778.

Valoczi, A., Hornyik, C., Varga, N., Burgyan, J., Kauppinen, S., and Havelda, Z. (2004). Sensitive and specific detection of microRNAs by Northern blot analysis using LNA-modified oligonucleotide probes. Nucleic Acids Res. *32*, e175.

Voinnet, O. (2001). RNA silencing as a plant immune system against viruses. Trends Genet. *17*, 449–459.

Wakita, T., Pietschmann, T., Kato, T., Date, T., Miyamoto, M., Zhao, Z., Murthy, K., Habermann, A., Krausslich, H.G., Mizokami, M., *et al.* (2005). Production of infectious hepatitis C virus in tissue culture from a cloned viral genome. Nat. Med. *11*, 791–796.

Wang, C., Gale, M., Jr., Keller, B.C., Huang, H., Brown, M.S., Goldstein, J.L., and Ye, J. (2005). Identification of FBL2 as a geranylgeranylated cellular protein required for hepatitis C virus RNA replication. Mol. Cell *18*, 425–434.

Wang, X.H., Aliyari, R., Li, W.X., Li, H.W., Kim, K., Carthew, R., Atkinson, P., and Ding, S.W. (2006a). RNA interference directs innate immunity against viruses in adult *Drosophila*. Science *312*, 452–454.

Wang, Y., Kato, N., Jazag, A., Dharel, N., Otsuka, M., Taniguchi, H., Kawabe, T., and Omata, M. (2006b). Hepatitis C virus core protein is a potent inhibitor of RNA silencing-based antiviral response. Gastroenterology *130*, 883–892.

Wang, Z., Ma, H.I., Li, J., Sun, L., Zhang, J., and Xiao, X. (2003). Rapid and highly efficient transduction by double-stranded adeno-associated virus vectors in vitro and *in vivo*. Gene Ther. *10*, 2105–2111.

Watanabe, T., Sudoh, M., Miyagishi, M., Akashi, H., Arai, M., Inoue, K., Taira, K., Yoshiba, M., and Kohara, M. (2006). Intracellular-diced dsRNA has enhanced efficacy for silencing HCV RNA and overcomes variation in the viral genotype. Gene Ther. *13*, 883–892.

Watanabe, T., Umehara, T., Yasui, F., Nakagawa, S., Yano, J., Ohgi, T., Sonoke, S., Satoh, K., Inoue, K., Yoshiba, M., *et al.* (2007). Liver target delivery of small interfering RNA to the HCV gene by lactosylated cationic liposome. J. Hepatol. 47, 744–750.

Watashi, K., Hijikata, M., Hosaka, M., Yamaji, M., and Shimotohno, K. (2003). Cyclosporin A suppresses replication of hepatitis C virus genome in cultured hepatocytes. Hepatology *38*, 1282–1288.

Watashi, K., Ishii, N., Hijikata, M., Inoue, D., Murata, T., Miyanari, Y., and Shimotohno, K. (2005). Cyclophilin B is a functional regulator of hepatitis C virus RNA polymerase. Mol. Cell *19*, 111–122.

Waterhouse, P.M., Wang, M.B., and Lough, T. (2001). Gene silencing as an adaptive defence against viruses. Nature *411*, 834–842.

Wellnitz, S., Klumpp, B., Barth, H., Ito, S., Depla, E., Dubuisson, J., Blum, H.E., and Baumert, T.F. (2002). Binding of hepatitis C virus-like particles derived from infectious clone H77C to defined human cell lines. J. Virol. *76*, 1181–1193.

Wen, W.H., Liu, J.Y., Qin, W.J., Zhao, J., Wang, T., Jia, L.T., Meng, Y.L., Gao, H., Xue, C.F., Jin, B.Q., *et al.* (2007). Targeted inhibition of HBV gene expression by single-chain antibody mediated small interfering RNA delivery. Hepatology *46*, 84–94.

Wilkins, C., Dishongh, R., Moore, S.C., Whitt, M.A., Chow, M., and Machaca, K. (2005). RNA interference is an antiviral defence mechanism in Caenorhabditis elegans. Nature *436*, 1044–1047.

Wilson, J.A., Jayasena, S., Khvorova, A., Sabatinos, S., Rodrigue-Gervais, I.G., Arya, S., Sarangi, F., Harris-Brandts, M., Beaulieu, S., and Richardson, C.D. (2003). RNA interference blocks gene expression and RNA synthesis from hepatitis C replicons propagated in human liver cells. Proc. Natl. Acad. Sci. U.S.A. *100*, 2783–2788.

Wilson, J.A., and Richardson, C.D. (2005). Hepatitis C virus replicons escape RNA interference induced by a short interfering RNA directed against the NS5b coding region. J. Virol. *79*, 7050–7058.

Witting, S.R., Brown, M., Saxena, R., Nabinger, S., and Morral, N. (2008). Helper-dependent adenovirus-mediated short hairpin RNA expression in the liver activates the interferon response. J. Biol. Chem. *283*, 2120–2128.

Wooddell, C.I., Van Hout, C.V., Reppen, T., Lewis, D.L., and Herweijer, H. (2005). Long-term RNA interference from optimized siRNA expression constructs in adult mice. Biochem. Biophys. Res. Commun. *334*, 117–127.

Yang, Y., Nunes, F.A., Berencsi, K., Furth, E.E., Gonczol, E., and Wilson, J.M. (1994). Cellular immunity to viral antigens limits E1-deleted adenoviruses for gene therapy. Proc. Natl. Acad. Sci. U.S.A. *91*, 4407–4411.

Ye, J., Wang, C., Sumpter, R., Jr., Brown, M.S., Goldstein, J.L., and Gale, M., Jr. (2003). Disruption of hepatitis C virus RNA replication through inhibition of host protein geranylgeranylation. Proc. Natl. Acad. Sci. U.S.A. *100*, 15865–15870.

Yi, R., Qin, Y., Macara, I.G., and Cullen, B.R. (2003). Exportin-5 mediates the nuclear export of pre-microRNAs and short hairpin RNAs. Genes Dev. *17*, 3011–3016.

Yi, Z., Fang, C., Pan, T., Wang, J., Yang, P., and Yuan, Z. (2006). Subproteomic study of hepatitis C virus replicon reveals Ras-GTPase-activating protein binding protein 1 as potential HCV RC component. Biochem. Biophys. Res. Commun. *350*, 174–178.

Yilmaz, N., Shiffman, M.L., Stravitz, R.T., Sterling, R.K., Luketic, V.A., Sanyal, A.J., Mills, A.S., Contos, M.J., Coterell, A., Maluf, D., *et al.* (2007). A prospective evaluation of fibrosis progression in patients with recurrent hepatitis C virus following liver transplantation. Liver Transpl. *13*, 975–983.

Yokota, T., Sakamoto, N., Enomoto, N., Tanabe, Y., Miyagishi, M., Maekawa, S., Yi, L., Kurosaki, M., Taira, K., Watanabe, M., *et al.* (2003). Inhibition of intracellular hepatitis C virus replication by synthetic and vector-derived small interfering RNAs. EMBO Rep. *4*, 602–608.

Zeuzem, S., Buti, M., Ferenci, P., Sperl, J., Horsmans, Y., Cianciara, J., Ibranyi, E., Weiland, O., Noviello, S., Brass, C., *et al.* (2006). Efficacy of 24 weeks treatment with peginterferon alfa-2b plus ribavirin in patients with chronic hepatitis C infected with genotype 1 and low pretreatment viremia. J. Hepatol. *44*, 97–103.

Zheng, A., Yuan, F., Li, Y., Zhu, F., Hou, P., Li, J., Song, X., Ding, M., and Deng, H. (2007). Claudin-6 and claudin-9 function as additional coreceptors for hepatitis C virus. J. Virol. *81*, 12465–12471.

Zhu, Q., Guo, J.T., and Seeger, C. (2003). Replication of hepatitis C virus subgenomes in nonhepatic epithelial and mouse hepatoma cells. J. Virol. *77*, 9204–9210.

Zhu, Z., Wilson, A.T., Mathahs, M.M., Wen, F., Brown, K.E., Luxon, B.A., and Schmidt, W.N. (2008). Heme oxygenase-1 suppresses hepatitis C virus replication and increases resistance of hepatocytes to oxidant injury. Hepatology *48*, 1430–1439.

# RNAi Applications to Defeat Respiratory Viral Infections

# 10

Sailen Barik

Abstract

Pathogenic respiratory viruses, exemplified by respiratory syncytial virus (RSV), influenza (Flu) and parainfluenza virus (PIV), are major disease agents that kill millions of humans worldwide. Other respiratory viruses that are also potential agents of bioterrorism include highly pathogenic avian flu virus, SARS coronavirus, and the henipaviruses. Respiratory infection by RSV and PIV is the most prevalent cause of paediatric hospitalization in industrialized Western nations. In the US alone, it leads to about 100,000 admissions per year at a cost of about US$300m. The lack of a reliable vaccine or antiviral against these viruses, in part due to the high mutation rate of the viral RNA genomes, has led to the adoption of the novel RNA interference (RNAi) strategy. This review summarizes significant recent progress in the development of RNAi therapeutics to defeat these viral respiratory diseases.

## Introduction: the antiviral basis of RNAi

RNA interference (RNAi) is a natural cellular pathway, evolutionarily conserved in essentially in all metazoans and triggered by double-stranded RNA (dsRNA) (Fire *et al.*, 1998; Mello *et al.*, 2004). All long dsRNA molecules are first cleaved by a specialized cytoplasmic RNase, called Dicer. The product is known as short interfering RNA or siRNA, and is a 20- to 22-basepair-long RNA that contains 2-nucleotide 3′-overhangs (Doench *et al.*, 2003; Elbashir *et al.*, 2001a,b; Zamore *et al.*, 2000). The antisense strand of the siRNA engages into the RNA-induced silencing complex (RISC), the active component of which is a slicer RNase of the Argonaute (Ago) family, mainly AGO2. The antisense strand essentially acts as a guide to recruit the RISC machinery to the complementary target mRNA, and the latter is then sliced by AGO2 (Liu *et al.*, 2004; Rivas *et al.*, 2005). The net result, therefore, is loss of the mRNA and hence the corresponding protein. As viruses are obligatory intracellular parasites and RNAi is an intracellular cytoplasmic pathway, the corollary is that properly designed RNAi can target viral mRNAs to inhibit viral replication.

A number of viruses have evolved strategies to evade RNAi that target various steps of the pathway; they include viral RNAs that compete for Dicer/RISC, viral proteins that interact with and inhibit Dicer, and dsRNA-binding viral proteins that sequester siRNAs, preventing the incorporation of the guide strand into RISC (Hasnoot *et al.*, 2007). The fortunately strong antiviral activity of dsRNA against respiratory viruses suggests that these viruses have not yet acquired a function to inhibit Dicer or RISC.

## The general failure of traditional antiviral strategies

We have chronicled the struggle of anti-RSV research in a recent review (Maggon and Barik, 2004). In brief, the main current treatment prescribed for RSV is Palivizumab (Synagis™), a humanized subunit antibody against the RSV fusion protein (F), marketed by MedImmune (Gaithersberg, MD, USA). It is highly expensive, costing about US$900 per infusion and requiring monthly injections throughout the RSV season due to its short-lasting protective effect. Prophylactic Synagis may be covered by health insurance, but only if the physician considers the infant high-risk. The most adverse reactions of Synagis are rare but include anaphylactic shock and acute hypersensitivity reactions. MedImmune also markets a live attenuated influenza vaccine formulation, named FluMist®. At this time, no specific treatment is available for PIV. All respiratory viruses described here not only have human strains but also animal strains that cause severe diseases in agriculturally important species such as cattle, pig, sheep, goat and chicken, at an enormous loss to the society. Thus, an effective antiviral is an unmet need in human as well as veterinary clinics and is destined to be a blockbuster drug.

Various features of the respiratory viruses are responsible for these difficulties of prevention and treatment (Bitko and Barik, 2007; Maggon and Barik, 2004; Nichol, 2008; Nokes and Cane, 2008; Simões, 2008). First of all, their RNA genomes are highly mutable, thus presenting a significant challenge in designing a reliable vaccine or antiviral. Second, they are relatively difficult to grow in cell cultures. Third, a number of viral proteins bind to specific cellular proteins, adding to the difficulty of obtaining cell-free viral material for vaccination. Finally, the immunopathology of these viruses is exceedingly complex, and this is acutely illustrated in RSV, for which vaccine-enhanced disease or 'immunopotentiation' was first discovered in clinical trials of formalin-inactivated virus as vaccine. In short, RSV exemplifies many intriguing challenges to an antiviral regimen. Essentially all pharmaceutical companies have cancelled their traditional RSV projects (Maggon and Barik, 2004) and some have recently focused on RNAi-based strategies that form the basis of this article.

## Short interfering RNA versus short hairpin RNA as antiviral

The discovery that synthetic siRNA introduced into cultured mammalian cell lines destroyed complementary cellular mRNA, ushered the era of RNAi as an easily adoptable gene silencing tool (Elbashir *et al.*, 2001a). Alternatively, one can clone and express short hairpin RNAs from recombinant plasmids or viral vectors, which are then intracellular cleaved by Dicer to generate siRNA (Rossi, 2008). The immense antiviral prospects of RNAi became apparent when it was demonstrated that siRNAs can be properly designed to destroy essential viral mRNAs and thereby inhibit virus replication (Bitko and Barik, 2001).

The choice between synthetic siRNA and recombinant shRNA is mainly determined by the nature of the targeted virus. In chronic viral infections, such as hepatitis and AIDS, a continuous and long-term RNAi therapy is called for, and therefore, recombinant expression of shRNA is preferred. Recombinant expression in general is also preferred in gene therapy (such as treatment of cystic fibrosis, Alzheimer's disease and cancer) but may cause adverse effects such as systemic infection by the viral vector, immune reaction against the viral vector even if it is attenuated, and carcinogenic potential of lentiviral vectors due to genomic integration (Verma and Weitzman, 2004). In happy contrast, all known respiratory viruses produce only an acute infection that lasts about a week (Bitko *et al.*, 2005; Maggon and Barik, 2004; Nichol, 2008); thus, synthetic siRNA therapy has been an effective strategy. Furthermore, the siRNA is safe, generally non-immunogenic and nontoxic, and can be easily and rapidly synthesized for quick manufacturing and deployment. Therapeutic speed is especially important for RNA viruses because the rapid mutation of their RNA genome may promote the generation and selection of RNAi-resistant viral strains with mismatch to the siRNA, as demonstrated for human immunodeficiency virus (Das *et al.*, 2004; Hasnoot *et al.*, 2007). In the relatively rare event of such a mutation, a second siRNA targeting another sequence of the same virus can be promptly designed and delivered.

## The benefits of intranasal siRNA to defeat respiratory viruses

The landmark first publication describing the antiviral efficacy of RNAi used synthetic siRNA in cell culture using RSV as the proof-of-concept virus (Bitko and Barik, 2001). For respiratory viruses, this was logically followed by intranasal application of the siRNA formulation in the mouse model of RSV infection (Bitko *et al.*, 2005).

Inhaled siRNA has the following advantages as a drug:

1. It is a strong antiviral agent with preventive as well as therapeutic value (Bitko *et al.*, 2005; Barik, 2004a,b; Chakraborty, 2007).
2. It can be designed and synthesized in hours, and new ones tested as needed (Barik, 2004a; Behlke, 2008).
3. It can be applied in combination with other siRNAs in a multidrug regimen (Bitko *et al.*, 2005).
4. It is non-invasive and easily administered via generic hand-held inhalers.
5. Dry siRNA in powder form is stable and can be shipped without refrigeration to remote areas and extreme climates; when needed, one can just add water, mix and inhale.
6. Many nasally applied drugs reach the central nervous system within minutes along olfactory and trigeminal neural pathways, bypassing the blood–brain barrier (Thorne *et al.*, 2004). If this occurs for inhaled siRNA, it should provide important therapy of viral infections of the brain and the CNS, such as meningitis, encephalitis and myelitis.
7. siRNA dissolved in sterile water exhibits substantial antiviral potency without transfection reagents (Bitko *et al.*, 2005), and available chemical modifications may further enhance bioavailability and potency (Bitko and Barik, 2007).

## Respiratory syncytial virus (RSV)

RSV continues to be the most common worldwide cause of lower respiratory tract infections (LRI) in neonates and infants (Maggon and Barik, 2004; Nokes and Cane, 2008; Simões, 2008). According to WHO, the global annual infection and mortality for RSV are roughly 64 million and 160,000 respectively (Maggon and Barik, 2004). The acute RSV infection is generally designated as 'croup' and the symptoms include pneumonia, bronchiolitis, rhinitis and often otitis media. Typically, it also leaves the individual with a predisposition to asthma throughout life. The high-risk group includes premature infants, senior citizens, those with chronic lung disease or congenital heart disease, and immunocompromised individuals such as transplant recipients, who use immunosuppressive drugs for the rest of their lives, and AIDS patients.

As mentioned before, RSV led the way as the first virus for which successful antiviral siRNA was demonstrated, first in cell culture (Bitko and Barik, 2001) and then in intranasal application in an authenticated Balb/c mouse model (Bitko *et al.*, 2005). In these studies, synthetic siRNA was used to target essential viral mRNAs coding for the viral transcription factor, P (phosphoprotein) and viral F (fusion) protein. An example of the effect of anti-P siRNA is shown in Fig. 10.1. RSV growth was inhibited, and specific siRNA-specific ablation of the corresponding protein and mRNA was clearly demonstrated. In the mouse study, anti-P mRNA was administered into mouse nostrils in two forms: carrier-free (naked siRNA, dissolved in sterile water) or complexed with TransIT-TKO® transfection reagent (Mirus Corporation). No special equipment was needed and the anesthetized animals inhaled the siRNA formulations through natural breathing. The optimal dosage was 3.5 mg siRNA per kg body weight at which nearly complete protection was noted with 95% survival rate. Besides nasal drops, the formulations were effective as aerosols in closed chambers. Of note, polyethyleneimine (PEI), a popular carrier for DNA and RNA (Ge *et al.*, 2004), was toxic when applied nasally (Bitko *et al.*, 2005); therefore, we do not recommended it for intranasal application of siRNA.

For both liquid and aerosol formulations, antiviral protection against was confirmed by a variety of physiological criteria (Fig. 10.2). First of all, there was very little virus left in the mouse lung, tested by indirect immunofluorescence; the histopathology was also essentially undetectable. Second, the respiratory rate of the siRNA-treated mice remained low, characteristic of normal, healthy animals (Bitko *et al.*, 2005). Third, cysteine leukotrienes, which are significant players in bronchoconstriction and elevated in RSV-infected children and asthma patients, were low in the siRNA-treated mice but elevated in the infected untreated animals (Bitko *et al.*, 2005). Thus, inhaled anti-P siRNA not only protected from RSV infection but prevented the development of respiratory symptoms characteristic of

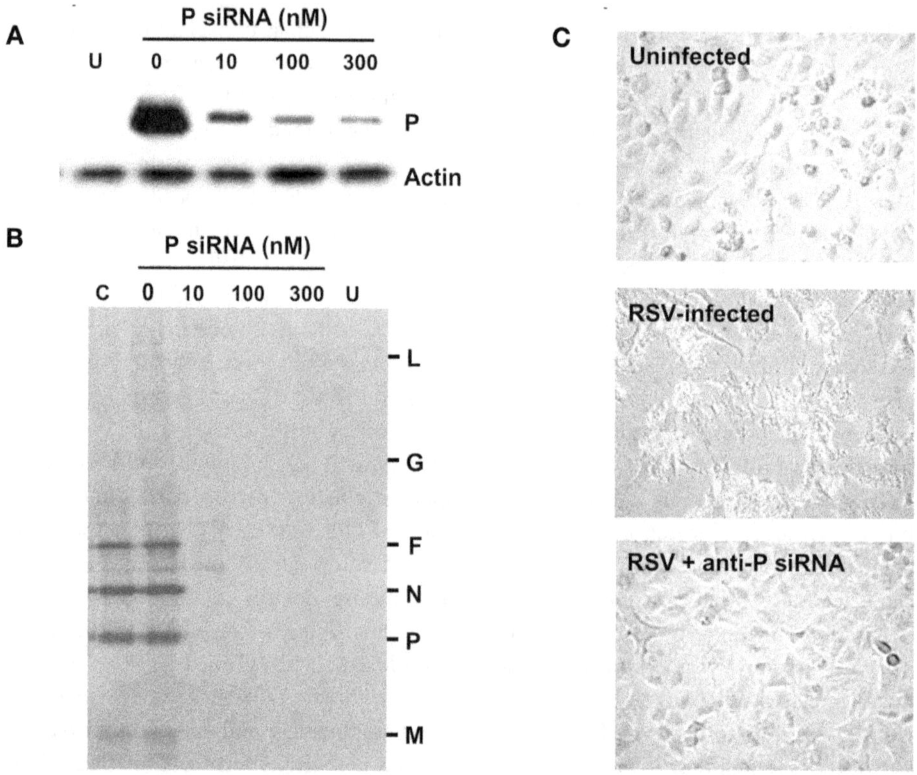


**Figure 10.1** The potent effect of anti-RSV-P siRNA (Bitko and Barik, 2001). (A) Immunoblot to show silencing of RSV P by anti-P siRNA used at the indicated concentrations in RSV-infected A549 cells in culture; U = uninfected cells. Actin serves as equal sample loading control, not affected by the siRNA. (B) Anti-P siRNA inhibited overall RSV growth in cell culture as shown by loss of all viral proteins in immunoblot (L, G etc.). C = siRNA-untreated control; U = uninfected cells. (C) Photomicrograph of A549 cells. The RSV-infected monolayer (middle panel) shows characteristic syncytia (fused cells), whereas the siRNA-treated cells (bottom) appear morphologically as healthy as the uninfected cells (top).

the infection. The inhaled siRNA did not activate dsRNA-dependent protein kinase, PKR, or interferon (IFN) pathway, confirming a specific RNAi mechanism and not a general inhibition of global translation (Bitko *et al.*, 2005; Heidel *et al.*, 2004).

In testing the therapeutic potency of the siRNA, the mice were first exposed to RSV and then the siRNA was nasally administered at different days. It was found that regardless of the time of administration, the siRNA prevented further growth of the virus and thus led to better prognosis (Fig. 10.3).

RSV also encodes two highly unique non-structural proteins, NS1 and NS2, which stimulate virus growth, mainly by suppressing the host's IFN response. In an interesting companion study, a complex of anti-NS1 siRNA and chitosan nanoparticles, administered intranasally, showed protection against RSV in the mouse model (Zhang *et al.*, 2005). This was unexpected, since deletion of either S gene strongly reduces RSV growth in the chimpanzee model (Teng *et al.*, 2000; Whitehead *et al.*, 1999). Together, these studies consolidated the use of siRNA as respiratory antiviral in both basic and clinical medicine.

The RSV breakthrough is now licensed and being developed for clinical use by Alnylam, a major RNAi pharmaceutical company (www.Alnylam.com). In their most recent progress reports, the candidate anti-RSV siRNA (dubbed ALN-RSV01), designed against RSV N mRNA, passed toxicity tests (DeVincenzo *et al.*, 2008) and Phase II clinical studies and now ready to enter Phase III trials. This is an historic milestone since it is the first anti-infective siRNA to reach the bedside.

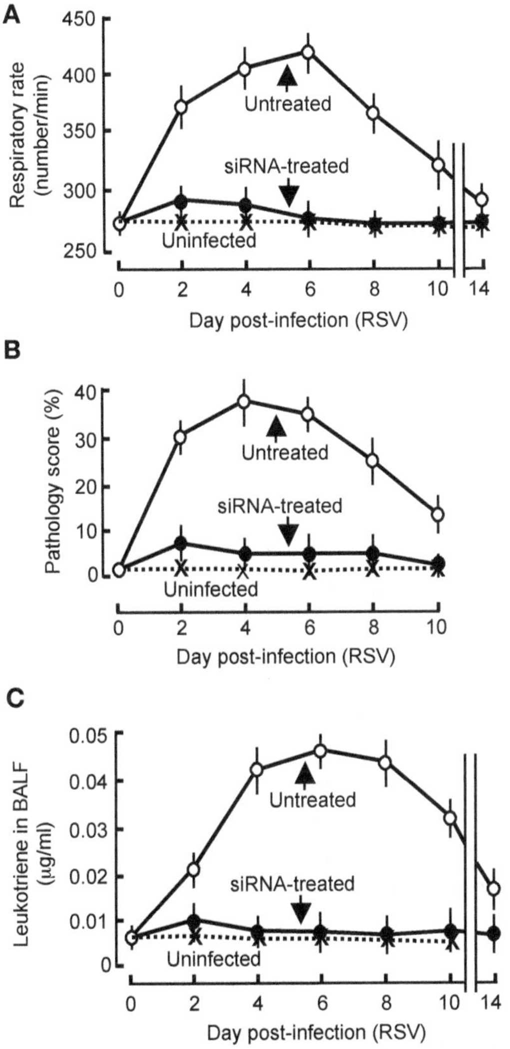


**Figure 10.2** Normal physiology of BALB/c mice exposed to RSV and anti-RSV-P siRNA (Bitko *et al,* 2005). For siRNA treatment, 3.5 mg siRNA per kg body weight was instilled intranasally. Overall, the treated mice (closed circles) appear similar to uninfected healthy controls (X), while the untreated RSV-infected mice (open circles) show the standard course of infection by all three criteria: (A) respiratory rate, (B) lung pathology score by H&E staining and (C) Cys-leukotrienes in bronchoalveolar lavage fluid.

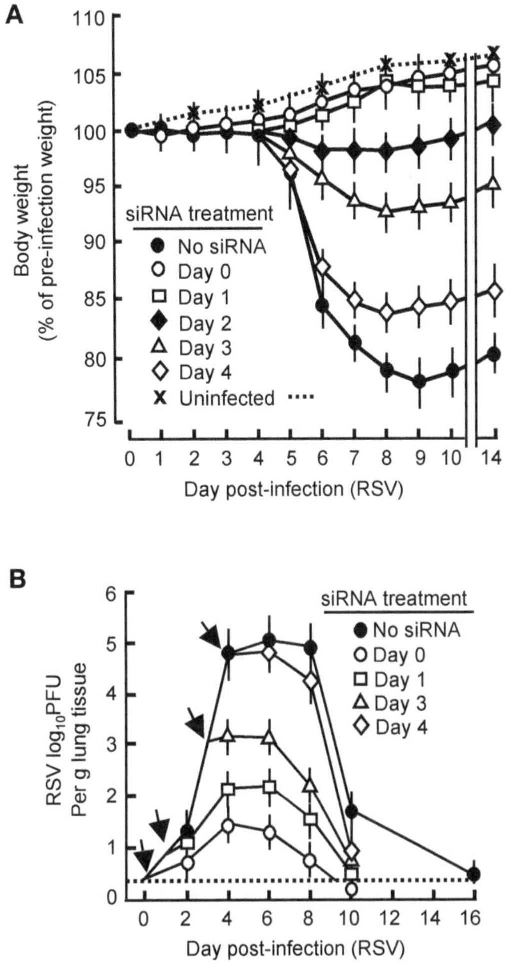


**Figure 10.3** Therapeutic effect of anti-P siRNA in mice already infected with RSV (Bitko *et al,* 2005). Changes in (A) body weight and (B) pulmonary viral titre during RSV infection in mice. Standard error bars are shown. The arrows indicate the day of siRNA administration. Note that mice treated with intranasal siRNA always fared better than the untreated ones, regardless of the day of treatment, although earlier treatment was more effective, as expected.

## Parainfluenza virus (PIV)

PIV is second only to RSV as a cause of paediatric respiratory disease and infects a slightly older age group. The intranasal moue RSV study mentioned above also included PIV, and strong antiviral effect was observed with using siRNA against P mRNA of PIV. The siRNA against PIV did not inhibit RSV and vice versa, showing specificity once more.

In a novel twist, a mixture of siRNAs against RSV and PIV was tested intranasally and it protected mice from joint challenge with both viruses (Bitko *et al.,* 2005). These results opened the door to cocktail-siRNA therapy against multiple respiratory viruses and are particularly relevant as multiple viral epidemics are common in winter seasons (Kim *et al.,* 2009), in part due to horizontal transmission among people crowded in close quarters. This experiment also revealed an upper limit of siRNA, possibly due to the saturation effect on the available cellular RISC machinery.

## Metapneumovirus

Human metapneumovirus (hMPV) is a relatively recent discovery; however, it often accompanies RSV, and like RSV, is a major respiratory viral pathogen in young children, elderly individuals and immunocompromised patients. Its avian cousin, aMPV, is the primary agent of severe rhinotracheitis in turkeys and swollen head syndrome in chickens, causing significant economic losses. Three recent studies examined siRNAs corresponding to various MPV genes for antiviral efficacy. In the first (Munir *et al.*, 2006), siRNA targeting avian metapneumovirus (aMPV) phosphoprotein (P) gene silenced P RNA and protein expression. In analogy to RSV, the P protein of MPV is a viral transcription factor, and therefore, anti-P siRNA broadly reduced the expression of other viral proteins in addition to P. Expectedly, as with RSV, the P-specific siRNAs led to inhibition of virus replication. In the second study (Ferreira *et al.*, 2007), siRNA was used to target aMPV nucleoprotein (N) and fusion (F) genes in infected CER (chicken embryo-related) cells. A marked decrease in virus replication was observed as determined by virus titration and real time RT-PCR assays. The production of viral N, F, and G mRNAs also decreased. It was proposed that a combination of siRNAs targeting the viral RNA polymerase complex may be used as a tool to study AMPV replication and/or antiviral therapy. In the most recent study (Deffrasnes *et al.*, 2008), a hMPV genome-wide search for siRNAs was conducted by computational analysis, which revealed 200 potentially effective 21-mer siRNAs. Initial screening with a luciferase assay identified 57 siRNAs of interest. Further evaluation of their inhibitory potential against the four hMPV subgroups by quantitative real-time reverse transcriptase PCR and plaque immunoassay identified two highly potent siRNAs with 50% inhibitory concentration (IC50) values in the subnanomolar range. Longer, 25/27-mer siRNAs, known as Dicer substrates were designed from the top two siRNA candidates and were at least as effective as their corresponding 21-mer siRNAs. In conclusion, appropriately designed siRNAs can be quite effective antagonists of metapneumovirus and should be advanced to animal and clinical studies.

## Influenza (flu) virus

Influenza A and B viruses are well known respiratory pathogens in adult humans and is a significant agent of morbidity and mortality in the elderly, with a healthcare tag of US$1–6m per million people in the US alone. Overall, flu epidemics affect 10–20% of the US population and cause roughly 100,000 hospitalizations annually (Nichol, 2008). Historically, flu vaccines have a success rate of just 60–80%, and the rate is even lower (<30–40%) in high-risk groups including infants, the elderly, pregnant women, and immunocompromised individuals. Anti-flu siRNAs targeting the viral nucleoprotein and RNA polymerase, essential functions for viral growth, had strong protective effect in combined intravenous and intranasal application of in an established murine model (Ge *et al.*, 2003, 2004; Tomkins *et al.*, 2004). Mixed siRNA administration led to essentially 100% survival of the infected mice. Intranasal delivery of plasmid DNA, expressing similar Flu-specific siRNAs also significantly decreased lung flu viral titres in infected mice (Ge *et al.*, 2004; Tomkins *et al.*, 2004).

## SARS coronavirus (SARS coV)

Severe acute respiratory syndrome (SARS), a highly lethal disease, is caused by the SARS coronavirus (SARS-CoV) (Peiris *et al.*, 2003; Ksiazek *et al.*, 2003), a member of the Nidovirales order that contains positive-stranded RNA genome (Marra *et al.*, 2003). There has been one SARS pandemic to date, in 2002–2003, with about 9000 known infections and 750 deaths reported worldwide, amounting to a fatality rate of roughly 9% (Periris *et al.*, 2004). For comparison, the case fatality rate for flu averages to about 0.6% (primarily among the elderly), although it can rise up to 33% in locally severe epidemics of new strains. Within a matter of weeks in early 2003, SARS spread from the Guangdong province of China to rapidly infect individuals in some 37 countries around the world (Periris *et al.*, 2004), emphasizing the importance of rapid deployment of therapy in remote areas, which is possible with siRNA.

Although treatment with conventional antiviral drugs showed some clinical success (Golda *et al.*, 2008; See *et al.*, 2009; Wong *et al.*, 2008) an effective and specific anti-SARS therapy is still needed. A recently established rhesus macaque SARS model replicated pathological features of the early stages of human SARS (Qin *et al.*, 2005). Potent anti-SARS siRNA screened against the virus in cell culture (Zheng *et al.*, 2004) showed strong protection in virus-infected macaques when delivered intranasally (Li *et al.*, 2005; Tang *et al.*, 2008). Suppression of SARS-CoV infection (Fig. 10.4) and siRNA-mediated viral RNA degradation in the upper airway epithelial cells were confirmed; negative control siRNA had no effect (Li *et al.*, 2005).

## Henipavirus

Hendra virus (HeV) and Nipah virus (NiV) are recently emerged zoonotic paramyxoviruses grouped within the new genus, *Henipavirus*. They cause fatal infections in a wide range of species, including humans, and continue to re-emerge sporadically in Australia and the Indian subcontinent. Both henipaviruses are classified as Biosafety Level 4 pathogen, and no therapeutics or vaccines are presently available against either. In a very recent but preliminary study (Mungall *et al.*, 2008), three out of four siRNAs designed against the RNA-dependent RNA polymerase gene of NiV effectively inhibited viral replication in minigenome assay. Loss of siRNA efficacy upon introduction of nucleotide mismatch validated the RNAi mechanism. Interestingly, only NiV N gene siRNAs were effective in reducing live NiV replication. It was conjectured that inhibition of early, abundantly expressed gene transcripts is more effective than later, less abundant transcripts.

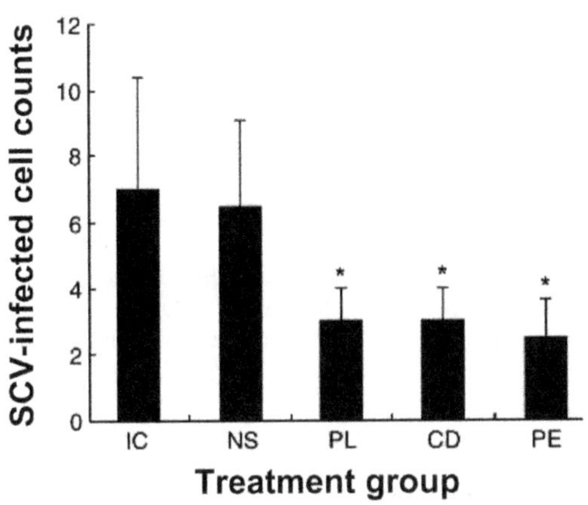


**Figure 10.4** Efficacy of siRNA against SARS coronavirus (SCV). Rhesus macaques (*Macaca mulatta*) were infected by intranasal inoculation of PUMC01, a natural isolate from China. Lungs were collected after necropsy of the macaques and the infected cells quantified by detecting viral antigens with immunohistochemistry. Asterisks (*) indicate significant effect ($P < 0.05$; $n = 4$ per group). IC = viral infection control (i.e. no siRNA); NS, non-specific siRNA control; PL, prophylactic treatment (siRNA before virus challenge); CD, co-delivery treatment (siRNA and virus at the same time); PE, post-exposure treatment (siRNA after virus). Adapted with permission of Dr Patrick Lu (Sirnaomics, MD, USA) (Li *et al.*, 2005).

## Future directions: second generation antiviral siRNA for improved efficacy

Finding a functional siRNA that silences a virus in tissue culture is no longer news. The hurdles actually begin thereafter in three main areas: specificity, stability, and delivery. These are obviously the areas where further improvement is needed to advance antiviral siRNAs to the clinic.

Regarding specificity, off-target effect (Anderson *et al.*, 2008; Svoboda, 2007; van Dongen *et al.*, 2008) and toxicity of siRNA (Grimm *et al.*, 2006) are possible, and a variety of approaches are being used to minimize it. First, design guidelines exist to find a siRNA sequence that works at low concentrations, such as low nanomolar or picomolar range (Birmingham *et al.*, 2007). Second, the delivery methods and reagents should be screened for an optimal balance of low toxicity and better delivery. As we mentioned, agents such as PEI can be toxic intranasally (Bitko *et al.*, 2005), and some delivery methods can actually modulate the off-target effect of the siRNA (Fedorov *et al.*, 2005). If possible, siRNA should be tested without any reagent as we have found them to be significantly active against RSV in the infected mouse lung (Bitko *et al.*, 2005). Using combinatorial chemistry, a large family of lipid-like transfection reagents has been developed recently (Akinc *et al.*, 2008) that can be tested on a case-by-case basis against various respiratory viruses. Lastly, a variety of chemical modifications of the siRNA have been tested recently in an attempt to improve the pharmacological parameters (Behlke, 2008; Jackson *et al.*, 2006; Manoharan, 2004; Watts *et al.*, 2008). As a rule, modifications to the sense strand often reduce off-target effects without inhibiting specific silencing but exhaustive substitutions of all positions appear to be detrimental. Synthetic chemical modifications include but are not limited to: aminolinker, biotin, fluorescein, 2′-deoxy, 2′-O-methyl, 2′-O-allyl, 2′-O-methyl, 2′-fluoro-2′-deoxyuridine, and 2′-fluoro-2′-deoxycytidine. There is currently no definitive guideline as to which chemistry will work best, and one must test them somewhat empirically in terms of number and placement of the modifications on the siRNA strands. Recently, siRNAs containing synthetic RNA-like nucleotide analogue, known as 'locked nucleic acid' (LNA), have received attention because of their high biostability and enhanced activity (Puri *et al.*, 2008). The other novel design is asymmetric siRNA, claimed to have significantly lower off-target effects (Sun *et al.*, 2008). These second generation, modified siRNAs are certainly worth testing in intranasal application against respiratory viruses.

## Acknowledgements

Research in the author's laboratory was supported by National Institutes of Health (NIH) grant AI059267 and was conducted in a facility constructed with support from Research Facilities Improvement Program Grant C06 RR11174 from the National Centre for Research Resources, NIH.

## References

Anderson, E., Boese, Q., Khvorova, A., and Karpilow, J. (2008). Identifying siRNA-induced off-targets by microarray analysis. Methods Mol. Biol. *442*, 45–63.

Akinc, A., Zumbuehl, A., Goldberg, M., Leshchiner, E.S., Busini, V., Hossain, N., Bacallado, S.A., Nguyen, D.N., Fuller, J., Alvarez, R., *et al.* (2008). A combinatorial library of lipid-like materials for delivery of RNAi therapeutics. Nat. Biotechnol. *26*, 561–569.

Barik, S. (2004a). Development of gene-specific double-stranded RNA drugs. Ann. Med. *36*, 540–551.

Barik, S. (2004b). Control of nonsegmented negative-strand RNA virus replication by siRNA. Virus Res. *102*, 27–35.

Barik, S. (2006). RNAi in moderation. Nat. Biotechnol. *24*, 796–797.

Behlke, M.A. (2008). Chemical modification of siRNAs for *in vivo* use. Oligonucleotides *18*, 305–319.

Birmingham, A., Anderson, E., Sullivan, K., Reynolds, A., Boese, Q., Leake, D., Karpilow, J., and Khvorova, A. (2007). A protocol for designing siRNAs with high functionality and specificity. Nat. Protoc. *2*, 2068–2078.

Bitko, V., and Barik, S. (2001). Phenotypic silencing of cytoplasmic genes using sequence-specific double-stranded short interfering RNA and its application in the reverse genetics of wild type negative-strand RNA viruses. BMC Microbiol. *1*, 34.

Bitko, V., and Barik, S. (2007). Intranasal antisense therapy: preclinical models with a clinical future? Curr. Opin. Mol. Ther. *9*, 119–125.

Bitko, V., Musiyenko, A., Shulyayeva, O., and Barik, S. (2005). Inhibition of respiratory viruses by nasally administered siRNA. Nat. Med. *11*, 50–55.

Chakraborty, C. (2007). Potentiality of small interfering RNAs (siRNA) as recent therapeutic targets for gene-silencing. Curr. Drug Targets *8*, 469–482.

Das, A.T., Brummelkamp, T.R., Westerhout, E.M., Vink, M., Madiredjo, M., Bernards, R., and Berkhout, B. (2004). Human immunodeficiency virus type 1 escapes from RNA interference-mediated inhibition. J. Virol. *78*, 2601–2605.

Deffrasnes, C., Cavanagh, M.H., Goyette, N., Cui, K., Ge, Q., Seth, S., Templin, M.V., Quay, S.C., Johnson, P.H., and Boivin, G. (2008). Inhibition of human metapneumovirus replication by small interfering RNA. Antiviral Ther. *13*, 821–832.

Doench, J.G., Petersen, C.P., and Sharp, P.A. (2003). siRNAs can function as miRNAs. Genes Dev. *17*, 438–442.

Elbashir, S.M., Harborth, J., Lendeckel, W., Yalcin, A., Weber, K., and Tuschl, T. (2001a). Duplexes of 21-nucleotide RNAs mediate RNA interference in cultured mammalian cells. Nature *411*, 494–498.

Elbashir, S.M., Lendeckel, W., and Tuschl, T. (2001b). RNA interference is mediated by 21- and 22-nucleotide RNAs. Genes Dev. *15*, 188–200.

Fedorov, Y., King, A., Anderson, E., Karpilow, J., Ilsley, D., Marshall, W., and Khvorova, A. (2005). Different delivery methods-different expression profiles. Nat. Methods. *2*, 241. Erratum in: Nat. Methods. (2005) *2*, 559.

Ferreira, H.L., Spilki, F.R., de Almeida, R.S., Santos, M.M., and Arns, C.W. (2007). Inhibition of avian metapneumovirus (AMPV) replication by RNA interference targeting nucleoprotein gene (N) in cultured cells. Antiviral Res. *74*, 77–81.

Fire, A., Xu, S., Montgomery, M.K., Kostas, S.A., Driver, S.E., and Mello, C.C. (1998). Potent and specific genetic interference by double-stranded RNA in *Caenorhabditis elegans*. Nature *391*, 806–811.

Ge, Q., Filip, L., Bai, A., Nguyen, T., Eisen, H.N., and Chen, J. (2004). Inhibition of influenza virus production in virus-infected mice by RNA interference. Proc. Natl. Acad. Sci. U.S.A. *101*, 8676–8681.

Ge, Q., McManus, M.T., Nguyen, T., Shen, C.H., Sharp, P.A., Eisen, H.N., and Chen, J. (2003). RNA interference of influenza virus production by directly targeting mRNA for degradation and indirectly inhibiting all viral RNA transcription. Proc. Natl. Acad. Sci. U.S.A. *100,* 2718–2723.

Golda, A., and Pyrc, K. (2008). Recent antiviral strategies against human coronavirus-related respiratory illnesses. Curr. Opin. Pulm. Med. *14,* 248–253.

Grimm, D., Streetz, K.L., Jopling, C.L., Storm, T.A., Pandey, K., Davis, C.R., Marion, P., Salazar, F., and Kay, M.A. (2006). Fatality in mice due to oversaturation of cellular microRNA/short hairpin RNA pathways. Nature *441,* 537–541.

Haasnoot, J., Westerhout, E.M., and Berkhout, B. (2007). RNA interference against viruses: strike and counterstrike. Nat. Biotechnol. *25,* 1435–1443.

Heidel, J.D., Hu, S., Liu, X.F., Triche, T.J., and Davis, M.E. (2004). Lack of interferon response in animals to naked siRNAs. Nat. Biotechnol. *22,* 1579–1582.

Jackson, A.L., Burchard, J., Leake, D., Reynolds, A., Schelter, J., Guo, J., Johnson, J.M., Lim, L., Karpilow, J., Nichols, K., Marshall, W., Khvorova, A., and Linsley, P.S. (2006). Position specific chemical modification of siRNAs reduces 'off-target' transcript silencing. RNA *12,* 1197–1205.

Kim, S.R., Ki, C.S., and Lee, N.Y. (2009). Rapid detection and identification of 12 respiratory viruses using a dual priming oligonucleotide system-based multiplex PCR assay. J. Virol. Methods *156,* 111–116.

Ksiazek, T.G., Erdman, D., Goldsmith, C.S., Zaki, S.R., Peret, T., Emery, S., Tong, S., Urbani, C., Comer, J.A., Lim, W., *et al.*; SARS Working Group. (2003). A novel coronavirus associated with severe acute respiratory syndrome. N. Engl. J. Med. *348,* 1953–1966.

Li, B.J., Tang, Q., Cheng, D., Qin, C., Xie, F.Y., Wei, Q., Xu, J., Liu, Y., Zheng, B.J., Woodle, M.C., Zhong, N., and Lu, P.Y. (2005). Using siRNA in prophylactic and therapeutic regimens against SARS coronavirus in Rhesus macaque. Nat. Med. *11,* 944–951.

Liu, J., Carmell, M.A., Rivas, F.V., Marsden, C.G., Thomson, J.M., Song, J.J., Hammond, S.M., Joshua-Tor, L., and Hannon, G.J. (2004). Argonaute2 is the catalytic engine of mammalian RNAi. Science *305,* 1437–1441.

Maggon, K., and Barik, S. (2004). New drugs and treatment for respiratory syncytial virus. Rev. Med. Virol. *14,* 149–168.

Manoharan, M. (2004).RNA interference and chemically modified small interfering RNAs. Curr. Opin. Chem. Biol. *8,* 570–579.

Marra, M.A., Jones, S.J., Astell, C.R., Holt, R.A., Brooks-Wilson, A., Butterfield, Y.S., Khattra, J., Asano, J.K., Barber, S.A., Chan, S.Y., *et al.* (2003). The genome sequence of the SARS-associated coronavirus. Science *300,* 1399–1404.

Mello, C.C., and Conte, D., Jr. (2004). Revealing the world of RNA interference. Nature *431,* 338–342.

Mungall, B.A., Schopman, N.C., Lambeth, L.S., and Doran, T.J. (2008). Inhibition of Henipavirus infection by RNA interference. Antiviral Res. *80,* 324–331.

Munir, S., Kaur, K., and Kapur, V. (2006). Avian metapneumovirus phosphoprotein targetedn RNA interference silences the expression of viral proteins and inhibits virus replication. Antiviral Res. *69,* 46–51.

Nichol, K.L. (2008). Efficacy and effectiveness of influenza vaccination. Vaccine *26* Suppl. *4,* D17–22.

Nokes, J.D., and Cane, P.A. (2008). New strategies for control of respiratory syncytial virus infection. Curr. Opin. Infect. Dis. *21,* 639–643.

Peiris, J.S., Lai, S.T., Poon, L.L., Guan, Y., Yam, L.Y., Lim, W., Nicholls, J., Yee, W.K., Yan, W.W., Cheung, M.T., *et al.*; SARS study group (2003). Coronavirus as a possible cause of severe acute respiratory syndrome. Lancet *361,* 1319–1325.

Peiris, J.S., Guan, Y., and Yuen, K.Y. (2004). Severe acute respiratory syndrome. Nat. Med. *10(12 Suppl),* S88–97.

Puri, N., Wang, X., Varma, R., Burnett, C., Beauchamp, L., Batten, D.M., Young, M., Sule, V., Latham, K., Sendera, T., Echeverri, C., Sachse, C., and Magdaleno, S. (2008). LNA incorporated siRNAs exhibit lower off-target effects compared to 2′-OMethoxy in cell phenotypic assays and microarray analysis. Nucleic Acids Symp. Ser. (Oxf). *52,* 25–26.

Qin, C., Wang, J., Wei, Q., She, M., Marasco, W.A., Jiang, H., Tu, X., Zhu, H., Ren, L., Gao, H., *et al.* (2005). An animal model of SARS produced by infection of Macaca mulatta with SARS coronavirus. J. Pathol. *206,* 251–259.

Rivas, F.V., Tolia, N.H., Song, J.J., Aragon, J.P., Liu, J., Hannon, G.J., and Joshua-Tor, L. (2005). Purified Argonaute2 and an siRNA form recombinant human RISC. Nat. Struct. Mol. Biol. *12,* 340–349.

Rossi, J.J. (2008). Expression strategies for short hairpin RNA interference triggers. Hum. Gene Ther. *19,* 313–317.

See, R.H., Petric, M., Lawrence, D.J., Mok, C.P., Rowe, T., Zitzow, L.A., Karunakaran, K.P., Voss, T.G., Brunham, R.C., Gauldie, J., Finlay, B.B., and Roper, R.L. (2008). Severe acute respiratory syndrome vaccine efficacy in ferrets: whole killed virus and adenovirus-vectored vaccines. J. Gen. Virol. *89,* 2136–2146.

Simões, E.A. (2008). RSV disease in the pediatric population: epidemiology, seasonal variability, and long-term outcomes. Manag. Care *17(11* Suppl. *12),* 3–6.

Sun, X., Rogoff, H.A., and Li, C.J. (2008). Asymmetric RNA duplexes mediate RNA interference in mammalian cells. Nat. Biotechnol. *26,* 1379–1382.

Svoboda, P. (2007). Off-targeting and other non-specific effects of RNAi experiments in mammalian cells. Curr. Opin. Mol. Ther. *9,* 248–257.

Tang, Q., Li, B., Woodle, M., and Lu, P.Y. (2008). Application of siRNA against SARS in the rhesus macaque model. Methods Mol. Biol. *442,* 139–158.

Teng, M.N., Whitehead, S.S., Bermingham, A., St Claire, M., Elkins, W.R., Murphy, B.R., and Collins, P.L. (2000). Recombinant respiratory syncytial virus that does not express the NS1 or M2-2 protein is highly attenuated and immunogenic in chimpanzees. J. Virol. *74,* 9317–9321.

Thorne, R.G., Pronk, G.J., Padmanabhan, V., and Frey, W.H., I.I. (2004). Delivery of insulin-like growth factor-I to the rat brain and spinal cord along olfactory and trigeminal pathways following intranasal administration. Neuroscience *127,* 481–496.

Tompkins, S.M., Lo, C.Y., Tumpey, T.M., and Epstein, S.L. (2004). Protection against lethal influenza virus challenge by RNA interference *in vivo*. Proc. Natl. Acad. Sci. U.S.A. *101,* 8682–8686.

Ui-Tei, K., Naito, Y., Nishi, K., Juni, A., and Saigo, K. (2008). Thermodynamic stability and Watson-Crick basepairing in the seed duplex are major determinants of the efficiency of the siRNA-based off-target effect. Nucleic Acids Res. *36,* 7100–7109.

van Dongen, S., Abreu-Goodger, C., and Enright, A.J. (2008). Detecting microRNA binding and siRNA off-target effects from expression data. Nat. Methods *5,* 1023–1025.

Verma, I.M., and Weitzman, M.D. (2004). Gene therapy: twenty-first century medicine. Annu. Rev. Biochem. *74,* 711–738.

Watts, J.K., Deleavey, G.F., and Damha, M.J. (2008). Chemically modified siRNA: tools and applications. Drug Discov. Today. *13,* 842–855.

Whitehead, S.S., Bukreyev, A., Teng, M.N., Firestone, C.Y., St Claire M., Elkins, W.R., Collins, P.L., and Murphy, B.R. (1999).Recombinant respiratory syncytial virus bearing a deletion of either the NS2 or SH gene is attenuated in chimpanzees. J. Virol. *73,* 3438–3442.

Wong, S.S., and Yuen, K.Y. (2008). The management of coronavirus infections with particular reference to SARS. J. Antimicrob. Chemother. *62,* 437–441.

Zamore, P.D., Tuschl, T., Sharp, P.A., and Bartel, D.P. (2000). RNAi: double-stranded RNA directs the ATP-dependent cleavage of mRNA at 21 to 23 nucleotide intervals. Cell *101,* 25–33.

Zhang, W., Yang, H., Kong, X., Mohapatra, S., San Juan-Vergara, H., Hellermann, G., Behera, S., Singam, R., Lockey, R.F., and Mohapatra, S.S. (2005). Inhibition of respiratory syncytial virus infection with intranasal siRNA nanoparticles targeting the viral NS1 gene. Nat. Med. *11,* 56–62. Erratum in: Nat. Med. (2005) *11,* 233.

Zheng, B.J., Guan, Y., Tang, Q., Du, C., Xie, F.Y., He, M.L., Chan, K.W., Wong, K.L., Lader, E., Woodle, M.C. *et al.* (2004). Prophylactic and therapeutic effects of small interfering RNA targeting SARS-coronavirus. Antivir. Ther. *9,* 365–374.v

# RNAi with Viral Vectors that Deliver Small Interference RNAs

11

Jovanna González-Rojas*, Xabier Abad* and Puri Fortes

Abstract

RNA interference (RNAi) has been successfully applied as a technology to inhibit gene expression for functional studies and offers great promise in therapeutic applications. The inhibitors that mediate interference are called siRNAs, small RNA molecules that can be expressed from viral vectors. Depending on the viral vector used and the route of administration, the inhibitors can be delivered in specific organs, for shorter or longer terms and with varying efficacy. In this chapter, we review the characteristics of the most relevant viral vectors for RNAi technology and the different cassettes that allow expression of the small RNA inhibitors. Further, we summarize the application of viral vectors delivering siRNAs in preclinical models of relevant diseases, such as cancer, neurological disorders and viral infections. Finally, we discuss several concerns that should be taken into account when considering viruses as vectors to mediate RNAi. We believe that the information provided in this chapter will help the successful design of experiments aimed to silence the expression of specific genes, both for gene function studies or for therapeutic purposes.

## Introduction

RNAi is a cellular regulatory mechanism that allows specific and powerful inhibition of gene expression. The inhibitors that mediate RNAi are small RNA molecules or small interfering RNAs (siRNAs) that guide the RNA silencing machinery to a target mRNA. Then, the siRNA basepairs with the mRNA, and gene expression is blocked. Thus, RNAi has been widely applied to the study of the functional role of individual genes. More recently, several libraries of inhibitory siRNAs have been constructed which allow high throughput RNAi screens, aimed at identifying the genes involved in specific functions. Gene function studies using RNAi have been predominantly performed in tissue cultured cells but also in animal models. Further development of the technology has allowed inducible or tissue-specific RNAi-mediated inhibitions. The success obtained so far with RNAi raises the attractive possibility of using this technology for clinical applications. Any disease caused by the expression of one or a few toxic genes could be treated by RNAi-mediated silencing of the disease-associated genes. Theoretically, RNAi could be successfully used in the treatment of cancer, viral infections, or dominant genetic disorders. With this in mind, siRNAs should be designed to target a gene of known sequence which has been clearly associated with a specific disease. Subsequently, functional siRNAs should be introduced into the patient without causing any harmful secondary effects. This is challenging as siRNA delivery is still a major issue. Also, several unwanted effects have been associated with the use of RNAi. Under certain circumstances, siRNA molecules are sensed by the cell leading to activation of the interferon response. Also, overexpression of siRNAs can saturate the cellular silencing machinery, which is required to control the expression of many genes involved in essential

* These authors have contributed equally to this work.

cellular processes. Besides, many siRNAs are not specific for their target and can inhibit the expression of other genes which could be required for proper cell functioning.

In this chapter, we review the use of viral vectors to deliver siRNAs. We discuss different expression constructs and strategies developed to express functional siRNAs. Also, we describe several viral vectors and their use to deliver siRNAs that target genes involved in several diseases. Finally, we list considerations that should be taken into account when using viral vectors to silence the expression of specific genes, both for gene function studies or for therapeutic purposes.

## RNAi mechanism and silencing technology

RNAi is a cellular posttranscriptional gene silencing process first described in plants and worms and later in many other organisms, including mammals. Regulation of gene expression is achieved by inducing DNA heterochromatinization, gene deletion, mRNA cleavage and destabilization and translation inhibition (Kim, 2005). In mammals, it has been clearly shown that the RNAi machinery can induce translation inhibition and/or mRNA degradation. This could lead to the RNAi-mediated regulation of the expression of 30% of mammalian genes, many involved in essential functions such as cell fate, proliferation and death (Lewis *et al.*, 2005).

RNAi is mediated in the cell by small non-coding RNAs called microRNAs (miRNAs). miRNAs are transcribed by polymerase II as long primary transcripts or pri-miRNAs (Lee *et al.*, 2004). Approximately 40% of human miRNA genes are in genomic clusters. Single or tandem pri-miRNAs are transcribed from non-protein coding genes or from the non-coding regions of protein coding genes, such as introns or 3′ untranslated sequences. Transcription can be constitutive, tissue specific or developmentally regulated leading to different miRNAs in different cells. After transcription, pri-miRNAs undergo a sequential maturation process (Fig. 11.1). First, pri-miRNAs bind to the microprocessor complex in the nucleus, which contains the double-stranded RNA binding protein di George critical region 8 (DGCR8) and Drosha, a nuclear RNAse III (Lee *et al.*, 2003; Han *et al.*, 2004; Denli *et al.*, 2004; Gregory *et al.*, 2004b). Drosha cleaves pri-miRNAs into pre-miRNAs, imperfectly paired stem–loop miRNA precursors of 60–80 nucleotides in length. Then, pre-miRNAs are exported to the cytoplasm by the RanGTP-dependent Exportin-5 transporter (Lund *et al.*, 2004; Yi *et al.*, 2003). There, they bind a cytoplasmic RNAse III enzyme named Dicer associated to the dsRNA-binding proteins TAR RNA binding protein (TRBP) (Chendrimada *et al.*, 2005) and protein activator of protein kinase (PACT) (Lee *et al.*, 2002b). Dicer processes pre-miRNAs into mature double-stranded miRNAs of 19–25 bp with 2 nucleotide 3′ overhangs (Lee *et al.*, 2002b) and helps coupling of miRNAs with the RNA-induced silencing complex (RISC) (Chendrimada *et al.*, 2005; Gregory *et al.*, 2005). This complex releases the passenger sense strand and retains a single intact antisense guide strand which will guide RISC to the target mRNA. Mature cellular miRNAs bind usually to 3′ untranslated regions of their targets which are not entirely complementary. Binding of the 'seed' sequence formed by nucleotides 2 to 7 of the 5′ end of the miRNA is enough for target recognition, even if larger interactions can also occur (Lewis *et al.*, 2005; Jackson *et al.*, 2006b). In fact, the number of targeted sequences and their extent of complementarity with the miRNA could determine the silencing efficiency and the molecular mechanism that leads to silencing. In plants, perfect pairing between the miRNA and the target leads to RISC-mediated cleavage of the target mRNA. In mammals, cellular miRNAs pair imperfectly to the 3′-UTR of target mRNAs leading RISC to inhibit protein translation and/or cause mRNA destabilization (Eulalio *et al.*, 2008). This inhibition requires P-bodies, where RISC brings targeted mRNAs and RNA degrading enzymes localize (Sen and Blau, 2005; Liu *et al.*, 2005). There, mRNAs could be degraded or could hide transiently from the translational machinery. Besides, other mechanisms have been proposed to explain translation inhibition of miRNA-targeted genes (Eulalio *et al.*, 2008) but more work is required to understand the exact molecular mechanisms that mediate silencing and the factors that lead either to translation inhibition, mRNA destabilization or both.

RNAi-based technology has been developed to use the cellular silencing machinery with

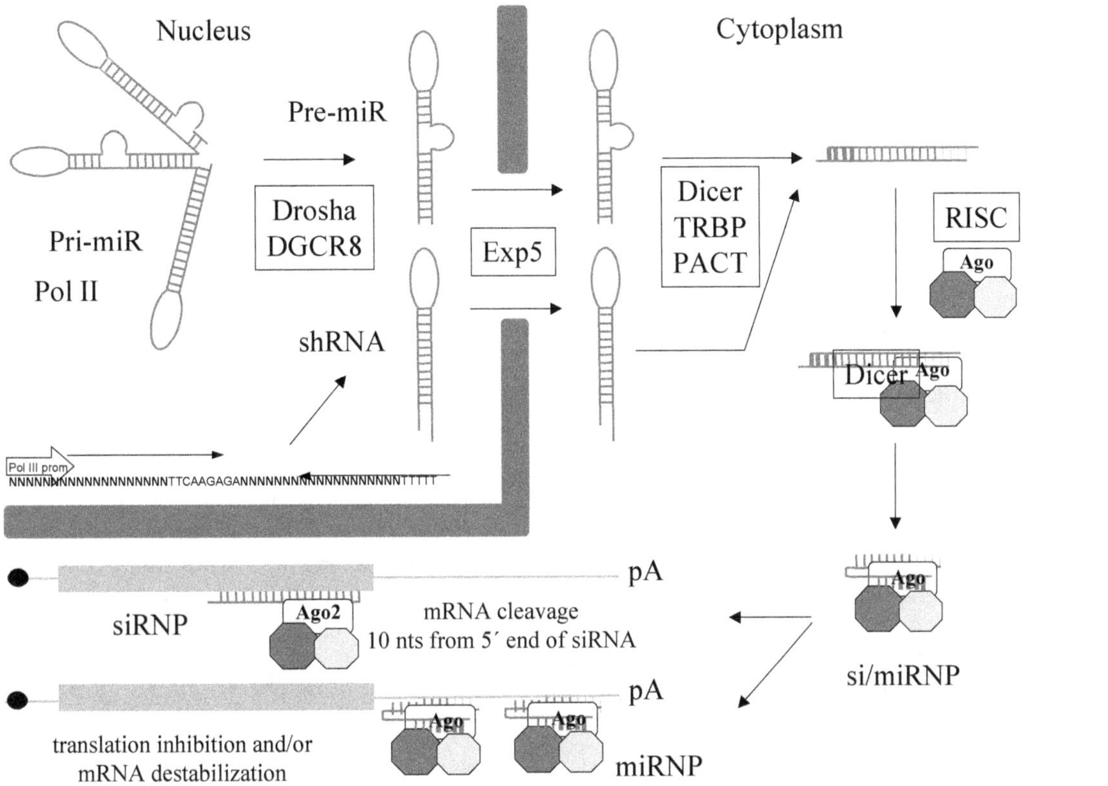


**Figure 11.1** Schematic of microRNA and siRNA processing and function. Precursor Pri-microRNA (Pri-miR) molecules are transcribed by polymerase II and processed by Drosha-DGCR8 to pre-microRNAs (pre-miR). shRNAs can be transcribed by polymerase III (as indicated) or by polymerase II. Pre-miRs and shRNAs are transported to the cytoplasm by Exp 5. There, Dicer–TRBP–PACT complex binds and processes the RNA hairpins to small dsRNA molecules. RISC binding results in retention of the guide or antisense strand, with lower thermodynamic stability at the 5′ end (indicated by thinner basepairings). The RISC complex loaded with antisense siRNAs can basepair with perfect complementarity with target mRNAs and induce mRNA cleavage 10 nucleotides upstream of the 5′ end of the siRNA. mRNA destabilization and/or translation inhibition occurs, when the loaded RISC complex binds the 3′-UTR of target mRNAs through nucleotides 2 to 7 of the 5′ end of the miRNA.

exogenous RNA inhibitors. Thus, siRNAs mimic double-stranded miRNAs, and can interact with the RISC complex to mediate gene silencing similarly to what has been described above for miRNAs (Hammond *et al.*, 2000; Elbashir *et al.*, 2001c). However, pairing between siRNAs and target sequences is generally designed to be perfect and induces the efficient site specific cleavage or slicing of the target mRNA by Argonaute 2 (Ago2), a component of RISC (Rivas *et al.*, 2005). Ago2 cleaves the target mRNA between bases 10 and 11 relative to the 5′ end of the guide strand (Elbashir *et al.*, 2001b) leading to degradation of the cleaved transcript by cellular exonucleases (Orban and Izaurralde, 2005). Then, the complex releases this cleaved transcript to act on another target molecule, so a single activated RISC can undergo multiple rounds of target mRNA cleavage (Hutvagner and Zamore, 2002). Thus, siRNAs can lead to stronger inhibitions (10 to 20 fold) than mammalian miRNAs, which silence the target gene less efficiently (2- to 3-fold) (Selbach *et al.*, 2008). Besides, siRNAs have been described to move to the nucleus and induce transcriptional gene silencing by histone methylation when they target promoter sequences (Morris *et al.*, 2004; Kim *et al.*, 2006; Janowski *et al.*, 2006; Ting *et al.*, 2005; Weinberg *et al.*, 2006;

Kawasaki and Taira, 2004; Fukagawa *et al.*, 2004; Berezhna *et al.*, 2006). The molecular mechanism by which this is achieved is poorly understood in mammalian cells and has been mostly studied in *Schizosaccharomyces pombe* (Volpe *et al.*, 2002; Verdel *et al.*, 2004).

siRNAs can be chemically synthesized and then introduced into the cell or can be expressed in the cell from an exogenous gene designed to be transcribed to siRNA precursor molecules similar to pre-miRNAs, called small or short-hairpin RNAs (shRNAs) (Brummelkamp *et al.*, 2002b). After transcription, shRNAs are transported to the cytoplasm by Exportin-5, where they are processed to siRNAs by Dicer and loaded into RISC, following a similar pathway to miRNAs (Lund *et al.*, 2004; Yi *et al.*, 2003) (Fig. 11.1). RISC will retain with higher frequency the strand of the siRNA whose 5′ end is less tightly base-paired, which should be antisense to the target (Schwarz *et al.*, 2003; Khvorova *et al.*, 2003).

Delivery of siRNAs and shRNAs into mammalian cells has been achieved by several means. In cell culture, the simplest method is the transfection of siRNAs or plasmids expressing shRNAs, which should lead to transient inhibitions of the target gene. For stable inhibitions, clones should be selected that have integrated the shRNA expressing gene into their genome. This is normally achieved by expressing the shRNA and a protein marker that allows selection, either from a plasmid that is transfected into the cell or from a virus used to infect the cell that integrates the viral genome into the cell genome. *In vivo*, silencing has been obtained by administration of chemically synthesized siRNAs, by physical-force techniques such as hydrodynamic injection of shRNA expressing plasmids or by transduction with viral vectors that express shRNAs. While synthetic siRNAs and shRNA expressing plasmids can only lead to transient inhibitions *in vivo*, durability of inhibition obtained with the viral vector will depend on the vector of choice. Virus tropism and infection efficiency should be also taken into account in the design of these experiments. For many applications, it is essential that most of the cells from the organ of interest can express the shRNA and inhibit the target gene, but efficient delivery of viral vectors in certain tissues is still a major issue in gene therapy applications.

## Alternatives to viral vectors expressing shRNAs

siRNAs can be synthesized chemically or can be produced enzymatically *in vitro* with active recombinant RNAses such as Dicer (Myers *et al.*, 2003). This strategy results in pools of siRNAs derived from a larger dsRNA which can be purified from the residual substrate. These siRNAs are effective inhibitors in transient transfections, as several highly efficient siRNAs are expected to be present in the pool. Advantages of this technique are the simplicity and the low cost. The major disadvantages are the reduced specificity and efficacy, the possibility of toxic effects and the difficult translation to clinical settings. Pools of siRNAs show decreased specificity of inhibition, mostly when only a member of a conserved family or a mutated locus needs to be silenced. Also, cocktails of siRNAs may compete against each other and with endogenous miRNAs for incorporation into RISC, leading to reduced efficacy and toxicity (Castanotto *et al.*, 2007). Besides, regulations require testing of individual siRNAs before pools of siRNAs can be used in clinical trials. Finally, enzymatically generated and shRNA-derived siRNAs are not easy substrates for chemical modifications which could improve siRNA functionality (see below).

The development of protocols that facilitate the chemical synthesis of RNA oligos has resulted in the production of high amounts of siRNAs of uniform composition and at a more affordable price. A great advantage of synthetic siRNAs is that they could be modified chemically to maintain efficacy and gain in stability, improve pharmacokinetics and cellular uptake and reduce toxic effects (Dykxhoorn and Lieberman, 2006). Terminal residues and internal nucleotides have been modified to increase stability against degradation by exonucleases and endonucleases. However, *in vivo* stability does not seem to be the major rate limiting step for *in vivo* functionality even after intravenous administration. Excretion of siRNAs from blood into urine occurs faster than degradation. Glomerular filtration can be avoided by increasing the size of siRNAs either by incorporating siRNAs into particles or by increasing siRNA affinity to serum proteins. Thus, pharmacokinetics have been improved with cholesterol-conjugated siRNAs, which enhance serum half-life by binding to serum albumin and improve cellular uptake by liver lipoprotein receptors (Soutschek *et al.*, 2004). After intravenous administration in rats, elimination half-life changes from the 6 min of the single siRNA to the 95 min of the cholesterol-conjugated molecule. However, very high doses are still required to detect siRNA silencing, suggesting that the distribution was widespread rather than liver specific.

In fact, one of the most difficult challenges of synthetic RNAi technology is the efficient and specific delivery of siRNAs. Cell-specific delivery would be the ideal choice for most applications as this would provide therapeutic effects with low doses of siRNAs and should avoid the adverse effects derived from unwanted delivery to non-targeted cells. However, at present even the non-selective delivery of siRNAs is still a challenge. siRNAs are negatively charged polymers, so they cannot penetrate the hydrophobic cell membrane to reach the cytoplasm, where they function. To enable non-specific intracellular delivery, several methods have been used with certain success. These include coupling to cationic complexes with lipids and polymers or physical force-based methods such as bombardment with siRNA-coated gold particles, electroporation or hydrodynamic injection, a rapid injection into the mouse tail vein of a large volume of aqueous solution containing the siRNA that creates a high blood pressure and results in nucleic acid delivery mostly into hepatocytes. Some of these delivery methods have limited potential applications in humans. Selective systemic delivery aims to target siRNAs to specific cell-surface receptors that will allow intracellular transport of the inhibitor through endocytosis. This has been obtained by coupling siRNAs to antibody fragments or RNA ligand aptamers that bind specific receptors, or by packaging siRNA into nanoparticles coated with receptor binding factors (de Fougerolles *et al.*, 2007; Kumar *et al.*, 2008).

Chemical modifications have been also performed that increase siRNA functionality and decrease toxic secondary effects once the siRNA is inside the cell. Modifications improve release of the passenger strand and decrease the risk of unspecific 'off-target' inhibitions and activation of the interferon response (Jackson *et al.*, 2006a). Modifications should also improve the limited persistence of siRNAs, which will require frequent dosing to maintain efficacy in many therapeutic approaches. Prolonged siRNA function is difficult to achieve as siRNAs dilute by cell division resulting in efficacies of 4–5 days in dividing cells. In non-dividing cells, siRNA silencing has been observed for 3 weeks (Song *et al.*, 2003; Omi *et al.*, 2004).

Great effort is still needed to improve chemical modifications or to develop novel protocols that could improve delivery, stability, functionality and decreased toxicity of synthetic siRNAs. Some of these issues have been partly solved with viral vectors that express shRNAs leading to tissue specific and long-lasting inhibitions. However, the use of viral vectors has other disadvantages such as induction of immune response which precludes repeated administration of the vector or the risk of insertional mutagenesis by viral genome integration. Besides, shRNAs expressed from viral vectors cannot be chemically modified to modulate or improve their function. Therefore synthetic siRNAs have advantages for some applications such as treatment of respiratory syncytial virus (RSV) infection by intranasal administration of antiviral siRNAs or treatment of macular degeneration by local delivery into the eye of siRNAs targeting the vascular endothelial growth factor (VEGF). Both treatments have shown efficacy in clinical trials (de Fougerolles *et al.*, 2007).

## Constructs expressing shRNAs

A mayor issue in the design of viral vectors expressing shRNAs is the choice of the expression cassette. Proper combinations of promoter and termination sequences should be selected that allow adequate expression of the transgen. Transgen can be a basic shRNA or a miRNA-like inhibitor, expressed in single or tandem versions. Furthermore, certain rules should be followed to increase the possibility of obtaining functional siRNAs.

### Promoters and terminators

#### *Polymerase III promoters and terminators*

U6 and H1 were the first polymerase III promoters described that express functional siRNAs. This was achieved by three different approaches: (i) expression of sense and antisense small

RNAs from separate expression cassettes (Yu *et al.*, 2002; Lee *et al.*, 2002a; Miyagishi and Taira, 2002), (ii) expression of the sense and antisense strand from a single expression cassette with convergent opposite promoters (Kaykas and Moon, 2004) and (iii) expression of shRNAs from a single cassette (Brummelkamp *et al.*, 2002b; Paddison *et al.*, 2002; Sui *et al.*, 2002; Lee *et al.*, 2002a) (Fig. 11.2). The separately expressed strands hybridize and produce functional siRNAs (Lee *et al.*, 2002a; Miyagishi and Taira, 2002). The small dsRNAs produced by either system are transported to the cytoplasm by Exportin 5. Then, siRNA duplexes interact with the RISC complex directly while the shRNAs are processed by Dicer to siRNAs. Currently pol III shRNA-expression is the method of choice for gene function studies in most laboratories. The pol III promoters most commonly used are U6 (Paddison *et al.*, 2002; Sui *et al.*, 2002) or H1 (Brummelkamp *et al.*, 2002b) but also 7SK (Czauderna *et al.*, 2003) or the strongest Epstein–Barr virus EBER promoter (Choy *et al.*, 2008) (Table 11.1). Even if all pol III promoters lead to functional shRNAs, some studies conclude that U6 promoters may be more efficient than H1 promoters in tissue culture cells and in animal models (Makinen *et al.*, 2006). Also, some protocols design the vector depending on the target sequence, as the U6 promoter strongly favours a 'G' at the first position of the shRNA transcript, while the H1 promoter has a weak preference for 'A' (Cullen, 2006). Independently of the choice of pol III promoter, shRNA transcription is initiated in a precise position outside the promoter sequence, continues with the 19- to 29-nucleotide-long top strand of the hairpin, the 4–19 nucleotide long hairpin loop and the bottom strand and terminates after the second, normally, or third residue of a track of 4–6 thymidines. Termination is designed to result in the 3′ 2 nucleotide overhang characteristic of RNAse III cleaved products such as pre-miRNAs.

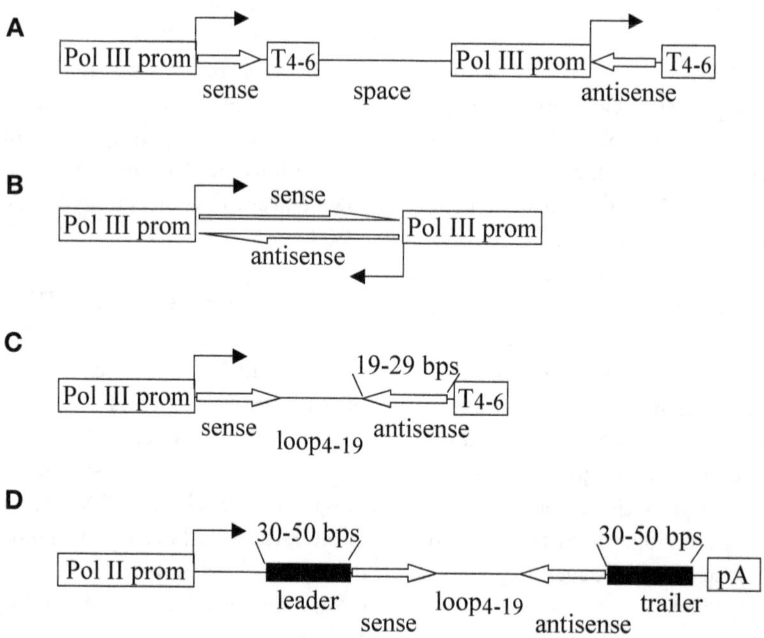


**Figure 11.2** Schematic of different siRNA expressing cassettes. (A) Expression of sense and antisense small RNAs from separate expression cassettes. (B) Expression of the sense and antisense strand from a single expression cassette with convergent opposite promoters. (C) Expression of shRNAs from a single cassette. (D) Expression of miRNA-like shRNAs from a polymerase II promoter. Promoter is indicated with a box, transcription start with an arrow, pA indicates polyadenylation sequences and the sense and antisense strands are shown by arrows in opposite directions. Sizes of some siRNA strands, terminators (T), loops, leader and trailer sequences are indicated in basepairs (bp).

**Table 11.1** Summary of cassettes used to express shRNAs and their characteristics

| Promoter | Modification | Characteristics | References |
|---|---|---|---|
| Pol III U6 promoter | None | Not inducible | Paddison *et al.* (2002), Sui *et al.* (2002), Yu *et al.* (2002), Lee *et al.* (2002a), Miyagishi and Taira (2002), Paul *et al.* (2002) |
| | Tet operator | Tetracycline-inducible silencing | Czauderna *et al.* (2003), Matsukura *et al.* (2003), Pluta *et al.* (2007), Aagaard *et al.* (2007), Chen *et al.* (2003) |
| | Tet operator | Tetracycline-inducible target expression | Pluta *et al.* (2007) |
| | 2× Tet operator | Tetracycline-inducible silencing | Lin *et al.* (2004) |
| | Gal 4 element | Ecdysone-inducible silencing | Gupta *et al.* (2004) |
| | LoxP insertions | Irreversible Cre-induced silencing | Kasim *et al.* (2004), Tiscornia *et al.* (2004), Coumoul *et al.* (2004, 2005), Ventura *et al.* (2004), Chang *et al.* (2004), Yu and McMahon (2006) |
| | LoxP insertions | Irreversible Cre-induced target expression | Ventura (2004), Meissner (2006) |
| Minimal pol III U6 promoter | Tet operator | Tetracycline inducible silencing | Amar *et al.* (2006) |
| Pol III H1 promoter | None | Not inducible | Brummelkamp *et al.* (2002b) |
| | Tet operator | Tetracycline-inducible silencing | van der Wetering *et al.* (2003), Aagaard *et al.* (2007), Grey *et al.* (2007), Hoeflich *et al.* (2006), Ngo *et al.* (2006), Matthess *et al.* (2005), Wiznerowicz and Trono (2003) |
| | Lac operator | IPTG-inducible silencing | Higuchi *et al.* (2004) |
| | LoxP insertions | Irreversible Cre-induced silencing | Heinonen *et al.* (2005), Fritsch *et al.* (2004) |
| Pol III 7SK promoter | None | Not inducible | Czauderna *et al.* (2003) |
| | Tet operator | Tetracycline-inducible silencing | Czauderna *et al.* (2003) |
| Pol III viral EBER promoter | None | Not inducible | Choy *et al.* (2008) |
| Pol II viral LTR promoter | None | Not inducible | Dickins *et al.* (2005) |
| Pol II U1 promoter | None | Not inducible | Denti *et al.* (2004) |
| Pol II viral CMV promoter | None | Not inducible | Stegmeier *et al.* (2005), Snyder *et al.* (2008) |
| AFP enh II + Albumin Pol II prom | Chimeric promoter | Tissue specific (liver) | Snyder *et al.* (2008) |

**Table 11.1** Continued

| Promoter | Modification | Characteristics | References |
|---|---|---|---|
| Pol II viral minimal CMV promoter | Tet operator | Tetracycline-inducible silencing | Grey *et al.* (2007), Shin *et al.* (2006), Dickins *et al.* (2005), Stegmeier *et al.* (2005) |
| Pol II viral minimal CMV promoter | Tet operator | Tetracycline-inducible target expression | Dickins *et al.* (2005), Stegmeier *et al.* (2005) |
| Pol II viral minimal CMV promoter | Tet operator + LoxP insertions | Tet-inducible silencing in specific loci | Wang *et al.* (2007) |
| Pol II viral HIV-1 LTR + hsp70 minimal promoter | Chimeric promoter | HIV-1 Tat-inducible silencing | Unwalla *et al.* (2004) |

Termination is not precise, but results in some heterogeneity that is not relevant for the functionality of the shRNA (Elbashir *et al.*, 2001c; Miyagishi and Taira, 2002).

*Modified polymerase III promoters: inducible systems*

Recent improvements in the expression cassette consist of modifications of pol III promoters to respond to inducers (Czauderna *et al.*, 2003; Gupta *et al.*, 2004, van de Wetering *et al.*, 2003; Higuchi *et al.*, 2004) (Table 11.1). Ideally, inducible promoters should control expression tightly to allow a fast, dose-dependent and reversible regulation of transcription in response to an inducer. The reality is that inducible promoters are leaky, but efficient to control gene expressing within a certain range. This is very important as pol III-based promoters are very strong, leading to high levels of shRNAs which could lead to toxicity due to saturation of the endogenous RNAi pathway, increased unspecific or 'off-target' inhibitions and activation of the immune response. Consequently, inducible promoters are of great interest as they allow expression levels to be fine tuned, thus decreasing toxicity, which is very important in stable inhibitions. Besides, induction allows the activation of silencing with the appropriate timing. This is crucial for the inhibition of essential genes in stable transfections, in tissue culture or in animal models.

Tetracycline inducible promoters have been first generated by inclusion of a binding site for a tetracycline repressor (Tet operator or tetO) between the TATA box and the transcription start site of an H1 promoter (van de Wetering *et al.*, 2003; Wiznerowicz and Trono, 2003; Matthess *et al.*, 2005; Ngo *et al.*, 2006; Hoeflich *et al.*, 2006). Binding of the repressor (tetR) to the tetO interferes with TATA function and blocks transcription of the shRNA and silencing. Addition of the inducer, tetracycline or doxycycline, leads to dissociation of the repressor, transcription and silencing. Similar strategies have been used to generate tetracycline inducible U6 (Czauderna *et al.*, 2003; Matsukura *et al.*, 2003) or 7SK (Czauderna *et al.*, 2003), promoters or IPTG inducible H1 promoters, which contain *lac* repressor binding sequences (Higuchi *et al.*, 2004) (Table 11.1).

Comparisons among existing methods have not been carefully carried out and the general belief is that they will perform very similarly. After addition of the inducer, silencing is fast and target gene expression is detected again 3–4 days after removal of doxycycline in inducible U6 promoters (Matsukura *et al.*, 2003) or IPTG in inducible H1 systems (Higuchi *et al.*, 2004). A general problem of these systems is that pol III promoters are weakened upon the insertion of regulatory sequences, leading to suboptimal levels of shRNAs and less efficient silencing (Berns *et al.*, 2004). Also, as tetracycline inducible promoters are a little leaky, certain down-regulation of target genes could be detected in the absence of inducer when working with potent shRNAs (Lin *et al.*, 2004).

Tighter repressions have been described with U6 promoters modified to include two tet operators (Lin *et al.*, 2004) or with ecdysone-inducible systems (Gupta *et al.*, 2004). For ecdysone induction, the U6 enhancer has been replaced with a GAL4 element that responds to a GAL4-Oct-2Q transactivator in cells expressing certain heterologous transcription factors (Gupta *et al.*, 2004). Addition of ecdysone results in dimerization of the transgenic transcription factors, activation of the transactivator and expression from the U6 promoter. Target gene can be expressed again 2–4 days after removal of ecdysone (Gupta *et al.*, 2004). The main disadvantage of this system is the high number of components that need to be expressed in the cell to allow induction. Ecdysone inducible systems work well in tissue cultured cells (Gupta *et al.*, 2004) and are generally less active than tetracycline-based strategies, which also can be used to control expression in animal models (Matsukura *et al.*, 2003; Pluta *et al.*, 2007; Amar *et al.*, 2006; Aagaard *et al.*, 2007).

shRNAs delivered from pol III promoters can also be regulated in mice in a cell specific manner when combined with the Cre-based system (Table 11.1). Cre allows recombination of loxP sequences that depending on the system used, can lead to silencing or target gene expression. To create a Cre inducible silencing system, a loxP flanked stuffer sequence has been inserted between the pol III promoter and the shRNA coding sequence, between the distal (DSE) and proximal (PSE) U6 promoter elements or between the PSE and the TATA box, so proper spacing for U6 transcription is disrupted (Kasim *et al.*, 2004; Tiscornia *et al.*, 2004; Coumoul *et al.*, 2004, 2005; Ventura *et al.*, 2004; Chang *et al.*, 2004; Yu and McMahon, 2006). Upon Cre induction, loxP recombination deletes the stuffer and leaves the promoter active or the shRNA sequence in a proper position for pol III transcription. This results in an efficient expression of shRNA both in cultured cells (Tiscornia *et al.*, 2004) and in mice (Ventura *et al.*, 2004). Also, the stuffer sequence could contain a reporter cassette, so reporter expression correlates with expression of the target gene. To create a Cre-mediated inactivation of silencing, the loxP sites flank a pol III-shRNA expression cassette that is removed upon Cre expression (Ventura *et al.*, 2004; Meissner and Jaenisch, 2006). If a reporter cassette is also included between the loxP sites, reporter expression should correlate with silencing (Ventura *et al.*, 2004). These approaches can work in mice as inducible and tissue specific systems, as a great collection of transgenic mice exists that express Cre in a conditional and/or tissue-specific manner. The major disadvantage of Cre-mediated mechanisms is the lack of reversibility, as it involves genomic rearrangements.

*Polymerase II promoters*

The use of pol III promoters has several advantages such as the small size of the entire expression cassette and their high activity in all cell types. This generally results in a strong silencing both in tissue cells and in animal models. Conversely, unless cre-lox systems are used, pol III promoters lack tissue-specificity, which is required for some applications *in vivo*. Also, some authors claim that pol III-expressed shRNAs could induce interferon response, as the triphosphate linked to their 5′ end could activate retinoic acid-inducible protein RIG-I, an inductor of interferon system (Hornung *et al.*, 2006; Nallagatla *et al.*, 2007). To circumvent these problems, pol II promoters have also been developed to express siRNAs (Table 11.1 and Fig. 11.2). This has not been straightforward as pol II promoters often have several transcription start sites and produce transcripts with a 5′ cap and a 3′ poly-A tail. However, shRNAs need to initiate precisely and terminate in two nucleotide overhangs so they can be efficiently recognized by Dicer for further processing. To generate precise ends, *cis*-acting hammerhead ribozymes have been used in combination with pol II promoters (Allen *et al.*, 2007). Alternatively, proper processing has been achieved with pol II synthesized miRNA-like transcripts (Dickins *et al.*, 2005) (Fig. 11.2). This approach has been successfully used as mature miRNAs, similar to siRNAs, result from pol II transcribed pri-miRNA precursors. pri-miRNAs should contain naturally occurring signal sequences that direct Drosha processing to yield pre-miRNAs, similar to shRNAs, that have the proper ends to be then cleaved by Dicer into siRNAs. Drosha binding protein DGCR8 recognizes the junction of the single-stranded regions of the pri-miRNA and the pre-miRNA sequences and Drosha cuts about one helical turn into the stem–loop structure (Han *et al.*, 2006). This indicates that the sequences flanking the pre-miRNAs and the hairpin structure, but not the exact stem sequence, are clearly important for correct miRNA processing (Zeng and Cullen, 2005). Thus, miRNA-based siRNAs could be expressed from pol II promoters that transcribe a pre-miRNA structure containing the siRNA sequences plus 30–50 nucleotides-long 5′ and 3′ end extensions derived from the wild-type miRNA gene (Zeng *et al.*, 2002) (Figs. 11.2 and 11.3). Even when these rules are followed, to date it is still unclear whether transcripts based on any miRNA can be efficiently processed to siRNAs in all cell types.

Several pol II promoters have been used to express siRNAs, including inducible (Unwalla *et al.*, 2004), cell specific (Snyder *et al.*, 2008) or strong promoters expressed in all cell types such us U1 promoter (Denti *et al.*, 2004) or cytomegalovirus (CMV) and other promoters of viral origin (Stegmeier *et al.*, 2005, Snyder *et al.*, 2008) (Table 11.1). Interestingly, an expression cassette containing a pol II tetracycline regulatable promoter has been inserted into a constitutively active region of the cell genome through Cre-mediated site-specific recombination (Wang *et al.*, 2007). The major advantage of this system is that shRNAs are inducibly expressed from a precise locus using a viral-free system to control the expression, for instance, in ES cells (Wang *et al.*, 2007). Conversely, this system is irreversible.

## Inhibitory RNAs

### *Standard shRNAs*

A shRNA is an RNA molecule that contains sense and antisense stem sequences of 19–29 nucleotides connected by a short loop of 4–19 nucleotides (Brummelkamp *et al.*, 2002b; Paddison *et al.*, 2002) (Fig. 11.3). The sequences of both the stem and the loop have to be chosen carefully to obtain efficient and specific silencing. A 4-nt loop (5′-UUCG-3′) supports significant silencing (Paul *et al.*, 2002), but 9-nt loops such as 5′-TTCAAGAGA-3′ tend to be better (Brummelkamp *et al.*, 2002b). Probably, the best option is to use natural miRNA loop sequences. Seven miRNA derived loop sequences were shown to have similar or greater silencing than similar constructs with standard 9-nt loops (Boden *et al.*, 2004). This seems to be more important when the copy number of the template is low. Recommended loops come from miR-23 (5′-CTTCCTGTCA-3′)

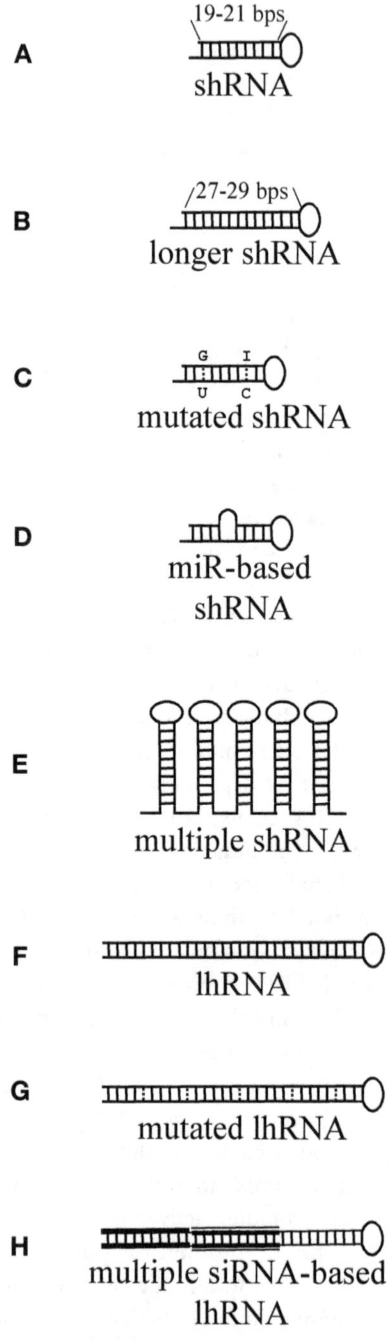


**Figure 11.3** Schematic of different classes of inhibitory hairpin RNAs. (A) Short hairpin RNAs with stems of 19–21 basepairs (bp). (B) Longer short hairpin RNAs with stems of 27–29 bp. (C) Short hairpin RNAs with mutations to decrease the thermodynamic stability. D) Short hairpin RNAs that mimic miRNAs. (E) Tandem short hairpin RNAs. (F) Long hairpin RNAs with stems longer than 30 bp. (G) Long hairpin RNAs with mutations to decrease the thermodynamic stability. (H) Tandem siRNAs that form a long hairpin RNA.

(Kawasaki and Taira, 2003) and miR-30 5′-CUGUGAAGCCACAGAUGGG-3′, which can be shortened to 5′-UAAGCCACAGG-3′ and still retain functionality (Zeng and Cullen, 2004). This may explain the better efficacy of miRNA-based shRNA inhibitors compared with first-generation shRNAs (see below) (Silva *et al.*, 2005).

More important than the loop sequence is the stem sequence. In fact, one of the most critical steps of RNAi technology is the selection of target sites. Normally, several target sites must be tested before an effective site is found. Even more, siRNAs with maximum potency and specificity should be identified, so they can be efficient silencers at low concentrations which will decrease negative side-effects. To identify target sequences with increased possibilities of being silenced by RNAi, several criteria should be followed. Early protocols suggested the choice of a target 19-bp unique sequence with a 45–55% G:C content after AA or NA nucleotides (where N represents any nucleotide). Hybridization of this sequence with the antisense sequence followed by 2-nt 3′ overhangs would form the stem of the shRNA (Elbashir *et al.*, 2001a). Currently, the probability of designing good shRNAs has increased with the use of more stringent algorithms built from the analysis of a large number of effective and ineffective siRNAs and a better understanding of the molecular mechanisms that drive RNAi (Reynolds *et al.*, 2004; Amarzguioui and Prydz, 2004; Takasaki *et al.*, 2004; Yuan *et al.*, 2004; Chalk *et al.*, 2004; Saetrom and Snove, 2004; Santoyo *et al.*, 2005; Boese *et al.*, 2005). Algorithms to design siRNAs and support for siRNA design can be found at http://www.dharmacon.com, http://www1.qiagen.com/siRNA, http://sfold.wadsworth.org, http://www.ambion.com/techlib/misc/siRNA finder.html, http://molecula.com/new/siRNAinquiry.html, http://web.mit.edu/mmcmanus/www/home1.2files/siRNAs.htm, http://hydra1.wistar.upenn.edu/Projects/siRNA/siRNAindex.htm, http://www.broad.mit, edu/genome_bio/trc/rnai.html, http://www.cshl.edu/public/SCIENCE/hannon.html, Deqor (Henschel *et al.*, 2004) and siDirect (Naito *et al.*, 2004). Alternatively, the following rules can be used to increase the probability of designing efficient siRNAs:

1. GC content of the target sequence should range between 30% and 52%, as siRNAs with higher GC content may have difficulties unwinding, while siRNAs with lower GC content may interact less well with the target sequence (Reynolds *et al.*, 2004; Boese *et al.*, 2005).
2. siRNAs should lack internal secondary structures that could interfere with the formation of the stable siRNA duplex or binding between the guide strand with RISC or with the target RNA (Reynolds *et al.*, 2004; Boese *et al.*, 2005). Conversely, target sequences should also lack high order structures, and should not be occluded by protein binding, so targets can be accessible to bind the RISC complex loaded with the guide siRNA (Heale *et al.*, 2005; Overhoff *et al.*, 2005; Westerhout *et al.*, 2005; Schubert *et al.*, 2005; Brown *et al.*, 2005; Luo and Chang, 2004; Reynolds *et al.*, 2004).
3. siRNAs should have lower thermodynamic stability at the 5′ end of the guide strand and higher thermodynamic stability at the 5′ end of the passenger strand, which will favour uptake of the guide strand into RISC complex (Schwarz *et al.*, 2003; Khvorova *et al.*, 2003). This has been observed for naturally occurring miRNAs. Also, effective siRNA tend to have at least 3 A/U residues at position 15–19 of the passenger strand (Schwarz *et al.*, 2003; Khvorova *et al.*, 2003; Reynolds *et al.*, 2004; Boese *et al.*, 2005). This increases the potency of siRNAs almost 100-fold, so it allows functionality at lower doses, which decreases off-target effects. Also, this decreases the possibility of loading into RISC the passenger strand, which could act as a guide sequence to silence unwanted genes.
4. siRNAs should have a U residue at position 10 of the sense strand, which is the favourite residue for Argonaute cleavage (Reynolds *et al.*, 2004; Boese *et al.*, 2005).
5. Efficient siRNAs tend to have an A residue at position 3 and 19 and should not have G at position 13 of the sense strand. Tracks of G:C seem deleterious while several A:U are favoured. This could be important for decreased probability of secondary structures,

target binding, cleavage or later recycling of RISC (Reynolds *et al.*, 2004; Boese *et al.*, 2005).

6 siRNAs should not contain sequences that activate the interferon response. Motifs such as 5′-UGUGU-3′ or 5′-GUCCUUCAA-3′ activate Toll-like receptor 7 (Yeung *et al.*, 2005; Karpala *et al.*, 2005; Judge *et al.*, 2005; Hornung *et al.*, 2005).

7 Target sequences should be unique. Thus, Blast searches should be performed to avoid sequences that are shared by other mRNAs. Ideally, the target sequence that binds nucleotides 2–7 of the guide strand, known as the 'seed' sequence, should also be unique or at least, should not be shared by the 3′-UTR of other mRNAs different than the target. The seed sequence could guide the siRNA-loaded RISC complex to the 3′-UTR of mRNAs different than the target and induce 'off target' inhibition by activation of a miRNA-type-mediated silencing (Sapru *et al.*, 2006; Leonard and Schaffer, 2006; Lewis *et al.*, 2005; Boese *et al.*, 2005; Jackson *et al.*, 2006b; Birmingham *et al.*, 2006).

8 Other restrictions apply for polymerase III transcribed shRNAs. No more than three consecutive T nucleotides can be present as they will constitute a premature termination signal. Also, the first nucleotide of the template should be a G (mostly for U6 promoters) or an A (for H1 promoters) as the polymerase requires initiation of transcription with a purine. This does not represent a serious limitation, since a G in the first position of the sense strand of siRNA correlates positively with siRNA functionality.

Unfortunately, siRNA design algorithms are imperfect and even when all the rules are followed there is no guarantee that efficient sequences will be identified. Also, identification of good target sequences for synthetic siRNAs cannot always be translated into good target shRNA sequences, which reduces the use of screens with synthetic siRNAs to identify good shRNA targets. Thus, many shRNAs should be tested per gene to identify those with high potency. For this, a PCR-based cloning strategy can be employed (Castanotto *et al.*, 2002). The PCR reaction is performed with a forward primer that contains polymerase III promoter and hairpin sequences and a reverse primer with the target sense and terminator sequences. The PCR product can be used directly to transfect cells and evaluate silencing.

Once the shRNA sequences have been chosen, they are translated to DNA and normally synthesized in the form of two complementary oligos that are annealed and cloned downstream of the elected promoter and regulatory sequences. Generally, oligos are also designed to create appropriate overhangs for directional cloning after hybridization. The resulting plasmids have a high chance of being mutated upon introduction into *E. coli* (Inoue *et al.*, 2006). The reason is that the DNA structure for shRNA expression has inverted repeats, which recombine easily in common bacterial strains. Therefore it is advisable to employ strains impaired in recombination such as *E. coli* SURE (recB recJ sbcC umuC uvrC) or GT116 (recA1ΔsbcC-sbcD). The use of longer loop sequences could facilitate cloning as it would separate the inverted repeats. However, a wiser approach has been to introduce mutations in the sense strand of the shRNA sequence (Inoue *et al.*, 2006; Miyagishi *et al.*, 2004) (Fig. 11.3). Recommended mutations can be mild C to T, A to G changes that will result in G–U wobbles or G to inosine substitutions that will yield I–C interactions, both with a similar energy to Watson–Crick A–U pairs. However, mutations that lead to mismatches or bulges in the shRNA could also be allowed, as these often occur in natural pre-miRNA structures (Fig. 11.3). This mutation approach has several advantages. On one hand the resulting plasmids decrease the rate of deletions in bacteria without losing silencing activity, because as the guide strand sequence is not altered, target specificity is not compromised. Moreover, when the mutations are located at nucleotides complementary to positions 2–4 of the 5′ end of the antisense strand, thermostability will be decreased and therefore efficiency increased. Finally, mismatches will diminish the double-strand nature of the structure decreasing possibilities of interferon induction.

Once a set of plasmids has been constructed that express shRNA inhibitors against a single target gene, silencing efficiency should be checked in tissue cultured cells. Several issues need to be considered such as transfection efficiency of the cells and stability of the protein encoded by the target gene. Ideally, transfection efficiency should be higher than 80–90% and inhibition should be evaluated 3 days after transfection by Western blot, RT-PCR or Northern blot. When the target gene encodes a very stable protein or the cells are not easily transfected, a plasmid expressing a reporter containing the target sequences or an epitope-tagged version of the protein encoded by the target gene, could be co-transfected with the shRNA expressing plasmid. As co-transfection efficiency is very high, following this protocol all cells that express the inhibitor will also express a recently synthesized target sequence. Target silencing can then be evaluated without considerations of target stability or transfection efficiency.

Several efficient inhibitors of the target gene should be used for gene function or therapeutic studies. Inhibitors with similar silencing potencies should give the same phenotypic outcomes; otherwise the response could be caused by inhibition of 'off-targets'. The best experiment is to reverse the phenotype upon expression of a functional version of the target gene, mutated to avoid silencing by the shRNA used. For therapeutic applications, high throughput experiments could be performed for the identification of off-targets and toxicity should be evaluated at different doses of the shRNA.

Other than for single gene function studies and therapeutic applications, shRNA expressing vectors can be used for high throughput RNAi screens, aimed at identify genes involved in specific functions. This is done using libraries of inhibitory shRNAs (Fewell and Schmitt, 2006; Moffat and Sabatini, 2006). Three major libraries have been described produced by Hannon-Elledge labs (Silva *et al.*, 2005; Chang *et al.*, 2006), the Netherlands Cancer Institute (NKI) (Berns *et al.*, 2004) and the RNAi consortium (TRC) (Moffat *et al.*, 2006; Root *et al.*, 2006). All express a resistance gene and one or more shRNAs (NKI, TCR) or miR30-based shRNAs (Hannon-Elledge) from H1 (NKI) or U6 (TCR, Hannon-Elledge) promoters against most human and/or mouse genes. These cassettes have been introduced in plasmids from retroviral (NKI, Hannon-Elledge) or lentiviral (TCR, Hannon-Elledge) background from which single copy insertions in transduced cells are enough to yield efficient knock downs. Viruses are produced to titres high enough so a single viral production can be used in many experiments, depending of the multiplicity of infection required for the cells of interest. Also, the vectors used in these libraries have been described to have low recombination rates. Libraries tend to have 5 inhibitors per gene and at least 40% of constructs result in inhibitions higher than 70%, leading to an average of 2 efficient inhibitors per gene. Other advantages of some of the libraries include easy transfer to other vectors with, for instance, tissue-specific or inducible promoters, which will be essential to produce the viruses that express functional shRNAs that target genes essential for cell viability or viral production. The libraries also include bar-codes to identify shRNAs against genes that perform specific functions using high throughput hybridization approaches.

## Alternative shRNAs

### *Longer shRNAs*

Synthetic siRNAs of 27–29 nt are 10- to 100-fold more efficient than the conventional 21-nt versions (Kim *et al.*, 2005; Siolas *et al.*, 2005). This has been explained by the fact that longer siRNAs could be processed by Dicer to generate several siRNAs and Dicer interaction could favour proper strand loading into RISC and RISC loading efficacy. Interestingly, some shRNAs with longer stems have been also found to be more efficient (Tan *et al.*, 2007) (Fig. 11.3). Lower concentrations of the shRNA are required to achieve maximal inhibition of target. The molecular mechanism under this result is still unclear. Dicer recognizes the free ends of the shRNA to cleave 21–22 nt further, so longer and shorter shRNAs should generate identical siRNAs (Siolas *et al.*, 2005). However, it has been suggested that Dicer could process longer stems more efficiently.

### *Multiple shRNAs*

There are several examples of pri-miRNAs that encode several miRNAs. The same strategy has been used to express simultaneously different shRNAs, against several regions of the same target or against different targets (Zhu *et al.*, 2007) (Fig. 11.3). As was the case with single miRNA-based shRNAs, the miRNA sequence is substituted by the siRNA sequence while the rest of the polycistron is kept identical. The best studied example involves modifications of the miR-17–92 polycistron, which expresses five different miRNAs, to yield five shRNAs against human immunodeficiency virus (HIV) (Liu *et al.*, 2008; Aagaard *et al.*, 2008). This strategy aims to prevent viral escape against single shRNAs by selection of point mutations in the target sequence.

*lhRNAs*

In plants, worms and flies, long hairpin RNAs (lhRNAs) are processed by Dicer to siRNAs and as a result they have been used to induce silencing. However, in most mammalian cells lhRNAs longer than 30 bp have been described to induce the IFN response, relegating the use of lhRNAs to mouse embryos, oocytes and, possibly, ES and tumour cells which lack interferon (IFN) response (Yang *et al.*, 2001). To avoid IFN response, lhRNAs have been modified to include multiple specific point mutations in the sense strand, similar to what has been done with shRNAs (Fig. 11.3). This allows lhRNAs longer than 60 bp to lead to RNAi without induction of the IFN system (Akashi *et al.*, 2005; Watanabe *et al.*, 2006a; Nishitsuji *et al.*, 2006; Weinberg *et al.*, 2007; Sano *et al.*, 2008). Target silencing was evaluated in lhRNAs of 62 basepairs containing multiple G–U wobbles (Weinberg *et al.*, 2007). The highest silencing efficacy was observed for targets that are complementary to sequences located at the base of the hairpin stem and this correlated with a higher concentration of siRNAs derived from this region of the lhRNA (Weinberg *et al.*, 2007). This work suggests that Dicer processing starts in the base of the stem and proceeds towards the loop with decreasing efficacy. Similar findings have been observed with lhRNAs resulting from the fusion of three efficient siRNAs that target non-continuous sites (Saayman *et al.*, 2008; Liu *et al.*, 2007). Silencing was primarily mediated by the siRNA processed from the base of the stem. Improvements that consisted of adjusting spacing at siRNAs junctions only resulted in functionality of siRNAs located in the first and second position (Saayman *et al.*, 2008). Surprisingly, unmodified lhRNAs longer than 50 bp have also been reported to activate RNAi without IFN stimulation (Konstantinova *et al.*, 2006; Konstantinova *et al.*, 2007; Barichievy *et al.*, 2007). However, it may be difficult to produce viruses that express lhRNAs due to high frequency of recombination.

## Viral vectors expressing siRNAs.

The use of viral gene therapy vectors for shRNA delivery provides a number of advantages over other siRNA delivery systems: (i) viral vectors can infect certain cells with high efficacy, leading to efficient and tissue-specific delivery of shRNAs, (ii) higher specificity can also be achieved as shRNAs can be expressed from inducible or tissue-specific promoters, (iii) depending on the choice of vector, shRNAs can be expressed transiently or stably, (iv) viral vectors are cost-effective, and (v) the majority of all available viral vectors have already been assessed clinically in phase I safety trials.

The first step in the design of protocols using viral vectors that express shRNAs is the selection of vector. The characteristics of several viral vectors are summarized in Table 11.2. The small length of a typical RNAi cassette permits shRNA expression even from the smallest vectors, and thus permits vector choice based on other criteria. Each type of viral vector has inherent specific properties that need to be taken into consideration depending on the required duration and location of shRNA expression. Different vectors will allow a broad range of possibilities that go from transient expression to stable integrations that can lead to germ-line transmission and to the creation of *in vivo* animal models. Also, different vectors will infect only proliferative or also quiescent cells and will have different host range from wide to more specific. Transduction efficiency will also depend on the vector. There are cells, such as hepatocytes, which are efficiently transduced by several vectors, while others can only be infected very poorly. However, in most cases, gene function inhibition is only achieved when most of the cells of a given tissue express the inhibitory shRNA. Strategies to increase cell transduction have shown limited success. Alternatively, selection of transduced cells has worked efficiently when using proliferative cells such as stem cells.

The two major classes of viral vectors used for shRNA delivery are based on: (i) RNA viruses such as retrovirus and lentivirus, which integrate

**Table 11.2** Summary of viral vectors used in gene therapy trials and their characteristics

| Viral vector | Genome | Packaging capacity | Host range | Integration | Duration of expression | Immunogenicity | Gene therapy trials* |
|---|---|---|---|---|---|---|---|
| *Enveloped* | | | | | | | |
| Retrovirus | RNA | 8 kb | Dividing cells | Random. Risk of insertional mutagenesis | Stable | Weak | 320 |
| | | | | Highly efficient | | | |
| Lentivirus | RNA | 8 kb | Dividing and non-dividing cells | Random. Risk of insertional mutagenesis | Stable | Weak | 18 |
| | | | Certain tropism by pseudotyping | | | | |
| HSV | dsDNA | 40 kb (replication defective) | Dividing and non-dividing cells | Episomal | Transient expression in cells other than neurons | Potential for innate and adaptive immunity | 47 |
| | | 150 kb (amplicon) | Broad: strong for neurons, but also stem and muscle cells | | | | |
| *Non-enveloped* | | | | | | | |
| Adenovirus | Linear dsDNA | 8 kb | Dividing and non-dividing cells | Episomal | Transient | Innate and adaptative immunity | 367 |
| | | | Liver and other organs by targeting | | | Inflammatory response | |
| Gutless-Ad | Linear dsDNA | 30 kb | Dividing and non-dividing cells | Episomal | Long-term | Weak | |
| AAV | Linear ssDNA/ dsDNA | <5 kb | Dividing and non-dividing cells | Episomal (>90%) | Long-term/transient | Mild. Minimal innate response | 60 |
| | | | | Low integration (<10%) | | Adaptive immune response to capsid | |
| SV40 | circular DNA | up to 2.5 kb | Dividing and non-dividing cells | Random | Long-term | Weak | 1 |

* Updated in September 2008. *The Journal of Gene Medicine* Clinical Trials Database.

into the host cell genome, and (ii) non-integrating DNA viruses such as adenovirus, adeno-associated virus (AAV) and herpesvirus, whose genome is maintained episomically in the transduced cell. The retroviral and lentiviral vectors are useful for long-term applications in proliferating cells such as hematopoietic stem cells. Non-integrating vectors such as AAV or gutless adenovirus can be used to sustain persistent expression in quiescent cells while others, such as first generation adenovirus, are optimal when transient expression is required (Rossi, 2008).

### Oncoretrovirus

Oncoretroviral vectors were the first class of viral vectors to be developed and are typically based on the Moloney murine leukaemia virus (MLV). Generally, the MLV envelope glycoprotein is substituted by the G protein of the vesicular stomatitis virus (VSV-G), resulting in a more stable viral particle that allows concentration of the viral stocks to high titres (Burns *et al.*, 1993). VSV-G pseudotyped MLV vectors have also an extremely broad host range compared with unmodified MLV vectors. When the target cell is infected by the vector, the viral genome can only reach the nucleus when the nuclear membrane breaks down upon division. Therefore, oncoretrovirus can only transduce dividing cells. Once in the cell, the viral genome is copied by reverse transcription to a proviral DNA that is then integrated into the host genome. These processes are mediated by proteins carried in the viral particle. Thus, viral production is performed in packaging cells able to express the viral Gag, Pol and Env proteins. In this way, all the viral genes can be removed from the genome which only retains the long terminal repeats (LTRs) and nearby signals required for packaging the RNA genome into the viral particle, reverse transcription and integration. Viral titres obtained are lower than with other viral vectors such as adenoviruses.

Integration of the proviral DNA into the host genome results in sustained heterologous gene expression for extended time periods and stable gene transmission to subsequent cell generations. This great advantage is counterbalanced by the increased risk of insertional mutagenesis that may lead to carcinogenesis. MLV genomes have been shown to integrate preferentially within a few kbp at either side of the transcription start site of active genes in human cells (Wu *et al.*, 2003). This integration can occur near cellular protooncogenes leading to their aberrant expression from the viral LTR, or alternatively could cause the disruption of a tumour suppressor gene (Ketzinel-Gilad *et al.*, 2006). In fact, several patients with X-SCID (X-linked severe immune deficiency syndrome) treated with retroviral vectors developed leukaemia (Hacein-Bey-Abina *et al.*, 2003).

Retroviral vectors have been used successfully in several clinical trials. Retroviruses have been the vectors of choice for the *ex vivo* transduction of repopulating haematopoietic stem cells and have demonstrated promising results in the treatment of inherited, monogenic blood disorders such as the haemoglobinopathies and SCIDs. These cells could also be targeted by retrovirus that expresses shRNAs against disease-associated RNAs such as HIV transcripts. In fact, retroviruses were among the first vectors used to express shRNAs and they have the capacity to encode lhRNAs or several shRNAs per viral genome (Brummelkamp *et al.*, 2002a; Barton and Medzhitov, 2002; Devroe and Silver, 2002). However, safety concerns due to integration mutagenesis have decreased the general employment of retroviral vectors in gene therapy.

### Lentivirus

Most of the lentiviral vectors used in gene therapy are based on the HIV (Vigna and Naldini, 2000) and equine infectious anaemia virus (EIAV). These viruses are members of the retrovirus family and share many characteristics with oncoretroviruses. Recombinant HIV production is also based on packaging cells that express in *trans* the Gag and Pol HIV genes with (Klages *et al.*, 2000) or without (Kotsopoulou *et al.*, 2000) the Rev gene, in the absence of any HIV virulence genes (Tat, Vif, Vpr, Vpu, and Nef) to avoid HIV pathogenicity (Kim *et al.*, 1998). This procedure results in low viral titres compared to other viral vectors such as adenoviruses. Similar to oncoretrovirus, lentivirus vectors are often pseudotyped with VSV-G, Mokola or Ebola virus glycoproteins. This allows viral entry into haematopoietic or embryonic stem cells (Siapati *et al.*, 2005), muscle (Gregory *et al.*, 2004a), hepatocytes (Follenzi *et al.*, 2002), neurons and glia (Azzouz *et al.*, 2002; Mitrophanous *et al.*, 1999; Wong *et al.*, 2004).

Once the viral RNA genome reaches the cytoplasm, reverse transcription generates a cDNA that forms the pre-initiation complex (PIC) complexed with other viral factors. A nuclear localization signal within the *gag* matrix protein (Bukrinsky *et al.*, 1993) and a 99-nt central polypurine track (cPPT) located in the viral polymerase gene contribute to the transport of the PIC into the nucleus. This is a major difference with oncoretroviruses and allows lentiviral transduction of non-dividing cells.

Similar to oncoretrovirus, lentiviral cDNA genome integrates into the cell genome allowing stable expression of transgenes. However, while retroviruses integrate preferentially around the transcription start site of active genes, HIV integration sites tend to be uniformly distributed within transcription units, but not upstream (Mitchell *et al.*, 2004). Detailed studies will be required to determine whether this difference in the integration results in lower risk of insertional mutagenesis by lentiviral vectors. Another issue that should be clearly addressed is whether the risk of insertional mutagenesis only affects frequent dividing cells and not non-dividing cells, such as neurons, which can be targeted by lentiviruses. Biosafety has also been addressed by the development of self-inactivating (SIN) vectors that contain deletions of the regulatory elements in the downstream LTR, eliminating the transcription of the packaging signal that is required for vector mobilization (Zufferey *et al.*, 1998). Similar strategies have been also implemented with oncoretroviral vectors.

Lentiviral vectors have been also used to express shRNAs in tissue cultured cells or *in vivo*. As lentiviral vectors can accommodate 8 kbp of exogenous DNA, there is plenty space to include cassettes that express shRNAs and reporter genes. Vectors have been developed that express single or tandem shRNAs from different promoters, including pol II and pol III, tissue-specific or inducible promoters (Ventura *et al.*, 2004; Westerhout *et al.*, 2006b; Wiznerowicz and Trono, 2003; Tiscornia *et al.*, 2004). Thus, lentiviruses have been shown to mediate stable knock-down of different genes both in isolated cells and in animal models. Interestingly, transgenic 'knockdown' mice have been developed with lentiviral vectors, which allow persistent expression of transgenes (Tiscornia *et al.*, 2003; Rubinson *et al.*, 2003). The resulting knock-down animals may show different phenotypes than transgenic animals in which the same gene has been knocked out. In fact, knock-downs may be better to mimic diseases in which pathogenesis is linked to gene-dosage variation (Chen *et al.*, 2006b). However, lentiviral-mediated shRNA expression may not be useful to generate large knock-down animals, where epigenetic regulation has been described to silence lentiviral expression (Hofmann *et al.*, 2006).

The intrinsic advantages of lentiviral vectors make them ideal for *ex vivo* transduction of repopulating haematopoietic stem cells without first inducing them to proliferate with cytokine stimulation, as cell division is not required for transduction. In fact, a clinical trial has been started that involves *ex vivo* lentiviral vector delivery of shRNAs targeting HIV in haematopoietic stem cells collected from HIV infected patients. The transduced cells are then to be reinfused into these patients for therapeutic benefit *in vivo* (Morris and Rossi, 2006). However, it must be kept in mind that the clinical experience with this vector is very limited.

### Adenovirus

Adenoviruses are non-enveloped viruses with a linear dsDNA which ends in inverted terminal repeats (ITRs). Viral entry requires the interaction between capsid proteins and the CAR receptor, present in the membrane of many cells, or factor X to transduce hepatocytes (Mentel *et al.*, 1997; Waddington *et al.*, 2008). The presence of these factors determines susceptibility to infection by wild-type viruses, but altered tropism has been obtained with capsid modifications. Viral particles enter the cell by endocytosis and are released into the cytoplasm where capsid proteins destabilize and allow transport of the viral genome to the nucleus. Therefore, adenovirus infects both dividing and non-dividing cells. In the nucleus, viral DNA is maintained episomically and starts transcription using the cell machinery. Transcription of the wild-type virus initiates with expression of the early E1genes, which activate synthesis of viral genes and viral replication. First generation adenoviral vectors have been created by deletion of E1 and the non-essential E3 genes. This generates

8 kb for cloning exogenous genes and makes the virus replication-deficient. Viral vector production is performed in packaging cells that express E1 proteins in *trans* and leads to viral stocks of high titres ($10^{11}$–$10^{15}$ pfu/ml). Transduction with adenoviral vectors leads to efficient expression of transgenes in different organs including the liver, lung and brain. Therefore such vectors are very popular in gene therapy trials. However, they induce potent innate and adaptative immune responses that result in transient transgene expression, which disappears in 2–3 weeks, unless the transduced cells are in immune privileged sites or the animals are immunocompromised (Yang and Wilson, 1995). To reduce immune responses, adenoviral vectors have been mixed with polyethyleneglycol (PEG), which decreases protein–protein interactions and, therefore, immunostimulation (Mok *et al.*, 2005). However, cells transduced with first generation adenoviral vectors can still express low amounts of viral genes able to induce cytotoxic T-cell responses that eliminate transgene expressing cells. To avoid this, high capacity, gutless or helper-dependent adenoviral vectors have been developed that are deleted of all viral sequences except for ITRs and packaging signals (Oka and Chan, 2005), and therefore show low immunogenicity and long-term expression. However, production of gutless vectors to high titres is still an issue. Finally, adenoviruses have been modified to conditionally replicating adenovirus (CRAd), able to lyse host cells that have been used therapeutically for the treatment of cancer. Viral replication amplifies the input dose of CRAd resulting in viral spread throughout the tumour. Tumour cell death may result from viral lysis or from the activation of beneficial cytotoxic immune responses. Safety is achieved using viruses that only replicate in tumours by deletion of essential viral genes required to induce cell proliferation.

First generation adenovirus, CRAds and high-capacity gutless adenovirus have been used to express shRNAs in tissue cultured cells and *in vivo*. Despite the success obtained with these vectors, some concerns have also been raised. First generation adenoviral vectors are not useful for the inhibition of chronic disease-related genes, as they will express inhibitory shRNAs only for short-term and readministered virus will be sequestered by the immune system. In fact, adenovirus-mediated RNAi is detected in mouse liver at 5 days post injection and functions to detectable levels until day 21 post injection (Narvaiza *et al.*, 2006). Also, first-generation adenovirus express viral genes to low levels, including virus associated (VA) RNAs. VA RNAs have been described to saturate the cellular silencing machinery, leading to toxicity (Cullen, 2006; Andersson *et al.*, 2005; Xu *et al.*, 2007). However, functional inhibitory exogenous shRNAs have been expressed from first-generation adenoviral vectors without saturation of the cellular silencing machinery (Narvaiza *et al.*, 2006), suggesting that the levels of VA RNAs express from these vectors are not high enough to saturate the RNAi pathway. Gutless vectors do not express VA RNAs and they can express shRNAs for long periods, but hairpin RNA structures could jeopardize genetic vector stability (Huang *et al.*, 2007a). In spite of these concerns, adenoviral-mediated target gene inhibition has been observed in several tissues including mice central nervous system after local stereotactic injection (Xia *et al.*, 2002), rat lung epithelial cells after intratracheal application (Gou *et al.*, 2004), lung alveolar epithelial type II cells after intrabronchial administration (Gou *et al.*, 2004) and liver after systemic administration (Xia *et al.*, 2002; Ragozin *et al.*, 2005; Narvaiza *et al.*, 2006; Carmona *et al.*, 2006; Uprichard *et al.*, 2005; Huang *et al.*, 2004; Morral *et al.*, 2002; Rauschhuber *et al.*, 2008; Ruiz *et al.*, 2009). First-generation adenoviral vectors and CRAds have also been used to target genes required for tumorigenesis, resulting in tumour growth reduction (Li *et al.*, 2005a; Zhang *et al.*, 2005; Yoo *et al.*, 2007). Also, gutless vectors expressing functional shRNAs have been produced that silence the expression of target genes for several weeks (Ruiz *et al.*, 2009; Morral *et al.*, 2002). Care should be taken as high doses of gutless vectors do not result in higher silencing and may lead to activation of IFN response (Witting *et al.*, 2008).

## Adenoassociated virus

Adenoassociated viruses (AAVs) are non-enveloped single-stranded DNA viruses that require adenovirus or herpesvirus helper virus functions to replicate. The AAV genome contains two ITRs and two viral genes – *rep* and *cap*

– which are deleted in AAV-based vectors. This generates 4.9 kbp for cloning exogenous genes (Samulski *et al.*, 1989). For viral entry, different AAV serotypes use different cell receptors and therefore infect different tissues with different efficiency. AAV5 has tropism for retina, brain and airway epithelial cells (Rabinowitz *et al.*, 2002; Zabner *et al.*, 2000), AAV1 for skeletal muscle and retina, AAV6 for skeletal muscle (Blankinship *et al.*, 2004), and AAV8 for liver (Davidoff *et al.*, 2005; Nathwani *et al.*, 2005). Once inside the cell, the single-stranded DNA genome has to be converted into double-stranded. This is a limiting step for viral transduction that has been overcome by the development of double-stranded vectors, able to express 10- to 100-fold higher levels of transgenes. The counterpart is that the space for cloning exogenous genes is reduced to 2.4 kbp (Mitani *et al.*, 1995). Viral genome can move to the nucleus where the rep viral protein mediates wild-type AAV virus integration in a specific position. AAV-based vectors do not encode Rep, so integration is very inefficient and occurs randomly at sites of dsDNA breaks in the host genome (Miller *et al.*, 2004a). Most of the genome persist as concatemers formed by head-to-tail recombination, fluxing between scarce linear and abundant circular forms (McCarty *et al.*, 2004). This results in AAV vectors that can transduce dividing and non-dividing cells and express transgenes long term.

AAV viruses are optimal vectors for RNAi delivery because of the small size of their viral genome, which is ideally suited for packaging a single or several shRNA expression cassettes, without the necessity of 'stuffer' sequences. Several AAV vectors expressing shRNAs have been used *in vivo* targeting brain and liver genes. Gene expression inhibition for several weeks has been reported after direct injection in the brain tissue (Davidson and Boudreau, 2007; Ketzinel-Gilad *et al.*, 2006), midbrain neurons (Leung and Whittaker, 2005; Hommel *et al.*, 2003), after bilateral injection into the paraventricular nucleus of the hypothalamus (Garza *et al.*, 2008), or after microinfusion into the rat hippocampus (Babcock *et al.*, 2005), among others. Also, AAV-based vectors have been used to express shRNAs against hepatic targets, including hepatitis B virus (HBV) and hepatitis C virus (HCV) (Grimm *et al.*, 2006; Carmona *et al.*, 2006; Uprichard *et al.*, 2005; Zhang *et al.*, 2004a; Moore *et al.*, 2005). Interestingly, AAV-mediated expression in mouse liver of 49 different shRNAs and longer shRNAs (with stems of 27 bp) against six individual targets, resulted in liver injury in 36 cases and animal death in 23 cases. Morbidity was associated with saturation of the cellular silencing machinery due to shRNA overexpression. Optimization of the shRNA dose and sequence lead to persistent and therapeutic silencing of HBV in mice (Grimm *et al.*, 2006).

### Herpes simplex virus

Herpes simplex virus (HSV) is an enveloped virus with a large (152 kbp) linear double-stranded DNA genome. HSV spreads by contact, infects and replicates in skin or mucous membrane. Viral particles released in the skin enter the axons of sensory neurons innervating the affected area. Then, viral components, including the genome and the nucleocapsid, are retrogradely transported towards the neuronal cell body. There, the nucleocapsid docks at the nuclear pore complex and viral DNA enters the nucleus. Thus, HSV vectors can transduce dividing and quiescent cells. Once in the nucleus, the virus may enter into a latent state for the lifetime of the host where viral DNA persists episomically, unless certain factors such as immunosuppression cause viral reactivation. Reactivation leads to a progeny that is anterogradely transported back to the site of initial inoculation. Viral particles are released in the axon to infect epithelial or mucosal cells causing cell lysis and a lesion at the site of initial inoculation (Frampton *et al.*, 2005).

Prior to viral replication, HSV transcribes immediate-early (IE) genes that activate first the expression of early genes, involved in viral replication, and then late genes that encode for structural proteins. Recombinant HSV (HSV-1) vectors have been created by deletion of multiple essential IE genes. This disruption impairs expression of early and late genes and viral replication. These recombinant viruses can be propagated to high titters in packaging cells that express *in trans* the required IE genes (Frampton *et al.*, 2005). Gutless or HSV amplicon vectors have also been developed where the viral genome only contains the HSV origin of replication and packaging signals (Saeki *et al.*, 2001). These HSV vectors retain the targeting properties and the ability for axonal transport found in the wild-type virus and have been widely used for long-term expression of therapeutic genes in the peripheral and central nervous system. Other cells have been also successfully transduced with HSV vectors such as skeletal muscle, stem cells or hepatocytes. Malignant gliomas and other tumours have also been targeted by oncolytic replication competent HSV vectors developed by deletion of viral essential genes with homology to cellular genes constitutively expressed in cancer cells (Young *et al.*, 2006).

HSV vectors have been also used to express shRNAs that target reporter or endogenous genes in the dorsal root ganglia or in hippocampus neurons in mice (Anesti *et al.*, 2008; Hong *et al.*, 2006). Interestingly, HSV delivered shRNAs have decreased amyloid-beta accumulation in mouse brain, which could be used as a therapy for Alzheimer's disease (Hong *et al.*, 2006).

### Others

Baculoviruses offer a number of advantages for gene delivery in mammalian cells. These include: (i) space for multiple exogenous genes or large inserts because of the huge and flexible viral genome (130kb), (ii) biosafety, as baculovirus do not infect humans, (iii) polyhedrin and p10 promoters allow high yields, (iv)) absence of toxicity and technical simplicity. However, cell entry of baculovirus could be a problem for *in vivo* gene delivery to polarized epithelial cells, since it has been seen that baculoviruses access hepatocytes through the basolateral surface. Virus pseudotyping has led to an increased binding of virus to cells, but not to an increased gene expression as compared to unmodified virions. Also, baculoviruses are rapidly inactivated by human serum complement (Hofmann and Strauss, 1998). This limitation can be overcome by coinjection of a soluble form of complement receptor type I with the baculovirus, by pseudotyping with VSV-G protein, or by using a route of administration that evades the complement system. For example, stereotaxic injection of virus into the striatum and vitreous body of rats resulted in transgene expression in neuronal cells (Huser and Hofmann, 2003). Other than to the brain, these vectors have been used for gene delivery to many primary cells, including primary rat chondrocytes, mouse primary kidney cells, human osteosarcoma cells, and human mesenchymal stem cells (Condreay *et al.*, 1999; Ho *et al.*, 2004; Aagaard and Rossi, 2007; Liang *et al.*, 2004; Ho *et al.*, 2005). Baculovirus have been used to express functional shRNAs against reporter genes in rat brain (Li *et al.*, 2006), shRNAs targeting HCV (Suzuki *et al.*, 2005) or lamin A/C in human cells lines and primary hepatic stellate cells (Nicholson *et al.*, 2005).

Simian Virus 40 (SV40) vectors have a small double-stranded circular DNA genome of 5.2 kb, allow long-term expression of transgenes and target a broad range of susceptible cell types, including resting and dividing cells such as haematopoietic cells, bone marrow cells, liver, kidney and lung (Vera and Fortes, 2004). SV40 avoids host immune response, entering into cells by caveolar endocytosis, followed by vesicular transport to the endoplasmic reticulum and not via the endosomal–lysosomal pathway used by most viruses. Thus, SV40 has been shown to be non-immunogenic in mice, allowing repeated administrations (Kimchi-Sarfaty *et al.*, 2002). SV40 vectors have been successfully used in several therapeutically relevant animal models but, to date, there has been no clinical experience with these vectors. SV40 vectors can be produced in cells or *in vitro* by packaging the DNA with recombinant capsids. Both systems have been used to produce SV40 vectors that express functional shRNAs and longer RNA hairpins (Kimchi-Sarfaty *et al.*, 2005; Strayer *et al.*, 2005, our unpublished results).

## Therapeutic applications of viral vectors expressing shRNAs

Many diseases are caused by the expression of one or few toxic genes whose silencing could lead to the cure of the disease. Thus, RNAi could be successfully used in the treatment of cancer, viral infections, dominant genetic disorders or other diseases where toxicity results from overexpression of a cellular gene or from expression of a modified version of the cellular gene.

## Cancer

Solid tumours are complex mixtures of different cells that result from the successful acquisition of uncontrolled growth. Thus, several pathways are activated that affect cell cycle regulation, decrease cell senescence and apoptosis and lead to proper interactions with the host that result in angiogenesis and immunosuppression. Further, tumour cells are quick to adopt modifications to resist toxic insults such as chemotherapy and radiotherapy. In this environment, the final goal of cancer treatments is not the reversion of the malignant state in each tumour cell, but the regression of the whole tumour. RNAi-based cancer therapies seem promising as they could lead to the inhibition of the expression of one or several genes that are essential for tumour growth. However, before a successful therapy is achieved, there are several issues that need to be addressed: target genes essential for tumour growth need to be identified and siRNAs against the desired targets need to be delivered to most tumour cells or, at least, to the tumour stem cells.

Many studies have used RNAi as an experimental tool to understand the cellular pathways that lead to tumour formation and growth. The genes involved in these pathways whose expression is altered in tumour cells are good candidates for RNAi-based therapies. These include oncogenes, which are involved in signal transduction pathways that lead to proliferation by decreased cell death and increased division. Examples of oncogenes whose down-regulation by viral vector-delivered siRNAs have led to decreased tumour growth in animal models include Ras and Raf (Table 11.3). Promising results have also been obtained in the preclinical evaluation of the antitumoral effect of viral-vector delivered siRNAs against genes involved in cell cycle regulation, apoptosis, senescence, and angiogenesis (Table 11.3).

However, in most cancers, inhibition of a single pathway is not sufficient to eradicate the disease, and it is necessary to block multiple genes simultaneously or combine gene silencing with other therapies. RNAi-based therapies can be developed to silence several targets in the same cell by combinatorial expression of different shRNAs from the same viral vector. Genes to be silenced have to be chosen so their combined down-regulation results in additive or synergistic therapeutic effects. An interesting alternative to this approach is the expression of tumour suppressor miRNAs. Some cellular miRNAs are important regulators of genes involved in cell proliferation and differentiation and can function as oncogenes (He *et al.*, 2005) or as tumour suppressors (Johnson *et al.*, 2005; Chan *et al.*, 2005). Expression of a single cellular tumour suppressor miRNA from a viral vector should lead to the down-regulation of one or several genes involved in tumour generation and maintenance. Also, combination of RNAi-mediated silencing of tumour genes and conventional therapies such as ionizing radiation or imatinib, has resulted in enhanced antitumoral effects (Table 11.3). Interestingly, the use of oncolytic viral vectors to deliver siRNAs is a combinatorial antitumoral approach on its own. Conditionally replicating oncolytic viruses should transduce tumour cells where they propagate leading to viral amplification, cell lysis, and immune activation. siRNAs expressed from oncolytic vectors could potentiate their antitumoral response when the siRNAs target cellular genes that result in increased viral amplification or immune activation.

In spite of the success obtained so far in preclinical studies with RNAi-based antitumour therapy, translation to the clinical arena remains distant. Most tumour growth inhibitions have been observed in tissue cultured cells. Efficient antitumoral responses have also been obtained when cells transduced *in vitro* are then inoculated *in vivo* in mice. However, therapeutic effect should be observed in already established tumours before clinical applications could be feasible. This is a difficult goal as vector delivery within solid tumours is not efficient. Surprisingly, intratumoral injection of viral vectors delivering siRNAs against single targets has also resulted in inhibition of tumour growth (Table 11.3).

## Neurological diseases

Many neurological diseases are caused by the expression of mutated or misfolded toxic proteins or by the hyperaccumulation of proteins that leads to neurotoxicity. These are ideal settings for the application of RNAi-based therapies which could target the toxic genes to decrease their expression. Importantly, specific siRNAs can be selected that

**Table 11.3** Summary of viral vector delivered siRNAs targeting genes relevant for tumour growth.

| Pathway | Target gene | Vector | Administration | *In vivo* result | References |
|---|---|---|---|---|---|
| Oncogenesis | K-ras | Retrovirus | *Ex vivo* transduction | Loss of tumorigenicity | Brummelkamp *et al.* (2002a) |
| | H-ras | Retrovirus | *Ex vivo* transduction | Inhibition of tumour growth | Yang *et al.* (2003) |
| | B-RAF | Lentivirus | *Ex vivo* transduction | Inhibition of tumour growth and invasive ability of melanoma | Sumimoto *et al.* (2004) |
| | PTTG1 | Adenovirus | *Ex vivo* transduction | Inhibition of tumour growth | Cho-Rok *et al.* (2006) |
| | TEL-PDGFβR | Retrovirus | *Ex vivo* transduction + imatinib or rapamycin | Inhibition of tumour growth | Chen *et al.* (2004) |
| | HIF-1α | Adenovirus | Intratumoral injection + ionizing radiation | Inhibition of tumour growth | Zhang *et al.* (2004b) |
| | C-met | Adenovirus | Intratumoral injection | Inhibition of tumour growth | Shinomiya *et al.* (2004) |
| | Gankyrin | Adenovirus | Intratumoral injection | Inhibition of tumour growth | Li *et al.* (2005a) |
| | K-ras | CRAd | Intratumoral injection | Inhibition of tumour growth | Zhang *et al.* (2006) |
| | EZH2 | Lentivirus | Intratumoral injection | Inhibition of tumour growth | Chen *et al.* (2007) |
| Apoptosis | Bcl-2 | Lentivirus | *Ex vivo* transduction+ S TRAIL | Induction of apoptosis | Kock *et al.* (2007) |
| | Survivin | Adenovirus | Intratumoral injection | Inhibition of tumour growth | Uchida *et al.* (2004) |
| Cell cycle regulation | E6, E7. | Lentivirus | *Ex vivo* transduction | Inhibition of tumour growth | Gu *et al.* (2006) |
| | Hec1 | Retrovirus | *Ex vivo* transduction | Inhibition of tumour growth | Gurzov and Izquierdo (2006) |
| | Hec1 | Adenovirus | Intratumoral injection | Inhibition of tumour growth | Gurzov and Izquierdo (2006) |
| | Hec1 | AAV2 | Intratumoral injection | Inhibition of tumour growth | Li *et al.* (2007) |
| Senescence | hTer | Lentivirus | *Ex vivo* transduction + expression of telomerase mutant | Inhibition of tumour growth | Li *et al.* (2004b) |
| Tumour–host interaction | VEGF | CRAd | *Ex vivo* transduction | Inhibition of tumour growth | Yoo *et al.* (2007) |

target the toxic mutant and not the wild-type essential form of the gene. A summary of the results obtained in this field is presented in Table 11.4.

Hyperphosphorylated tau protein in neurofibrillary tangles and accumulation of beta amyloid peptide (Aβ), a cleavage product of amyloid precursor protein (APP), are characteristics of Alzheimer's disease (AD), the most common form of dementia. Decreased in Aβ can result from down-regulation of APP but also from decreasing BACE1, the beta-secretase involved in the cleavage of APP. Thus, tau, APP and BACE1 have been targeted by RNAi in different settings. RNAi-mediated down-regulation of mutated tau protein (TauV337M) has been shown to work *in vitro*, but the effects *in vivo* still need to be demonstrated (Miller *et al.*, 2004b). Down-regulation of BACE1 or APP reduced the formation of amyloid plaques in the injected brain regions (Singer *et al.*, 2005; Hong *et al.*, 2006).

The accumulation of human α-synuclein (α-syn) in neurons and glia leads to several neurodegenerative disorders known as synocleinopathies. Parkinson's disease (PD) is a synocleinopathy where overexpression of α-syn causes the selective loss of dopaminergic neurons in the substantia nigra pars compacta. Thus, genetic variability in α-syn promoter is associated with idiopathic Parkinson and familial Parkinson has been linked to mutations, duplications and triplications of α-syn. RNAi-mediated down-regulation of α-syn has been shown in a human neuroblastoma dopaminergic cell line and in rat brain (Sapru *et al.*, 2006). Interestingly, decrease of α-syn *in vitro* protects against MPP+, a potent and selective dopaminergic neurotoxin, but studies showing a behavioural phenotype after α-syn down-regulation are still lacking (Fountaine and Wade-Martins, 2007). A convincing animal model for behavioural studies could result from down-regulation of tyrosine hydrolase, a rate-limiting enzyme in dopamine synthesis whose decrease leads to behavioural deficits similar to neurotoxin-induced models of Parkinson (Hommel *et al.*, 2003).

Accumulation of $PrP^{sc}$, a protease-resistant isoform of the cellular prion protein ($PrP^{c}$), results in a transmissible neurodegenerative disease know as Prion disease or transmissible spongiform encephalopathies (TSEs). $PrP^{c}$ is required for $PrP^{sc}$ replication and pathogenesis so reducing $PrP^{c}$ levels extends survival time after $PrP^{sc}$ infection. RNAi-mediated down-regulation of PrP gene has been achieved in a murine neuroblastoma cell line persistently infected with a mouse $PrP^{sc}$ strain (Pfeifer *et al.*, 2006) and in mice with established prion disease (White *et al.*, 2008; Pfeifer *et al.*, 2006). Further, chimeric mice and goat fetus have been obtained with RNAi-decreased PrP (Pfeifer *et al.*, 2006; Golding *et al.*, 2006). Down-regulation of PrP gene in prion mice models has correlated with increased survival time (White *et al.*, 2008; Pfeifer *et al.*, 2006), prevention of behavioural deficits associated with early prion disease, reduced spongiform degeneration and protection against neuronal loss (White *et al.*, 2008).

Mutations in Cu-Zn superoxide dismutase (SOD1) and a three-nucleotide deletion in the torsin A (TOR1A) gene may cause amyotrophyc lateral sclerosis (ALS) and dystonia, respectively. In dystonia, the deletion eliminates a glutamic acid residue in torsin A that results in a dominant negative accumulation of wild-type torsin A in oligomeric structures of the nuclear envelope. RNAi-mediated down-regulation of the mutated form of the protein results in the normal distribution of torsin A in mammalian neurons (Gonzalez-Alegre *et al.*, 2005). ALS is a neurodegenerative disease characterized by the selective and progressive loss of motoneurons in spinal cord, brainstem and cerebral cortex that lead to weakness, skeletal muscle atrophy and death. One-fifth of familial ALS is linked to dominant mutations in the essential and ubiquitously expressed gene SOD1 that result in novel toxic properties rather than altered dismutase activity. Down-regulation of mutant SOD1 by RNAi could be used therapeutically for the treatment of ALS, but siRNA inhibitors should be very specific to affect the mutant SOD1 and not the essential wild-type protein. This is not easy as many SOD1 mutations are single nucleotide changes. However, siRNAs have been described that only target SOD1 mutant G39A and lead to increased survival of motor neurons, delayed appearance of the first signs of the disease and prolonged survival in ALS mice (Raoul *et al.*, 2005; Ralph *et al.*, 2005; Azzouz and Mazarakis, 2004; Miller *et al.*, 2005; Towne *et al.*, 2008).

**Table 11.4** Summary of viral vector delivered siRNAs targeting genes relevant for neurological disorders

| Disease model | Protein target | Delivery | Target/model | References |
|---|---|---|---|---|
| Alzheimer's disease (AD) | APP | HSV | Brain (hippocampus) | Hong *et al.* (2006) |
| | BACE-1 | Lentivirus | Brain (striatum or hippocampus) | Singer *et al.* (2005) |
| Parkinson's disease (PD) | α-Synuclein A53T | Lentivirus | Brain | Sapru *et al.* (2006) |
| | Tyr hydroxylase | AAV | Mid-brain | Hommel *et al.* (2003) |
| Prion disease | PrP | Lentivirus | Brain (striatum or hippocampus) | Pfeifer *et al.* (2006) |
| | PrP | Lentivirus | Brain (hippocampus) | White *et al.* (2008) |
| | PrP | Lentivirus | Goat fetus KD | Golding *et al.* (2005) |
| Dystonia (DYT1) | Torsin A (TOR1A) | Lentivirus (allele specific) | Brain (neurons) | Gonzalez-Alegre *et al.* (2005) |
| | Torsin A (TOR1A) | FIV | Brain (neurons) | Gonzalez-Alegre *et al.* (2005) |
| Amyotrophyc lateral sclerosis (ALS) | $SOD^{G93A}$ | Lentivirus | Brain (motoneurons) | Raoul *et al.* (2005) |
| | $SOD^{G93A}$ | AAV6 | Brain (motoneurons) intravenous injection in SOD fALS mice | Towne *et al.* (2008) |
| Spinocerebellar ataxia type (SCA) | Ataxin-$1^{82Q}$ | AAV | Brain (midline lobules) (intracerebellar injection in SCA1 mice) | Xia *et al.* (2004) |
| Huntington's disease (HD) | $Huntingtin^{N171-82Q}$ | AAV | Brain (striatum) | Harper *et al.* (2005) |
| | Mutant Htt | AAV5 + siHUNT | Brain (striatum) in R6/1 mouse | Rodriguez-Lebron *et al.* (2005) |
| | Mutant Htt | AAV | Brain (striatum) | Franich *et al.* (2008) |

Some neurological diseases are caused by the expansion of the trinucleotide CAG. CAG repeats lead to a polyglutamine track that alters protein function and result in the formation of intranuclear and cytoplasmic aggregates in different neuronal cell types. PolyQ tracks in ataxin-1 lead to degeneration of Purkinje cerebellar neurons, spinocerebellar tracts and some brain stem nuclei which result in spinocerebellar ataxia (SCA). A polyQ extension in the N-terminus of huntingtin protein leads to a toxic gain-of-function that provokes the dysfunction and death of striatal medium spiny GABAergic neurons and later cortical neurons and results in Huntington's disease (HD). The age of onset and the rate of progression of these diseases are inversely correlated to the size of the polyQ expansion. RNAi-mediated silencing of CAG extensions has been studied for the treatment of these diseases. This is not easy as siRNA inhibitors of CAG repeats need to be gene specific because these repeats are common to many normal transcripts. Delivery of specific siRNAs to decreased toxic ataxin-1 in mice results in improvement of the motor performance and attenuation of cerebellar pathology (Xia *et al.*, 2004). Down-regulation of expanded huntingtin prevents the formation of nuclear protein aggregates, improved motor performance and ameliorated the behavioural phenotype in transgenic mice (Harper *et al.*, 2005; Franich *et al.*, 2008; Machida *et al.*, 2006; Rodriguez-Lebron *et al.*, 2005).

Thus, promising results have been obtained by RNAi treatment of neurological diseases in animal models. However, much work needs to be done to improve delivery of shRNAs to different neuronal populations. Also, it should be addressed whether damaged neurons also have functional RNAi machinery that can mediate down-regulation of the toxic genes. Finally, it will be crucial to determine whether silencing of neurological toxic genes by RNAi is beneficial at late stages of the disease, so they can be used therapeutically both for the prevention and the reversion of neurological diseases.

### Genetic and immunological diseases

RNAi-based therapies could be used to treat any genetic disease, besides the neurological genetic pathologies already described. Thus, preclinical examples exist of viral vectors delivering siRNAs for the treatment of sickle cell anaemia (Samakoglu *et al.*, 2006) and Duchenne muscular dystrophy (DMD) (Ghahramani Seno *et al.*, 2008). Also, immunological diseases such as asthma could be treated with RNAi. In fact, lentiviral vectors have been used for the delivery of siRNAs targeting GATA-3, the key transcriptional factor for Th2 commitment in T cells strongly associated with asthma and allergic disease. Intratracheal administration of the vectors in a murine model of asthma attenuated antigen-induced airway inflammation and hyperresponsiveness (Lee *et al.*, 2008).

### Infections

Infections by bacteria, virus and parasites could also be treated with RNAi-mediated silencing. Most development has been done using RNAi as an antiviral. Bacterial genes are not silenced by siRNAs because they are expressed outside of the host cell and bacteria lack RNAi machinery (Lieberman *et al.*, 2003). However it is still possible to affect bacterial infections by down-regulation of host genes required for bacterial invasions or for the immune response that leads to damaging secondary effects in some bacterial infections. Unlike in bacteria, RNAi machinery has been described in parasites such as malaria and in the principal malaria vector in Africa, *Anopheles gambiae*. Therefore, malaria could be treated by silencing essential parasite genes or by down-regulation of non-essential mosquito genes required for parasite infection. Development of an RNAi-based therapy for malaria still needs proper delivery of siRNAs that target the parasite or the mosquito cells.

RNAi-based techniques have tremendous potential as antivirals. Genome sequences from several strains of most pathogenic viruses are publicly available. Also, new emerging pathogenic viruses can be rapidly identified and sequenced, as was recently the case with severe acute respiratory syndrome (SARS). When time consuming detailed studies of viral life cycle have not been performed, the sequence is enough to design antiviral RNAi inhibitors that block viral gene expression. However, when more information about the virus and their interaction with the host cell is available, antiviral siRNAs can also target non-essential cellular genes whose function is important for virus viability. For RNA viruses, siRNA inhibitors could also target the viral genome or the replicating intermediates. However, in some cases the viral genome is protected by the viral capsid or is inaccessible to the RNAi machinery in specific cellular compartments.

Every class of human virus has been experimentally targeted with RNAi. Therapeutic antiviral siRNAs have been described for acute and chronic viral infections. Generally, chemically synthesized siRNAs or plasmid expressing shRNAs have been used for the pre-clinical treatment of acute infections. Chronic infections require long-term expression of the siRNA inhibitors using viral vectors. Examples exist in which the viral vector that delivers shRNAs derives from the target virus. Thus, both parental and recombinant viruses transduce the same cells. However, in general, viral delivery of siRNAs to specific target cells is still a problem. Also, several viruses have evolved to escape from the cellular silencing machinery by different molecular mechanisms. Many viruses could counteract the RNAi machinery by expression of viral silencing suppressors. In RNA viruses that lead to chronic infections, viral escape has also been reported by mutation of the target sequence.

### Human immunodeficiency virus type 1 (HIV-1)

HIV-1 is a lentivirus with a positive single-stranded RNA genome widely used as a gene therapy vector (see above). The wild-type virus recognizes the CD4 receptor and the chemokine coreceptor CCR5 or CXCR4 to infect CD4+ T cells. The CD4-CCR5 route also serves to infect other immune cells such as dendritic cells, which can also be infected through DC-SIGN. After binding, the viral genome and several enzymes, including reverse transcriptase and integrase, are injected into the cell. There, the RNA genome is retrotranscribed into double-stranded DNA that integrates into the cell genome. The viral genome contains several regulatory elements, such as the LTRs and encodes for three structural proteins (Gag, Pol and Env) and six regulatory factors (Tat, Rev, Nef, Vif, Vpr and Vpu) that allow viral replication and production of new viral particles. Viral infection leads to the decrease of immune cells which results in failure of the immune system that cannot cope with opportunistic infections. It has been estimated that HIV infection has killed more than 25 million people and 0.6% of the world's population is infected. Nowadays, HIV-1 infection is treated with the highly active antiretroviral therapy (HAART). Where available, HAART has improved the lifespan of the patients resulting in long-term treatments that lead to life-threatening secondary effects, emergence of virus resistance and high cost. Thus, new therapies for HIV-1 treatment are required and RNAi-based HIV-1 silencing could be the most promising alternative.

Development of therapies for HIV-1 silencing needs to face the lack of proper animal models and efficient and specific systemic siRNA-delivery vectors. HIV-1 naturally infects human and a small number of nonhuman primate species. This limitation in host range restricts enormously the animal models suitable for the study of HIV-1 infection. Many groups have tested the efficiency of RNAi in suppressing HIV-1 replication in cell lines, and in primary human T cells and macrophages (Novina *et al.*, 2002; Lee *et al.*, 2005). Only in a few cases have studies been carried out in mouse chimeras with human haematopoietic cells (Berges *et al.*, 2006; Anderson *et al.*, 2007) or in primates infected with SIV virus, which is very similar to HIV-1. The delivery of siRNAs to the HIV-infected cells has been accomplished with viral vectors, such as lentivirus, retrovirus or AAV, that guarantee long-term expression of the transgene. In clinical settings and to increase transduction efficiency, T cells should be isolated from patients, transduced *ex vivo* with the shRNA expressing vector, expanded and reinfused into the same patient. Alternatively, HIV-resistant progeny T cells and macrophages can be generated by transducing haematopoietic stem cells (HSC) with vectors that express shRNAs against HIV-1 (Li *et al.*, 2005b). As in all cases the cells will be expanded, the use of lentiviral or retroviral vectors that will integrate the shRNA-expression cassette in the cell genome, so the inhibitor can be expressed in all daughter cells, is mandatory.

Many experiments *in vitro* have shown inhibition of HIV-1 after transient or stable expression of siRNAs targeting HIV-1 genes such as Rev, Tat, Nef, Env, Gag, and Vif (Table 11.5).

However, inhibition of HIV-1 replication with a single siRNA inhibitor is limited to the short term. Over longer times, the high mutation rate of RNA viruses results in the rapid emergence of HIV-1 mutants resistant to inhibition (Boden *et al.*, 2003; Das *et al.*, 2004; Sabariegos *et al.*, 2006). This has been described even when the targets are highly conserved sequences. Resistant viruses result from point mutations or deletions within the HIV-1 target that overcome RNAi-mediated inhibition. Given the specificity of RNAi, any single point mutation in the target sequence is enough to decrease silencing, and single point mutations in the central region of the target abolished silencing completely (Senserrich *et al.*, 2008; von Eije *et al.*, 2008; Pusch *et al.*, 2003). Also, mutations outside the target sequence have been described that lead to an altered RNA secondary structure of the target area which becomes inaccessible to inhibitory siRNAs (Westerhout *et al.*, 2005).

A way to avoid viral escape from silencing of viral genes, is the use of RNAi-mediated down-regulation of cellular cofactors required for HIV infection. It is expected that such an approach will result in a major obstacle to viral evolution. There are many cellular proteins important for HIV-1 viability. In fact, a recent siRNA library screen identified more than 250 cellular proteins required for HIV-1 replication (Brass *et al.*, 2008). Inhibition of HIV replication has been observed by silencing of cellular cofactors such as the HIV receptor CD4; co-receptors CXCR4 and CCR5; Arp2/3 complex, important for the actin polymerization required for HIV transport towards the nucleus (Komano *et al.*, 2004); Rab6A, a protein that regulates retrograde transport from Golgi to endoplasmic reticulum and is required for an early step in HIV-1 infection; PARP-1, required for HIV integration (Kameoka *et al.*, 2004); the integrase binding protein LEDGF/p75 (Vandekerckhove *et al.*, 2006); NFκB (Surabhi and Gaynor, 2002); SPT5 or P-TEFb, composed of cyclin T1, CDK9 and Tat, which function in HIV elongation of RNA polymerase II on the HIV LTR (Chiu *et al.*, 2004; Ping *et al.*, 2004; Li *et al.*, 2005c; Leonard and Schaffer, 2005); Sam68 or hRIP, required for HIV Rev-mediated nuclear export of unspliced viral genomic RNA (Modem *et al.*, 2005; Yu *et al.*, 2005); human deoxyhypusine synthase, required to activate the cofactor for rev transactivation eIF-5A (Hauber *et al.*, 2005); cyclophilin A (Sokolskaja *et al.*, 2004); DC-SIGN (Arrighi *et al.*, 2004); cdk2 (Ammosova *et al.*, 2005); etc. (Table 11.5). Some of these factors, such as NFκB, CD4 and CXCR4 are important proteins for the cell and therefore their down-regulation is not appropriate as a therapy. However, other factors such as CCR5, are not essential for normal cell function. Targeting this cell receptor could protect cells from the initial step of infection and this may force the virus to stay longer in the extracellular space where it is more susceptible to the immune system. Besides, the virus would need to make a challenging evolutionary jump to be able to cope with the loss of a receptor. However, even when down-regulation of CCR5 resulted in a 3 to 7 fold reduction in viral infectivity relative to controls, the cells were still susceptible to infection by T-tropic CXCR4 utilizing virus.

The best therapeutic approach for HIV treatment could be the combination of siRNAs that target conserved viral sequences and host genes important for viral viability. Theoretically, with the combination of four effective siRNAs, viral escape is impossible as long as viral suppression is complete. In fact, expression of three different shRNAs from a single vector resulted in delayed HIV-1 escape (ter Brake *et al.*, 2006). There are several ways of expressing different shRNAs from a single vector. When the same repeated promoter is used to transcribe independent shRNAs, vector production will lead to recombination of the repeated sequences and reduced inhibitory activity. To avoid recombination, different promoters with similar activities have been successfully used (ter Brake *et al.*, 2008). An alternative is the use of a single promoter to transcribe tandem miRNA-based shRNAs (Liu *et al.*, 2008) or extended shRNAs (Liu *et al.*, 2007). Finally, lhRNAs could be used as a method to generate several siRNAs (Konstantinova *et al.*, 2006; Konstantinova *et al.*, 2007). The down side of these approaches is that the use of multiple siRNAs can increase the chance of off-target effects and toxicity by saturating the cellular RNAi machinery. Thus, the optimal alternative could be the combination of RNAi-mediated silencing with other approaches to inhibit viral replication, such as HAART. A potent anti-HIV-1 therapy has

**Table 11.5** Summary of viral vector delivered siRNAs targeting genes relevant for HIV

| Target gene | siRNA type | Promoter | Vector | Host | Fold inhibition | Reference |
|---|---|---|---|---|---|---|
| **Single viral gene** | | | | | | |
| *Rev* | shRNA | | Lentivirus | T lymphocytes differentiated in SCID- hu mice and CD34+ progenitor cell-derived macrophages | | Banerjea *et al.* (2003) |
| *Tat* | shRNA | H1 and MTD | AAV2 | H9 cells and primary human lymphocytes | >9 | Boden *et al.* (2003, 2004) |
| *Nef* | shRNA | H1 | Retrovirus | SupT1 | >10 | Das *et al.* (2004) |
| *Env* | shRNA | U6 | Lentivirus | COS, MT-4 | >10 | Hayafune *et al.* (2006) |
| *Nef* | shRNA | U6 | Lentivirus | MDMs | >9 | Yamamoto *et al.* (2006) |
| *Nef* | shRNA | | Lentivirus | Human haematopoietic stem cells, CD4(+) T cells | | ter Brake *et al.* (2006) |
| *Gag*, *nef* or *vir* | shRNA | U6 | Lentivirus | CD4 T-cells HeLa-CD4 | Different | Lee *et al.* (2005) |
| *Nef* | lhRNA | – | Lentivirus | SupT1 | >7 | Konstantinova *et al.* (2007) |
| **Single cellular gene** | | | | | | |
| *DC-SIGN* or *rL-SIGN* | shRNA | H1 | Lentivirus | DC cells | | Arrighi *et al.* (2004) |
| LEDGF/p75 | shRNA | mU6 | Lentivirus | HeLaP4-derived cell lines | 3–5 | Vandekerckhove *et al.* (2006) |
| **Combinations** | | | | | | |
| Cellular *CXCR4* and *CCR5* | shRNA | H1 and U6 | Lentivirus | Magi and Ghost C334 haematopoietic cells (2005) | | Anderson and Akkina (2005) |
| Viral Gag, Pol and env | shRNA (21pb) | H1 | Lentivirus | 293T, Magi, GHOST, hi5, CEM-A, Molt-4, PBMC | >20 | Chang *et al.* (2005) |
| Viral *Tat* and *int* | shRNA, lhRNA | H1 | Lentivirus | Human CD4 T cells MT4 cells | >9 | Nishitsuji *et al.* (2006) |
| Viral *Tat* and *rev* | shRNA | U6 | Lentivirus | Human haematopoietic stem cells CD4(+) T cells | >20 | Li *et al.* (2005b), Anderson *et al.* (2007) |
| Viral *Gag* and *pol* | miRNA polycistron | | Lentivirus | SupT1 | >10 | Liu *et al.* (2008) |
| Viral *Gag* and *pol* | Multiple shRNAs | pol III: U6, H1; pol II: U1, 7SK | Lentivirus | SupT1 | >10 | ter Brake *et al.* (2008) |

been obtained with lentiviral vectors that express three different molecules: a shRNA targeting Tat and Rev, a nucleolar decoy to block HIV-1 Tat function and a hammerhead ribozyme to cleave the cellular HIV-1 co-receptor CCR5. This triple vector has shown a complete protection against HIV-1 in experimental models and is being assayed in a clinical trial by administration *ex vivo* in haematopoietic stem cells of HIV-1 infected patients (Li *et al.*, 2005b).

## Hepatitis C virus (HCV)

HCV is a member of the Flaviridae family with a 9.6-kb single-stranded RNA genome of positive polarity. The genome contains a single open reading frame that encodes a polyprotein cleaved post-transcriptionally into four structural proteins (core, p7, E1 and E2) and six non-structural proteins (NS2, NS3, NS4A, NS4B, NS5A and NS5B). The open reading frame is flanked by untranslated regions (UTR) that contain several well-conserved domains important for the HCV life cycle. The 5′-UTR includes a highly conserved region important for replication that functions as an internal ribosome entry site (IRES) to initiate polyprotein translation. Replication occurs attached to cytoplasmic membranes and requires an RNA-dependent RNA polymerase that lacks proof reading and leads to imperfect copies of the viral genome. This results in distinct but related HCV genomes that coexist in the infected individual and are known as quasispecies.

Approximately 170 million people (2–3% of the world population) are chronically infected with HCV. 85% will develop a chronic liver disease, such as liver fibrosis, cirrhosis and hepatocellular carcinoma. Current therapy for HCV is the combination of pegylated (peg)IFN-α plus ribavirin (Chander *et al.*, 2002). This therapy is effective in 40–80% of the cases of HCV, depending on the HCV serotype that needs to be treated. There is no vaccine that protects against HCV infection. Therefore, the development of new therapies for chronic HCV is mandatory.

Novel therapies should be easier to evaluate with the recent development of HCV replicons that persistently replicate in stably transfected human hepatoma Huh-7 cell lines (Lohmann *et al.*, 1999; Blight *et al.*, 2000). Replicons can be full-length or subgenomic, which lack the structural sequences and may express reporter genes. Functional inhibitors evaluated in the replicon system have to be assayed *in vivo*. *In vivo* models for HCV infection are limited to chimpanzee (Wieland and Chisari, 2005) and chimeric immunodeficient mice that are grafted with human hepatocytes (Mercer *et al.*, 2001). These models should also be used for the development of RNAi-mediated silencing of HCV. Besides, it is essential to design the best delivery vector that targets HCV infected cells and expresses siRNAs stably. Retrovirus, lentivirus, adenovirus and AAV have been used to deliver shRNAs targeting HCV (Table 11.6).

HCV RNA was predicted to be highly susceptible to RNAi because replication occurs in the cytoplasm and the viral genome resembles an mRNA. The results have confirmed the prediction. The replicon system has been used to evaluate siRNA-targeting the conserved 5′-UTR (Seo *et al.*, 2003; Randall *et al.*, 2003; Yokota *et al.*, 2003; Kanda *et al.*, 2007), core (Randall *et al.*, 2003), E2 (Prabhu *et al.*, 2005), NS3 (Takigawa *et al.*, 2004; Kapadia *et al.*, 2003; Prabhu *et al.*, 2005), NS4B (Randall *et al.*, 2003), NS5A (Sen *et al.*, 2003), NS5B (Takigawa *et al.*, 2004; Wilson *et al.*, 2003; Kapadia *et al.*, 2003; Prabhu *et al.*, 2005; McCaffrey *et al.*, 2002). NS5B and IRES have also been targeted by RNAi in adult mice (McCaffrey *et al.*, 2002; Wang *et al.*, 2005). These results indicate that most regions of HCV RNA are accessible to the RNAi machinery and lead to down-regulation of HCV protein expression and to a decline in both genomic and antigenomic HCV RNA (Wilson *et al.*, 2003). Moreover, siRNA-mediated inhibition is more effective in reducing HCV RNA levels than high doses of IFN (Kapadia *et al.*, 2003).

Given the high rate of mutation of HCV genome, it is not surprising that after treatment with anti-HCV siRNAs, HCV escape mutants emerge with several point mutations in the siRNA targeted sequence (Wilson and Richardson, 2005). An alternative to delay viral escape might be to select target sequences that are conserved because they are essential for viral replication. In the case of HCV, well-conserved sequences are in 5′-UTR IRES, in the region of the AUG translation initiation codon, where mutations may lead to loss of function. Also RNAi-mediated inhibition of

**Table 11.6** Summary of viral vector delivered siRNAs targeting genes relevant for HCV

| Target gene | siRNA type | Promoter | Vector | Host | Fold inhibition | References |
|---|---|---|---|---|---|---|
| Viral 5′UTR, C, NS4B, NS5A, NS5B | esiRNA/ shRNA | U6 | Retrovirus | Huh-7 | 100 | Kronke *et al.* (2004) |
| Viral 5′UTR, C, NS3, NS5B | shRNA | U6 | Lentivirus | Huh-7 | 4–9 | Takigawa *et al.* (2004) |
| Cellular La, PTB, eIF2βγ | shRNA (19pb) | U6 | Adenovirus | Huh-7 | 13 | Zhang *et al.* (2004a) |
| Cellular Cyp-A, B, C | shRNA | U6 | Retrovirus | Huh-7 | 10 | Nakagawa *et al.* (2005) |
| Cellular + viral 5′-UTR, 3′-UTR PSMA7, HuR | shRNA (19–21pb) | U6 | Retrovirus | Huh-7 | >2 | Korf *et al.* (2005) |
| Cellular + viral CD81, IRES, NS5B | shRNA (19–21pb) | H1 | Lentivirus | Huh-7 | >32 | Henry *et al.* (2006) |

cellular targets and combinations of different siRNAs targeting HCV should limit the emergence of resistant viruses (Henry *et al.*, 2006; Korf *et al.*, 2005). Several cellular factors have been identified that are required for HCV infection. HCV replication has been decreased by RNAi-mediated inhibition of host cell receptor CD81 (Henry *et al.*, 2006); the proteosome a-subunit 7 (PSMA7), which modulates HCV IRES activity (Kruger *et al.*, 2001); Hu antigen R (HuR), which interacts with the 3′-UTR (Spangberg *et al.*, 2000); polypyrimidine track-binding protein (PTB), which interacts with the IRES (Chang and Luo, 2006); RNA helicases p68 and A (RHA) (Isken *et al.*, 2007); La autoantigen (Domitrovich *et al.*, 2005); cyclophilins (Nakagawa *et al.*, 2005; Watashi *et al.*, 2005); human VAMP-associated protein of 33 kDa (hVAP-33) (Zhang *et al.*, 2004a) and 62 host genes targeted with a systematic RNAi screen that encode proteins that interact with the HCV genome or belong to cellular pathways that may modulate HCV infection (Randall *et al.*, 2007). Interestingly, Dicer and other components of the RNAi pathway also inhibit HCV replication (Randall *et al.*, 2007). This is surprising as it has been described that the HCV genome forms dsRNA regions that can be targeted by Dicer, suggesting that Dicer inhibition should improve HCV replication. However, the RNAi machinery is required for production of miR-122, a liver specific miRNA required for HCV replication. In fact, miR-122 blockade with antisense oligos impairs HCV replication (Randall *et al.*, 2007). In spite of the successful inhibition of HCV targeting genes that affect HCV viability, treatment of chronic HCV will require the long-term silencing of these genes, which could result in unwanted secondary effects.

Combinations of different siRNAs that target HCV replication have been shown to decrease viral escape and have revealed additive HCV inhibition (Henry *et al.*, 2006). This can be achieved by expression of independent shRNAs from a single vector (Takigawa *et al.*, 2004) or by expression of lhRNAs that are processed to different siRNAs (Romano *et al.*, 2006; Akashi *et al.*, 2005; Hamazaki *et al.*, 2004; Watanabe *et al.*, 2006a). HCV infection has been successfully targeted by lhRNAs with multiple point mutations within the sense strand designed to decrease IFN response (Akashi *et al.*, 2005; Watanabe *et al.*, 2006a). As expression of multiple siRNAs could increase the chance of silencing non-targeted genes, combination therapies have also been tested for HCV inhibition. Thus, the optimal therapy could be the combination of RNAi-mediated silencing with IFN and ribavirin or with hairpin ribozymes targeting the 3′-UTR of the HCV genome, as has been already shown.

## Hepatitis B virus (HBV)

HBV is the sole human pathogen in the family of Hepadnaviridae that predominantly infects the liver. HBV is a non-cytopathic enveloped virus with a partly double-stranded DNA genome of 3.2 kbp. In the infected hepatocyte, the viral genome is converted to a covalently closed circular DNA (cccDNA) that serves as a template for expression of viral genes, formation of pregenomic RNA and production of progeny. The viral genome contains regulatory regions embedded in four overlapping open reading frames that encode core, polymerase, surface and X proteins. In addition, eight HBV genotypes have been described which have distinct clinical characteristics and geographical distribution. In general, HBV infection may counteract the immune response leading to chronic infections. Though vaccines against HBV are available, about 2 billion people are infected with HBV and 400 million are chronic HBV carriers, at high risk of developing liver cirrhosis or liver cancer (Romano *et al.*, 2006). Currently, there are two treatments for antiviral therapy of chronic HBV infection: IFN-α, with limited efficacy (Lee, 1997) and nucleoside analogues such as adefovir or lamivudine, which are effective only in the short term (Liaw, 2002) and fail to decrease HBV antigen load, which leads to chronic inflammation and development of liver disease.

Novel therapies are required for HBV treatment and HBV RNAi-mediated silencing could be a promising option. Therapy efficiency should be tested in any of the several tissue culture cells that support viral replication, such as 2.2.15, and HBV transgenic mice, whose hepatocytes support viral replication but not reinfection by HBV as they lack the unidentified HBV receptor. Besides, a good experimental model results from woodchucks naturally infected with woodchuck hepatitis virus, which is very similar to HBV.

HBV expression and replication has been down-regulated *in vitro* and *in vivo* with siRNAs (Table 11.7). Interestingly, unlike lamivudine, HBV antiviral RNAi can down-regulate surface antigen expression, the major cause of liver damage, and can target lamivudine-resistance YMDD polymerase gene. There are examples of efficient anti-HBV siRNAs that target sequences from core, polymerase, surface and X genes (Uprichard *et al.*, 2005; Wu *et al.*, 2005; Shlomai and Shaul, 2003; Chen *et al.*, 2003; Liu *et al.*, 2004; Ren *et al.*, 2005; Ni *et al.*, 2004; Yang *et al.*, 2005; Hamasaki *et al.*, 2003; Ying *et al.*, 2007). In fact, the HBV genome is so compact that single siRNAs can silence all viral genes. Anti-HBV siRNAs have been expressed from regular shRNAs, from miR-like transcripts (Ely *et al.*, 2008; Snyder *et al.*, 2008) and from lhRNAs containing multiple G:U

**Table 11.7** Summary of viral vector delivered siRNAs targeting genes relevant for HBV

| Target gene | siRNA type | Promoter | Vector | Host | Fold inhibition | Reference |
|---|---|---|---|---|---|---|
| SAg, Pol, X | shRNA | U6 | Adenovirus | HBV transgenic mice | 9 | Uprichard *et al.* (2005) |
| SAg | shRNA | H1 | PFV, AAV | 293T, 2.2.15 | 4–9 | Moore *et al.* (2005) |
| SAg | shRNA | H1 | dsAAV2/8 | HBV transgenic mice | >9 | Chen *et al.* (2007) |
| SAg | shRNA | Pol II Apo E | dsAAV2/8 | HBV transgenic mice | 6 | Giering *et al.* (2008) |
| HBV genome | shRNA | U6 | Gutless adenovirus | HBV transgenic mice | 9 | Rauschhuber *et al.* (2008) |
| X | shRNA | U6 | Adenovirus | Hepatocytes HBV transgenic mice | >9 | Carmona *et al.* (2006) |
| SAg, pol | 2 shRNA | U6 | Retrovirus | 2.2.15 cells | 7 | Jia *et al.* (2007) |
| Pol, core, SAg, X | shRNA | U6 | AAV2/8 | HBV transgenic mice | >9 | Grimm *et al.* (2006) |
| S Ag and X | shRNA | U6 | HFV | HepG2, 2.2.15 | 9 | Sun *et al.* (2007) |

pairings (Weinberg *et al.*, 2007). Also, simultaneous delivery of two shRNAs that target different regions resulted in synergistic antiviral effects. The same result has been obtained by combining RNAi and lamivudine (Chen *et al.*, 2003). This is important as HBV polymerase lacks proofreading activity resulting in rapid mutagenesis of HBV and emergence of lamivudine resistant variants. However, viral escape of HBV against single siRNA targets has not been observed in experimental models. Thus, sustained inhibition of HBV has been observed in tissue culture for longer than 30 passages (Ren *et al.*, 2006) and in transgenic mice for 5 months after a single injection of viral vectors expressing the antiviral siRNAs (Carmona *et al.*, 2006; Uprichard *et al.*, 2005). It is still possible that siRNAs may force selection pressure on pre-existent viral mutants (Wu *et al.*, 2005). Anti-HBV RNAi therapy has to face the issue of proper delivery. Viral vectors need to be developed that target hepatocytes efficiently and express non-toxic antiviral siRNAs. In fact, for *in vivo* delivery, mostly adenovirus (Romano *et al.*, 2006; Uprichard *et al.*, 2005; Carmona *et al.*, 2006) and AAV viruses have been used in experimental models (Grimm *et al.*, 2006; Suzuki *et al.*, 2005; Chen *et al.*, 2007). Adenovirus has the disadvantage of short-term expression and HBV treatment is expected to require inhibitors that are expressed for long periods of time. Therefore AAV has been the vector of choice in most *in vivo* studies. AAV8 has an exquisite preference and efficiency to transduce mouse hepatocytes and has resulted in efficient inhibition of HBV replication in transgenic models. Studies performed with AAV8 delivered shRNAs have also highlighted the importance of optimizing the dose of inhibitor. Vectors expressing 25-mer anti-HBV shRNAs consistently caused toxicity in the treated mice, while low doses of 19-mer inhibitors lead to safe and long-term antiviral activity (Grimm *et al.*, 2006). This result demonstrates the potential of RNAi application in anti-HBV therapy when all the relevant considerations are taken into account.

## Other viruses

In summary, 25 different RNA viruses, and 11 different DNA viruses have been efficiently targeted by RNAi (Haasnoot and Berkhout, 2006). Besides HIV, HCV and HBV, these include important human pathogens such as dengue (DEN) virus, SARS coronavirus, RSV, poliovirus, influenza A virus, HSV, human papillomavirus (HPV), West Nile virus (WNV), coxsackievirus and CMV. Some of these viruses lead to acute infections that have been mostly targeted by synthetic siRNAs or plasmid expressing shRNAs. This is the case of influenza virus, HSV-2, DEN, Marburg virus and RSV. Viral vectors expressing siRNAs have shown efficient inhibition of infection by dsDNA viruses such as CMV (Dittmer and Bogner, 2006) or papillomavirus (Gu *et al.*, 2006) or positive stranded RNA viruses such as SARS (Zhang *et al.*, 2007) or Coxsackievirus (Fechner *et al.*, 2008; Kim *et al.*, 2008). Most of the experiments have been performed in cell culture infected cells where shRNA delivery has been accomplished with first generation adenoviruses (Zhang *et al.*, 2008; Chen *et al.*, 2006a), lentivirus (Kim *et al.*, 2008; Gu *et al.*, 2006; Kumar *et al.*, 2006), retrovirus (Dittmer and Bogner, 2006) or AAV (Fechner *et al.*, 2008). In some cases antiviral effects have also been observed in animal models. Thus, AAV or lentivirus delivered shRNAs protect mice against coxsackievirus infection by decreasing viral titres and myocarditis (Fechner *et al.*, 2008; Kim *et al.*, 2008). Detailed studies showed attenuation of cardiac dysfunction after intravenous injection of cardiotropic AAV that delivers antiviral shRNAs (Fechner *et al.*, 2008). In addition, a single intracranial administration in mice of shRNA expressing lentivirus was sufficient to protect against lethal encephalitis after challenge with WNV or Japanese encephalitis (JEV) virus (Kumar *et al.*, 2006). Interestingly, as this shRNA was targeting a conserved region among all flaviviruses, the same vector was able to protect against WNV or JEV (Kumar *et al.*, 2006). This demonstrates that RNAi can be used as a broad spectrum antiviral agent to suppress related viruses across species. Guinea pigs and swine have also been protected against foot and mouth disease virus (FMDV) by adenoviral delivered shRNAs (Chen *et al.*, 2006a). This shows the feasibility of antiviral RNAi in larger animals.

## Potential limitations of RNAi therapy

The use of siRNAs *in vivo* still faces some important problems that could lead to toxicity due to activation of IFN response, lack of specificity, or inhibition of the endogenous miRNA machinery.

### Activation of IFN response

Many viruses produce long dsRNA molecules that are formed, in most cases, as replication intermediates. As long dsRNAs are not natural products in healthy cells, they are sensed by the cell as an indicator of danger that leads to the activation of the antiviral IFN response. The most relevant cellular factors that sense dsRNA are the dsRNA-dependent protein kinase (PKR), 2′–5′-oligoadenylate synthetase (OAS), and retinoic acid inducible gene-I (RIG-I) in the cytoplasm and Toll-like receptors 3 (TLR3), which mostly locate in cell membranes. Activation of PKR and OAS by cytoplasmic dsRNA has been well-studied and results in a general inhibition of gene expression. The activation of OAS results in the formation of 2′–5′-oligoadenylates that bind to and activate RNase L, which degrades cellular and viral RNAs (Player and Torrence, 1998; Wreschner *et al.*, 1981; Dauber *et al.*, 2004). Activated PKR phosphorylates eukaryotic initiation factor 2$\alpha$ leading to translation inhibition and initiates a signalling cascade that results in the secretion of IFN (Gil and Esteban, 2000). RIG-I binding of dsRNA unwinds the double strand and activates IFN expression (Marques *et al.*, 2006). In neighbouring cells, IFN activates different pathways with antiviral, antiproliferative and pro-apoptotic activities. In fact, IFN activates expression of OAS and PKR so cells exposed to IFN are sensitive to very low levels of dsRNA, which eventually lead to cell death. dsRNA released from lysed cells can be sensed by TLR3, present in the membrane of many cells, which also activates the expression of inflammatory cytokines and IFN (Alexopoulou *et al.*, 2001; Doyle *et al.*, 2003).

Strategies that require expression of dsRNA molecules in the cell, such as RNAi, need to take into consideration the IFN response. Initial experiments demonstrated that siRNAs, unlike long dsRNAs, do not activate the IFN response (Elbashir *et al.*, 2001b). Consequently, it was widely accepted that dsRNA molecules shorter than 30 bp are not sensed by the antiviral defence machinery. However, it was soon discovered that siRNAs and vectors expressing shRNAs can activate PKR, OAS, TLRs, RIG-I and IFN (Elbashir *et al.*, 2001b; Kariko *et al.*, 2004; Sledz *et al.*, 2003; Bridge *et al.*, 2003; Jackson and Linsley, 2004; Fish and Kruithof, 2004). Currently, examples also exist of IFN activation by viral delivered siRNAs both in tissue culture cells and in model animals (Witting *et al.*, 2008). However, siRNAs may lead to activation of components that overlap but are not identical to those activated in a regular IFN response (Persengiev *et al.*, 2004).

The understanding of the way the cell reacts against exogenous dsRNAs has helped the design of dsRNA that are less immunostimulatory. In the case of PKR, activation by dsRNA is complex. PKR has at least two dsRNA binding regions, both of which may be occupied with a single dsRNA larger than 30 bp (Carlson *et al.*, 2003). PKR activity increases with the length of the dsRNA up to 85 bp (Manche *et al.*, 1992). However, each individual site can bind dsRNAs of only 11–16 bp (Bevilacqua and Cech, 1996). PKR activation depends on the site elected (Carlson *et al.*, 2003). Moreover, the sequence and secondary structure of the dsRNA may also affect PKR activity which can also be altered depending on the tissue, growth and health status of the cell (Carlson *et al.*, 2003; Zamanian-Daryoush *et al.*, 1999; Manche *et al.*, 1992). In fact, the best minimal motif for PKR activation is a 16-bp dsRNA stem flanked by 11- to 15-nt single-stranded tails (Zheng *et al.*, 2005). Unlike PKR, RIG-I seems to sense the ends of the dsRNA molecules. T7 RNA polymerase and polymerase III transcribed RNAs start with a 5′ triphosphate able to activate RIG-I (Kim *et al.*, 2004). RIG-I also reacts to blunt-ended siRNAs lacking the 2-nt 3′overhangs characteristic of Dicer processing (Marques *et al.*, 2006). Thus, RIG-I is insensitive to polymerase II transcribed dsRNAs, which are capped at the 5′ end, and Dicer processed endogenous miRNAs. Finally, TLR3, TLR7 and TLR8 sense danger motifs in the cell membrane and in endosomes, which is the arrival organelle for most methods of non-viral delivery of siRNAs. There, TLR3 reacts to dsRNA and TLR7/8 to ssRNA with GU rich sequences (5′-GUCCUUCAA-3′ and 5′-UGUGU-3′) (Judge *et al.*, 2005; Hornung *et al.*, 2005). Several chemical modifications of siRNAs have been described which attenuate TLR-induced immunostimulatory effects (Judge *et al.*, 2006).

The knowledge of the mechanism of recognition of exogenous dsRNAs by cellular factors can help to design viral vector delivered siRNAs with decreased possibilities of IFN activation. Delivery with viral vectors does not locate siRNAs in the endosome, so this method avoids IFN activation through this pathway. However, it is still wise to avoid GU-rich danger sequences in the siRNA sequence. Also, pol II promoters are favoured against pol III, which start transcription with a 5′ triphosphate. In the case of U6 promoters, higher activation of IFN genes has correlated with the presence of an AA dinucleotide near the transcription start site (Pebernard and Iggo, 2004; Scacheri *et al.*, 2004). Preserving the wild-type C/G sequence at position −1/+1 prevents this effect. Besides, activation of IFN has a strong dose effect. As a result, siRNA expression has to be kept to the minimal functional dose. Finally, the length and structure of the dsRNAs have to be taken into consideration. Reducing the dsRNA nature by introducing mismatches should lead to decreased risk of IFN activation. Use of longer (27–29 bp) siRNAs has shown efficient specific inhibitions in some reports and IFN activation in others (Reynolds *et al.*, 2006; Kim *et al.*, 2005; Siolas *et al.*, 2005). However, expression of lhRNAs seems to be well tolerated and does not induce IFN activation either *in vitro* or *in vivo* (Akashi *et al.*, 2005; Weinberg *et al.*, 2007; Strat *et al.*, 2006; Watanabe *et al.*, 2006a). This could reflect the fact that endogenous expression of low amounts of nuclear lhRNAs may mimic natural miRNA precursors that are not detected as dangerous.

Even if these results are promising, care should be taken in the evaluation of any novel combination of viral vector and siRNA expressing cassette. Also, the results obtained may reflect defective interferon responses in the cells used, as has been described for many commonly used tumour cells. In fact, different cells and cell states could react differently to the same dsRNA stimuli. In any case, it should also be taken into consideration that in some treatments activation of IFN response could be beneficial. This is the case of HCV inhibition or macular degeneration therapy, where control siRNAs showed therapeutic effects by activation of TLR3 (Kleinman *et al.*, 2008).

### Off-target inhibition

Initially, RNAi was described as a highly specific technique for gene expression inhibition. Several studies with various model systems supported this idea. Thus, siRNAs can be designed to silence only a pathogenic allele that differs in a single nucleotide from the wild-type non-targeted gene (Brummelkamp *et al.*, 2002a; Miller *et al.*, 2004b; Ding *et al.*, 2003). Also, single nucleotide changes in the target sequence allow viral escape from RNAi (Boden *et al.*, 2003; Das *et al.*, 2004; Gitlin *et al.*, 2002). However, experiments performed with point mutations in the siRNA or the target sequence lead to the conclusion that some mismatches can be tolerated without disrupting silencing (Jacque *et al.*, 2002; Czauderna *et al.*, 2003; Yu *et al.*, 2002; Pusch *et al.*, 2003; Vickers *et al.*, 2003). Changes at the 3′ end of the guide siRNA strand are relatively better tolerated, and there is higher tolerance for G:U wobble pairing between the RNA oligo and the target RNA (Gitlin *et al.*, 2002; Das *et al.*, 2004; Boden *et al.*, 2003; Elbashir *et al.*, 2001a; Amarzguioui *et al.*, 2003; Saxena *et al.*, 2003). This can lead to inhibition of genes different than the target gene, which contains target sequences with few differences. These off-targets can be generally predicted with regular blast searches.

Microarray analysis has been also used to evaluate inhibition of genes different than the target. The results have been contradictory. Some studies found high specificity of siRNAs (Chi *et al.*, 2003; Semizarov *et al.*, 2004) while others did not (Jackson *et al.*, 2003; Persengiev *et al.*, 2004). When off target inhibition was found, some authors find that it cannot be eliminated by decreasing the concentration of the siRNA (Semizarov *et al.*, 2004; Persengiev *et al.*, 2004) while others do not (Jackson *et al.*, 2003). All studies agree that the magnitude of silencing of off target transcripts is lower than that of the on target gene. All of the transcripts that were regulated with similar kinetics to the target transcript, and therefore, could not be explained as a secondary effect of on-target down-regulation, displayed partial sequence complementarity to the siRNA. Initial analysis demonstrated that contiguous complementarity of only 11 nucleotides is enough to induce off-target silencing (Jackson *et al.*, 2003). More recent work indicates that the majority of experimentally verified off-targets have in their 3'UTRs a 6–7 nucleotide complementarity with the siRNA in the 'seed' region (Jackson *et al.*, 2006b; Birmingham *et al.*, 2006; Lim *et al.*, 2005; Lin *et al.*, 2005; Scacheri *et al.*, 2004). This has lead to the widely accepted conclusion that siRNAs behave as miRNAs. miRNAs regulate gene expression by inhibiting translation and/or activating mRNA degradation. Only when the latter mechanism operates will the decrease in the mRNA levels be detected by microarray analysis. Therefore, there is a general belief that microarray studies underestimate off target effects as they do not account for differences in protein expression resulting for translation inhibition. This issue needs to be solved by high throughput proteomic analysis capable of distinguishing low differences in protein levels such as SILAC.

Several methods have been proposed to decrease off-target effects. For synthetic siRNAs, certain modifications have been described that decrease the number of off-targets and the degree of their silencing (Jackson *et al.*, 2006a). This strategy cannot be applied to siRNAs delivered from shRNAs. When both the sense and the antisense strand of the siRNA can be loaded into RISC, both can cause off-targeting effects. Proper design of the siRNA will allow RISC loading of only the antisense strand, and decrease the total amount of off-targets. Off-target effects are specific for the siRNA sequence and not for the target gene, so different siRNAs against the same gene will have different off-targets with varying toxicity (Jackson *et al.*, 2003). In principle, off-targets can be predicted *in silico* by screening for complementarity between the siRNA and all the 3'-UTRs of the transcriptome. The results will not be definitive as large numbers of genes with good seed matches remain unaffected (Birmingham *et al.*, 2006). It has been suggested that miRNA-like inhibition of targets will depend on accessibility, the presence of certain sequences close to the target or the action of other inhibitory miRNAs bound to the same target (Saetrom *et al.*, 2007). As miRNA expression is cell specific, it is possible that off-target effects will also be different in different cells. Given the complexity of predictions, it is advisable to monitor off-targets of a given siRNA by microarray, Northern blot and Western blot analysis. Alternatively, several siRNAs can be used that need to yield the same result for gene function studies. This is not an option for therapeutic siRNAs, whose off-target toxicity should be evaluated in careful preclinical studies. The outcome of these analyses is difficult to predict. On one hand, studies of transfected cells *in vitro* have shown that one third of randomly chosen siRNAs affect cell viability (Fedorov *et al.*, 2006). On the other hand, obvious side-effects have not been reported *in vivo* with many functional siRNAs (Grimm *et al.*, 2006).

### Saturation of the cellular silencing machinery

Exogenous siRNAs and endogenous miRNAs share the same cellular machinery, whose functionality is essential for cell viability. When cellular siRNAs are expressed over a certain threshold, the limiting silencing machinery can be saturated leading to cell toxicity. This has been shown in mice after inoculation of high doses of AAV recombinant vectors expressing siRNAs with different stem lengths (Grimm *et al.*, 2006). In this case, shorter stems of 19 bp were less harmful than longer duplexes of 25 bp, which lead to decreased miRNA levels in targeted hepatocytes, liver failure and death. Saturation occurred at the level of Exp 5, which transports shRNAs and pre-miRNAs to the cytoplasm. Overexpression of Exp 5 relieved saturation and restored decreased miRNA levels. However, other components of the machinery, such as Dicer and RISC, have been shown to be limited and therefore could be also saturable.

Saturation of the cellular RNAi machinery can be prevented when low doses of siRNAs are expressed in the cell. The best option is to first identify those siRNA inhibitors that are functional at very low concentrations. When using viral vectors to deliver shRNAs, the lowest functional titre should be employed. Ideally, promoters that drive the expression of shRNAs should be inducible so siRNA expression can be decreased from the outside with decreasing doses of inductor.

Strong promoters such as U6, should be avoided. H1 promoters seem to be less problematic (An *et al.*, 2006). Following these rules it has been shown that efficient inhibition can be observed *in vivo* without saturation of the endogenous machinery (Grimm *et al.*, 2006; Narvaiza *et al.*, 2006).

## Difficulties in the production of viral vectors expressing siRNAs

There are several reasons to predict that viral vectors expressing siRNAs will be more difficult to produce than standard vectors. Decreased viral production or amplification could result from reorganization of the viral genome or from specific or unspecific effects of the siRNAs expressed by the vector.

Recombinant virus production requires that the viral genome enters a packaging cell where viral replication is favoured. Recombinant viral progeny, in some cases, can be used to transduce fresh packaging cells where new rounds of replication occur. Thus, a single viral genome is replicated and amplified to very high titres. When viral genomes contain repetitive sequences that are not essential for the virus, recombination of these sequences may easily occur, resulting in viral genomes with faster replication capacity. Therefore, defective viral sequences are favourably amplified over the original genome. The efficiency of generation of defective viruses should increase with the length of the repetitive sequence and should decrease when the complementarily of repetitive sequences decreases. Thus, viral vectors that express lhRNAs should be more difficult to produce than vectors that express shRNAs. Also, introduction of mutations in the sense strand of the hairpin may help to increase viral production.

Packaging cells are dedicated to recombinant virus production. Thus, these cells need to cope with many copies of viral genomes that may express high levels of transgenes. When recombinant vectors deliver shRNAs, packaging cells should express high amounts of shRNAs, high enough to be able to lead to toxicity due to IFN activation, off-target effects and saturation of the silencing machinery. These effects could go undetected as packaging cells are normally lysed a few days after viral infection to release viral progeny. However toxicity due to high levels of siRNAs could affect viral production. In fact, expression in packaging cells of antagonists that block IFN and the RNAi machinery increases virus production (de Vries *et al.*, 2008; Poluri and Sutton, 2008). Also, Dicer inhibition with siRNAs in packaging cells leads to higher production of shRNA expressing lentivirus and retrovirus (Poluri and Sutton, 2008). The Dicer-dependent titter decreased could be due to induction of off target inhibitions that affect virus production. Finally, examples exist of siRNAs anti-HIV expressed from HIV recombinant vectors. When the target region is in the recombinant vector, viral production should be blocked. Also, viral production is blocked when the antiviral siRNA targets a cellular gene required for production of the recombinant vector (Poluri and Sutton, 2008). In these cases alternative vectors should be used. Also, point mutations in the target region of the recombinant vector could be performed that do not affect viral production but render the vector insensitive to the antiviral siRNA.

## Viral effects on RNAi machinery

### *miRNAs expressed upon viral infection*

Mammalian viruses have been shown to use the RNAi machinery of the infected cell for their own benefit. This is achieved by expression of viral miRNAs important for virus viability and by changing the cellular miRNA signature. Thus, cellular miRNAs important for virus production are expressed while detrimental miRNAs decrease their function by several means.

Genome-wide screens by computational approaches and high-throughput validation have identified more than 100 putative microRNA precursors encoded by viruses. Viral miRNAs have been mostly validated in viruses that establish latent infections, where miRNAs may represent an effective way of modulating host cell gene expression in the long term. Thus, many viral miRNAs have been described in herpesviruses, including HSV (Cui *et al.*, 2006), human CMV (Pfeffer *et al.*, 2005), and the oncogenic Epstein–Barr virus (17 miRNA) (Pfeffer *et al.*, 2005; Xing and Kieff, 2007; Grundhoff *et al.*, 2006) or Kaposi's sarcoma-associated herpesvirus (12 miRNA) (Samols *et al.*, 2005; Pfeffer *et al.*, 2005). Other smaller DNA viruses such as adenovirus and SV40 also encode miRNAs (Andersson *et al.*, 2005; Aparicio *et al.*, 2006; Xu *et al.*, 2007; Sullivan *et al.*,

2005). Controversy exists over miRNA production by retroviruses such as HIV. Viral miRNAs are generally not conserved between different viral strains or with their host miRNAs (Dolken *et al.*, 2007; Buck *et al.*, 2007; Cai *et al.*, 2006; Schafer *et al.*, 2007), indicating rapid evolution of viral miRNAs. Results indicate that viral miRNAs control expression of both viral (Sullivan *et al.*, 2005; Grey *et al.*, 2007; Murphy *et al.*, 2008; Lo *et al.*, 2007) and cellular genes (Samols *et al.*, 2007; Skalsky *et al.*, 2007; Gottwein *et al.*, 2007) resulting in increased efficacy of infection and regulation of cell adhesion, migration, angiogenesis and immune response.

The best-characterized example of cellular miRNA that helps viral infection is miR-122. This liver-specific miRNA binds the 5′ end of HCV genome to increase viral replication, which depends on miR-122 function (Jopling *et al.*, 2005). Thus, the region bound by miR-122 in the 5′-UTR is a good therapeutic target to guide anti-HCV siRNAs.

Little is known about miRNA expression from viral vectors. Upon transduction, viral vectors could also modify the cellular miRNA signature or could express viral miRNAs that affect cell function. In fact, modification of cellular miRNAs is the result of saturation of the RNAi machinery by overexpression of exogenous shRNAs (see above). Also, first generation adenoviruses have the potential to express VA RNA derived viral miRNAs (Andersson *et al.*, 2005; Aparicio *et al.*, 2006; Xu *et al.*, 2007).

## Viral silencing suppressors

RNAi could work as an antiviral defence mechanism in infected cells. This has been well described for plant viruses and there is increasing evidence that the same could be true for mammalian viruses (Voinnet, 2005; Sullivan and Ganem, 2005; Schutz and Sarnow, 2006). There are several reasons why viruses could be targeted by the RNAi machinery. Viral infection courses in many cases with generation of dsRNA molecules that could be substrates for Dicer, which should produce antiviral siRNAs (Bennasser *et al.*, 2005; Pfeffer *et al.*, 2005). Also, cellular miRNAs could target viral expression and therefore affect viral production. Thus, miR-32 affects PFV-1 replication (Lecellier *et al.*, 2005) and a cluster of miRNAs including miR-28, miR-125b, miR-150, miR-223 and miR-382 redundantly target HIV expression (Huang *et al.*, 2007b). HIV production is also compromised by overexpression of miR-17–5p and miR17-20a, which could target PCAF (p300/CBP associated factor), a cellular cofactor for recruiting Tat to TAR and relieving the block of transcriptional elongation upon HIV integration (Triboulet *et al.*, 2007). Other small non-coding RNAs called piRNAs have been shown to suppress replication of mammalian endogenous retroviruses in germ and somatic cells (Watanabe *et al.*, 2006b; Carmell *et al.*, 2007).

Viruses have evolved to respond to antiviral RNAi by protection against RNAi, modulation of the cellular miRNA profile, mutational escape from RNAi and suppression of RNAi. As stated above, viruses can shield their genomes to protect them against RNAi (Westerhout *et al.*, 2006a), alter the miRNA signature in the infected cell and mutate their primary sequences or secondary structure to evade siRNA recognition (Das *et al.*, 2004; Westerhout *et al.*, 2005). Furthermore, several viruses have been described to express viral factors that decrease the functionality of the RNAi machinery (Fig. 11.4).

Silencing suppressors that target the RNAi machinery at different steps were first described in plant viruses. The p19 protein from tombosvirus sequesters siRNAs (Silhavy *et al.*, 2002), HC-Pro from polyvirus targets RISC (Llave *et al.*, 2000; Kasschau *et al.*, 2003), Cmv2b from mosaic virus blocks siRNA spreading and silencing by DNA methylation (Guo and Ding, 2002; Lucy *et al.*, 2000). Later, functional homologues of silencing suppressors were described in mammalian viruses (Fig. 11.4). E3L protein from vaccinia virus, NS1 protein from influenza virus and VP35 from Ebola are dsRNA binding proteins that may act as silencing suppressors by sequestering siRNAs and miRNAs and their precursors (Li *et al.*, 2004a; Haasnoot *et al.*, 2007; Delgadillo *et al.*, 2004; Bucher *et al.*, 2004; de Vries and Berkhout, 2008; Haasnoot *et al.*, 2007). Core protein from HCV blocks Dicer (Chen *et al.*, 2008; Wang *et al.*, 2006). HIV Tat protein can suppress RNAi by partially repressing the processing activity of Dicer (Bennasser *et al.*, 2005). Also, TAR RNA from HIV could sequester TRBP, rendering it unavailable for Dicer–RISC complexes (Bennasser

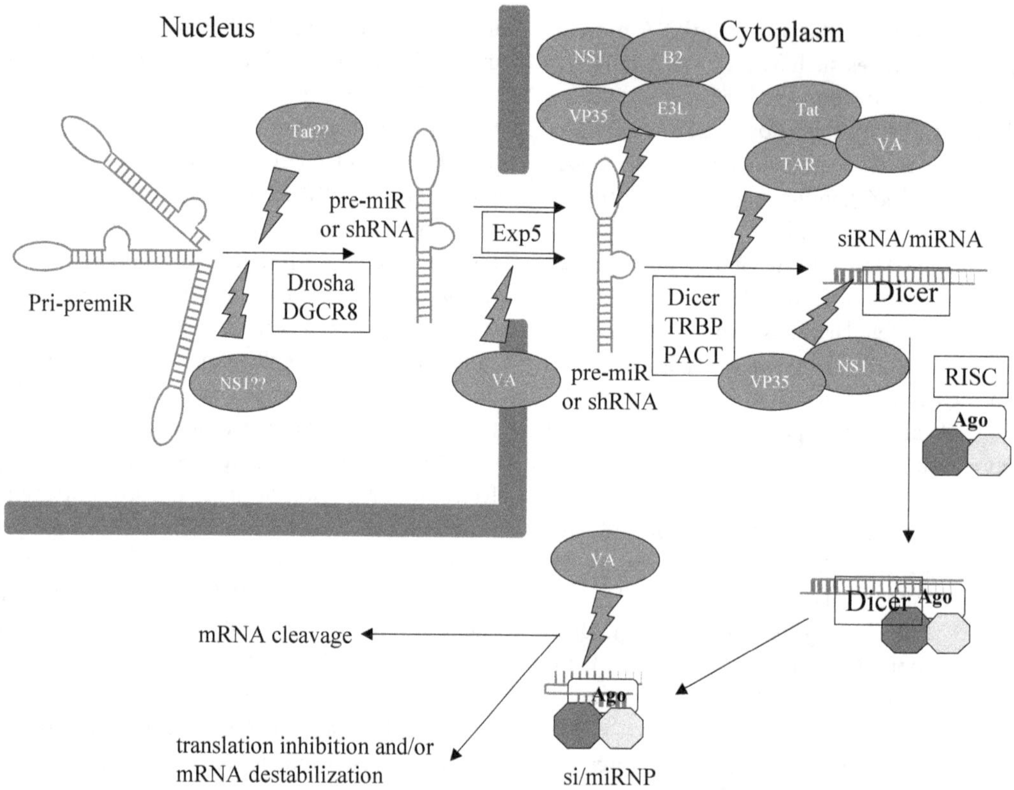


**Figure 11.4** Schematic of the interaction of viral silencing suppressors with the RNA interference machinery. HIV Tat and Influenza NS1 proteins could potentially sequester nuclear miRNA and siRNA precursors. The export to the cytoplasm of pre-miRNAs or shRNAs can also be inhibited by saturation of exportin5 by adenoviral VA RNAs. Once in the cytoplasm shRNAs and pre-miRNAs can be sequestered by NS1, Nodamura B2, Ebola VP35 and Vaccinia E3L proteins. VA RNAs can block Dicer and Tar RNA could sequester TRBP to affect Dicer processing. miRNAs and siRNAs can also be sequestered by VP35 and NS1. VA RNAs processed to viral miRNAs can saturate RISC complex leading to inhibition of RNAi. For details of the RNAi machinery see Fig. 11.1.

*et al.*, 2006). PFV-1 Tas protein blocks the RNAi pathway to overcome viral inhibition by cellular miR-32 (Lecellier *et al.*, 2005). Adenovirus VA RNAs can block the RNAi machinery and at the same time use the same pathway to be processed to viral miRNAs. VA RNAs are dsRNA molecules expressed to high amounts in the cell ($10^8$ molecules per infected cell) that act as a decoy substrate for Exportin 5 and bind to and inhibit Dicer (Lu and Cullen, 2004; Andersson *et al.*, 2005)). Three to five per cent of Dicer-bound VA RNAs can be processed to viral miRNAs important for viral viability that can saturate RISC complex (Andersson *et al.*, 2005; Aparicio *et al.*, 2006; Xu *et al.*, 2007).

In spite of these results, controversy still exists and definitive experiments should be performed to show that silencing suppression is required for the viability of mammalian viruses. Antiviral RNAi in plants could have been evolutionarily substituted in mammals by the antiviral interferon response. In fact, there is certain cross-talk between the RNAi machinery and the IFN pathway. Thus, IFN induction in HCV infected cells leads to the expression of antiviral cellular miRNAs (Pedersen *et al.*, 2007). IFN induced ADAR protein can alter or block miRNA functionality (Ohman, 2007). The Dicer binding proteins TRBP and PACT have IFN effector functions. Furthermore, many viral silencing suppressors are antagonists of the IFN response.

The multiple overlaps between IFN and RNAi pathways suggest that both mechanisms could act cooperatively to control virus infection.

Viral silencing suppressors could interfere with the activity of RNAi-based therapeutics. This could happen when the suppressors are expressed in the infected cell to be treated or when they are expressed from the viral vector used to deliver the therapeutic shRNAs, such as VA RNAs from first generation adenoviruses. However, viruses have been efficiently inhibited by RNAi and RNAi has been described *in vivo* in shRNAs delivered by first generation adenoviruses. This indicates that silencing suppressors are not very effective. However, RNAi in the presence of silencing suppressors may not be optimal and therefore virus inhibition by targeting the suppressor function may have better therapeutic effects.

## Concluding remarks

Two key challenges in developing RNAi as a therapy are avoiding toxic effects and ensuring efficient delivery. Viral vectors offer several advantages for shRNA delivery, but they still need to undergo major improvements concerning viral vector production, target cell specificity, long-term expression in humans and lack of toxicity associated with vector integration or with the expression of viral factors. Toxicity associated to shRNA expression has been related to activation of the IFN response, off-target effects and saturation of the cellular silencing machinery. All of these are dependent on siRNA concentration. Thus, it is important to identify the most potent siRNAs against the target gene of interest to obtain functional inhibitions at the lowest concentration of siRNA possible.

As RNAi is a novel technique, it is possible that novel issues regarding efficiency or secondary effects will arise in the future. However, given the great efforts made in the scientific community to understand silencing and to develop this technology further, it is conceivable that therapeutic RNA could be proven successful in clinical settings. There is great hope that in a few years, RNAi technology can revolutionize the treatment of many common and rare diseases currently with no effective cure, in the same way that silencing has changed completely the field of gene function analysis in basic science.

## References

Aagaard, L., Amarzguioui, M., Sun, G., Santos, L.C., Ehsani, A., Prydz, H., and Rossi, J.J. (2007). A facile lentiviral vector system for expression of doxycycline-inducible shRNAs: knockdown of the pre-miRNA processing enzyme Drosha. Mol. Ther. *15*, 938–945.

Aagaard, L., and Rossi, J.J. (2007). RNAi therapeutics: principles, prospects and challenges. Adv. Drug Deliv. Rev. *59*, 75–86.

Aagaard, L.A., Zhang, J., von Eije, K.J., Li, H., Saetrom, P., Amarzguioui, M., and Rossi, J.J. (2008). Engineering and optimization of the miR-106b cluster for ectopic expression of multiplexed anti-HIV RNAs. Gene Ther. *15*, 1536–1549.

Akashi, H., Miyagishi, M., Yokota, T., Watanabe, T., Hino, T., Nishina, K., Kohara, M., and Taira, K. (2005). Escape from the interferon response associated with RNA interference using vectors that encode long modified hairpin-RNA. Mol. Biosyst. *1*, 382–390.

Alexopoulou, L., Holt, A.C., Medzhitov, R., and Flavell, R.A. (2001). Recognition of double-stranded RNA and activation of NF-kappaB by Toll-like receptor 3. Nature *413*, 732–738.

Allen, D., Kenna, P.F., Palfi, A., McMahon, H.P., Millington-Ward, S., O'Reilly, M., Humphries, P., and Farrar, G.J. (2007). Development of strategies for conditional RNA interference. J. Gene Med. *9*, 287–298.

Amar, L., Desclaux, M., Faucon-Biguet, N., Mallet, J., and Vogel, R. (2006). Control of small inhibitory RNA levels and RNA interference by doxycycline induced activation of a minimal RNA polymerase III promoter. Nucleic Acids Res. *34*, e37.

Amarzguioui, M., Holen, T., Babaie, E., and Prydz, H. (2003). Tolerance for mutations and chemical modifications in a siRNA. Nucleic Acids Res. *31*, 589–595.

Amarzguioui, M., and Prydz, H. (2004). An algorithm for selection of functional siRNA sequences. Biochem. Biophys. Res. Commun. *316*, 1050–1058.

Ammosova, T., Berro, R., Kashanchi, F., and Nekhai, S. (2005). RNA interference directed to CDK2 inhibits HIV-1 transcription. Virology *341*, 171–178.

An, D.S., Qin, F.X., Auyeung, V.C., Mao, S.H., Kung, S.K., Baltimore, D., and Chen, I.S. (2006). Optimization and functional effects of stable short hairpin RNA expression in primary human lymphocytes via lentiviral vectors. Mol. Ther. *14*, 494–504.

Anderson, J., Li, M.J., Palmer, B., Remling, L., Li, S., Yam, P., Yee, J.K., Rossi, J., Zaia, J., and Akkina, R. (2007). Safety and efficacy of a lentiviral vector containing three anti-HIV genes--CCR5 ribozyme, tat-rev siRNA, and TAR decoy--in SCID-hu mouse-derived T cells. Mol. Ther. *15*, 1182–1188.

Andersson, M.G., Haasnoot, P.C., Xu, N., Berenjian, S., Berkhout, B., and Akusjarvi, G. (2005). Suppression of RNA interference by adenovirus virus-associated RNA. J. Virol. *79*, 9556–9565.

Anderson, J., and Akkina, R. (2005). CXCR4 and CCR5 shRNA transgenic CD34+ cell derived macrophages are functionally normal and resist HIV-1 infection. Retrovirology *2*, 53.

Anesti, A.M., Peeters, P.J., Royaux, I., and Coffin, R.S. (2008). Efficient delivery of RNA Interference to peripheral neurons *in vivo* using herpes simplex virus. Nucleic Acids Res. *36*, e86.

Aparicio, O., Razquin, N., Zaratiegui, M., Narvaiza, I., and Fortes, P. (2006). Adenovirus virus-associated RNA is processed to functional interfering RNAs involved in virus production. J. Virol. *80*, 1376–1384.

Arrighi, J.F., Pion, M., Wiznerowicz, M., Geijtenbeek, T.B., Garcia, E., Abraham, S., Leuba, F., Dutoit, V., Ducrey-Rundquist, O., van Kooyk, Y., *et al.* (2004). Lentivirus-mediated RNA interference of DC-SIGN expression inhibits human immunodeficiency virus transmission from dendritic cells to T cells. J. Virol. *78*, 10848–10855.

Azzouz, M., Martin-Rendon, E., Barber, R.D., Mitrophanous, K.A., Carter, E.E., Rohll, J.B., Kingsman, S.M., Kingsman, A.J., and Mazarakis, N.D. (2002). Multicistronic lentiviral vector-mediated striatal gene transfer of aromatic L-amino acid decarboxylase, tyrosine hydroxylase, and GTP cyclohydrolase I induces sustained transgene expression, dopamine production, and functional improvement in a rat model of Parkinson's disease. J. Neurosci *22*, 10302–10312.

Azzouz, M., and Mazarakis, N. (2004). Non-primate EIAV-based lentiviral vectors as gene delivery system for motor neuron diseases. Curr. Gene Ther. *4*, 277–286.

Babcock, A.M., Standing, D., Bullshields, K., Schwartz, E., Paden, C.M., and Poulsen, D.J. (2005). In vivo inhibition of hippocampal Ca2+/calmodulin-dependent protein kinase II by RNA interference. Mol. Ther. *11*, 899–905.

Banerjea, A., Li, M., Bauer, G., Remling, L., Lee, N., Rossi, J., and Akkina, R. (2003). Inhibition of HIV-1 by lentiviral vector-transduced siRNAs in T lymphocytes differentiated in SCID-hu mice and CD34+ progenitor cell-derived macrophages. Mol. Ther. *8*, 62–71.

Barichievy, S., Saayman, S., von Eije, K.J., Morris, K.V., Arbuthnot, P., and Weinberg, M.S. (2007). The inhibitory efficacy of RNA POL III-expressed long hairpin RNAs targeted to untranslated regions of the HIV-1 5′ long terminal repeat. Oligonucleotides *17*, 419–431.

Barton, G.M., and Medzhitov, R. (2002). Retroviral delivery of small interfering RNA into primary cells. Proc. Natl. Acad. Sci. U.S.A. *99*, 14943–14945.

Bennasser, Y., Le, S.Y., Benkirane, M., and Jeang, K.T. (2005). Evidence that HIV-1 encodes an siRNA and a suppressor of RNA silencing. Immunity *22*, 607–619.

Bennasser, Y., Yeung, M.L., and Jeang, K.T. (2006). HIV-1 TAR RNA subverts RNA interference in transfected cells through sequestration of TAR RNA-binding protein, TRBP. J. Biol. Chem. *281*, 27674–27678.

Berezhna, S.Y., Supekova, L., Supek, F., Schultz, P.G., and Deniz, A.A. (2006). siRNA in human cells selectively localizes to target RNA sites. Proc. Natl. Acad. Sci. U.S.A. *103*, 7682–7687.

Berges, B.K., Wheat, W.H., Palmer, B.E., Connick, E., and Akkina, R. (2006). HIV-1 infection and CD4 T cell depletion in the humanized Rag2$^{-/-}$gamma c$^{-/-}$ (RAG-hu) mouse model. Retrovirology 3, 76.

Berns, K., Hijmans, E.M., Mullenders, J., Brummelkamp, T.R., Velds, A., Heimerikx, M., Kerkhoven, R.M., Madiredjo, M., Nijkamp, W., Weigelt, B., *et al.* (2004). A large-scale RNAi screen in human cells identifies new components of the p53 pathway. Nature *428*, 431–437.

Bevilacqua, P.C., and Cech, T.R. (1996). Minor-groove recognition of double-stranded RNA by the double-stranded RNA-binding domain from the RNA-activated protein kinase PKR. Biochemistry *35*, 9983–9994.

Birmingham, A., Anderson, E.M., Reynolds, A., Ilsley-Tyree, D., Leake, D., Fedorov, Y., Baskerville, S., Maksimova, E., Robinson, K., Karpilow, J., *et al.* (2006). 3′-UTR seed matches, but not overall identity, are associated with RNAi off-targets. Nat. Methods *3*, 199–204.

Blankinship, M.J., Gregorevic, P., Allen, J.M., Harper, S.Q., Harper, H., Halbert, C.L., Miller, D.A., and Chamberlain, J.S. (2004). Efficient transduction of skeletal muscle using vectors based on adeno-associated virus serotype 6. Mol. Ther. *10*, 671–678.

Blight, K.J., Kolykhalov, A.A., and Rice, C.M. (2000). Efficient initiation of HCV RNA replication in cell culture. Science *290*, 1972–1974.

Boden, D., Pusch, O., Lee, F., Tucker, L., and Ramratnam, B. (2003). Human immunodeficiency virus type 1 escape from RNA interference. J. Virol. *77*, 11531–11535.

Boden, D., Pusch, O., Silbermann, R., Lee, F., Tucker, L., and Ramratnam, B. (2004). Enhanced gene silencing of HIV-1 specific siRNA using microRNA designed hairpins. Nucleic Acids Res. *32*, 1154–1158.

Boese, Q., Leake, D., Reynolds, A., Read, S., Scaringe, S.A., Marshall, W.S., and Khvorova, A. (2005). Mechanistic insights aid computational short interfering RNA design. Methods Enzymol *392*, 73–96.

Brass, A.L., Dykxhoorn, D.M., Benita, Y., Yan, N., Engelman, A., Xavier, R.J., Lieberman, J., and Elledge, S.J. (2008). Identification of host proteins required for HIV infection through a functional genomic screen. Science *319*, 921–926.

Bridge, A.J., Pebernard, S., Ducraux, A., Nicoulaz, A.L., and Iggo, R. (2003). Induction of an interferon response by RNAi vectors in mammalian cells. Nat. Genet. *34*, 263–264.

Brown, K.M., Chu, C.Y., and Rana, T.M. (2005). Target accessibility dictates the potency of human RISC. Nat. Struct. Mol. Biol. *12*, 469–470.

Brummelkamp, T.R., Bernards, R., and Agami, R. (2002a). Stable suppression of tumorigenicity by virus-mediated RNA interference. Cancer Cell *2*, 243–247.

Brummelkamp, T.R., Bernards, R., and Agami, R. (2002b). A system for stable expression of short interfering RNAs in mammalian cells. Science *296*, 550–553.

Buck, A.H., Santoyo-Lopez, J., Robertson, K.A., Kumar, D.S., Reczko, M., and Ghazal, P. (2007). Discrete clusters of virus-encoded micrornas are associated with complementary strands of the genome and the 7.2-kilobase stable intron in murine cytomegalovirus. J. Virol. *81*, 13761–13770.

Bucher, E., Hemmes, H., de Haan, P., Goldbach, R., and Prins, M. (2004). The influenza A virus NS1 protein binds small interfering RNAs and suppresses RNA silencing in plants. J. Gen. Virol. *85*, 983–991.

Bukrinsky, M.I., Haggerty, S., Dempsey, M.P., Sharova, N., Adzhubel, A., Spitz, L., Lewis, P., Goldfarb, D., Emerman, M., and Stevenson, M. (1993). A nuclear localization signal within HIV-1 matrix protein that governs infection of non-dividing cells. Nature *365*, 666–669.

Burns, J.C., Friedmann, T., Driever, W., Burrascano, M., and Yee, J.K. (1993). Vesicular stomatitis virus G glycoprotein pseudotyped retroviral vectors: concentration to very high titer and efficient gene transfer into mammalian and nonmammalian cells. Proc. Natl. Acad. Sci. U.S.A. *90*, 8033–8037.

Cai, X., Schafer, A., Lu, S., Bilello, J.P., Desrosiers, R.C., Edwards, R., Raab-Traub, N., and Cullen, B.R. (2006). Epstein–Barr virus microRNAs are evolutionarily conserved and differentially expressed. PLoS Pathog. *2*, e23.

Carlson, C.B., Stephens, O.M., and Beal, P.A. (2003). Recognition of double-stranded RNA by proteins and small molecules. Biopolymers *70*, 86–102.

Carmell, M.A., Girard, A., van de Kant, H.J., Bourc'his, D., Bestor, T.H., de Rooij, D.G., and Hannon, G.J. (2007). MIWI2 is essential for spermatogenesis and repression of transposons in the mouse male germline. Dev. Cell *2*, 503–514.

Carmona, S., Ely, A., Crowther, C., Moolla, N., Salazar, F.H., Marion, P.L., Ferry, N., Weinberg, M.S., and Arbuthnot, P. (2006). Effective inhibition of HBV replication *in vivo* by anti-HBx short hairpin RNAs. Mol. Ther. *13*, 411–421.

Castanotto, D., Li, H., and Rossi, J.J. (2002). Functional siRNA expression from transfected PCR products. Rna *8*, 1454–1460.

Castanotto, D., Sakurai, K., Lingeman, R., Li, H., Shively, L., Aagaard, L., Soifer, H., Gatignol, A., Riggs, A., and Rossi, J.J. (2007). Combinatorial delivery of small interfering RNAs reduces RNAi efficacy by selective incorporation into RISC. Nucleic Acids Res. *35*, 5154–5164.

Chalk, A.M., Wahlestedt, C., and Sonnhammer, E.L. (2004). Improved and automated prediction of effective siRNA. Biochem. Biophys. Res. Commun. *319*, 264–274.

Chan, J.A., Krichevsky, A.M., and Kosik, K.S. (2005). MicroRNA-21 is an antiapoptotic factor in human glioblastoma cells. Cancer Res. *65*, 6029–6033.

Chang, L., Liu, X., and He, J. (2005). Lentiviral siRNAs targeting multiple highly conserved RNA sequences of human immunodeficiency virus type 1. Gene Ther. *12*, 1133–1144.

Chander, G., Sulkowski, M.S., Jenckes, M.W., Torbenson, M.S., Herlong, H.F., Bass, E.B., and Gebo, K.A. (2002). Treatment of chronic hepatitis C: a systematic review. Hepatology *36*, S135–144.

Chang, H.S., Lin, C.H., Chen, Y.C., and Yu, W.C. (2004). Using siRNA technique to generate transgenic animals with spatiotemporal and conditional gene knockdown. Am J Pathol *165*, 1535–1541.

Chang, K., Elledge, S.J., and Hannon, G.J. (2006). Lessons from Nature: microRNA-based shRNA libraries. Nat. Methods *3*, 707–714.

Chang, K.S., and Luo, G. (2006). The polypyrimidine tract-binding protein (PTB) is required for efficient replication of hepatitis C virus (HCV) RNA. Virus Res. *115*, 1–8.

Chen, C.C., Ko, T.M., Ma, H.I., Wu, H.L., Xiao, X., Li, J., Chang, C.M., Wu, P.Y., Chen, C.H., Han, J.M., *et al.* (2007). Long-term inhibition of hepatitis B virus in transgenic mice by double-stranded adeno-associated virus 8-delivered short hairpin RNA. Gene Ther. *14*, 11–19.

Chen, W., Liu, M., Jiao, Y., Yan, W., Wei, X., Chen, J., Fei, L., Liu, Y., Zuo, X., Yang, F., *et al.* (2006a). Adenovirus-mediated RNA interference against foot-and-mouth disease virus infection both in vitro and *in vivo*. J. Virol. *80*, 3559–3566.

Chen, W., Zhang, Z., Chen, J., Zhang, J., Zhang, J., Wu, Y., Huang, Y., Cai, X., and Huang, A. (2008). HCV core protein interacts with Dicer to antagonize RNA silencing. Virus Res. *133*, 250–258.

Chen, Y., Du, D., Wu, J., Chan, C.P., Tan, Y., Kung, H.F., and He, M.L. (2003). Inhibition of hepatitis B virus replication by stably expressed shRNA. Biochem. Biophys. Res. Commun. *311*, 398–404.

Chen, Z., Stockton, J., Mathis, D., and Benoist, C. (2006b). Modeling CTLA4-linked autoimmunity with RNA interference in mice. Proc. Natl. Acad. Sci. U.S.A. *103*, 16400–16405.

Chen, J., Wall, N., Kocher, K., Duclos, N., Fabbro, D., Neuberg, D., Griffin, J., Shi, Y., and Gilliland, D. (2004). Stable expression of small interfering RNA sensitizes TEL-PDGFbetaR to inhibition with imatinib or rapamycin. J. Clin. Invest. *113*, 1784–1791.

Chendrimada, T.P., Gregory, R.I., Kumaraswamy, E., Norman, J., Cooch, N., Nishikura, K., and Shiekhattar, R. (2005). TRBP recruits the Dicer complex to Ago2 for microRNA processing and gene silencing. Nature *436*, 740–744.

Chi, J.T., Chang, H.Y., Wang, N.N., Chang, D.S., Dunphy, N., and Brown, P.O. (2003). Genomewide view of gene silencing by small interfering RNAs. Proc. Natl. Acad. Sci. U.S.A. *100*, 6343–6346.

Chiu, Y.L., Cao, H., Jacque, J.M., Stevenson, M., and Rana, T.M. (2004). Inhibition of human immunodeficiency virus type 1 replication by RNA interference directed against human transcription elongation factor P-TEFb (CDK9/CyclinT1). J. Virol. *78*, 2517–2529.

Cho-Rok, J., Yoo, J., Jang, Y.J., Kim, S., Chu, I.S., Yeom, Y.I., Choi, J.Y., and Im, D.S. (2006). Adenovirus-mediated transfer of siRNA against PTTG1 inhibits liver cancer cell growth in vitro and *in vivo*. Hepatology *43*, 1042–1052.

Choy, E.Y., Kok, K.H., Tsao, S.W., and Jin, D.Y. (2008). Utility of Epstein–Barr virus-encoded small RNA promoters for driving the expression of fusion transcripts harboring short hairpin RNAs. Gene Ther. *15*, 191–202.

Condreay, J.P., Witherspoon, S.M., Clay, W.C., and Kost, T.A. (1999). Transient and stable gene expression in mammalian cells transduced with a recombinant baculovirus vector. Proc. Natl. Acad. Sci. U.S.A. *96*, 127–132.

Coumoul, X., Li, W., Wang, R.H., and Deng, C. (2004). Inducible suppression of Fgfr2 and Survivin in ES cells using a combination of the RNA interference (RNAi) and the Cre-LoxP system. Nucleic Acids Res. *32*, e85.

Coumoul, X., Shukla, V., Li, C., Wang, R.H., and Deng, C.X. (2005). Conditional knockdown of Fgfr2 in mice using Cre-LoxP induced RNA interference. Nucleic Acids Res. *33*, e102.

Cui, C., Griffiths, A., Li, G., Silva, L.M., Kramer, M.F., Gaasterland, T., Wang, X.J., and Coen, D.M. (2006). Prediction and identification of herpes simplex virus 1-encoded microRNAs. J. Virol. *80*, 5499–5508.

Cullen, B.R. (2006). Induction of stable RNA interference in mammalian cells. Gene Ther. *13*, 503–508.

Czauderna, F., Santel, A., Hinz, M., Fechtner, M., Durieux, B., Fisch, G., Leenders, F., Arnold, W., Giese, K., Klippel, A., and Kaufmann, J. (2003). Inducible shRNA expression for application in a prostate cancer mouse model. Nucleic Acids Res. *31*, e127.

Das, A.T., Brummelkamp, T.R., Westerhout, E.M., Vink, M., Madiredjo, M., Bernards, R., and Berkhout, B. (2004). Human immunodeficiency virus type 1 escapes from RNA interference-mediated inhibition. J. Virol. *78*, 2601–2605.

Dauber, B., Heins, G., and Wolff, T. (2004). The influenza B virus nonstructural NS1 protein is essential for efficient viral growth and antagonizes beta interferon induction. J. Virol. *78*, 1865–1872.

Davidoff, A.M., Gray, J.T., Ng, C.Y., Zhang, Y., Zhou, J., Spence, Y., Bakar, Y., and Nathwani, A.C. (2005). Comparison of the ability of adeno-associated viral vectors pseudotyped with serotype 2, 5, and 8 capsid proteins to mediate efficient transduction of the liver in murine and nonhuman primate models. Mol. Ther. *11*, 875–888.

Davidson, B.L., and Boudreau, R.L. (2007). RNA interference: a tool for querying nervous system function and an emerging therapy. Neuron *53*, 781–788.

de Fougerolles, A., Vornlocher, H.P., Maraganore, J., and Lieberman, J. (2007). Interfering with disease: a progress report on siRNA-based therapeutics. Nat Rev Drug Discov *6*, 443–453.

de Vries, W., and Berkhout, B. (2008). RNAi suppressors encoded by pathogenic human viruses. Int. J. Biochem. Cell Biol. *40*, 2007–2012.

de Vries, W., Haasnoot, J., van der Velden, J., van Montfort, T., Zorgdrager, F., Paxton, W., Cornelissen, M., van Kuppeveld, F., de Haan, P., and Berkhout, B. (2008). Increased virus replication in mammalian cells by blocking intracellular innate defense responses. Gene Ther. *15*, 545–552.

Delgadillo, M.O., Saenz, P., Salvador, B., Garcia, J.A., and Simon-Mateo, C. (2004). Human influenza virus NS1 protein enhances viral pathogenicity and acts as an RNA silencing suppressor in plants. J. Gen. Virol. *85*, 993–999.

Denli, A.M., Tops, B.B., Plasterk, R.H., Ketting, R.F., and Hannon, G.J. (2004). Processing of primary microRNAs by the Microprocessor complex. Nature *432*, 231–235.

Denti, M.A., Rosa, A., Sthandier, O., De Angelis, F.G., and Bozzoni, I. (2004). A new vector, based on the PolII promoter of the U1 snRNA gene, for the expression of siRNAs in mammalian cells. Mol Ther. *10*, 191–199.

Devroe, E., and Silver, P.A. (2002). Retrovirus-delivered siRNA. BMC Biotechnol *28*, 15.

Dickins, R.A., Hemann, M.T., Zilfou, J.T., Simpson, D.R., Ibarra, I., Hannon, G.J., and Lowe, S.W. (2005). Probing tumor phenotypes using stable and regulated synthetic microRNA precursors. Nat. Genet. *37*, 1289–1295.

Ding, H., Schwarz, D.S., Keene, A., Affar el, B., Fenton, L., Xia, X., Shi, Y., Zamore, P.D., and Xu, Z. (2003). Selective silencing by RNAi of a dominant allele that causes amyotrophic lateral sclerosis. Aging Cell 2, 209–217.

Dittmer, A., and Bogner, E. (2006). Specific short hairpin RNA-mediated inhibition of viral DNA packaging of human cytomegalovirus. FEBS Lett. *580*, 6132–6138.

Dolken, L., Perot, J., Cognat, V., Alioua, A., John, M., Soutschek, J., Ruzsics, Z., Koszinowski, U., Voinnet, O., and Pfeffer, S. (2007). Mouse cytomegalovirus microRNAs dominate the cellular small RNA profile during lytic infection and show features of posttranscriptional regulation. J. Virol. *81*, 13771–13782.

Domitrovich, A.M., Diebel, K.W., Ali, N., Sarker, S., and Siddiqui, A. (2005). Role of La autoantigen and polypyrimidine tract-binding protein in HCV replication. Virology *335*, 72–86.

Doyle, S.E., O'Connell, R., Vaidya, S.A., Chow, E.K., Yee, K., and Cheng, G. (2003). Toll-like receptor 3 mediates a more potent antiviral response than Toll-like receptor 4. J. Immunol. *170*, 3565–3571.

Dykxhoorn, D.M., and Lieberman, J. (2006). Running interference: prospects and obstacles to using small interfering RNAs as small molecule drugs. Annu. Rev. Biomed. Eng. *8*, 377–402.

Elbashir, S.M., Harborth, J., Lendeckel, W., Yalcin, A., Weber, K., and Tuschl, T. (2001a). Duplexes of 21-nucleotide RNAs mediate RNA interference in cultured mammalian cells. Nature *411*, 494–498.

Elbashir, S.M., Lendeckel, W., and Tuschl, T. (2001b). RNA interference is mediated by 21- and 22-nucleotide RNAs. Genes Dev. *15*, 188–200.

Elbashir, S.M., Martinez, J., Patkaniowska, A., Lendeckel, W., and Tuschl, T. (2001c). Functional anatomy of siRNAs for mediating efficient RNAi in *Drosophila melanogaster* embryo lysate. EMBO J. *20*, 6877–6888.

Ely, A., Naidoo, T., Mufamadi, S., Crowther, C., and Arbuthnot, P. (2008). Expressed anti-HBV primary microRNA shuttles inhibit viral replication efficiently in vitro and *in vivo*. Mol. Ther. *16*, 1105–1112.

Eulalio, A., Huntzinger, E., and Izaurralde, E. (2008). Getting to the root of miRNA-mediated gene silencing. Cell *132*, 9–14.

Fechner, H., Sipo, I., Westermann, D., Pinkert, S., Wang, X., Suckau, L., Kurreck, J., Zeichhardt, H., Muller, O., Vetter, R., *et al.* (2008). Cardiac-targeted RNA interference mediated by an AAV9 vector improves cardiac function in coxsackievirus B3 cardiomyopathy. J. Mol. Med. *86*, 987–997.

Fedorov, Y., Anderson, E.M., Birmingham, A., Reynolds, A., Karpilow, J., Robinson, K., Leake, D., Marshall, W.S., and Khvorova, A. (2006). Off-target effects by siRNA can induce toxic phenotype. Rna *12*, 1188–1196.

Fewell, G.D., and Schmitt, K. (2006). Vector-based RNAi approaches for stable, inducible and genome-wide screens. Drug Discov. Today *11*, 975–982.

Fish, R.J., and Kruithof, E.K. (2004). Short-term cytotoxic effects and long-term instability of RNAi delivered using lentiviral vectors. BMC Mol. Biol. *5*, 9.

Follenzi, A., Sabatino, G., Lombardo, A., Boccaccio, C., and Naldini, L. (2002). Efficient gene delivery and targeted expression to hepatocytes *in vivo* by improved lentiviral vectors. Hum. Gene Ther. *13*, 243–260.

Fountaine, T.M., and Wade-Martins, R. (2007). RNA interference-mediated knockdown of alpha-synuclein protects human dopaminergic neuroblastoma cells from MPP(+) toxicity and reduces dopamine transport. J. Neurosci. Res. *85*, 351–363.

Frampton, A.R., Jr., Goins, W.F., Nakano, K., Burton, E.A., and Glorioso, J.C. (2005). HSV trafficking and development of gene therapy vectors with applications in the nervous system. Gene Ther. *12*, 891–901.

Franich, N.R., Fitzsimons, H.L., Fong, D.M., Klugmann, M., During, M.J., and Young, D. (2008). AAV vector-mediated RNAi of mutant huntingtin expression is neuroprotective in a novel genetic rat model of Huntington's disease. Mol. Ther. *16*, 947–956.

Fritsch, L., Martinez, L.A., Sekhri, R., Naguibneva, I., Gerard, M., Vandromme, M., Schaeffer, L., and Harel-Bellan, A. (2004). Conditional gene knock-down by CRE-dependent short interfering RNAs. EMBO Rep. *5*, 178–182.

Fukagawa, T., Nogami, M., Yoshikawa, M., Ikeno, M., Okazaki, T., Takami, Y., Nakayama, T., and Oshimura, M. (2004). Dicer is essential for formation of the heterochromatin structure in vertebrate cells. Nat. Cell Biol. *6*, 784–791.

Garza, J.C., Kim, C.S., Liu, J., Zhang, W., and Lu, X.Y. (2008). Adeno-associated virus-mediated knockdown of melanocortin-4 receptor in the paraventricular nucleus of the hypothalamus promotes high-fat diet-induced hyperphagia and obesity. J. Endocrinol. *197*, 471–482.

Ghahramani Seno, M.M., Graham, I.R., Athanasopoulos, T., Trollet, C., Pohlschmidt, M., Crompton, M.R., and Dickson, G. (2008). RNAi-mediated knockdown of dystrophin expression in adult mice does not lead to overt muscular dystrophy pathology. Hum. Mol. Genet. *17*, 2622–2632.

Giering, J., Grimm, D., Storm, T., and Kay, M. (2008). Expression of shRNA from a tissue-specific pol II promoter is an effective and safe RNAi therapeutic. Mol. Ther. *16*, 1630–1636.

Gil, J., and Esteban, M. (2000). Induction of apoptosis by the dsRNA-dependent protein kinase (PKR): mechanism of action. Apoptosis *5*, 107–114.

Gitlin, L., Karelsky, S., and Andino, R. (2002). Short interfering RNA confers intracellular antiviral immunity in human cells. Nature *418*, 430–434.

Golding, M.C., Long, C.R., Carmell, M.A., Hannon, G.J., and Westhusin, M.E. (2006). Suppression of prion protein in livestock by RNA interference. Proc. Natl. Acad. Sci. U.S.A. *103*, 5285–5290.

Gonzalez-Alegre, P., Bode, N., Davidson, B.L., and Paulson, H.L. (2005). Silencing primary dystonia: lentiviral-mediated RNA interference therapy for DYT1 dystonia. J. Neurosci. *25*, 10502–10509.

Gottwein, E., Mukherjee, N., Sachse, C., Frenzel, C., Majoros, W.H., Chi, J.T., Braich, R., Manoharan, M., Soutschek, J., Ohler, U., and Cullen, B.R. (2007). A viral microRNA functions as an orthologue of cellular miR-155. Nature *450*, 1096–1099.

Gou, D., Narasaraju, T., Chintagari, N.R., Jin, N., Wang, P., and Liu, L. (2004). Gene silencing in alveolar type II cells using cell-specific promoter in vitro and *in vivo*. Nucleic Acids Res. *32*, e134.

Gray, D.C., Hoeflich, K.P., Peng, L., Gu, Z., Gogineni, A., Murray, L.J., Eby, M., Kljavin, N., Seshagiri, S., Cole, M.J., and Davis, D.P. (2007). pHUSH: a single vector system for conditional gene expression. BMC Biotechnol. *7*, 61.

Gregory, L.G., Waddington, S.N., Holder, M.V., Mitrophanous, K.A., Buckley, S.M., Mosley, K.L., Bigger, B.W., Ellard, F.M., Walmsley, L.E., Lawrence, L., *et al.* (2004a). Highly efficient EIAV-mediated in utero gene transfer and expression in the major muscle groups affected by Duchenne muscular dystrophy. Gene Ther. *11*, 1117–1125.

Gregory, R.I., Chendrimada, T.P., Cooch, N., and Shiekhattar, R. (2005). Human RISC couples microRNA biogenesis and posttranscriptional gene silencing. Cell *123*, 631–640.

Gregory, R.I., Yan, K.P., Amuthan, G., Chendrimada, T., Doratotaj, B., Cooch, N., and Shiekhattar, R. (2004b). The Microprocessor complex mediates the genesis of microRNAs. Nature *432*, 235–240.

Grey, F., Meyers, H., White, E.A., Spector, D.H., and Nelson, J. (2007). A human cytomegalovirus-encoded microRNA regulates expression of multiple viral genes involved in replication. PLoS Pathog. *3*, e163.

Grimm, D., Streetz, K.L., Jopling, C.L., Storm, T.A., Pandey, K., Davis, C.R., Marion, P., Salazar, F., and Kay, M.A. (2006). Fatality in mice due to oversaturation of cellular microRNA/short hairpin RNA pathways. Nature *441*, 537–541.

Grundhoff, A., Sullivan, C.S., and Ganem, D. (2006). A combined computational and microarray-based approach identifies novel microRNAs encoded by human gamma-herpesviruses. Rna *12*, 733–750.

Gu, W., Putral, L., Hengst, K., Minto, K., Saunders, N.A., Leggatt, G., and McMillan, N.A. (2006). Inhibition of cervical cancer cell growth in vitro and *in vivo* with lentiviral-vector delivered short hairpin RNA targeting human papillomavirus E6 and E7 oncogenes. Cancer Gene Ther. *13*, 1023–1032.

Guo, H.S., and Ding, S.W. (2002). A viral protein inhibits the long range signaling activity of the gene silencing signal. EMBO J. *21*, 398–407.

Gupta, S., Schoer, R.A., Egan, J.E., Hannon, G.J., and Mittal, V. (2004). Inducible, reversible, and stable RNA interference in mammalian cells. Proc. Natl. Acad. Sci. U.S.A. *101*, 1927–1932.

Gurzov, E.N., and Izquierdo, M. (2006). RNA interference against Hec1 inhibits tumor growth *in vivo*. Gene Ther. *13*, 1–7.

Haasnoot, J., and Berkhout, B. (2006). RNA interference: its use as antiviral therapy. Handb. Exp. Pharmacol. 117–150.

Haasnoot, J., de Vries, W., Geutjes, E.J., Prins, M., de Haan, P., and Berkhout, B. (2007). The Ebola virus VP35 protein is a suppressor of RNA silencing. PLoS Pathog. *3*, e86.

Hacein-Bey-Abina, S., von Kalle, C., Schmidt, M., Le Deist, F., Wulffraat, N., McIntyre, E., Radford, I., Villeval, J.L., Fraser, C.C., Cavazzana-Calvo, M., and Fischer, A. (2003). A serious adverse event after successful gene therapy for X-linked severe combined immunodeficiency. N. Engl. J. Med. *348*, 255–256.

Hamasaki, K., Nakao, K., Matsumoto, K., Ichikawa, T., Ishikawa, H., and Eguchi, K. (2003). Short interfering RNA-directed inhibition of hepatitis B virus replication. FEBS Lett. *543*, 51–54.

Hamazaki, H., Ujino, S., Abe, E., Miyano-Kurosaki, N., Shimotohno, K., and Takaku, H. (2004). RNAi expression mediated inhibition of HCV replication. Nucleic Acids Symp Ser (Oxf), 307–308.

Hammond, S.M., Bernstein, E., Beach, D., and Hannon, G.J. (2000). An RNA-directed nuclease mediates post-transcriptional gene silencing in *Drosophila* cells. Nature *404*, 293–296.

Han, J., Lee, Y., Yeom, K.H., Kim, Y.K., Jin, H., and Kim, V.N. (2004). The Drosha–DGCR8 complex in primary microRNA processing. Genes Dev. *18*, 3016–3027.

Han, J., Lee, Y., Yeom, K.H., Nam, J.W., Heo, I., Rhee, J.K., Sohn, S.Y., Cho, Y., Zhang, B.T., and Kim, V.N. (2006). Molecular basis for the recognition of primary microRNAs by the Drosha–DGCR8 complex. Cell *125*, 887–901.

Harper, S.Q., Staber, P.D., He, X., Eliason, S.L., Martins, I.H., Mao, Q., Yang, L., Kotin, R.M., Paulson, H.L., and Davidson, B.L. (2005). RNA interference improves motor and neuropathological abnormalities in a Huntington's disease mouse model. Proc. Natl. Acad. Sci. U.S.A. *102*, 5820–5825.

Hauber, I., Bevec, D., Heukeshoven, J., Kratzer, F., Horn, F., Choidas, A., Harrer, T., and Hauber, J. (2005). Identification of cellular deoxyhypusine synthase as a novel target for antiretroviral therapy. J. Clin. Invest. *115*, 76–85.

Hayafune, M., Miyano-Kurosaki, N., Park, W., Moori, Y., and Takaku, H. (2006). Silencing of HIV-1 gene expression by two types of siRNA expression systems. Antivir. Chem. Chemother. *17*, 241–249.

He, L., Thomson, J.M., Hemann, M.T., Hernando-Monge, E., Mu, D., Goodson, S., Powers, S., Cordon-Cardo, C., Lowe, S.W., Hannon, G.J., and Hammond, S.M. (2005). A microRNA polycistron as a potential human oncogene. Nature *435*, 828–833.

Heale, B.S., Soifer, H.S., Bowers, C., and Rossi, J.J. (2005). siRNA target site secondary structure predictions using local stable substructures. Nucleic Acids Res. *33*, e30.

Heinonen, J.E., Mohamed, A.J., Nore, B.F., and Smith, C.I. (2005). Inducible H1 promoter-driven lentiviral siRNA expression by Stuffer reporter deletion. Oligonucleotides *15*, 139–144.

Henry, S.D., van der Wegen, P., Metselaar, H.J., Tilanus, H.W., Scholte, B.J., and van der Laan, L.J. (2006). Simultaneous targeting of HCV replication and viral binding with a single lentiviral vector containing multiple RNA interference expression cassettes. Mol. Ther. *14*, 485–493.

Henschel, A., Buchholz, F., and Habermann, B. (2004). DEQOR: a web-based tool for the design and quality control of siRNAs. Nucleic Acids Res. *32*, W113–120.

Higuchi, M., Tsutsumi, R., Higashi, H., and Hatakeyama, M. (2004). Conditional gene silencing utilizing the lac repressor reveals a role of SHP-2 in cagA-positive Helicobacter pylori pathogenicity. Cancer Sci. *95*, 442–447.

Ho, Y.C., Chen, H.C., Wang, K.C., and Hu, Y.C. (2004). Highly efficient baculovirus-mediated gene transfer into rat chondrocytes. Biotechnol. Bioeng. *88*, 643–651.

Ho, Y.C., Chung, Y.C., Hwang, S.M., Wang, K.C., and Hu, Y.C. (2005). Transgene expression and differentiation of baculovirus-transduced human mesenchymal stem cells. J. Gene Med. *7*, 860–868.

Hoeflich, K.P., Gray, D.C., Eby, M.T., Tien, J.Y., Wong, L., Bower, J., Gogineni, A., Zha, J., Cole, M.J., Stern, H.M., *et al.* (2006). Oncogenic BRAF is required for tumor growth and maintenance in melanoma models. Cancer Res. *66*, 999–1006.

Hofmann, A., Kessler, B., Ewerling, S., Kabermann, A., Brem, G., Wolf, E., and Pfeifer, A. (2006). Epigenetic regulation of lentiviral transgene vectors in a large animal model. Mol. Ther. *13*, 59–66.

Hofmann, C., and Strauss, M. (1998). Baculovirus-mediated gene transfer in the presence of human serum or blood facilitated by inhibition of the complement system. Gene Ther. *5*, 531–536.

Hommel, J.D., Sears, R.M., Georgescu, D., Simmons, D.L., and DiLeone, R.J. (2003). Local gene knockdown in the brain using viral-mediated RNA interference. Nat. Med. *9*, 1539–1544.

Hong, C.S., Goins, W.F., Goss, J.R., Burton, E.A., and Glorioso, J.C. (2006). Herpes simplex virus RNAi and neprilysin gene transfer vectors reduce accumulation of Alzheimer's disease-related amyloid-beta peptide *in vivo*. Gene Ther. *13*, 1068–1079.

Hornung, V., Ellegast, J., Kim, S., Brzozka, K., Jung, A., Kato, H., Poeck, H., Akira, S., Conzelmann, K.K., Schlee, M., *et al.* (2006). 5′-Triphosphate RNA is the ligand for RIG-I. Science *314*, 994–997.

Hornung, V., Guenthner-Biller, M., Bourquin, C., Ablasser, A., Schlee, M., Uematsu, S., Noronha, A., Manoharan, M., Akira, S., de Fougerolles, A., *et al.* (2005). Sequence-specific potent induction of IFN-alpha by short interfering RNA in plasmacytoid dendritic cells through TLR7. Nat. Med. *11*, 263–270.

Huang, A., Chen, Y., Wang, X., Zhao, S., Su, N., and White, D.W. (2004). Functional silencing of hepatic microsomal glucose-6-phosphatase gene expression *in vivo* by adenovirus-mediated delivery of short hairpin RNA. FEBS Lett. *558*, 69–73.

Huang, B.J., Liu, R.Y., Huang, J.L., Liang, Z.H., Gao, G.F., Wu, J.X., and Huang, W. (2007a). Long-Term toxicity studies in Canine of E10A, an adenoviral vector for human endostatin gene. Hum. Gene Ther. *18*, 207–221.

Huang, J., Wang, F., Argyris, E., Chen, K., Liang, Z., Tian, H., Huang, W., Squires, K., Verlinghieri, G., and Zhang, H. (2007b). Cellular microRNAs contribute to HIV-1 latency in resting primary CD4+ T lymphocytes. Nat. Med. *13*, 1241–1247.

Huser, A., and Hofmann, C. (2003). Baculovirus vectors: novel mammalian cell gene-delivery vehicles and their applications. Am J Pharmacogenomics 3, 53–63.

Hutvagner, G., and Zamore, P.D. (2002). A microRNA in a multiple-turnover RNAi enzyme complex. Science *297*, 2056–2060.

Inoue, A., Sawata, S.Y., and Taira, K. (2006). Molecular design and delivery of siRNA. J. Drug Target. *14*, 448–455.

Isken, O., Baroth, M., Grassmann, C.W., Weinlich, S., Ostareck, D.H., Ostareck-Lederer, A., and Behrens, S.E. (2007). Nuclear factors are involved in hepatitis C virus RNA replication. Rna *13*, 1675–1692.

Jackson, A.L., Bartz, S.R., Schelter, J., Kobayashi, S.V., Burchard, J., Mao, M., Li, B., Cavet, G., and Linsley, P.S. (2003). Expression profiling reveals off-target gene regulation by RNAi. Nat. Biotechnol. *21*, 635–637.

Jackson, A.L., Burchard, J., Leake, D., Reynolds, A., Schelter, J., Guo, J., Johnson, J.M., Lim, L., Karpilow, J., Nichols, K., *et al.* (2006a). Position-specific chemical modification of siRNAs reduces 'off-target' transcript silencing. Rna *12*, 1197–1205.

Jackson, A.L., Burchard, J., Schelter, J., Chau, B.N., Cleary, M., Lim, L., and Linsley, P.S. (2006b). Widespread siRNA 'off-target' transcript silencing mediated by seed region sequence complementarity. Rna *12*, 1179–1187.

Jackson, A.L., and Linsley, P.S. (2004). Noise amidst the silence: off-target effects of siRNAs? Trends Genet. *20*, 521–524.

Jacque, J.M., Triques, K., and Stevenson, M. (2002). Modulation of HIV-1 replication by RNA interference. Nature *418*, 435–438.

Janowski, B.A., Huffman, K.E., Schwartz, J.C., Ram, R., Nordsell, R., Shames, D.S., Minna, J.D., and Corey, D.R. (2006). Involvement of AGO1 and AGO2 in mammalian transcriptional silencing. Nat. Struct. Mol. Biol. *13*, 787–792.

Jia, F., Zhang, Y.Z., and Liu, C.M. (2007). Stable inhibition of hepatitis B virus expression and replication in HepG2.2.15 cells by RNA interference based on retrovirus delivery. J. Biotechnol. *128*, 32–40.

Johnson, S.M., Grosshans, H., Shingara, J., Byrom, M., Jarvis, R., Cheng, A., Labourier, E., Reinert, K.L., Brown, D., and Slack, F.J. (2005). RAS is regulated by the let-7 microRNA family. Cell *120*, 635–647.

Jopling, C.L., Yi, M., Lancaster, A.M., Lemon, S.M., and Sarnow, P. (2005). Modulation of hepatitis C virus RNA abundance by a liver-specific MicroRNA. Science *309*, 1577–1581.

Judge, A.D., Bola, G., Lee, A.C., and MacLachlan, I. (2006). Design of noninflammatory synthetic siRNA mediating potent gene silencing *in vivo*. Mol. Ther. *13*, 494–505.

Judge, A.D., Sood, V., Shaw, J.R., Fang, D., McClintock, K., and MacLachlan, I. (2005). Sequence-dependent stimulation of the mammalian innate immune response by synthetic siRNA. Nat. Biotechnol. *23*, 457–462.

Kameoka, M., Nukuzuma, S., Itaya, A., Tanaka, Y., Ota, K., Ikuta, K., and Yoshihara, K. (2004). RNA interference directed against Poly(ADP-Ribose) polymerase 1 efficiently suppresses human immunodeficiency virus type 1 replication in human cells. J. Virol. *78*, 8931–8934.

Kanda, T., Steele, R., Ray, R., and Ray, R.B. (2007). Small interfering RNA targeted to hepatitis C virus 5′ non-translated region exerts potent antiviral effect. J. Virol. *81*, 669–676.

Kapadia, S.B., Brideau-Andersen, A., and Chisari, F.V. (2003). Interference of hepatitis C virus RNA replication by short interfering RNAs. Proc. Natl. Acad. Sci. U.S.A. *100*, 2014–2018.

Kariko, K., Bhuyan, P., Capodici, J., and Weissman, D. (2004). Small interfering RNAs mediate sequence-independent gene suppression and induce immune activation by signaling through toll-like receptor 3. J. Immunol. *172*, 6545–6549.

Karpala, A.J., Doran, T.J., and Bean, A.G. (2005). Immune responses to dsRNA: implications for gene silencing technologies. Immunol Cell Biol *83*, 211–216.

Kasim, V., Miyagishi, M., and Taira, K. (2004). Control of siRNA expression using the Cre-loxP recombination system. Nucleic Acids Res. *32*, e66.

Kasschau, K.D., Xie, Z., Allen, E., Llave, C., Chapman, E.J., Krizan, K.A., and Carrington, J.C. (2003). P1/HC-Pro, a viral suppressor of RNA silencing, interferes with *Arabidopsis* development and miRNA function. Dev. Cell *4*, 205–217.

Kawasaki, H., and Taira, K. (2003). Short hairpin type of dsRNAs that are controlled by tRNA(Val) promoter significantly induce RNAi-mediated gene silencing in the cytoplasm of human cells. Nucleic Acids Res. *31*, 700–707.

Kawasaki, H., and Taira, K. (2004). Induction of DNA methylation and gene silencing by short interfering RNAs in human cells. Nature *431*, 211–217.

Kaykas, A., and Moon, R.T. (2004). A plasmid-based system for expressing small interfering RNA libraries in mammalian cells. BMC Cell Biol 5, 16.

Ketzinel-Gilad, M., Shaul, Y., and Galun, E. (2006). RNA interference for antiviral therapy. J. Gene Med. *8*, 933–950.

Khvorova, A., Reynolds, A., and Jayasena, S.D. (2003). Functional siRNAs and miRNAs exhibit strand bias. Cell *115*, 209–216.

Kim, D.H., Behlke, M.A., Rose, S.D., Chang, M.S., Choi, S., and Rossi, J.J. (2005). Synthetic dsRNA Dicer substrates enhance RNAi potency and efficacy. Nat. Biotechnol. *23*, 222–226.

Kim, D.H., Longo, M., Han, Y., Lundberg, P., Cantin, E., and Rossi, J.J. (2004). Interferon induction by siRNAs and ssRNAs synthesized by phage polymerase. Nat. Biotechnol. *22*, 321–325.

Kim, D.H., Villeneuve, L.M., Morris, K.V., and Rossi, J.J. (2006). Argonaute-1 directs siRNA-mediated transcriptional gene silencing in human cells. Nat. Struct. Mol. Biol. *13*, 793–797.

Kim, V.N. (2005). MicroRNA biogenesis: coordinated cropping and dicing. Nat Rev Mol. Cell Biol. *6*, 376–385.

Kim, V.N., Mitrophanous, K., Kingsman, S.M., and Kingsman, A.J. (1998). Minimal requirement for a lentivirus vector based on human immunodeficiency virus type 1. J. Virol. *72*, 811–816.

Kim, Y.J., Ahn, J., Jeung, S.Y., Kim, D.S., Na, H.N., Cho, Y.J., Yun, S.H., Jee, Y., Jeon, E.S., Lee, H., and Nam, J.H. (2008). Recombinant lentivirus-delivered short hairpin RNAs targeted to conserved coxsackievirus sequences protect against viral myocarditis and improve survival rate in an animal model. Virus Genes *36*, 141–146.

Kimchi-Sarfaty, C., Ben-Nun-Shaul, O., Rund, D., Oppenheim, A., and Gottesman, M.M. (2002). In vitro-packaged SV40 pseudovirions as highly efficient vectors for gene transfer. Hum. Gene Ther. *13*, 299–310.

Kimchi-Sarfaty, C., Brittain, S., Garfield, S., Caplen, N.J., Tang, Q., and Gottesman, M.M. (2005). Efficient delivery of RNA interference effectors via in vitro-packaged SV40 pseudovirions. Hum. Gene Ther. *16*, 1110–1115.

Klages, N., Zufferey, R., and Trono, D. (2000). A stable system for the high-titer production of multiply attenuated lentiviral vectors. Mol. Ther. *2*, 170–176.

Kleinman, M.E., Yamada, K., Takeda, A., Chandrasekaran, V., Nozaki, M., Baffi, J.Z., Albuquerque, R.J., Yamasaki, S., Itaya, M., Pan, Y., *et al.* (2008). Sequence- and target-independent angiogenesis suppression by siRNA via TLR3. Nature *452*, 591–597.

Kock, N., Kasmieh, R., Weissleder, R., and Shah, K. (2007). Tumor therapy mediated by lentiviral expression of shBcl-2 and S-TRAIL. Neoplasia *9*, 435–442.

Komano, J., Miyauchi, K., Matsuda, Z., and Yamamoto, N. (2004). Inhibiting the Arp2/3 complex limits infection of both intracellular mature vaccinia virus and primate lentiviruses. Mol. Biol. Cell *15*, 5197–5207.

Konstantinova, P., de Vries, W., Haasnoot, J., ter Brake, O., de Haan, P., and Berkhout, B. (2006). Inhibition of human immunodeficiency virus type 1 by RNA interference using long-hairpin RNA. Gene Ther. *13*, 1403–1413.

Konstantinova, P., ter Brake, O., Haasnoot, J., de Haan, P., and Berkhout, B. (2007). Trans-inhibition of HIV-1 by a long hairpin RNA expressed within the viral genome. Retrovirology *4*, 15.

Korf, M., Jarczak, D., Beger, C., Manns, M.P., and Kruger, M. (2005). Inhibition of hepatitis C virus translation and subgenomic replication by siRNAs directed against highly conserved HCV sequence and cellular HCV cofactors. J. Hepatol. *43*, 225–234.

Kotsopoulou, E., Kim, V.N., Kingsman, A.J., Kingsman, S.M., and Mitrophanous, K.A. (2000). A Rev-independent human immunodeficiency virus type 1 (HIV-1)-based vector that exploits a codon-optimized HIV-1 gag-pol gene. J. Virol. *74*, 4839–4852.

Krönke, J., Kittler, R., Buchholz, F., Windisch, M., Pietschmann, T., Bartenschlager, R., and Frese, M. (2004). Alternative approaches for efficient inhibition of hepatitis C virus RNA replication by small interfering RNAs. J. Virol. *78*, 3436–3446.

Kruger, M., Beger, C., Welch, P.J., Barber, J.R., Manns, M.P., and Wong-Staal, F. (2001). Involvement of proteasome alpha-subunit PSMA7 in hepatitis C virus internal ribosome entry site-mediated translation. Mol. Cell Biol. *21*, 8357–8364.

Kumar, P., Ban, H.S., Kim, S.S., Wu, H., Pearson, T., Greiner, D.L., Laouar, A., Yao, J., Haridas, V., Habiro, K., *et al.* (2008). T cell-specific siRNA delivery suppresses HIV-1 infection in humanized mice. Cell *134*, 577–586.

Kumar, P., Lee, S.K., Shankar, P., and Manjunath, N. (2006). A single siRNA suppresses fatal encephalitis induced by two different flaviviruses. PLoS Med *3*, e96.

Lecellier, C.H., Dunoyer, P., Arar, K., Lehmann-Che, J., Eyquem, S., Himber, C., Saib, A., and Voinnet, O. (2005). A cellular microRNA mediates antiviral defense in human cells. Science *308*, 557–560.

Lee, C.C., Huang, H.Y., and Chiang, B.L. (2008). Lentiviral-mediated GATA-3 RNAi decreases allergic airway inflammation and hyperresponsiveness. Mol. Ther. *16*, 60–65.

Lee, N.S., Dohjima, T., Bauer, G., Li, H., Li, M.J., Ehsani, A., Salvaterra, P., and Rossi, J. (2002a). Expression of small interfering RNAs targeted against HIV-1 rev transcripts in human cells. Nat. Biotechnol. *20*, 500–505.

Lee, S.K., Dykxhoorn, D.M., Kumar, P., Ranjbar, S., Song, E., Maliszewski, L.E., Francois-Bongarcon, V., Goldfeld, A., Swamy, N.M., Lieberman, J., and Shankar, P. (2005). Lentiviral delivery of short hairpin RNAs protects CD4 T cells from multiple clades and primary isolates of HIV. Blood *106*, 818–826.

Lee, W.M. (1997). Hepatitis B virus infection. N. Engl. J. Med. *337*, 1733–1745.

Lee, Y., Ahn, C., Han, J., Choi, H., Kim, J., Yim, J., Lee, J., Provost, P., Radmark, O., Kim, S., and Kim, V.N. (2003). The nuclear RNase III Drosha initiates microRNA processing. Nature *425*, 415–419.

Lee, Y., Jeon, K., Lee, J.T., Kim, S., and Kim, V.N. (2002b). MicroRNA maturation: stepwise processing and subcellular localization. EMBO J. *21*, 4663–4670.

Lee, Y., Kim, M., Han, J., Yeom, K.H., Lee, S., Baek, S.H., and Kim, V.N. (2004). MicroRNA genes are transcribed by RNA polymerase II. EMBO J. *23*, 4051–4060.

Leonard, J.N., and Schaffer, D.V. (2005). Computational design of antiviral RNA interference strategies that resist human immunodeficiency virus escape. J. Virol. *79*, 1645–1654.

Leonard, J.N., and Schaffer, D.V. (2006). Antiviral RNAi therapy: emerging approaches for hitting a moving target. Gene Ther. *13*, 532–540.

Leung, R.K., and Whittaker, P.A. (2005). RNA interference: from gene silencing to gene-specific therapeutics. Pharmacol Ther *107*, 222–239.

Lewis, B.P., Burge, C.B., and Bartel, D.P. (2005). Conserved seed pairing, often flanked by adenosines, indicates that thousands of human genes are microRNA targets. Cell *120*, 15–20.

Li, C.X., Parker, A., Menocal, E., Xiang, S., Borodyansky, L., and Fruehauf, J.H. (2006). Delivery of RNA interference. Cell Cycle *5*, 2103–2109.

Li, H., Fu, X., Chen, Y., Hong, Y., Tan, Y., Cao, H., Wu, M., and Wang, H. (2005a). Use of adenovirus-delivered siRNA to target oncoprotein p28GANK in hepatocellular carcinoma. Gastroenterology *128*, 2029–2041.

Li, M.J., Kim, J., Li, S., Zaia, J., Yee, J.K., Anderson, J., Akkina, R., and Rossi, J.J. (2005b). Long-term inhibition of HIV-1 infection in primary hematopoietic cells by lentiviral vector delivery of a triple combination of anti-HIV shRNA, anti-CCR5 ribozyme, and a nucleolar-localizing TAR decoy. Mol. Ther. *12*, 900–909.

Li, W.X., Li, H., Lu, R., Li, F., Dus, M., Atkinson, P., Brydon, E.W., Johnson, K.L., Garcia-Sastre, A., Ball, L.A., *et al.* (2004a). Interferon antagonist proteins of influenza and vaccinia viruses are suppressors of RNA silencing. Proc. Natl. Acad. Sci. U.S.A. *101*, 1350–1355.

Li, Z., Xiong, Y., Peng, Y., Pan, J., Chen, Y., Wu, X., Hussain, S., Tien, P., and Guo, D. (2005c). Specific inhibition of HIV-1 replication by short hairpin RNAs targeting human cyclin T1 without inducing apoptosis. FEBS Lett. *579*, 3100–3106.

Li, G.Q., Gu, H.X., Li, D., and Xu, W.Z. (2007). Inhibition of hepatitis B virus cccDNA replication by siRNA. Biochem. Biophys. Res. Commun. *355*, 404–408.

Li, S., Rosenberg, J.E., Donjacour, A.A., Botchkina, I.L., Hom, Y.K., Cunha, G.R., and Blackburn, E.H. (2004b). Rapid inhibition of cancer cell growth induced by lentiviral delivery and expression of mutant-template telomerase RNA and anti-telomerase short-interfering RNA. Cancer Res. *64*, 4833–4840.

Liang, C.Y., Wang, H.Z., Li, T.X., Hu, Z.H., and Chen, X.W. (2004). High efficiency gene transfer into mammalian kidney cells using baculovirus vectors. Arch Virol *149*, 51–60.

Liaw, Y.F. (2002). Management of patients with chronic hepatitis B. J. Gastroenterol. Hepatol. *17*, 406–408.

Lieberman, J., Song, E., Lee, S.K., and Shankar, P. (2003). Interfering with disease: opportunities and roadblocks to harnessing RNA interference. Trends Mol. Med *9*, 397–403.

Lim, L.P., Lau, N.C., Garrett-Engele, P., Grimson, A., Schelter, J.M., Castle, J., Bartel, D.P., Linsley, P.S., and Johnson, J.M. (2005). Microarray analysis shows that some microRNAs down-regulate large numbers of target mRNAs. Nature *433*, 769–773.

Lin, X., Yang, J., Chen, J., Gunasekera, A., Fesik, S.W., and Shen, Y. (2004). Development of a tightly regulated U6 promoter for shRNA expression. FEBS Lett. *577*, 376–380.

Lin, X., Ruan, X., Anderson, M.G., McDowell, J.A., Kroeger, P.E., Fesik, S.W., and Shen, Y. (2005). siRNA-mediated off-target gene silencing triggered by a 7 nt complementation. Nucleic Acids Res. *33*, 4527–4535.

Liu, J., Guo, Y., Xue, C.F., Li, Y.H., Huang, Y.X., Ding, J., Gong, W.D., and Zhao, Y. (2004). Effect of vector-expressed siRNA on HBV replication in hepatoblastoma cells. World J Gastroenterol *10*, 1898–1901.

Liu, J., Valencia-Sanchez, M.A., Hannon, G.J., and Parker, R. (2005). MicroRNA-dependent localization of targeted mRNAs to mammalian P-bodies. Nat. Cell Biol. *7*, 719–723.

Liu, Y.P., Haasnoot, J., and Berkhout, B. (2007). Design of extended short hairpin RNAs for HIV-1 inhibition. Nucleic Acids Res. *35*, 5683–5693.

Liu, Y.P., Haasnoot, J., ter Brake, O., Berkhout, B., and Konstantinova, P. (2008). Inhibition of HIV-1 by multiple siRNAs expressed from a single microRNA polycistron. Nucleic Acids Res. *36*, 2811–2824.

Lo, A.K., To, K.F., Lo, K.W., Lung, R.W., Hui, J.W., Liao, G., and Hayward, S.D. (2007). Modulation of LMP1 protein expression by EBV-encoded microRNAs. Proc. Natl. Acad. Sci. U.S.A. *104*, 16164–16169.

Lohmann, V., Korner, F., Koch, J., Herian, U., Theilmann, L., and Bartenschlager, R. (1999). Replication of subgenomic hepatitis C virus RNAs in a hepatoma cell line. Science *285*, 110–113.

Lu, S., and Cullen, B.R. (2004). Adenovirus VA1 noncoding RNA can inhibit small interfering RNA and MicroRNA biogenesis. J. Virol. *78*, 12868–12876.

Lucy, A.P., Guo, H.S., Li, W.X., and Ding, S.W. (2000). Suppression of post-transcriptional gene silencing by a plant viral protein localized in the nucleus. EMBO J. *19*, 1672–1680.

Lund, E., Guttinger, S., Calado, A., Dahlberg, J.E., and Kutay, U. (2004). Nuclear export of microRNA precursors. Science *303*, 95–98.

Luo, K.Q., and Chang, D.C. (2004). The gene-silencing efficiency of siRNA is strongly dependent on the local structure of mRNA at the targeted region. Biochem. Biophys. Res. Commun. *318*, 303–310.

Llave, C., Kasschau, K.D., and Carrington, J.C. (2000). Virus-encoded suppressor of posttranscriptional gene silencing targets a maintenance step in the silencing pathway. Proc. Natl. Acad. Sci. U.S.A. *97*, 13401–13406.

McCaffrey, A.P., Meuse, L., Pham, T.T., Conklin, D.S., Hannon, G.J., and Kay, M.A. (2002). RNA interference in adult mice. Nature *418*, 38–39.

McCarty, D.M., Young, S.M., Jr., and Samulski, R.J. (2004). Integration of adeno-associated virus (AAV) and recombinant AAV vectors. Annu. Rev. Genet. *38*, 819–845.

Machida, Y., Okada, T., Kurosawa, M., Oyama, F., Ozawa, K., and Nukina, N. (2006). rAAV-mediated shRNA ameliorated neuropathology in Huntington disease model mouse. Biochem. Biophys. Res. Commun. *343*, 190–197.

Makinen, P.I., Koponen, J.K., Karkkainen, A.M., Malm, T.M., Pulkkinen, K.H., Koistinaho, J., Turunen, M.P., and Yla-Herttuala, S. (2006). Stable RNA interference: comparison of U6 and H1 promoters in endothelial cells and in mouse brain. J. Gene Med. *8*, 433–441.

Manche, L., Green, S.R., Schmedt, C., and Mathews, M.B. (1992). Interactions between double-stranded RNA regulators and the protein kinase DAI. Mol. Cell Biol. *12*, 5238–5248.

Marques, J.T., Devosse, T., Wang, D., Zamanian-Daryoush, M., Serbinowski, P., Hartmann, R., Fujita, T., Behlke, M.A., and Williams, B.R. (2006). A structural basis for discriminating between self and nonself double-stranded RNAs in mammalian cells. Nat. Biotechnol. *24*, 559–565.

Matsukura, S., Jones, P.A., and Takai, D. (2003). Establishment of conditional vectors for hairpin siRNA knockdowns. Nucleic Acids Res. *31*, e77.

Matthess, Y., Kappel, S., Spankuch, B., Zimmer, B., Kaufmann, M., and Strebhardt, K. (2005). Conditional inhibition of cancer cell proliferation by tetracycline-responsive, H1 promoter-driven silencing of PLK1. Oncogene *24*, 2973–2980.

Meissner, A., and Jaenisch, R. (2006). Generation of nuclear transfer-derived pluripotent ES cells from cloned Cdx2-deficient blastocysts. Nature *439*, 212–215.

Mentel, R., Dopping, G., Wegner, U., Seidel, W., Liebermann, H., and Dohner, L. (1997). Adenovirus–receptor interaction with human lymphocytes. J. Med. Virol. *51*, 252–257.

Mercer, D.F., Schiller, D.E., Elliott, J.F., Douglas, D.N., Hao, C., Rinfret, A., Addison, W.R., Fischer, K.P., Churchill, T.A., Lakey, J.R., *et al.* (2001). Hepatitis C virus replication in mice with chimeric human livers. Nat. Med. *7*, 927–933.

Miller, D.G., Petek, L.M., and Russell, D.W. (2004a). Adeno-associated virus vectors integrate at chromosome breakage sites. Nat. Genet. *36*, 767–773.

Miller, T.M., Kaspar, B.K., Kops, G.J., Yamanaka, K., Christian, L.J., Gage, F.H., and Cleveland, D.W. (2005). Virus-delivered small RNA silencing sustains strength in amyotrophic lateral sclerosis. Ann Neurol *57*, 773–776.

Miller, V.M., Gouvion, C.M., Davidson, B.L., and Paulson, H.L. (2004b). Targeting Alzheimer's disease genes with RNA interference: an efficient strategy for silencing mutant alleles. Nucleic Acids Res. *32*, 661–668.

Mitani, K., Graham, F.L., Caskey, C.T., and Kochanek, S. (1995). Rescue, propagation, and partial purification of a helper virus-dependent adenovirus vector. Proc. Natl. Acad. Sci. U.S.A. *92*, 3854–3858.

Mitchell, R.S., Beitzel, B.F., Schroder, A.R., Shinn, P., Chen, H., Berry, C.C., Ecker, J.R., and Bushman, F.D. (2004). Retroviral DNA integration: ASLV, HIV, and MLV show distinct target site preferences. PLoS Biol. *2*, E234.

Mitrophanous, K., Yoon, S., Rohll, J., Patil, D., Wilkes, F., Kim, V., Kingsman, S., Kingsman, A., and Mazarakis, N. (1999). Stable gene transfer to the nervous system using a non-primate lentiviral vector. Gene Ther. *6*, 1808–1818.

Miyagishi, M., Sumimoto, H., Miyoshi, H., Kawakami, Y., and Taira, K. (2004). Optimization of an siRNA-expression system with an improved hairpin and its significant suppressive effects in mammalian cells. J. Gene Med. *6*, 715–723.

Miyagishi, M., and Taira, K. (2002). U6 promoter-driven siRNAs with four uridine 3′ overhangs efficiently suppress targeted gene expression in mammalian cells. Nat. Biotechnol. *20*, 497–500.

Modem, S., Badri, K.R., Holland, T.C., and Reddy, T.R. (2005). Sam68 is absolutely required for Rev function and HIV-1 production. Nucleic Acids Res. *33*, 873–879.

Moffat, J., Grueneberg, D.A., Yang, X., Kim, S.Y., Kloepfer, A.M., Hinkle, G., Piqani, B., Eisenhaure, T.M., Luo, B., Grenier, J.K., *et al.* (2006). A lentiviral RNAi library for human and mouse genes applied to an arrayed viral high-content screen. Cell *124*, 1283–1298.

Moffat, J., and Sabatini, D.M. (2006). Building mammalian signalling pathways with RNAi screens. Nat. Rev. Mol. Cell. Biol. *7*, 177–187.

Mok, H., Palmer, D.J., Ng, P., and Barry, M.A. (2005). Evaluation of polyethylene glycol modification of first-generation and helper-dependent adenoviral vectors to reduce innate immune responses. Mol. Ther. *11*, 66–79.

Moore, M.D., McGarvey, M.J., Russell, R.A., Cullen, B.R., and McClure, M.O. (2005). Stable inhibition of hepatitis B virus proteins by small interfering RNA expressed from viral vectors. J. Gene Med. *7*, 918–925.

Morral, N., McEvoy, R., Dong, H., Meseck, M., Altomonte, J., Thung, S., and Woo, S.L. (2002). Adenovirus-mediated expression of glucokinase in the liver as an adjuvant treatment for type 1 diabetes. Hum. Gene Ther. *13*, 1561–1570.

Morris, K.V., Chan, S.W., Jacobsen, S.E., and Looney, D.J. (2004). Small interfering RNA-induced transcriptional gene silencing in human cells. Science *305*, 1289–1292.

Morris, K.V., and Rossi, J.J. (2006). Lentiviral-mediated delivery of siRNAs for antiviral therapy. Gene Ther. *13*, 553–558.

Murphy, E., Vanicek, J., Robins, H., Shenk, T., and Levine, A.J. (2008). Suppression of immediate-early viral gene expression by herpesvirus-coded microRNAs: implications for latency. Proc. Natl. Acad. Sci. U.S.A. *105*, 5453–5458.

Myers, J.W., Jones, J.T., Meyer, T., and Ferrell, J.E., Jr. (2003). Recombinant Dicer efficiently converts large dsRNAs into siRNAs suitable for gene silencing. Nat. Biotechnol. *21*, 324–328.

Naito, Y., Yamada, T., Ui-Tei, K., Morishita, S., and Saigo, K. (2004). siDirect: highly effective, target-specific siRNA design software for mammalian RNA interference. Nucleic Acids Res. *32*, W124–129.

Nakagawa, M., Sakamoto, N., Tanabe, Y., Koyama, T., Itsui, Y., Takeda, Y., Chen, C.H., Kakinuma, S., Oooka, S., Maekawa, S., *et al.* (2005). Suppression of hepatitis C virus replication by cyclosporin a is mediated by blockade of cyclophilins. Gastroenterology *129*, 1031–1041.

Nallagatla, S.R., Hwang, J., Toroney, R., Zheng, X., Cameron, C.E., and Bevilacqua, P.C. (2007). 5′-triphosphate-dependent activation of PKR by RNAs with short stem-loops. Science *318*, 1455–1458.

Narvaiza, I., Aparicio, O., Vera, M., Razquin, N., Bortolanza, S., Prieto, J., and Fortes, P. (2006). Effect of adenovirus-mediated RNA interference on endogenous microRNAs in a mouse model of multidrug resistance protein 2 gene silencing. J. Virol. *80*, 12236–12247.

Nathwani, A.C., McIntosh, J., and Davidoff, A.M. (2005). An update on gene therapy for hemophilia. Curr Hematol Rep *4*, 287–293.

Ngo, V.N., Davis, R.E., Lamy, L., Yu, X., Zhao, H., Lenz, G., Lam, L.T., Dave, S., Yang, L., Powell, J., and Staudt, L.M. (2006). A loss-of-function RNA interference screen for molecular targets in cancer. Nature *441*, 106–110.

Ni, Q., Chen, Z., Yao, H.P., Yang, Z.G., Liu, K.Z., and Wu, L.L. (2004). Inhibition of human La protein by RNA interference down-regulates hepatitis B virus mRNA in 2.2.15 cells. World J Gastroenterol *10*, 2050–2054.

Nicholson, L.J., Philippe, M., Paine, A.J., Mann, D.A., and Dolphin, C.T. (2005). RNA interference mediated in human primary cells via recombinant baculoviral vectors. Mol. Ther. *11*, 638–644.

Nishitsuji, H., Kohara, M., Kannagi, M., and Masuda, T. (2006). Effective suppression of human immunodeficiency virus type 1 through a combination of short- or long-hairpin RNAs targeting essential sequences for retroviral integration. J. Virol. *80*, 7658–7666.

Novina, C.D., Murray, M.F., Dykxhoorn, D.M., Beresford, P.J., Riess, J., Lee, S.K., Collman, R.G., Lieberman, J., Shankar, P., and Sharp, P.A. (2002). siRNA-directed inhibition of HIV-1 infection. Nat. Med. *8*, 681–686.

Ohman, M. (2007). A-to-I editing challenger or ally to the microRNA process. Biochimie *89*, 1171–1176.

Oka, K., and Chan, L. (2005). Construction and characterization of helper-dependent adenoviral vectors for sustained *in vivo* gene therapy. Methods Mol Med *108*, 329–350.

Omi, K., Tokunaga, K., and Hohjoh, H. (2004). Long-lasting RNAi activity in mammalian neurons. FEBS Lett. *558*, 89–95.

Orban, T.I., and Izaurralde, E. (2005). Decay of mRNAs targeted by RISC requires XRN1, the Ski complex, and the exosome. Rna *11*, 459–469.

Overhoff, M., Alken, M., Far, R.K., Lemaitre, M., Lebleu, B., Sczakiel, G., and Robbins, I. (2005). Local RNA target structure influences siRNA efficacy: a systematic global analysis. J. Mol. Biol. *348*, 871–881.

Paddison, P.J., Caudy, A.A., Bernstein, E., Hannon, G.J., and Conklin, D.S. (2002). Short hairpin RNAs (shRNAs) induce sequence-specific silencing in mammalian cells. Genes Dev. *16*, 948–958.

Paul, C.P., Good, P.D., Winer, I., and Engelke, D.R. (2002). Effective expression of small interfering RNA in human cells. Nat. Biotechnol. *20*, 505–508.

Pebernard, S., and Iggo, R.D. (2004). Determinants of interferon-stimulated gene induction by RNAi vectors. Differentiation *72*, 103–111.

Pedersen, I.M., Cheng, G., Wieland, S., Volinia, S., Croce, C.M., Chisari, F.V., and David, M. (2007). Interferon modulation of cellular microRNAs as an antiviral mechanism. Nature *449*, 919–922.

Persengiev, S.P., Zhu, X., and Green, M.R. (2004). Nonspecific, concentration-dependent stimulation and repression of mammalian gene expression by small interfering RNAs (siRNAs). Rna *10*, 12–18.

Pfeffer, S., Sewer, A., Lagos-Quintana, M., Sheridan, R., Sander, C., Grasser, F.A., van Dyk, L.F., Ho, C.K., Shuman, S., Chien, M., *et al.* (2005). Identification of microRNAs of the herpesvirus family. Nat Methods *2*, 269–276.

Pfeifer, A., Eigenbrod, S., Al-Khadra, S., Hofmann, A., Mitteregger, G., Moser, M., Bertsch, U., and Kretzschmar, H. (2006). Lentivector-mediated RNAi efficiently suppresses prion protein and prolongs survival of scrapie-infected mice. J. Clin. Invest. *116*, 3204–3210.

Ping, Y.H., Chu, C.Y., Cao, H., Jacque, J.M., Stevenson, M., and Rana, T.M. (2004). Modulating HIV-1 replication by RNA interference directed against human transcription elongation factor SPT5. Retrovirology *1*, 46.

Player, M.R., and Torrence, P.F. (1998). The 2–5A system: modulation of viral and cellular processes through acceleration of RNA degradation. Pharmacol Ther *78*, 55–113.

Pluta, K., Diehl, W., Zhang, X.Y., Kutner, R., Bialkowska, A., and Reiser, J. (2007). Lentiviral vectors encoding tetracycline-dependent repressors and transactivators for reversible knockdown of gene expression: a comparative study. BMC Biotechnol. 7, 41.

Poluri, A., and Sutton, R.E. (2008). Titers of HIV-based vectors encoding shRNAs are reduced by a dicer-dependent mechanism. Mol. Ther. *16*, 378–386.

Prabhu, R., Vittal, P., Yin, Q., Flemington, E., Garry, R., Robichaux, W.H., and Dash, S. (2005). Small interfering RNA effectively inhibits protein expression and negative strand RNA synthesis from a full-length hepatitis C virus clone. J. Med. Virol. *76*, 511–519.

Pusch, O., Boden, D., Silbermann, R., Lee, F., Tucker, L., and Ramratnam, B. (2003). Nucleotide sequence homology requirements of HIV-1-specific short hairpin RNA. Nucleic Acids Res. *31*, 6444–6449.

Rabinowitz, J.E., Rolling, F., Li, C., Conrath, H., Xiao, W., Xiao, X., and Samulski, R.J. (2002). Cross-packaging of a single adeno-associated virus (AAV) type 2 vector genome into multiple AAV serotypes enables transduction with broad specificity. J. Virol. *76*, 791–801.

Ragozin, S., Niemeier, A., Laatsch, A., Loeffler, B., Merkel, M., Beisiegel, U., and Heeren, J. (2005). Knockdown of hepatic ABCA1 by RNA interference decreases plasma HDL cholesterol levels and influences post-prandial lipemia in mice. Arterioscler. Throm.b Vasc .Biol. *25*, 1433–1438.

Ralph, G.S., Radcliffe, P.A., Day, D.M., Carthy, J.M., Leroux, M.A., Lee, D.C., Wong, L.F., Bilsland, L.G., Greensmith, L., Kingsman, S.M., *et al.* (2005). Silencing mutant SOD1 using RNAi protects against neurodegeneration and extends survival in an ALS model. Nat. Med. *11*, 429–433.

Randall, G., Grakoui, A., and Rice, C.M. (2003). Clearance of replicating hepatitis C virus replicon RNAs in cell culture by small interfering RNAs. Proc. Natl. Acad. Sci. U.S.A. *100*, 235–240.

Randall, G., Panis, M., Cooper, J.D., Tellinghuisen, T.L., Sukhodolets, K.E., Pfeffer, S., Landthaler, M., Landgraf, P., Kan, S., Lindenbach, B.D., *et al.* (2007). Cellular cofactors affecting hepatitis C virus infection and replication. Proc. Natl. Acad. Sci. U.S.A. *104*, 12884–12889.

Raoul, C., Abbas-Terki, T., Bensadoun, J.C., Guillot, S., Haase, G., Szulc, J., Henderson, C.E., and Aebischer, P. (2005). Lentiviral-mediated silencing of SOD1 through RNA interference retards disease onset and progression in a mouse model of ALS. Nat. Med. *11*, 423–428.

Rauschhuber, C., Xu, H., Salazar, F.H., Marion, P.L., and Ehrhardt, A. (2008). Exploring gene-deleted adenoviral vectors for delivery of short hairpin RNAs and reduction of hepatitis B virus infection in mice. J. Gene Med. *10*, 878–889.

Ren, X., Luo, G., Xie, Z., Zhou, L., Kong, X., and Xu, A. (2006). Inhibition of multiple gene expression and virus replication of HBV by stable RNA interference in 2.2.15 cells. J Hepatol *44*, 663–670.

Ren, X.R., Zhou, L.J., Luo, G.B., Lin, B., and Xu, A. (2005). Inhibition of hepatitis B virus replication in 2.2.15 cells by expressed shRNA. J Viral Hepat *12*, 236–242.

Reynolds, A., Anderson, E.M., Vermeulen, A., Fedorov, Y., Robinson, K., Leake, D., Karpilow, J., Marshall, W.S., and Khvorova, A. (2006). Induction of the interferon response by siRNA is cell type- and duplex length-dependent. Rna *12*, 988–993.

Reynolds, A., Leake, D., Boese, Q., Scaringe, S., Marshall, W.S., and Khvorova, A. (2004). Rational siRNA design for RNA interference. Nat. Biotechnol. *22*, 326–330.

Rivas, F.V., Tolia, N.H., Song, J.J., Aragon, J.P., Liu, J., Hannon, G.J., and Joshua-Tor, L. (2005). Purified Argonaute2 and an siRNA form recombinant human RISC. Nat. Struct. Mol. Biol. *12*, 340–349.

Rodriguez-Lebron, E., Denovan-Wright, E.M., Nash, K., Lewin, A.S., and Mandel, R.J. (2005). Intrastriatal rAAV-mediated delivery of anti-huntingtin shRNAs induces partial reversal of disease progression in R6/1 Huntington's disease transgenic mice. Mol. Ther. *12*, 618–633.

Romano, P.R., McCallus, D.E., and Pachuk, C.J. (2006). RNA interference-mediated prevention and therapy for hepatocellular carcinoma. Oncogene *25*, 3857–3865.

Root, D.E., Hacohen, N., Hahn, W.C., Lander, E.S., and Sabatini, D.M. (2006). Genome-scale loss-of-function screening with a lentiviral RNAi library. Nat Methods *3*, 715–719.

Rossi, J.J. (2008). Expression strategies for short hairpin RNA interference triggers. Hum. Gene Ther. *19*, 313–317.

Rubinson, D.A., Dillon, C.P., Kwiatkowski, A.V., Sievers, C., Yang, L., Kopinja, J., Rooney, D.L., Zhang, M., Ihrig, M.M., McManus, M.T., *et al.* (2003). A lentivirus-based system to functionally silence genes in primary mammalian cells, stem cells and transgenic mice by RNA interference. Nat. Genet. *33*, 401–406.

Ruiz, R., Witting, S.R., Saxena, R., and Morral, N. (2009). Robust hepatic gene silencing for functional studies using helper-dependent adenovirus vectors. Hum. Gene Ther. *20*, 87–94.

Saayman, S., Barichievy, S., Capovilla, A., Morris, K.V., Arbuthnot, P., and Weinberg, M.S. (2008). The efficacy of generating three independent anti-HIV-1 siRNAs from a single U6 RNA Pol III-expressed long hairpin RNA. PLoS ONE 3, e2602.

Sabariegos, R., Gimenez-Barcons, M., Tapia, N., Clotet, B., and Martinez, M.A. (2006). Sequence homology required by human immunodeficiency virus type 1 to escape from short interfering RNAs. J. Virol. *80*, 571–577.

Saeki, Y., Fraefel, C., Ichikawa, T., Breakefield, X.O., and Chiocca, E.A. (2001). Improved helper virus-free packaging system for HSV amplicon vectors using an ICP27-deleted, oversized HSV-1 DNA in a bacterial artificial chromosome. Mol. Ther. *3*, 591–601.

Saetrom, P., Heale, B.S., Snove, O., Jr., Aagaard, L., Alluin, J., and Rossi, J.J. (2007). Distance constraints between microRNA target sites dictate efficacy and cooperativity. Nucleic Acids Res. *35*, 2333–2342.

Saetrom, P., and Snove, O., Jr. (2004). A comparison of siRNA efficacy predictors. Biochem. Biophys. Res. Commun. *321*, 247–253.

Samakoglu, S., Lisowski, L., Budak-Alpdogan, T., Usachenko, Y., Acuto, S., Di Marzo, R., Maggio, A., Zhu, P., Tisdale, J.F., Riviere, I., and Sadelain, M. (2006). A genetic strategy to treat sickle cell anemia by coregulating globin transgene expression and RNA interference. Nat. Biotechnol. *24*, 89–94.

Samols, M.A., Hu, J., Skalsky, R.L., and Renne, R. (2005). Cloning and identification of a microRNA cluster within the latency-associated region of Kaposi's sarcoma-associated herpesvirus. J. Virol. *79*, 9301–9305.

Samols, M.A., Skalsky, R.L., Maldonado, A.M., Riva, A., Lopez, M.C., Baker, H.V., and Renne, R. (2007). Identification of cellular genes targeted by KSHV-encoded microRNAs. PLoS Pathog. *3*, e65.

Samulski, R.J., Chang, L.S., and Shenk, T. (1989). Helper-free stocks of recombinant adeno-associated viruses: normal integration does not require viral gene expression. J. Virol. *63*, 3822–3828.

Sano, M., Li, H., Nakanishi, M., and Rossi, J.J. (2008). Expression of long anti-HIV-1 hairpin RNAs for the generation of multiple siRNAs: advantages and limitations. Mol. Ther. *16*, 170–177.

Santoyo, J., Vaquerizas, J.M., and Dopazo, J. (2005). Highly specific and accurate selection of siRNAs for high-throughput functional assays. Bioinformatics *21*, 1376–1382.

Sapru, M.K., Yates, J.W., Hogan, S., Jiang, L., Halter, J., and Bohn, M.C. (2006). Silencing of human alpha-synuclein in vitro and in rat brain using lentiviral-mediated RNAi. Exp Neurol *198*, 382–390.

Saxena, S., Jonsson, Z.O., and Dutta, A. (2003). Small RNAs with imperfect match to endogenous mRNA repress translation. Implications for off-target activity of small inhibitory RNA in mammalian cells. J. Biol. Chem. *278*, 44312–44319.

Scacheri, P.C., Rozenblatt-Rosen, O., Caplen, N.J., Wolfsberg, T.G., Umayam, L., Lee, J.C., Hughes, C.M., Shanmugam, K.S., Bhattacharjee, A., Meyerson, M., and Collins, F.S. (2004). Short interfering RNAs can induce unexpected and divergent changes in the levels of untargeted proteins in mammalian cells. Proc. Natl. Acad. Sci. U.S.A. *101*, 1892–1897.

Schafer, A., Cai, X., Bilello, J.P., Desrosiers, R.C., and Cullen, B.R. (2007). Cloning and analysis of microRNAs encoded by the primate gamma-herpesvirus rhesus monkey rhadinovirus. Virology *364*, 21–27.

Schubert, S., Grunweller, A., Erdmann, V.A., and Kurreck, J. (2005). Local RNA target structure influences siRNA efficacy: systematic analysis of intentionally designed binding regions. J. Mol. Biol. *348*, 883–893.

Schutz, S., and Sarnow, P. (2006). Interaction of viruses with the mammalian RNA interference pathway. Virology *344*, 151–157.

Schwarz, D.S., Hutvagner, G., Du, T., Xu, Z., Aronin, N., and Zamore, P.D. (2003). Asymmetry in the assembly of the RNAi enzyme complex. Cell *115*, 199–208.

Selbach, M., Schwanhausser, B., Thierfelder, N., Fang, Z., Khanin, R., and Rajewsky, N. (2008). Widespread changes in protein synthesis induced by microRNAs. Nature *455*, 58–63.

Semizarov, D., Kroeger, P., and Fesik, S. (2004). siRNA-mediated gene silencing: a global genome view. Nucleic Acids Res. *32*, 3836–3845.

Sen, A., Steele, R., Ghosh, A.K., Basu, A., Ray, R., and Ray, R.B. (2003). Inhibition of hepatitis C virus protein expression by RNA interference. Virus Res. *96*, 27–35.

Sen, G.L., and Blau, H.M. (2005). Argonaute 2/RISC resides in sites of mammalian mRNA decay known as cytoplasmic bodies. Nat. Cell Biol. *7*, 633–636.

Senserrich, J., Pauls, E., Armand-Ugon, M., Clotet-Codina, I., Moncunill, G., Clotet, B., and Este, J.A. (2008). HIV-1 resistance to the anti-HIV activity of a shRNA targeting a dual-coding region. Virology *372*, 421–429.

Seo, M.Y., Abrignani, S., Houghton, M., and Han, J.H. (2003). Small interfering RNA-mediated inhibition of hepatitis C virus replication in the human hepatoma cell line Huh-7. J. Virol. *77*, 810–812.

Shinomiya, N., Gao, C.F., Xie, Q., Gustafson, M., Waters, D.J., Zhang, Y.W., and Vande Woude, G.F. (2004). RNA interference reveals that ligand-independent met activity is required for tumor cell signaling and survival. Cancer Res. *64*, 7962–7970.

Shlomai, A., and Shaul, Y. (2003). Inhibition of hepatitis B virus expression and replication by RNA interference. Hepatology *37*, 764–770.

Siapati, E.K., Bigger, B.W., Miskin, J., Chipchase, D., Parsley, K.L., Mitrophanous, K., Themis, M., Thrasher, A.J., and Bonnet, D. (2005). Comparison of HIV- and EIAV-based vectors on their efficiency in transducing murine and human hematopoietic repopulating cells. Mol. Ther. *12*, 537–546.

Silhavy, D., Molnar, A., Lucioli, A., Szittya, G., Hornyik, C., Tavazza, M., and Burgyan, J. (2002). A viral protein suppresses RNA silencing and binds silencing-generated, 21- to 25-nucleotide double-stranded RNAs. EMBO J. *21*, 3070–3080.

Silva, J.M., Li, M.Z., Chang, K., Ge, W., Golding, M.C., Rickles, R.J., Siolas, D., Hu, G., Paddison, P.J., Schlabach, M.R., *et al.* (2005). Second-generation shRNA libraries covering the mouse and human genomes. Nat. Genet. *37*, 1281–1288.

Singer, O., Marr, R.A., Rockenstein, E., Crews, L., Coufal, N.G., Gage, F.H., Verma, I.M., and Masliah, E. (2005). Targeting BACE1 with siRNAs ameliorates Alzheimer disease neuropathology in a transgenic model. Nat. Neurosci. *8*, 1343–1349.

Siolas, D., Lerner, C., Burchard, J., Ge, W., Linsley, P.S., Paddison, P.J., Hannon, G.J., and Cleary, M.A. (2005). Synthetic shRNAs as potent RNAi triggers. Nat. Biotechnol. *23*, 227–231.

Skalsky, R.L., Samols, M.A., Plaisance, K.B., Boss, I.W., Riva, A., Lopez, M.C., Baker, H.V., and Renne, R. (2007). Kaposi's sarcoma-associated herpesvirus encodes an ortholog of miR-155. J. Virol. *81*, 12836–12845.

Sledz, C.A., Holko, M., de Veer, M.J., Silverman, R.H., and Williams, B.R. (2003). Activation of the interferon system by short-interfering RNAs. Nat. Cell Biol. *5*, 834–839.

Snyder, L.L., Esser, J.M., Pachuk, C.J., and Steel, L.F. (2008). Vector design for liver-specific expression of multiple interfering RNAs that target hepatitis B virus transcripts. Antiviral Res. *80*, 36–44.

Sokolskaja, E., Sayah, D.M., and Luban, J. (2004). Target cell cyclophilin A modulates human immunodeficiency virus type 1 infectivity. J. Virol. *78*, 12800–12808.

Song, E., Lee, S.K., Dykxhoorn, D.M., Novina, C., Zhang, D., Crawford, K., Cerny, J., Sharp, P.A., Lieberman, J., Manjunath, N., and Shankar, P. (2003). Sustained small interfering RNA-mediated human immunodeficiency virus type 1 inhibition in primary macrophages. J. Virol. *77*, 7174–7181.

Soutschek, J., Akinc, A., Bramlage, B., Charisse, K., Constien, R., Donoghue, M., Elbashir, S., Geick, A., Hadwiger, P., Harborth, J., *et al.* (2004). Therapeutic silencing of an endogenous gene by systemic administration of modified siRNAs. Nature *432*, 173–178.

Spangberg, K., Wiklund, L., and Schwartz, S. (2000). HuR, a protein implicated in oncogene and growth factor mRNA decay, binds to the 3′ ends of hepatitis C virus RNA of both polarities. Virology *274*, 378–390.

Stegmeier, F., Hu, G., Rickles, R.J., Hannon, G.J., and Elledge, S.J. (2005). A lentiviral microRNA-based system for single-copy polymerase II-regulated RNA interference in mammalian cells. Proc. Natl. Acad. Sci. U.S.A. *102*, 13212–13217.

Strat, A., Gao, L., Utsuki, T., Cheng, B., Nuthalapaty, S., Mathis, J.M., Odaka, Y., and Giordano, T. (2006). Specific and nontoxic silencing in mammalian cells with expressed long dsRNAs. Nucleic Acids Res. *34*, 3803–3810.

Strayer, D.S., Feitelson, M., Sun, B., and Matskevich, A.A. (2005). Paradigms for conditional expression of RNA interference molecules for use against viral targets. Methods Enzymol. *392*, 227–241.

Sui, G., Soohoo, C., Affar el, B., Gay, F., Shi, Y., Forrester, W.C., and Shi, Y. (2002). A DNA vector-based RNAi technology to suppress gene expression in mammalian cells. Proc. Natl. Acad. Sci. U.S.A. *99*, 5515–5520.

Sullivan, C.S., and Ganem, D. (2005). MicroRNAs and viral infection. Mol. Cell *20*, 3–7.

Sullivan, C.S., Grundhoff, A.T., Tevethia, S., Pipas, J.M., and Ganem, D. (2005). SV40-encoded microRNAs regulate viral gene expression and reduce susceptibility to cytotoxic T cells. Nature *435*, 682–686.

Sumimoto, H., Miyagishi, M., Miyoshi, H., Yamagata, S., Shimizu, A., Taira, K., and Kawakami, Y. (2004). Inhibition of growth and invasive ability of melanoma by inactivation of mutated BRAF with lentivirus-mediated RNA interference. Oncogene 23, 6031–6039.

Sun, Y., Li, Z., Li, L., Li, J., Liu, X., and Li, W. (2007). Effective inhibition of hepatitis B virus replication by small interfering RNAs expressed from human foamy virus vectors. Int. J. Mol. Med. *19*, 705–711.

Surabhi, R.M., and Gaynor, R.B. (2002). RNA interference directed against viral and cellular targets inhibits human immunodeficiency Virus Type 1 replication. J. Virol. *76*, 12963–12973.

Suzuki, H., Kaneko, H., Tamai, N., Miyano-Kurosaki, N., Hashimoto, K., Shimotohno, K., and Takaku, H. (2005). Suppression of HCV RNA replication by baculovirus-mediated shRNA expression. Nucleic Acids Symp Ser. (Oxf.), 339–340.

Takasaki, S., Kotani, S., and Konagaya, A. (2004). An effective method for selecting siRNA target sequences in mammalian cells. Cell Cycle *3*, 790–795.

Takigawa, Y., Nagano-Fujii, M., Deng, L., Hidajat, R., Tanaka, M., Mizuta, H., and Hotta, H. (2004). Suppression of hepatitis C virus replicon by RNA interference directed against the NS3 and NS5B regions of the viral genome. Microbiol. Immunol. *48*, 591–598.

Tan, E.L., Tan, T.M., Chow, V.T., and Poh, C.L. (2007). Enhanced potency and efficacy of 29-mer shRNAs in inhibition of Enterovirus 71. Antiviral Res. *74*, 9–15.

ter Brake, O., Konstantinova, P., Ceylan, M., and Berkhout, B. (2006). Silencing of HIV-1 with RNA interference: a multiple shRNA approach. Mol. Ther. *14*, 883–892.

ter Brake, O., t Hooft, K., Liu, Y.P., Centlivre, M., von Eije, K.J., and Berkhout, B. (2008). Lentiviral vector design for multiple shRNA expression and durable HIV-1 inhibition. Mol. Ther. *16*, 557–564.

Ting, A.H., Schuebel, K.E., Herman, J.G., and Baylin, S.B. (2005). Short double-stranded RNA induces transcriptional gene silencing in human cancer cells in the absence of DNA methylation. Nat. Genet. *37*, 906–910.

Tiscornia, G., Singer, O., Ikawa, M., and Verma, I.M. (2003). A general method for gene knockdown in mice by using lentiviral vectors expressing small interfering RNA. Proc. Natl. Acad. Sci. U.S.A. *100*, 1844–1848.

Tiscornia, G., Tergaonkar, V., Galimi, F., and Verma, I.M. (2004). CRE recombinase-inducible RNA interference mediated by lentiviral vectors. Proc. Natl. Acad. Sci. U.S.A. *101*, 7347–7351.

Towne, C., Raoul, C., Schneider, B.L., and Aebischer, P. (2008). Systemic AAV6 delivery mediating RNA interference against SOD1: neuromuscular transduction does not alter disease progression in fALS mice. Mol. Ther. *16*, 1018–1025.

Triboulet, R., Mari, B., Lin, Y.L., Chable-Bessia, C., Bennasser, Y., Lebrigand, K., Cardinaud, B., Maurin, T., Barbry, P., Baillat, V., *et al.* (2007). Suppression of microRNA-silencing pathway by HIV-1 during virus replication. Science *315*, 1579–1582.

Uchida, H., Tanaka, T., Sasaki, K., Kato, K., Dehari, H., Ito, Y., Kobune, M., Miyagishi, M., Taira, K., Tahara, H., and Hamada, H. (2004). Adenovirus-mediated transfer of siRNA against survivin induced apoptosis and attenuated tumor cell growth in vitro and *in vivo*. Mol. Ther. *10*, 162–171.

Unwalla, H.J., Li, M.J., Kim, J.D., Li, H.T., Ehsani, A., Alluin, J., and Rossi, J.J. (2004). Negative feedback inhibition of HIV-1 by TAT-inducible expression of siRNA. Nat. Biotechnol. *22*, 1573–1578.

Uprichard, S.L., Boyd, B., Althage, A., and Chisari, F.V. (2005). Clearance of hepatitis B virus from the liver of transgenic mice by short hairpin RNAs. Proc. Natl. Acad. Sci. U.S.A. *102*, 773–778.

van de Wetering, M., Oving, I., Muncan, V., Pon Fong, M.T., Brantjes, H., van Leenen, D., Holstege, F.C., Brummelkamp, T.R., Agami, R., and Clevers, H. (2003). Specific inhibition of gene expression using a stably integrated, inducible small-interfering-RNA vector. EMBO Rep. *4*, 609–615.

Vandekerckhove, L., Christ, F., Van Maele, B., De Rijck, J., Gijsbers, R., Van den Haute, C., Witvrouw, M., and Debyser, Z. (2006). Transient and stable knockdown of the integrase cofactor LEDGF/p75 reveals its role in the replication cycle of human immunodeficiency virus. J. Virol. *80*, 1886–1896.

Ventura, A., Meissner, A., Dillon, C.P., McManus, M., Sharp, P.A., Van Parijs, L., Jaenisch, R., and Jacks, T. (2004). Cre-lox-regulated conditional RNA interference from transgenes. Proc. Natl. Acad. Sci. U.S.A. *101*, 10380–10385.

Vera, M., and Fortes, P. (2004). Simian virus-40 as a gene therapy vector. DNA Cell Biol. *23*, 271–282.

Verdel, A., Jia, S., Gerber, S., Sugiyama, T., Gygi, S., Grewal, S.I., and Moazed, D. (2004). RNAi-mediated targeting of heterochromatin by the RITS complex. Science *303*, 672–676.

Vickers, T.A., Koo, S., Bennett, C.F., Crooke, S.T., Dean, N.M., and Baker, B.F. (2003). Efficient reduction of target RNAs by small interfering RNA and RNase H-dependent antisense agents. A comparative analysis. J. Biol. Chem. *278*, 7108–7118.

Vigna, E., and Naldini, L. (2000). Lentiviral vectors: excellent tools for experimental gene transfer and promising candidates for gene therapy. J. Gene Med. *2*, 308–316.

Voinnet, O. (2005). Induction and suppression of RNA silencing: insights from viral infections. Nat. Rev. Genet. *6*, 206–220.

Volpe, T.A., Kidner, C., Hall, I.M., Teng, G., Grewal, S.I., and Martienssen, R.A. (2002). Regulation of heterochromatic silencing and histone H3 lysine-9 methylation by RNAi. Science *297*, 1833–1837.

von Eije, K.J., ter Brake, O., and Berkhout, B. (2008). Human immunodeficiency virus type 1 escape is restricted when conserved genome sequences are targeted by RNA interference. J. Virol. *82*, 2895–2903.

Waddington, S.N., McVey, J.H., Bhella, D., Parker, A.L., Barker, K., Atoda, H., Pink, R., Buckley, S.M., Greig, J.A., Denby, L., *et al.* (2008). Adenovirus serotype 5 hexon mediates liver gene transfer. Cell *132*, 397–409.

Wang, J., Theunissen, T.W., and Orkin, S.H. (2007). Site-directed, virus-free, and inducible RNAi in embryonic stem cells. Proc. Natl. Acad. Sci. U.S.A. *104*, 20850–20855.

Wang, Q., Contag, C.H., Ilves, H., Johnston, B.H., and Kaspar, R.L. (2005). Small hairpin RNAs efficiently inhibit hepatitis C IRES-mediated gene expression in human tissue culture cells and a mouse model. Mol. Ther. *12*, 562–568.

Wang, Y., Kato, N., Jazag, A., Dharel, N., Otsuka, M., Taniguchi, H., Kawabe, T., and Omata, M. (2006). Hepatitis C virus core protein is a potent inhibitor of RNA silencing-based antiviral response. Gastroenterology *130*, 883–892.

Watanabe, T., Sudoh, M., Miyagishi, M., Akashi, H., Arai, M., Inoue, K., Taira, K., Yoshiba, M., and Kohara, M. (2006a). Intracellular-diced dsRNA has enhanced efficacy for silencing HCV RNA and overcomes variation in the viral genotype. Gene Ther. *13*, 883–892.

Watanabe, T., Takeda, A., Tsukiyama, T., Mise, K., Okuno, T., Sasaki, H., Minami, N., and Imai, H. (2006b). Identification and characterization of two novel classes of small RNAs in the mouse germline: retrotransposon-derived siRNAs in oocytes and germline small RNAs in testes. Genes Dev. *20*, 1732–1743.

Watashi, K., Ishii, N., Hijikata, M., Inoue, D., Murata, T., Miyanari, Y., and Shimotohno, K. (2005). Cyclophilin B is a functional regulator of hepatitis C virus RNA polymerase. Mol. Cell *19*, 111–122.

Weinberg, M.S., Ely, A., Barichievy, S., Crowther, C., Mufamadi, S., Carmona, S., and Arbuthnot, P. (2007). Specific inhibition of HBV replication in vitro and *in vivo* with expressed long hairpin RNA. Mol. Ther. *15*, 534–541.

Weinberg, M.S., Villeneuve, L.M., Ehsani, A., Amarzguioui, M., Aagaard, L., Chen, Z.X., Riggs, A.D., Rossi, J.J., and Morris, K.V. (2006). The antisense strand of small interfering RNAs directs histone methylation and transcriptional gene silencing in human cells. Rna *12*, 256–262.

Westerhout, E.M., Ooms, M., Vink, M., Das, A.T., and Berkhout, B. (2005). HIV-1 can escape from RNA interference by evolving an alternative structure in its RNA genome. Nucleic Acids Res. *33*, 796–804.

Westerhout, E.M., ter Brake, O., and Berkhout, B. (2006a). The virion-associated incoming HIV-1 RNA genome is not targeted by RNA interference. Retrovirology 3, 57.

Westerhout, E.M., Vink, M., Haasnoot, P.C., Das, A.T., and Berkhout, B. (2006b). A conditionally replicating HIV-based vector that stably expresses an antiviral shRNA against HIV-1 replication. Mol. Ther. *14*, 268–275.

White, M.D., Farmer, M., Mirabile, I., Brandner, S., Collinge, J., and Mallucci, G.R. (2008). Single treatment with RNAi against prion protein rescues early neuronal dysfunction and prolongs survival in mice with prion disease. Proc. Natl. Acad. Sci. U.S.A. *105*, 10238–10243.

Wieland, S.F., and Chisari, F.V. (2005). Stealth and cunning: hepatitis B and hepatitis C viruses. J. Virol. *79*, 9369–9380.

Wilson, J.A., Jayasena, S., Khvorova, A., Sabatinos, S., Rodrigue-Gervais, I.G., Arya, S., Sarangi, F., Harris-Brandts, M., Beaulieu, S., and Richardson, C.D. (2003). RNA interference blocks gene expression and RNA synthesis from hepatitis C replicons propagated in human liver cells. Proc. Natl. Acad. Sci. U.S.A. *100*, 2783–2788.

Wilson, J.A., and Richardson, C.D. (2005). Hepatitis C virus replicons escape RNA interference induced by a short interfering RNA directed against the NS5b coding region. J. Virol. *79*, 7050–7058.

Witting, S.R., Brown, M., Saxena, R., Nabinger, S., and Morral, N. (2008). Helper-dependent adenovirus-mediated short hairpin RNA expression in the liver activates the interferon response. J. Biol. Chem. *283*, 2120–2128.

Wiznerowicz, M., and Trono, D. (2003). Conditional suppression of cellular genes: lentivirus vector-mediated drug-inducible RNA interference. J. Virol. *77*, 8957–8961.

Wong, L.F., Azzouz, M., Walmsley, L.E., Askham, Z., Wilkes, F.J., Mitrophanous, K.A., Kingsman, S.M., and Mazarakis, N.D. (2004). Transduction patterns of pseudotyped lentiviral vectors in the nervous system. Mol. Ther. *9*, 101–111.

Wreschner, D.H., James, T.C., Silverman, R.H., and Kerr, I.M. (1981). Ribosomal RNA cleavage, nuclease activation and 2–5A(ppp(A2'p)nA) in interferon-treated cells. Nucleic Acids Res. *9*, 1571–1581.

Wu, H.L., Huang, L.R., Huang, C.C., Lai, H.L., Liu, C.J., Huang, Y.T., Hsu, Y.W., Lu, C.Y., Chen, D.S., and Chen, P.J. (2005). RNA interference-mediated control of hepatitis B virus and emergence of resistant mutant. Gastroenterology *128*, 708–716.

Wu, X., Li, Y., Crise, B., and Burgess, S.M. (2003). Transcription start regions in the human genome are favored targets for MLV integration. Science *300*, 1749–1751.

Xia, H., Mao, Q., Eliason, S.L., Harper, S.Q., Martins, I.H., Orr, H.T., Paulson, H.L., Yang, L., Kotin, R.M., and Davidson, B.L. (2004). RNAi suppresses polyglutamine-induced neurodegeneration in a model of spinocerebellar ataxia. Nat. Med. *10*, 816–820.

Xia, H., Mao, Q., Paulson, H.L., and Davidson, B.L. (2002). siRNA-mediated gene silencing in vitro and *in vivo*. Nat. Biotechnol. *20*, 1006–1010.

Xing, L., and Kieff, E. (2007). Epstein–Barr virus BHRF1 micro- and stable RNAs during latency III and after induction of replication. J. Virol. *81*, 9967–9975.

Xu, N., Segerman, B., Zhou, X., and Akusjarvi, G. (2007). Adenovirus virus-associated RNAII-derived small RNAs are efficiently incorporated into the rna-induced silencing complex and associate with polyribosomes. J. Virol. *81*, 10540–10549.

Yamamoto, T., Miyoshi, H., Yamamoto, N., Yamamoto, N., Inoue, J., and Tsunetsugu-Yokota, Y. (2006). Lentivirus vectors expressing short hairpin RNAs against the U3-overlapping region of HIV nef inhibit HIV replication and infectivity in primary macrophages. Blood *108*, 3305–3312.

Yang, S., Tutton, S., Pierce, E., and Yoon, K. (2001). Specific double-stranded RNA interference in undifferentiated mouse embryonic stem cells. Mol. Cell Biol. *21*, 7807–7816.

Yang, Y., and Wilson, J.M. (1995). Clearance of adenovirus-infected hepatocytes by MHC class I-restricted CD4+ CTLs *in vivo*. J. Immunol. *155*, 2564–2570.

Yang, Z.G., Chen, Z., Ni, Q., Xu, N., Shao, J.B., and Yao, H.P. (2005). Inhibition of hepatitis B virus surface antigen expression by small hairpin RNA in vitro. World J Gastroenterol *11*, 498–502.

Yang, G., Thompson, J.A., Fang, B., and Liu, J. (2003). Silencing of H-ras gene expression by retrovirus-mediated siRNA decreases transformation efficiency and tumor growth in a model of human ovarian cancer. Oncogene *22*, 5694–5701.

Yeung, M.L., Bennasser, Y., Le, S.Y., and Jeang, K.T. (2005). siRNA, miRNA and HIV: promises and challenges. Cell Res. *15*, 935–946.

Yi, R., Qin, Y., Macara, I.G., and Cullen, B.R. (2003). Exportin-5 mediates the nuclear export of pre-microRNAs and short hairpin RNAs. Genes Dev. *17*, 3011–3016.

Ying, R.S., Zhu, C., Fan, X.G., Li, N., Tian, X.F., Liu, H.B., and Zhang, B.X. (2007). Hepatitis B virus is inhibited by RNA interference in cell culture and in mice. Antiviral Res. *73*, 24–30.

Yokota, T., Sakamoto, N., Enomoto, N., Tanabe, Y., Miyagishi, M., Maekawa, S., Yi, L., Kurosaki, M., Taira, K., Watanabe, M., and Mizusawa, H. (2003). Inhibition of intracellular hepatitis C virus replication by synthetic and vector-derived small interfering RNAs. EMBO Rep. *4*, 602–608.

Yoo, J.Y., Kim, J.H., Kwon, Y.G., Kim, E.C., Kim, N.K., Choi, H.J., and Yun, C.O. (2007). VEGF-specific short hairpin RNA-expressing oncolytic adenovirus elicits potent inhibition of angiogenesis and tumor growth. Mol. Ther. *15*, 295–302.

Young, L.S., Searle, P.F., Onion, D., and Mautner, V. (2006). Viral gene therapy strategies: from basic science to clinical application. J. Pathol. *208*, 299–318.

Yu, J., and McMahon, A.P. (2006). Reproducible and inducible knockdown of gene expression in mice. Genesis *44*, 252–261.

Yu, J.Y., DeRuiter, S.L., and Turner, D.L. (2002). RNA interference by expression of short-interfering RNAs and hairpin RNAs in mammalian cells. Proc. Natl. Acad. Sci. U.S.A. *99*, 6047–6052.

Yu, Z., Sanchez-Velar, N., Catrina, I.E., Kittler, E.L., Udofia, E.B., and Zapp, M.L. (2005). The cellular HIV-1 Rev cofactor hRIP is required for viral replication. Proc. Natl. Acad. Sci. U.S.A. *102*, 4027–4032.

Yuan, B., Latek, R., Hossbach, M., Tuschl, T., and Lewitter, F. (2004). siRNA Selection Server: an automated siRNA oligonucleotide prediction server. Nucleic Acids Res. *32*, W130–134.

Zabner, J., Seiler, M., Walters, R., Kotin, R.M., Fulgeras, W., Davidson, B.L., and Chiorini, J.A. (2000). Adeno-associated virus type 5 (AAV5) but not AAV2 binds to the apical surfaces of airway epithelia and facilitates gene transfer. J. Virol. *74*, 3852–3858.

Zamanian-Daryoush, M., Der, S.D., and Williams, B.R. (1999). Cell cycle regulation of the double stranded RNA activated protein kinase, PKR. Oncogene *18*, 315–326.

Zeng, Y., and Cullen, B.R. (2004). Structural requirements for pre-microRNA binding and nuclear export by Exportin 5. Nucleic Acids Res. *32*, 4776–4785.

Zeng, Y., and Cullen, B.R. (2005). Efficient processing of primary microRNA hairpins by Drosha requires flanking nonstructured RNA sequences. J. Biol. Chem. *280*, 27595–27603.

Zeng, Y., Wagner, E.J., and Cullen, B.R. (2002). Both natural and designed micro RNAs can inhibit the expression of cognate mRNAs when expressed in human cells. Mol. Cell *9*, 1327–1333.

Zhang, H.B., Wu, J., Xian, J., Pei, L., and Wang, J. (2008). Two approaches to construct mammalian expression vector of shRNA to reduce expression and replication of HBV in vitro. Mol. Biol. Rep. *35*, 465–472.

Zhang, J., Yamada, O., Sakamoto, T., Yoshida, H., Iwai, T., Matsushita, Y., Shimamura, H., Araki, H., and Shimotohno, K. (2004a). Down-regulation of viral replication by adenoviral-mediated expression of siRNA against cellular cofactors for hepatitis C virus. Virology *320*, 135–143.

Zhang, S.Z., Pan, F.Y., Xu, J.F., Yuan, J., Guo, S.Y., Dai, G., Xue, B., Shen, W.G., Wen, C.J., Zhao, D.H., and Li, C.J. (2005). Knockdown of c-Met by adenovirus-delivered small interfering RNA inhibits hepatocellular carcinoma growth in vitro and *in vivo*. Mol. Cancer Ther. *4*, 1577–1584.

Zhang, X., Wu, K., Yue, X., Zhu, Y., and Wu, J. (2007). Inhibition of SARS-CoV gene expression by adenovirus-delivered small hairpin RNA. Intervirology *50*, 63–70.

Zhang, X., Kon, T., Wang, H., Li, F., Huang, Q., Rabbani, Z.N., Kirkpatrick, J.P., Vujaskovic, Z., Dewhirst, M.W., and Li, C.Y. (2004b). Enhancement of hypoxia-induced tumor cell death in vitro and radiation therapy *in vivo* by use of small interfering RNA targeted to hypoxia-inducible factor-1alpha. Cancer Res. *64*, 8139–8142.

Zhang, Y.A., Nemunaitis, J., Samuel, S.K., Chen, P., Shen, Y., and Tong, A.W. (2006). Antitumor activity of an oncolytic adenovirus-delivered oncogene small interfering RNA. Cancer Res. *66*, 9736–9743.

Zheng, Z.M., Tang, S., and Tao, M. (2005). Development of resistance to RNAi in mammalian cells. Ann. N.Y. Acad. Sci. *1058*, 105–118.

Zhu, X., Santat, L.A., Chang, M.S., Liu, J., Zavzavadjian, J.R., Wall, E.A., Kivork, C., Simon, M.I., and Fraser, I.D. (2007). A versatile approach to multiple gene RNA interference using microRNA-based short hairpin RNAs. BMC Mol Biol *8*, 98.

Zufferey, R., Dull, T., Mandel, R.J., Bukovsky, A., Quiroz, D., Naldini, L., and Trono, D. (1998). Self-inactivating lentivirus vector for safe and efficient *in vivo* gene delivery. J. Virol. *72*, 9873–9880.

# Index

www.ingramcontent.com/pod-product-compliance
Lightning Source LLC
LaVergne TN
LVHW062122200926
848484LV00007B/1737
*9781904455561*